Turbulent Shear Flows 7

Selected Papers from the Seventh International Symposium on Turbulent Shear Flows, Stanford University, USA, August 21–23, 1989

Editors:

F. Durst B. E. Launder W. C. Reynolds
F. W. Schmidt J. H. Whitelaw

With 241 Figures

Springer-Verlag
Berlin Heidelberg New York London Paris
Tokyo Hong Kong Barcelona

Franz Durst
Lehrstuhl für Strömungsmechanik, Universität Erlangen-Nürnberg
Cauerstraße 4, 8520 Erlangen, FRG

Brian E. Launder
Department of Mechanical Engineering, University of Manchester,
Institute of Science and Technology,
Manchester M60 1QD, UK

William C. Reynolds
Mechanical Engineering Department, Thermosciences Division,
Stanford University, Stanford, CA 94305-3030, USA

Frank W. Schmidt
Mechanical Engineering Department, The Pennsylvania State University,
University Park, PA 16802, USA

James H. Whitelaw
Department of Mechanical Engineering, Imperial College of Science and Technology,
Exhibition Road, London SW7 2BX, England

Library of Congress Cataloging-in-Publication Data.
International Symposium on Turbulent Shear Flows (7th: 1989: Stanford University)
Turbulent shear flows 7: selected papers from the Seventh International Symposium on Turbulent Shear Flows, Stanford University, USA, August 21–23, 1989 editors: F. Durst ... [et al.]. p. cm.
Includes index.

ISBN-13: 978-3-642-76089-1 e-ISBN-13: 978-3-642-76087-7
DOI: 10.1007/978-3-642-76087-7

1. Turbulence—Congresses. 2. Shear flow—Congresses. 3. Turbulent boundary layer—Congresses. I. Durst, F. II. Title. TA357.5.T87I58 1989 620.1'064 dc20 90-25961 CIP

Softcover reprint of the hardcover 1st edition 1991

Typesetting: Thomson Press (India) Ltd, New Delhi
Offsetprinting: Mercedes-Druck, Berlin.
2125/3020-543210—Printed on acid-free paper

Preface

The Seventh Symposium was held on the campus of Stanford University with a combination of facilities and weather which made it possible to add open-air poster sessions and coffee breaks to the programme. This was particularly convenient as the call for papers attracted close to three hundred abstracts and a total number of participants well in excess of this number. Some one hundred and thirty papers were presented in carefully phased parallel sessions and thirty six further contributions were made available in the form of posters. In addition, a lively open-forum session allowed additional speakers to make brief presentations. The staff of the Thermo-Sciences Division of the Department of Mechanical Engineering at Stanford undertook the local arrangements with evident success and their extensive record of contributions to Turbulent Shear Flows made the venue particularly appropriate. Also, the Centre for Turbulence Studies, based on the faculty of the University and the NASA Ames Research Center, provided a considerable body of expertise with emphasis on direct numerical stimulation.

From the papers presented at the Symposium, we have selected twenty-seven for inclusion in this volume and each of these has been improved and extended in accordance with recommendations of the Editors. We have chosen four subject areas which accord with the most exciting topics of the Symposium, namely wall flows, free flows, reacting flows and numerical simulations and, as in previous volumes, we have asked eminent authorities to prepare introductory articles for each topic so as to put the individual contributions in context with each other and with related research.

Financial support for the Seventh Symposium was provided by the Boeing Airplane Company, General Motors, Rockwell International and the National Science Foundation.

It was offered in cooperation with the Fluid Engineering and Heat Transfer Divisions of the ASME.

The technical programme of the Symposium was the responsibility of our Papers Committee guided by at least two reviews of each extended abstract. These reviews were provided by members of the Advisory Committee who also served as Chairmen at the Symposium. This Advisory Committee comprised:

R. J. Adrian (USA)
J. C. Andre (France)
R. A. Antonia (Australia)
G. Bergeles (Greece)
R. W. Bilger (Australia)
R. F. Blackwelder (USA)
L. J. S. Bradbury (UK)
P. Bradshaw (USA)
M. Coanic (France)
J. Coustlex (France)
D. F. G. Durao (Portugal)
R. Friedrich (Germany)
I. S. Gartshore (Canada)
W. K. George (USA)
C. H. Gibson (USA)
M. M. Gibson (UK)
V. W. Goldschmidt (USA)
H. Ha Minh (France)
K. Hanjalic (Yugoslavia)
T. I. Hanratty (USA)
J. R. Herring (USA)
R. Houdeville (France)
J. A. C. Humphrey (USA)
F. Hussain (USA)
N. Kasagi (Japan)
W. Kollmann (USA)
E. Krause (Germany)
A. Leonard (USA)
M. Lesleur (France)
J. L. Lumley (USA)
P. Moin (USA)
Y. Nagano (Japan)
I. Nakamura (Japan)
W. Rodi (Germany)
G. S. Samuelsen (USA)
U. Schumann (Germany)
M. Sokolov (Israel)
R. L. Street (USA)
K. Suzuki (Japan)
M. Wolfshiein (Israel)

It is with considerable pleasure that we record our thanks to the staff of Springer-Verlag for producing this book in accordance with the requirements of the Symposium.

London 1990

Editors

Contents

Part I Scalar and Stratified Flows

Introductory Remarks. By C. H. Gibson 3

Reconnection of Two Antiparallel Vortex Tubes: A New Cascade Mechanism. By M. V. Melander and F. Hussain 9

High Resolution Three-Dimensional (256^3) Spatio-Temporal Measurements of the Conserved Scalar Field in Turbulent Shear Flows. By W. J. A. Dahm and K. A. Buch 17

Direct Simulation of Homogeneous Turbulence and Gravity Waves in Sheared and Unsheared Stratified Flows. By T. Gerz and U. Schumann 27

Turbulence Model for Triple Velocity and Scalar Correlations. By Y. Nagano and M. Tagawa 47

Fossil Two-Dimensional Turbulence in the Ocean. By C. H. Gibson . . . 63

Part II Wall Flows

Introductory Remarks. By R. M. C. So and Y. G. Lai 81

The Structure of Pressure Fluctuations in Turbulent Shear Flows. By J. Kim and M. J. Lee 87

The Structure of Sheared Turbulence Near a Plane Boundary. By M. J. Lee and J. C. R. Hunt 101

Study on a Turbulent Boundary Layer Disturbed by a Cylinder—Effect of Cylinder Size and Position. By K. Suzuki, H. Suzuki, Y. Kikkawa and Y. Kawaguchi 119

Some Characteristics of Bypass Transition in a Heated Boundary Layer. By K. H. Sohn, J. E. O'Brien and E. Reshotko 137

Turbulent Characteristics Inside a Turbulent Spot in a Plane Poiseuille Flow. By D. S. Henningson and J. Kim 155

Part III Free Shear Flows

Introductory Remarks. By L. J. S. Bradbury 169

Three-Dimensional Aspects and Transition of the Wake of a Circular Cylinder. By C. H. K. Williamson 173

Experimental and Numerical Analysis of the Three-Dimensional Evolution of an Axisymmetric Jet. By E. Meiburg, J. C. Lasheras and J. E. Martin 195

Turbulent Plane Jet Excited Mechanically by an Oscillating Thin Plate in the Potential Core. By M. Miyata, N. Kurita and I. Nakamura 209

Structure in the Near Field of the Transverse Jet. By T. F. Fric and A. Roshko 225

Reynolds Stress Evolution in Curved Two-Stream Turbulent Mixing Layers. By M. W. Plesniak and J. P. Johnston 239

The Effect of Extra Strain Rates of Streamline Curvature and Divergence on Mixing Layers. By A. E. Johnson and P. E. Hancock 253

Part IV Reacting Flows

Introductory Remarks. By W. P. Jones 271

Mixing Models for Turbulent Flows with Exothermic Reactions. By J.-Y. Chen and W. Kollmann 277

A Lagrangian Intermittent Model for Turbulent Combustion Theoretical Basis and Comparisons with Experiments. By M. Gonzalez and R. Borghi 293

Direct Simulation of a Passive Diffusion Flame in Sheared and Unsheared Homogeneous Turbulence. By S. Elghobashi and K. K. Nomura 313

Large Scale Structures in Reacting Mixing Layers. By C. Lee, R. W. Metcalfe and F. Hussain 331

Modelling and Numerical Simulation of Premixed Turbulent Combustion in a Boundary Layer. By B. Rogg 345

Investigation of the Combustion-Turbulence Interaction in Premixed Stagnation Flames of H_2-CH_4 Mixtures. By Y. Liu, B. Lenze and W. Leuckel 357

Coherent Flame Model for Non-Uniformly Premixed Turbulent Flames. By D. Veynante, F. Lacas, E. Maistret and S. M. Candel 367

Index of Contributors 379

Part I

Scalar and Stratified Flows

Introduction to Scalar and Stratified Flows

Carl H. Gibson

Departments of Applied Mechanics and Engineering Sciences, and Scripps Institution of Oceanography, University of California at San Diego, La Jolla, CA 92093, USA

Perhaps the most important practical aspect of turbulent shear flows is their dominant effect on scalar fields such as temperature, density or chemical species. When turbulence exists, it tends to completely determine the mixing and diffusion of such quantities. Industrial flows with chemical reactions, combustion, and natural flows in the ocean and atmosphere usually involve turbulence constrained by forces, and complicated by factors, that laboratory studies often suppress; for example, stratification, rotation and shear. Unstratified, nonrotating, unsheared turbulence as described by the Batchelor (1967) classic book is so complex and poorly understood that many fluid dynamicists, and most undergraduate fluid mechanics textbooks, manage to avoid the subject completely. Papers in the present chapter on Scalar and Stratified flows confront many of the awkward uncertainties attending turbulence in the "real world". Some excellent new tools exist today that were not available to Batchelor, and their impact is reflected in the following articles.

Digital computers can now convincingly simulate the full details of viscous vortex dynamics, as shown by the paper of Melander and Hussain. No longer can one assume that the immutable vortex lines of potential flows move with the fluid and maintain constant vorticity magnitude (root enstrophy) values. Such constant "enstrophy" lines are diffusively unstable and break up into isolated points of maximum vorticity magnitude, with diffusive drift velocities, that are connected by topological lines of minimal enstrophy gradient (Gibson 1988, 1990). These topological "ridge lines" can "cut-and-connect" in a complex dance of viscous vorticity annihilation, 2D-dipole pumping, local self-induction and vortex stretching described in the Melander and Hussain paper. We know from other numerical simulations such as Ashurst et al. (1987) and Gibson et al. (1988) that the response of various scalar field "dancing partners" are equally intricate, and that the small scale mixing depends crucially on the interaction of the rate-of-strain tensor, the rotation tensor and the scalar field topologies.

Equally impressive new laboratory experimental tools are emerging, as demonstrated by the Dahm and Buch paper on a nonintrusive, high resolution, high frequency planar imaging technique that optically detects 65K samples of dye concentration in water on a 1 cm^2 plane 100 times a second. This can resolve the 3D field to the Batchelor viscous-diffusive length scale of the turbulent dye mixing $L_B \equiv (D/\gamma)^{1/2}$, where D is the molecular diffusivity of the dye, $\gamma \equiv (\varepsilon/\nu)^{1/2}$ is the rate-of-strain parameter of Batchelor (1959), ε is the viscous dissipation rate per unit mass,

Turbulent Shear Flows 7

and ν is the kinematic viscosity. Thus, a 256^3 set of 44 micron elements ($1.4\,cm^3$) could be resolved for a flow at a velocity normal to the image plane of $0.44\,cm\,s^{-1}$. Assuming the diffusivity of fluorescein is about $10^{-6}\,cm^2\,s^{-1}$ gives $\gamma \approx 0.052\,rad\,s^{-1}$, and $\varepsilon \approx 3.2 \times 10^{-5}\,cm^2\,s^{-3}$. Consequently, a Reynolds number of 10^4 could be achieved in a water flow field with a length scale of 2.2 meters. This is probably within laboratory capabilities, although care must be taken to see that buoyancy fluxes from heat flow and evaporation do not overcome the indicated turbulence mechanical power input for the tank of only 34 microwatts. Comparisons of Batchelor scale normalized spectra in the viscous convective and diffusive subranges with the Batchelor (1959) theoretical spectrum from this instrument will be interesting, since such high Schmidt number scalar spectra to diffusive scales have not yet been presented. The probability laws for the dissipation rate of scalar variance $\chi \equiv 2D(\nabla C)^2$ are also of interest, where C is the dye concentration.

Direct numerical simulation of the effects of stratification on turbulence are presented by Gerz and Schumann for a moderate Reynolds number $Re_\lambda \approx 25$ and seven Froude numbers from 0.3 to ∞, and show the wide range of statistical parameters that can be generated with relative ease by computer simulations compared to laboratory studies. Their result that

$$0.4L_R \geqq L_E \geqq 6.0L_K \tag{1}$$

confirms the criteria for the beginning of fossilization (onset of buoyancy effects) at the buoyant-inertial transition and completion of fossilization (extinction of turbulence) at the buoyant-inertial-viscous transition predicted by Gibson (1981)

$$1.2L_R \geqq \lambda \geqq 15L_K \tag{2}$$

where $L_R \equiv (\varepsilon/N^3)^{1/2}$ is the Ozmidov scale, $L_K \equiv (\nu^3/\varepsilon)^{1/4} = (\nu/\gamma)^{1/2}$ is the Kolmogorov scale and λ is the wavelength of a streamwise Fourier velocity mode, assuming that the Ellison scale $L_E \equiv \rho'/\overline{\rho, z}$ is equal to $\lambda/3$, which seems very reasonable. The predictions of Eq. (2) are also confirmed by the laboratory studies of Stillinger et al. (1983) and Itsweire et al. (1986) assuming $L_E \approx \lambda/2$.

Another important result, but not pointed out by the authors, was that after the onset of fossilization, as indicated by the departure of L_E in curve B of Fig. 9a from the unstratified case shown by curve A, the overturn scale L_E remains relatively constant at its onset fossil value L_{E0} while the corresponding Ozmidov scale L_R shown by curve B in Fig. 9b decreases rapidly with time. This shows that the density microstructure after the inertial-buoyancy transition is fossil turbulence in the range of scales $L_{E0} \geqq L \geqq 0.6L_R$ because the microstructure scales caused by the previous turbulence persist after the flow ceases to be turbulent, consistent with the laboratory studies of Stillinger et al. (1983) and Itsweire et al. (1986) and the fossil turbulence theory of Gibson (1980). The result is inconsistent with the Gargett (1988) prediction that the vertical turbulent energy scale $(L_O)_z \approx L_E$ and the horizontal turbulent energy scale $(L_O)_{x,y} \approx L_R$ for laboratory turbulence (and numerical simulations) with such low Reynolds numbers.

Gargett's prediction says that the numerical and laboratory studies just cited should be buoyancy-affected and highly anisotropic from the beginning, but Gerz and Schumann show their velocity fields are not anisotropic by computing anisotropy tensor values (Fig. 10a, curves B, C, D) which start from near zero and remain rather small even after fossilization begins. The Gargett (1988) scaling also predicts that no fossil turbulence, with $L_E \gg L_R$, will be formed, but this is also contradicted by the Gerz and Schumann numerical results (and the laboratory studies of Stillinger et al. (1983); Itsweire et al. (1986) and others). Both findings have important implications for oceanography, since they show that laboratory studies and numerical simulations have adequately large Reynolds numbers to simulate the stratified turbulence and mixing processes in the ocean, and that probably fossil turbulence also exists in the ocean (this is not generally accepted by oceanographers). Rather unexpected persistent countergradient fluxes of heat and momentum are detected by the numerical simulations of Gerz and Schumann when shear flow is included.

Nagano and Tagawa develop a new model for triple correlations of velocity and scalar fluctuations that is more appropriate for near-wall turbulence than the usual models that assume gradient-type diffusion. The model is based on the observations from numerical simulations that the triple correlations are determined primarily by organized structures such as bursts and sweeps. The new model involves the skewness factors for the scalar alone, or for the individual velocity components, and gives good agreement with measurements in pipes, flat plate boundary layers, jets, and backward facing step flow, as shown in Fig. 2a–d, much better than gradient diffusion models shown in Figs. 2d, 4, 5a, b, 6, and 7.

Gibson extends the Gibson (1980) theory of stratified fossil turbulence to include the Coriolis forces of a rotating coordinate system, as applied to the ocean where *2D* turbulence and *2D* fossil turbulence may exist. When not only viscous forces but also buoyancy and Coriolis forces may constrain the growth of turbulence, as in the ocean and atmosphere, it is important to define turbulence carefully to distinguish it from internal waves and Coriolis-inertial waves. As pointed out by Phillips (1969, p. 198):

"Not all of the random motions found in the ocean ... can be described properly as turbulence. The characteristic properties of turbulent motions are that they possess a random distribution of vorticity in which there is no unique relation between the frequency and wave-number of the Fourier modes; that they are diffusive and dissipative. A distinction is drawn ... between turbulence in a stably stratified fluid on the one hand and a random field of internal gravity waves on the other. This differentiation is useful, not only conceptually but also observationally since the mechanisms of energy transfer (in both physical and Fourier space) are essentially different."

Eddy-like motions of turbulence arise and grow by entrainment to form a patch when the inertial-vortex forces of a shear layer dominate the damping effects of viscosity, buoyancy and Coriolis forces. This happens rarely in the ocean, which is mostly nonturbulent. For eddy motions to be considered turbulent, by definition, the Reynolds, Froude, and Rossby numbers must be larger than critical values. Turbulence may be *2D* if motion in one Cartesian direction is prevented, perhaps by walls or buoyancy forces. These criteria may be expressed in terms of transition length scales L_K, L_R, and L_H at the inertial-viscous, inertial-buoyancy and inertial-Coriolis transition points, respectively. In order for an eddy of scale L to be *3D* turbulence

$$L_R \text{ and } L_H > L > L_K \tag{3}$$

where universal proportionality constants have been omitted, $L_H \equiv (\varepsilon/f^3)^{1/2}$ is the Hopfinger scale and f is the Coriolis parameter equal to twice the component of the rotation vector perpendicular to the coordinate plane of the $2D$ turbulence. When the energy scale L_O of growing $3D$ turbulence eddies reaches a constraining scale, either L_R or L_H, then the patch ceases to grow, the turbulent kinetic energy is converted to internal wave or Coriolis-inertial wave energy, and partially mixed patches of scalar fluctuations with $L_E \approx L_{E0} \approx L_{R0}$ or L_{H0} persist as fossil turbulence remnants of the previous turbulence. Unless further turbulent energy is supplied, say by a shear flow, the turbulence decays and ceases to exist entirely when L_R or $L_H \approx L_K = (\nu/\gamma)^{1/2}$. This occurs at either the Coriolis-inertial-viscous transition, with $\varepsilon_f = \nu f^2$, or at the buoyancy-inertial-viscous transition, with $\varepsilon_F = \nu N^2$, at the fossil Kolmogorov length scales $L_{Kf} = (\nu/f)^{1/2}$ or $L_{KF} = (\nu/N)^{1/2}$, respectively. The fossil turbulence remnants in the scalar and vorticity fields preserve information about previous turbulence events, and may be useful in the ocean to avoid undersampling errors in estimates of turbulence mixing rates and exchange coefficients. Hydrodynamic phase diagrams are described to permit classification, mostly by comparisons of turbulence and mixing length scales, of microstructure and mesostructure according to hydrodynamic states; that is, active:active-fossil:fossil ($3D$ or $2D$). Most oceanic microstructure appears to be active-fossil. The laboratory and numerical simulation evidence for 3 fossil turbulence is strong, and preliminary laboratory evidence for fossil $2D$ turbulence is encouraging.

Acknowledgements: This work was supported by ONR and NSF.

References

Ashurst, W. T., Kerstein, A. R., Kerr, R. M., Gibson, C. H. (1987): Alignment of vorticity and scalar gradient with strain rate in simulated Navier–Stokes turbulence. Phys. Fluids **30(8)**, 2343–2353

Batchelor, G. K. (1967): *Theory of Homogeneous Turbulence*. Cambridge University Press

Batchelor, G. K. (1959): Small-scale variation of convected quantities like temperature in turbulent fluid. J. Fluid Mech. **5**, 113–139

Dahm, W. J. A., Buch, K. A.: High resolution three-dimensional (256^3) spatio-temporal measurements of the conserved scalar field in turbulent shear flows, this chapter

Gargett, A. E. (1988): The scaling of turbulence in the presence of stable stratification. J. Geophys. Res. **93**, 5021–5036

Gerz, T., Schumann, U.: Direct simulation of homogeneous turbulence and gravity waves in sheared and unsheared stratified flows, this chapter

Gibson, C. H.: Fossil Two-dimensional turbulence in the ocean, this chapter

Gibson, C. H. (1980): Fossil temperature, salinity, and vorticity turbulence in the ocean. In *Marine Turbulence*, J. Nihoul, ed. Elsevier Publishing Co., Amsterdam, 221–257, 1980.

Gibson, C. H. (1981): Fossil turbulence and internal waves. In *American Institute of Physics Conference Proceedings No 76: Nonlinear Properties of Internal Waves*, Bruce West, ed.. American Institute of Physics, 159–179

Gibson, C. H. (1988): Isoenstrophy points and surfaces in turbulent flow and mixing. Fluid Dynamics Research **3**, 331–336

Gibson, C. H., Ashurst, W. T., Kerstein, A. R. (1988): Mixing of strongly diffusive passive scalars like temperature by turbulence. J. Fluid Mech. **194**, 261–293

Gibson, C. H. (1990): Scalar field topology in turbulent mixing, in Topological Fluid Mechanics, Proceedings of the IUTAM Symposium, Cambridge 1989, H. K. Moffatt and A. Tsinober (Eds.), Cambridge University Press, 85–94

Itsweire, E. C., Helland, K. N., Van Atta, C. W. (1986): The evolution of grid-generated turbulence in a stably stratified fluid. J. Fluid Mech. **162**, 299–338

Melander, M. V., Hussain, F. Cut-and-connect of two antiparallel vortex tubes: a new cascade mechanism, this chapter

Nagano, Y., Tagawa, M. Turbulence model for triple velocity and scalar correlations, this publication

Phillips, O. M. (1969): *The Dynamics of the Upper Ocean.* Cambridge University Press

Stillinger, D. C., Helland, K. N., Van Atta, C. W. (1983): Experiments on the transition of homogeneous turbulence to internal waves in a stratified fluid. J. Fluid Mech. **131**, 91–122

Reconnection of Two Antiparallel Vortex Tubes: A New Cascade Mechanism

M. V. Melander[1] and F. Hussain[2]

[1] Southern Methodist University, Dallas, Texas, USA
[2] University of Houston, Texas, USA

Abstract

Numerical simulations of two antiparallel vortices reveal a new and crucial mechanism, which accumulates the reconnected vortex lines into characteristic 'bridges'. The simultaneous occurrence of viscous vorticity annihilation, $2D$-dipole pumping, local self-induction and vortex stretching results in a characteristic curvature reversal, and also causes the reconnection to be only partial.

Nomenclature

A	area of dipole head in π_s
$\mathscr{C}$	contact zone
$\mathscr{E}$	external vorticity distribution
Re	Reynolds number
S_B, S_T	stagnation points
t	time
u, v, w	velocity components
x, y, z	space coordinates
z_0	z-coordinate of π_d-intercept
π_s	the xy-plane
π_d	the yz-plane
Γ_s	circulation in π_s $(x > 0)$
Γ_d	circulation in π_d $(z > 0)$
ν	kinematic viscosity
$\boldsymbol{\omega} = (\omega_1, \omega_2, \omega_3)$	vorticity vector
$\omega_m(t)$	peak vorticity in π_s

Introduction

Motivated by our early conjectures that reconnection (also known as cut-and-connect or crosslinking) of viscous vortices are frequent occurrences in turbulent flows and that they play important roles in mixing and chemical reaction, and generation of turbulence, helicity density and aerodynamic noise (Takaki and Hussain 1985; Hussain 1986), we have explored the heart of the phenomenon

Turbulent Shear Flows 7

through analysis of direct numerical simulations. Reconnection is intrinsically interesting as an example of nonpreserving topology in fluid mechanics, as well as an example to illustrate differences between inviscid and viscous vortex dynamics, and differences between scalar and vorticity transports even at unity Schmidt number. Recently the reconnection process has become the focus of intense research activity, primarily because the viscous relinking of vortex lines occurs on a convective timescale. Based on a sequence of 64^3-spectral simulations (Melander and Hussain 1988, 1989), we explain the physics causing the observed characteristics of the reconnection phenomenon. A striking finding is that the process is incomplete; we explain *why* and *why* it is interesting.

Initial Condition and Implications of Symmetry

To identify the effect of viscosity clearly, we use a symmetric initial condition (Fig. 1) consisting of two antiparallel vortices with inclined sinusoidal perturbations (Melander and Hussain 1988, 1989). Perfect symmetry is physically unrealistic, but mathematically it is of paramount importance, as we explain below. For clarity, the xy-plane is called the symmetric plane or π_s, while the yz-plane is the dividing plane or π_d. In both symmetry planes, the normal velocity component vanishes identically, that is $u(0, y, z) = 0$ and $w(x, y, 0) = 0$. Thus, in the absence of viscosity the symmetry planes are material planes (e.g. a particle initially in π_s will remain in π_s for all times, although it is free to move around in π_s). Since π_d is initially irrotational and is a material plane, it would remain irrotational were it not for the presence of viscosity (in an inviscid flow irrotational fluid remains irrotational). Hence a vortex line can not be turned to intercept π_d solely by convective effects.

Another consequence of the symmetry is that the tangential vorticity vanishes in both symmetry planes at all times ($\omega_1(x, y, 0) = \omega_2(x, y, 0) = \omega_2(0, y, z) = \omega_3(0, y, z) = 0$). The y-axis is therefore always irrotational. Furthermore, the circulations in the symmetry half-planes would be constant in the absence of viscosity. For later use we define Γ_s to be the circulation in the half-plane π_s^+ $(x > 0)$, while Γ_d is the circulation in π_d^+ $(z > 0)$. It is also important to observe that $\omega_3 \geqq 0$ in π_s^- and $\omega_3 \leqq 0$ in π_s^+, as this implies that Γ_s can only change through viscous vorticity annihilation along the irrotational y-axis. An almost similar argument can be made for π_d, the difference being that the vorticity in π_d is created along the y-axis rather than annihilated. A simple relation exists between the circulations in the symmetry half-planes: $d\Gamma_d/dt = -d\Gamma_s/dt$. This is a consequence of the fact that an arbitrary vortex line must intercept either π_s or π_d (possibly ending on the y-axis where $\boldsymbol{\omega} = \mathbf{0}$). Thus circulation is viscously transferred from π_s to π_d via the y-axis.

The Mechanism of Reconnection

The evolution of the vortices is shown in Fig. 1, while the vorticity distributions in π_s and π_d are shown in Figs. 2 and 3. The Reynolds number $\mathrm{Re} \equiv \Gamma/\nu = 1000$. Figure 4i shows Γ_s vs. time t (here t is nondimensionalized by the initial peak vorticity $\omega_m(0)$,

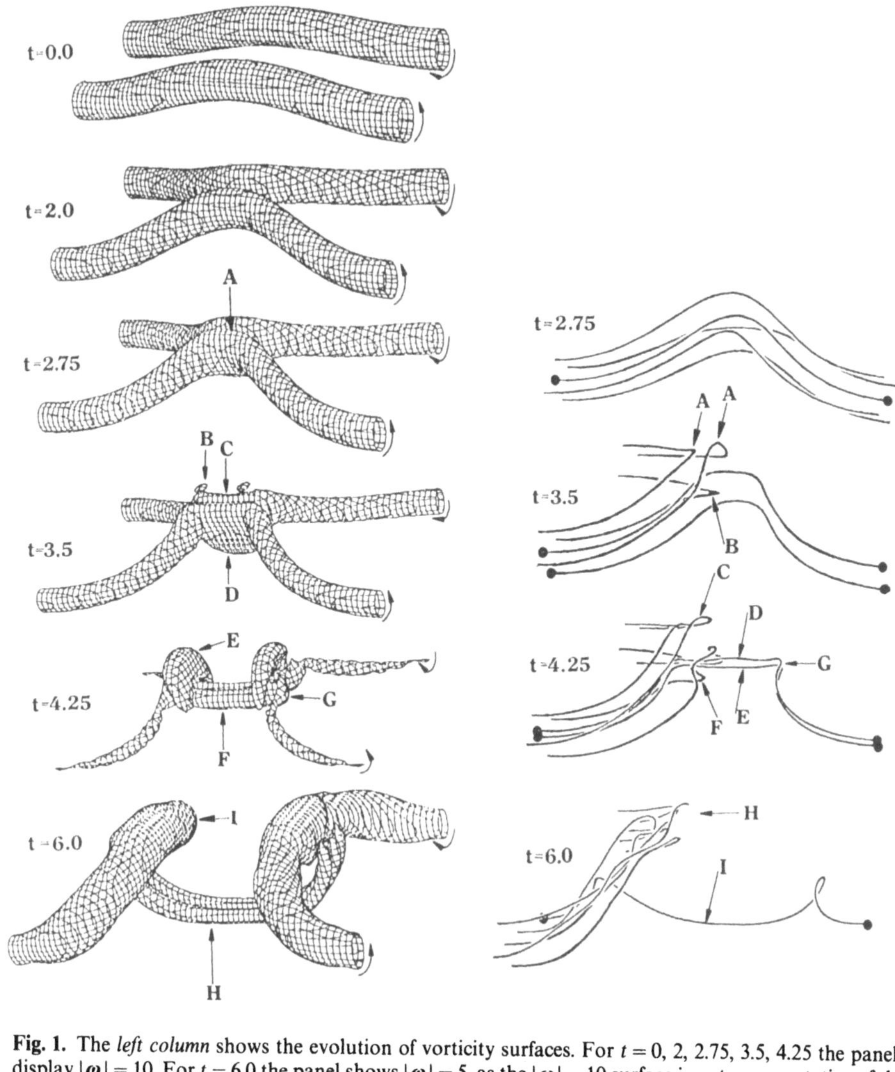

Fig. 1. The *left column* shows the evolution of vorticity surfaces. For $t = 0, 2, 2.75, 3.5, 4.25$ the panels display $|\boldsymbol{\omega}| = 10$. For $t = 6.0$ the panel shows $|\boldsymbol{\omega}| = 5$, as the $|\boldsymbol{\omega}| = 10$ surface is not representative of the overall vorticity configuration. The letters in the figure point to characteristic features. A: the formation of a contact zone and flattened vortex cores. B: the early stage of the formation of a bridge. Note that the bridge forms in front of the annihilating vortex pair. C, D: head-tail structure in the contact zone. E: a fully developed bridge at the end of the reconnection process. F: the "threads". G: sharp bends in the non-reconnected part of the vortex-pair (see also *G* in the right column). H: threads which may undergo a secondary reconnection a later point in time. I: reconnected vortex.

The *right column* shows typical samples of actual vortex lines. A: reconnected vortex lines in the bridge. B: a vortex line which has recently reconnected at *y*-axis. C: vortex lines in the bridge. D, E: Vortex lines in the threads. F: a newly reconnected vortex line. G: Sharp bends in the non-reconnected vortex lines emerging from the contact zone. H: The vortex lines reveal a complicated internal structure in the reconnected vortex. I: Vortex line in the threads.

Coordinates (x, y, z) can be inferred from Fig. 4h

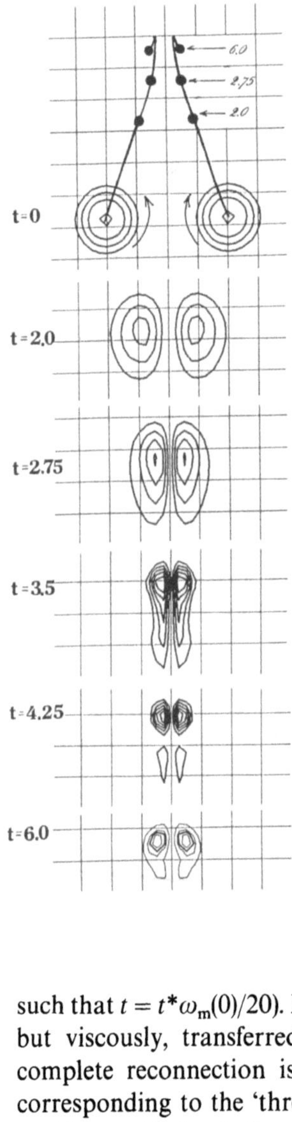

Fig. 2. Vorticity distribution in π_s. The heavy lines show contours of $|\boldsymbol{\omega}|$ in increments of 5. The displayed grid is four times coarser than the computational grid and is shown only for the reader's orientation. The panel at $t = 0.0$ also presents the trajectory of the vorticity peak in π_s, with the positions at $t = 2$, 2.75, and 6.0 marked. At $t = 2.75$ the formation of a contact zone is evident, and at $t = 3.5$ the characteristic head-tail structure can be seen. Later ($t = 4.25$ and $t = 6.$) the head has developed into long lasting "threads", which can be seen to be slowly pulling away from each other at $t = 6.0$. The thin lines shown at $t = 6.0$ are vorticity contours in increments of 2

such that $t = t^*\omega_m(0)/20$). During the reconnection 80% of the circulation is rapidly, but viscously, transferred from π_s into π_d (Fig. 4i). This transfer stops before complete reconnection is achieved, and 20% of the circulation remains in π_s, corresponding to the 'threads' seen in Fig. 1 (see H in the left column of Fig. 1).

The rapid circulation transfer from π_s to π_d begins when the vortices come in close contact near $\pi_s \cap \pi_d$ (see A in the left column of Fig. 1) and form a contact zone $\mathscr{C}(t)$ (Fig. 4a) with steep vorticity gradients. As the flow is highly viscous around $\mathscr{C}$ and nearly inviscid elsewhere, we may think of the vortex lines as material lines far away from $\mathscr{C}$. Vortex lines originating from this inviscid region have unique continuations into and through $\mathscr{C}$. Therefore, it is possible to qualitatively explore the evolution of vortex lines, even though they are not associated with material particles in the viscous region $\mathscr{C}$. This idea is applied in Fig. 4a–f to explain the consequences of viscous vorticity annihilation, which occurs along the y-axis where $\boldsymbol{\omega} \equiv \mathbf{0}$. Schematically, each vortex is represented by three vortex lines, one line from each vortex intercepts $\mathscr{C}$ (Fig. 4a), and these shortly afterwards annihilate each other

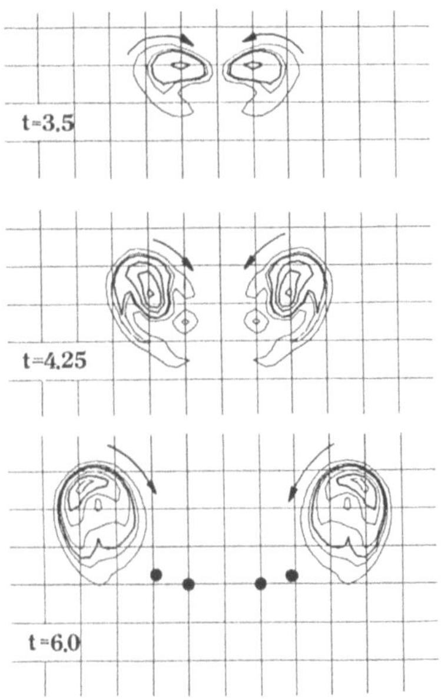

Fig. 3. Vorticity distribution in π_d. The heavy contours show $|\omega|$ in increments of 5, while the thin lines show increments of 2. Initially there is no vorticity in π_d; therefore we have not shown panels prior to $t = 3.5$. In the panel at $t = 6.0$ we show the trajectory of the vorticity peak in π_d

at $x = z = 0$ by cross-diffusion (Fig. 4b). This produces two reconnected lines intercepting π_d at $z = \pm z_0$ (Fig. 4c). The π_d-intercept z_0 quickly retreats away from $x = z = 0$, because of strong diffusion and nearly antiparallel vorticity vectors (in the limit of antiparallel vectors z_0 moves infinitely fast). This fast viscous transport of vortex lines stops at the edge of $\mathscr{C}$, where the vorticity vectors cease to be nearly antiparallel.

A cross section AB (Fig. 4e) at the edge of $\mathscr{C}$ reveals the 2D-dipole structure of the original vortices. In a co-translating frame the streamline pattern becomes the well known recirculation bubble of a vortex-dipole (Fig. 4g); although this frame is not precisely defined for an unsteady dipole, the existence of two stagnation points S_T and S_B is clear. The large velocity, occurring between S_T and S_B, advects z_0 towards S_T, ahead of the dipole. At S_T the relative upward advection of z_0 stops, and our reconnected vortex line unfolds along the diverging separatrix a–a (Fig. 4g). In this location, the vortex line continuously stretches and begins to wrap around the dipolar vortex pair (Fig. 4f). Meanwhile, the circulatory motion in the recirculation bubble continuously pumps fresh non-reconnected lines into $\mathscr{C}$ near S_B (Fig. 4g). Because of this mechanism, reconnected vortex lines accumulate ahead of the antiparallel vortex pair (Fig. 4f). Thus, the circulatory motion in the recirculation bubble creates characteristic 'humps' on top of the antiparallel vortices in Fig. 1 (see B, E, I in the left column of Fig. 1). This 'bridging' process is illustrated in the perspective view (Fig. 5) of one half of the vortices ($z > 0$); this should be compared with the actual vortex lines shown in Fig. 1.

Looking like curved hairpins (Fig. 1, left column E, I), the bridges move upward and away from π_s (Fig. 3) due to self-induction. Meanwhile, new vortex lines are continuously drawn into the swirling bridges, and as the accumulation progresses,

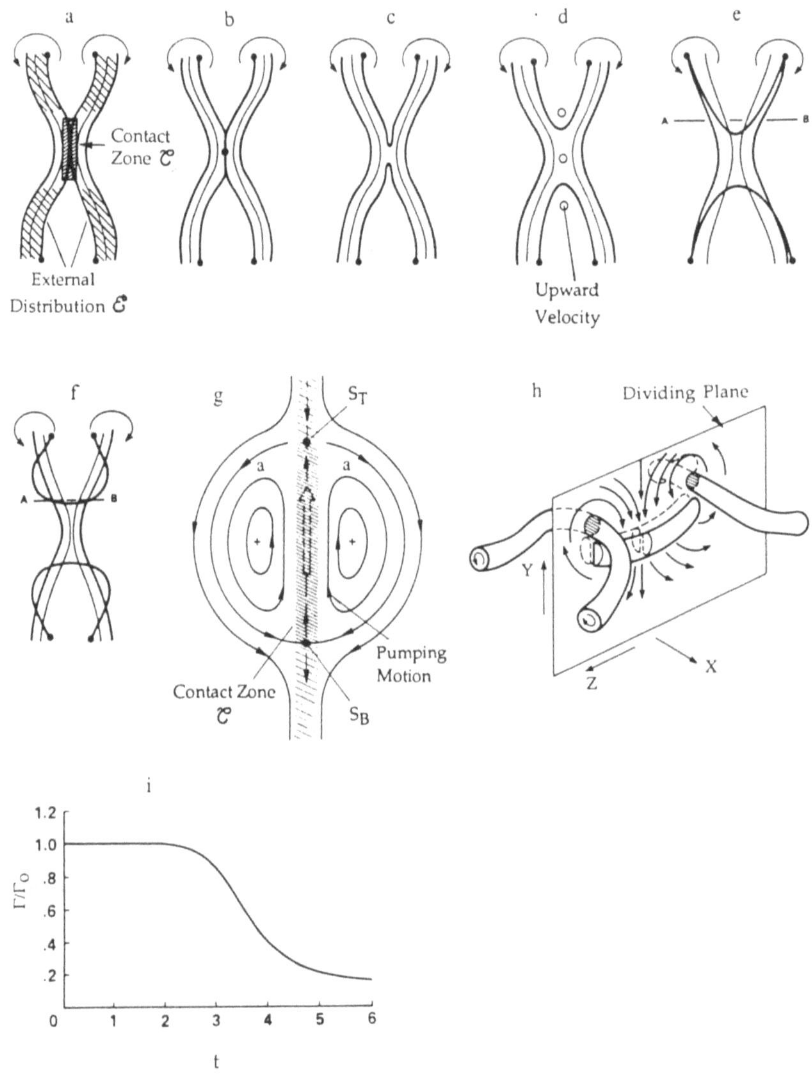

Fig. 4a–i. Annihilation and reconnection sequence; each vortex is denoted by three vortex lines; (**g**) streamline pattern in the symmetric plane; (**h**) induced motion of the bridges and the downwash; (**i**) circulation in π_s^+ $(x > 0)$

the cores of the bridges become rounded. However, the vorticity distribution inside these cores is heavily skewed in the direction of the bridges' propagation. Near a bridge's intersection with π_d, the helicity production is large since reconnected and non-reconnected lines must wrap around each other.

During the bridging process, a complex interaction of competing effects governs the rate of annihilation, by changing the vorticity gradients in $\mathscr{C}$. These effects are diffusion, stretching in the z-direction, and local 'self-induction'. The diffusion

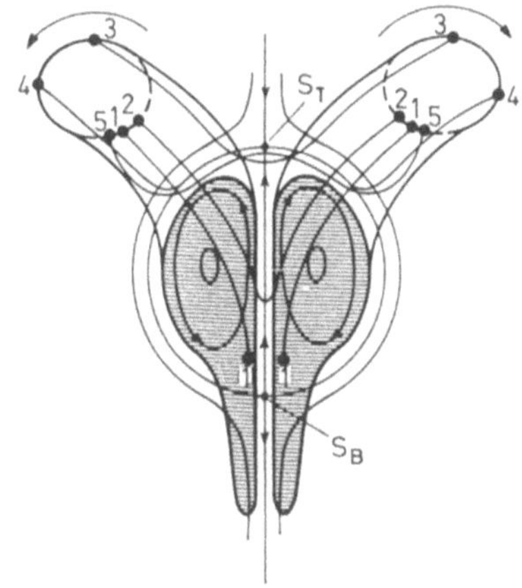

Fig. 5. Advection and accumulation of vortex lines during bridging. The streamline pattern of Fig. 2(g) is superimposed on the vorticity contour (hatched area). Lines 1–1, 2–2, 3–3, 4–4 and 5–5 represent successive positions of vortex lines

counteracts the growth of gradients and can only decrease the peak vorticity ω_m in π_s. On the contrary, any increase in ω_m is due to stretching normal to π_s. This 'axial' stretching is mainly generated by the vorticity distribution $\mathscr{E}(t)$ away from $\mathscr{C}$ ($\mathscr{E}$ is shown in Fig. 4a, note $\mathscr{E}$ includes the bridges), but also in part by self-induced lengthening of vortex lines near $\mathscr{C}$ (analogous to the axisymmetric collision of vortex rings). Depending on the relative positions of $\mathscr{E}$ and the dipole in π_s, the external stretching is either positive (vortex stretching) or negative (vortex compression): positive if the dipole has fallen behind $\mathscr{E}$ (Fig. 4h), and negative if the dipole has advanced ahead of $\mathscr{E}$. The self-induction on each side of the sheetlike region $\mathscr{C}$ is normal to π_d, and is directed toward π_d whenever the non-reconnected vortex lines in $\mathscr{C}$ curve upward, and away from π_d when they curve downward. The motion toward π_d increases gradients across the y-axis by squashing the two vortex cores into a characteristic head-tail structure (Fig. 2), while motion away from π_d decreases the gradients by separating the vortex cores in π_s. The above three effects—diffusion, stretching and self-induction—dominate the present evolution (Re = 1000) in the following order. At first the vortex lines in $\mathscr{C}$ curve upward, and the gradients across π_d increase because self-induction toward π_d presses the dipole cores into a head-tail shape; meanwhile, the external stretching is negative, hence the increase in ω_m is small. However, as the dipole in π_s falls behind $\mathscr{E}$ (for reasons discussed later) such that the vortex lines in $\mathscr{C}$ curve downward, the axial stretching picks up and produces a large increase in ω_m, and consequently larger gradients. Finally, toward the end of the rapid circulation transfer from π_s to π_d (Fig. 4i), the local self-induction away from π_d arrests the cross-diffusion by producing a near balance between axial stretching, diffusion, and separation of the dipole cores in π_s.

The curvature reversal, which causes the incomplete reconnection, is a characteristic and crucial event. It occurs, because the upward motion of the dipole in π_s slows down relative to that of $\mathscr{E}$ (Fig. 1)—but why does the dipole in π_s slow down? A downward velocity field in $\mathscr{C}$—the downwash—opposes the upward motion in π_s (Fig. 4h). Furthermore, the dipole loses circulation through vorticity annihilation, and also relaxes into a head-tail shape. The downwash is induced by the ever growing bridges as well as by non-reconnected vortex lines in $\mathscr{E}$, so that a downwash will be present even at $Re = \infty$. The relaxation into a head-tail shape, an inviscid process, virtually decreases the dipole circulation as the dipole propagation velocity is determined almost exclusively by the head. Although this inviscid effect is weak at $Re = 1000$, it may well be the chief mechanism for curvature reversal at higher Re. An indication of this would be that the area A of the head decreases with higher Re, such that $\omega_m A^{1/2}$ is bounded as $Re \to \infty$. This indicates the possibility that the amount of circulation transferred from π_s to π_d tends to zero as $Re \to \infty$ while the transfer is still occurring within a convective timescale.

Conclusion

In our bridging mechanism the well known viscous vorticity annihilation is working in conjunction with viscous vorticity transport, dipole pumping, local self-induction and axial stretching such as to accumulate the reconnected vortex lines into bridges, which characteristically appear ahead of the original vortices and are aligned orthogonal to the latter. A curvature reversal in $\mathscr{C}$ arrests the rapid viscous annihilation by producing a near balance between diffusion, axial stretching, and core separation. Consequently, the reconnection is only partial. This partial reconnection points to a cascade of partial reconnections as the resolution of the old paradox concerning the reconnection process in the limit $Re \to \infty$, and also suggests a new mechanism for producing enhanced mixing and cascade of vorticity and helicity.

Acknowledgements: This and other simulations were done during 1988 Summer workshop of NASA-Stanford Center for Turbulence Research (CTR), during which we had fruitful discussions with K. Shariff. We also acknowledge the support of ONR Grant N00014-87-K-0126.

References

Hussain, F. (1986): J. Fluid Mech. **173**, 303
Melander, M. V., Hussain, F. (1989): Phys. of Fluids A, **1**, 633
Melander, M. V., Hussain, F. (1988): Report CTR-S88, 257, Stanford U
Takaki, R., Hussain, F. (1985): Turb. Shear Flow V, 3.19, Cornell U

High Resolution Three-Dimensional (256^3) Spatio-Temporal Measurements of the Conserved Scalar Field in Turbulent Shear Flows

Werner J. A. Dahm and Kenneth A. Buch

Department of Aerospace Engineering, University of Michigan, Ann Arbor, MI 48109-2140, USA

Abstract

We present results from highly resolved three-dimensional spatio-temporal measurements of the conserved scalar field $\zeta(\mathbf{x}, t)$ in a turbulent shear flow. Each of these experiments consists of 256^3 individual point measurements of the local instantaneous conserved scalar value in the flow. The spatial and temporal resolution of these measurements reach beyond the local Kolmogorov scale and resolve the local strain-limited molecular diffusion scale in the flow. This resolution capability, together with the signal quality achieved, allows spatial derivatives involved in determining the instantaneous local molecular mixing rate field $\nabla\zeta \cdot \nabla\zeta(\mathbf{x}, t)$ to be accurately obtained at each point in the resulting data volume. These results clearly show molecular mixing occurring in thin strained laminar diffusion layers in a turbulent flow.

Introduction

In the mixing of a conserved scalar $\zeta(\mathbf{x}, t)$ in a turbulent flow, the associated scalar energy per unit mass $\frac{1}{2}\zeta^2(\mathbf{x}, t)$ follows a transport equation in which the corresponding local instantaneous rate of scalar energy dissipation per unit mass, $(\mathrm{ReSc})^{-1}\nabla\zeta \cdot \nabla\zeta(\mathbf{x}, t)$, gives the rate at which non-uniformities in the scalar field are reduced by the molecular diffusivity. In this context, the scalar dissipation gives the local instantaneous rate of molecular mixing in the flow. The concepts of scalar energy and its dissipation rate offer a useful perspective for understanding mixing and chemical reaction in turbulent shear flows, and play a central role in many approaches for modeling turbulent reacting flows.

A number of investigations in recent years have obtained measurements of conserved scalar fields in turbulent flows. Virtually all of these have made use of various planar imaging techniques; a good review of such methods is given by Hanson (1986). However, most of these investigations have been primarily aimed at the topologically simple large scale features of transitional flows, rather than the comparatively complex fine scale structure of the conserved scalar field in fully developed turbulent flows. Furthermore, and related to this, many of these measurements have had a spatial and temporal resolution far too coarse to resolve the finest scales on which gradients exist and molecular mixing occurs in fully developed turbulent flows. Accurate measurements of the conserved scalar field, with a resolution sufficient to distinguish the finest spatial and temporal scales involved in the scalar gradient field $\nabla\zeta(\mathbf{x}, t)$ have been difficult to obtain and have limited progress in this approach to mixing and reaction in turbulent shear flows.

Here we present results from highly resolved three-dimensional (256^3) spatio-temporal measurements of the conserved scalar field $\zeta(\mathbf{x}, t)$ in the fully-developed self-similar far field of a turbulent shear flow. Our experiments are based on high-speed acquisition of 256 successive 256 × 256 spatial data planes of laser induced fluorescence in the far field of an axisymmetric coflowing turbulent jet in water. The resulting 256^3 data volumes consist of the temporal dimension and two of the three spatial dimensions in $\zeta(\mathbf{x}, t)$. The spatial and temporal resolution of these measurements exceeds the local Kolmogorov scale and reaches the local strain-limited molecular diffusion scale in the flow, allowing spatial derivatives involved in determining the local instantaneous scalar gradient field $\nabla\zeta(\mathbf{x}, t)$ and the corresponding scalar dissipation field $\nabla\zeta\cdot\nabla\zeta(\mathbf{x}, t)$ to be accurately obtained from the resulting data volumes.

Experiment Configuration

The experiments described here are conducted in the far field of an axisymmetric turbulent jet issuing into the coflowing stream formed in a low turbulence level, gravity driven, actively-controlled Bernoulli tunnel. High resolution three-dimensional spatio-temporal data volumes are obtained for the resulting space and time varying conserved scalar field by designing the experiments so that spatial structure in the scalar field smaller than the local strain-limited molecular diffusion scale in the flow can be resolved between adjacent pixels in each individual spatial data plane as well as by the laser sheet thickness, and by successively acquiring such data planes at a rate exceeding the local molecular diffusion-scale passage time in the flow. The jet fluid mixture fraction, in this case determined by the concentration of a passive laser fluorescent dye (disodium fluorescein) carried by the jet fluid, is a conserved scalar in the flow. Here the mixture fraction is determined by measuring the intensity of laser induced fluorescence from dye-containing fluid in a thin laser sheet intersecting the flow and containing the jet axis. The fluorescence intensity from the imaged portion of the flow, centered on the jet axis, was optimized by determining the fluorescent species concentration at which the product of the typical local scalar concentration in the flow and the laser sheet power, which decreases as it propagates through the flow due to absorption and reëmission of energy by the dye, reaches its maximum. Following Dahm (1985) and Dahm and Dimotakis (1986), the local instantaneous fluorescence intensity $F(x, y)$ measured at any point in the plane is then related to the corresponding conserved scalar value $\zeta(x, y, t)$ through a simple attenuation integral and the overall image transfer function $h(x, y)$ as

$$\frac{\zeta(x,y,t)}{\zeta(\hat{x},\hat{y},t)}=\frac{F(x,y,t)/F(\hat{x},\hat{y},t)}{h(x,y)/h(\hat{x},\hat{y})}\exp\int_{\hat{y}}^{y} I(c_0\cdot\zeta(x,\xi,t))\,\mathrm{d}\xi \qquad (1)$$

where $(\hat{x}, \hat{y})$ denotes an arbitrary reference location and c_0 denotes the dye concentration corresponding to $\zeta = 1$. The transfer function $h(x, y)$ is determined from a measurement of $F(x, y)$ at a uniform dye concentration $\bar{c}$ as

$$\frac{h(x,y)}{h(\hat{x},\hat{y})}=\frac{F(x,y)}{F(\hat{x},\hat{y})}\exp\{I(\bar{c})\cdot(y-\hat{y})\} \tag{2}$$

where $I(\bar{c})$ in Eqs. (1) and (2) is the measured fractional attenuation per unit beam propagation distance y at the dye concentration $\bar{c}$. The attenuation integral in Eq. (1) is necessary to precisely determine the conserved scalar field $\zeta(x, y, t)$ from the measured fluorescence intensity data. However, the principal objective in these experiments is in determining the scalar *gradient* field from such fluorescence data, and in that case only the attenuation between adjacent pixels is relevant. For the present measurements, this was less than 0.09% and therefore neglected, so that

$$\nabla\left(\frac{\zeta(x,y,t)}{\zeta(\hat{x},\hat{y},t)}\right)\approx\nabla\left(\frac{F(x,y,t)/F(\hat{x},\hat{y},t)}{h(x,y)/h(\hat{x},\hat{y})}\right) \tag{3}$$

The key elements of the data acquisition system used to measure the space and time varying laser induced fluorescence intensity $F(x, y, t)$ in these experiments are shown in Fig. 1. Briefly, the fluorescence is optically projected onto a 256×256 photodiode array, having a pixel spacing of $40\,\mu$m. This array is driven from an external clock at pixel rates up to 11 MHz, and thus can continuously measure successive 256^2 data planes at rates up to 140 planes per second. Of particular importance here, this variable clocking capability allows the data rate *relative to the flow time scales* to be kept constant among measurements at different flow conditions, thereby allowing comparisons at the same relative temporal resolution. The fluorescence signal from the array is serially acquired through a programmable digital port interface slaved to the array clock, digitized to 8-bits digital resolution and ported directly into high-speed memory. Programmable gain and offset permit the data to span the entire 8 bits of digital resolution, allowing fully 256 different conserved scalar levels to be distinguished. The rms pixel noise in the resulting measured fluorescence signal was in all cases less than ± 3 digital signal levels out of

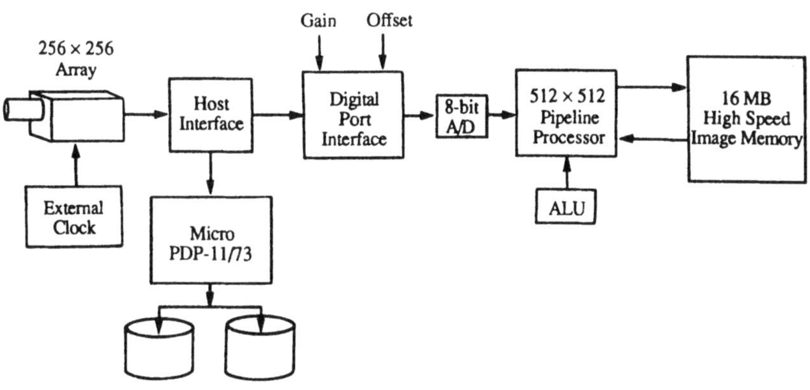

Fig. 1. Schematic showing key elements of the high-speed variable rate data acquisition and processing system for acquiring 256^3 spatio-temporal data volumes of the conserved scalar field $\zeta(\mathbf{x}, t)$ in turbulent flows

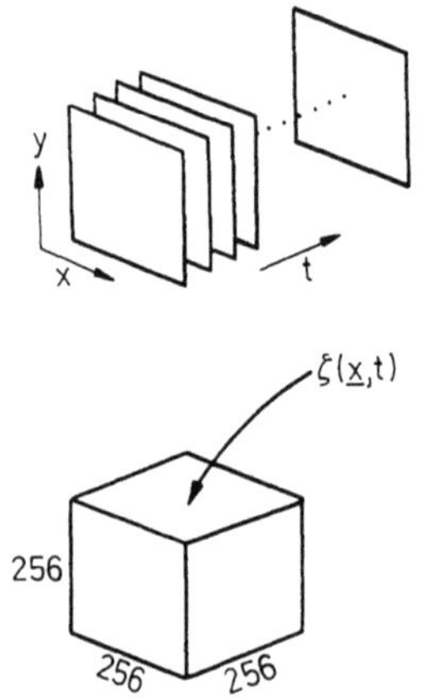

Fig. 2. Structure of the measured spatio-temporal data volume $\zeta(\mathbf{x},t)$ in the conserved scalar field as 256 successive 256 × 256 spatial data planes. Note that scales smaller than the local strain-limited molecular diffusion scale are spatially resolved within each data plane, and are temporally resolved from one plane to the next. Spatial gradients involved in determining the local instantaneous rate of molecular mixing in the flow can therefore be meaningfully determined from these data

the 256. The 16 MB high-speed memory can accommodate 256 successive individual data planes, producing the 256^3 spatio-temporal data volume shown in Fig. 2.

For the pixel clock rate used in the present measurements, the time between acquisition of successive data planes was in all cases less than 10 msec. The 1/e waist thickness of the laser sheet was measured as 130 μm, while at the edges of the imaged portion the sheet thickness was approximately 280 μm. As a result, for the image ratios used in these measurements (as low as 1.1:1) the volume in the flow imaged onto each pixel had dimensions as small as 44 μm × 44 μm × 130 μm.

To determine the resulting measurement resolution capabilities, this pixel image volume and time must be compared with the smallest local spatial and temporal scales of the conserved scalar field in the flow. Following Maczynski (1962), in the far field of such an axisymmetric coflowing turbulent jet, the growth of the local flow width δ and the decay of the local mean centerline excess velocity $u \equiv (U - U_\infty)$ where U is the local mean centerline velocity, with increasing downstream distance x follow non-algebraic similarity scalings of the form

$$(\delta/\vartheta) \sim f_1(x/\vartheta) \tag{4a}$$

$$(u/U_\infty)^{-1} \sim f_2(x/\vartheta) \tag{4b}$$

where $\vartheta^2 = J_0/(\pi\rho_\infty U_\infty^2)$ is the 'momentum radius' of the flow, with J_0 denoting the jet source momentum flux and U_∞ the speed of the coflowing stream. Measurements by Reichardt (1965) and Biringen (1975) confirm this similarity scaling and provide the scaling functions f_1 and f_2, which ultimately set the measurement resolution requirements as described below.

Following Burgers (1948) and Corcos and Sherman (1984), for local diffusion of vorticity in the presence of a time-varying strain rate $\sigma(t)$, the competing effects of strain and diffusion establish an equilibrium strain-limited vorticity diffusion layer thickness $\lambda_\nu \sim (\nu/\sigma)^{1/2}$, closely related to the Kolmogorov scale, giving the finest scale on which spatial gradients in the strain rate and vorticity fields can be sustained in the flow. A similar competition between the effects of strain and molecular diffusion of the conserved scalar establishes a local strain-limited molecular

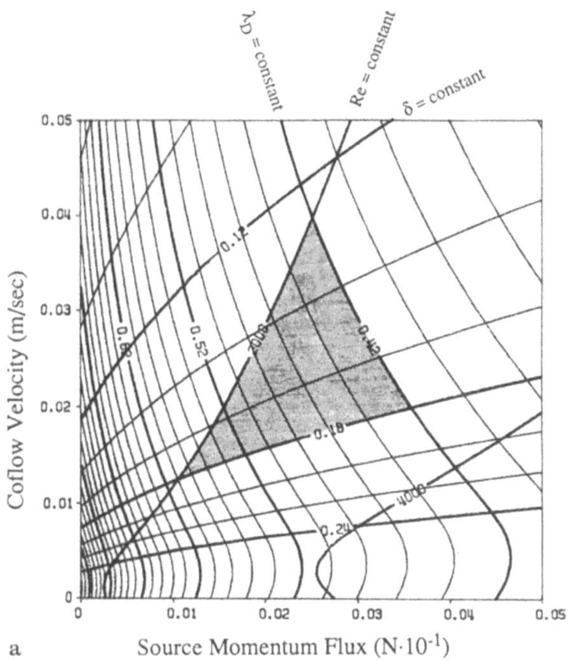

a

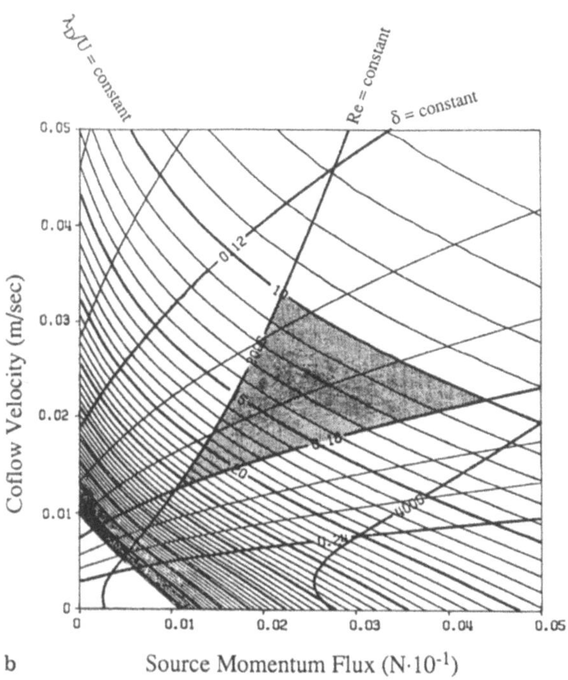

b

Fig. 3 a. Spatial resolution capabilities of the present measurements, giving experimental conditions for which the local strain-limited molecular diffusion scale in the flow can be spatially resolved. Shown are contours of the local Reynolds number *Re*, the local flow width δ (in mm), and the local molecular diffusion scale λ_D (in mm). The shaded area gives the range of conditions in the present experiments for which λ_D can be spatially resolved by better than a factor of three. At the actual measurement conditions, the volume in the flow imaged onto each pixel was more than two orders of magnitude smaller than λ_D^3. **b**. Temporal resolution achieved in the present measurements, giving the experimental conditions for which the local strain-limited molecular diffusion scale in the flow is temporally resolvable. Shown are contours of the local Reynolds number Re, the local flow width δ (in mm), and the local molecular diffusion scale passage time λ_D/U (in msec). The shaded area gives the range of conditions in the present experiments for which λ_D can be temporally resolved in the present measurements diffusion layer thickness $\lambda_D \sim (D/\sigma)^{1/2}$, related to the Batchelor scale and giving the smallest scale on which spatial gradients in the conserved scalar field can be sustained by the flow. The ratio of the vorticity and scalar diffusivities, ν and D respectively, establishes the relation between these two scales as $\lambda_D \sim \lambda_\nu \cdot Sc^{-1/2}$, where $Sc \equiv (\nu/D)$ is the Schmidt number. The similarity of the two diffusion processes indicates that the proportionality constant should be approximately one. The highest strain rates occurring locally in the flow scale as $\sigma \sim (u/\delta) \cdot Re^{1/2}$, giving the finest local scale in the conserved scalar field as $(\lambda_D/\delta) \sim Sc^{-1/2} \cdot Re^{-3/4}$. Measurements by Dowling (1988) and Dowling and Dimotakis (1988) give indications that the resulting proportionality constant is roughly 25. Here $Re \equiv u\delta/\nu$ is the local Reynolds number of the flow, where $\delta(x)$ and $u(x)$ are determined by Eqs. (4a) and (4b). The scaling functions $f_1(x/\vartheta)$ and $f_2(x/\vartheta)$ then give the results shown in Fig. 3a and 3b for the local molecular diffusion scale λ_D and the local molecular diffusion scale passage time λ_D/U at the downstream location $x = 1.0\,\text{m}$ at which these measurements were conducted. The shaded regions denote the range of conditions for which the local molecular diffusion scale could be resolved by better than a factor of three, together with the further constraints that the local Reynolds number exceed 2,000 and the local full width of the flow is no more than one half the tunnel width. Note that, for the values of J_0 and U_∞ at which the measurements were actually conducted, the volume in the flow imaged onto each pixel was approximately 235 times smaller than the resulting local strain-limited molecular diffusion scale volume λ_D^3, with its maximum dimension approximately 2 times smaller than λ_D, and the data acquisition rate was sufficient to resolve the local molecular diffusion scale between successive spatial data planes.

Results and Discussion

Figures 4a and 5a show digital reconstructions of the data in two typical 256 × 256 spatial data planes. Each data volume is composed of 256 successive data planes such as these. The 256 different color levels denote the local conserved scalar value $\zeta(\mathbf{x}, t)$, with pure blue denoting $\zeta = 0$ and increasing uniformly to pure red, corresponding to the highest scalar values in the data volume. The local Reynolds number for these measurements is approximately 3,500, for which the data shown span approximately one strain-limited vorticity diffusion scale λ_ν and approximately 25 strain-limited molecular diffusion scales λ_D in both the downstream and radial directions. In terms of the local flow width δ, the data span approximately $\frac{1}{18}$ of the total flow width in both directions. In the temporal dimension, the 256 data planes in the volume span approximately eight local vorticity scale passage times λ_ν/U, or about one-half of the local large scale passage time δ/U.

While data such as those in Figs. 4a and 5a permit detailed numerical examination of the conserved scalar field $\zeta(\mathbf{x}, t)$, the principal interest here is in the scalar gradient field $\nabla\zeta(\mathbf{x}, t)$, and in particular in the corresponding scalar dissipation field $(\text{ReSc})^{-1}\nabla\zeta \cdot \nabla\zeta(\mathbf{x}, t)$, giving the instantaneous local molecular mixing rate in the

a

b

c

Fig. 4 a. A typical 256 × 256 spatial data plane showing the instantaneous conserved scalar values denoted by pseudocolor assignments, with pure blue denoting $\zeta(\mathbf{x}, t) = 0$ and increasing uniformly to pure red, denoting the highest values observed in the data volume. **b.** Scalar dissipation $\nabla\zeta \cdot \nabla\zeta$ obtained from the scalar values in the data plane shown in Fig. 4a. Data of this type, together with that shown in Fig. 4a, allow the instantaneous values of the conserved scalar and the scalar dissipation to be determined at each point in the data volume. Only non-zero values of the dissipation are shown here. Note that relatively high dissipation rates are comparatively rare, and appear to be localized along thin interfaces in the flow. **c.** The logarithm of the scalar dissipation rate, $\log_e \nabla\zeta \cdot \nabla\zeta(\mathbf{x}, t)$, obtained by a logarithmic mapping of the color assignments in Fig. 4b to allow increased constrast at low dissipation values. The data here show that, not only are high dissipation rates confined to thin diffusion layers in the flow, but indeed essentially all the dissipation occurs in such layers

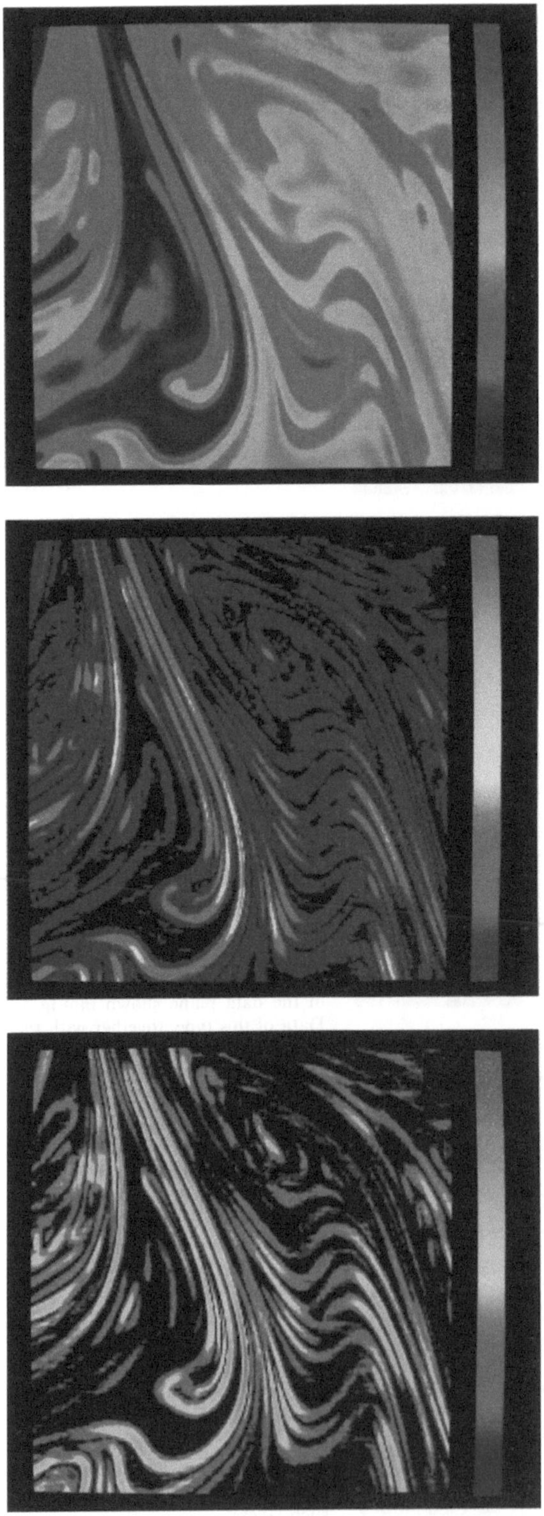

a

b

c

Fig. 5 a. Another 256 × 256 spatial data plane for slightly different conditions, showing the instantaneous conserved scalar values denoted by pseudocolor assignments, with pure blue denoting $\zeta(\mathbf{x},t)=0$, and with mixed fluid values increasing uniformly from blue to red. **b.** The scalar dissipation $\nabla\zeta\cdot\nabla\zeta$ obtained from the scalar values in the data plane shown in Fig. 5. Note again that molecular mixing of the scalar is essentially confined to thin strained diffusion layers in the flow. The measurement resolution was sufficient to allow internal structure within these diffusion layers to be examined. **c.** The logarithm of the scalar dissipation rate, $\log_e \nabla\zeta\cdot\nabla\zeta(\mathbf{x},t)$ corresponding to the data in Figs. 5a and 5b. Again, essentially all the molecular mixing appears to occur within thin strained laminar diffusion layers in the turbulent flow

flow. These data are sufficiently resolved to allow accurate determination of the spatial derivatives within each data plane. Of course, the out-of-plane component of the scalar gradient vector cannot be determined from such spatio-temporal data. Nevertheless, derivatives within each spatial data plane give the projection of the true scalar gradient vector into the plane. Here this projection is determined by direct differentiation of the raw data within each data plane using linear central difference approximations. No smoothing or filtering is needed. The magnitude squared of the resulting projection vector then gives the two-dimensional approximation for the scalar dissipation $\nabla\zeta \cdot \nabla\zeta(\mathbf{x}, t)$. For the two data planes shown in Figs. 4a and 5a, the resulting scalar dissipation fields are presented in Figs. 4b and 5b. The quality of these scalar dissipation fields is perhaps the single strongest testimony to the resolution of the measurements. The temporal evolution of the scalar and dissipation fields can be investigated by comparing data in the successive planes.

Although a wide variety of statistical information could be derived from data such as those in Figs. 4 and 5, our primary interest is in identifying topological and structural features associated with the molecular mixing in turbulent flows. To this end, we note first that the scalar dissipation rate is rather low throughout most of the data volume. Notice that in Figs. 4b and 5b, only non-zero values of the dissipation are shown. Large values of the dissipation occur rather infrequently, and it is generally assumed that the scalar dissipation is lognormally distributed. This does not say, however, that most of the molecular mixing occurs at low values of the dissipation. To answer this first requires integrating the scalar dissipation field obtained from these measurements. This question has significant implications for theoretical descriptions of extinction in turbulent reacting flows.

To increase the contrast at low values of the scalar dissipation, the same data in Figs. 4b and 5b are shown in Figs. 4c and 5c after a logarithmic mapping of the color assignments. In this case, the colors denote increasing values of $\log_e \nabla\zeta \cdot \nabla\zeta(\mathbf{x}, t)$. Notice that these results now show that essentially all values of the dissipation occur along thin interfaces in the flow. These data give direct evidence of molecular mixing occurring in thin strained laminar diffusion layers in a turbulent flow. The resolution is sufficient to allow numerical experiments to be performed on these data volumes to investigate internal structure within these layers. Preliminary results from such experiments suggest good agreement with asymptotic solutions for strain-limited molecular diffusion layer structure.

Last, note that within each data plane, and in comparing among different planes within the data volume, these molecular diffusion layers all appear to have about the same thickness, and are of the order of the local strain-limited molecular diffusion scale estimates λ_D in Fig. 3a. On the one hand, this is consistent with the notion of the strain rate being essentially uniform over regions smaller than the local vorticity diffusion scale λ_ν, as is the case within each data plane. However, the temporal dimension of our data volume spans approximately eight local λ_ν/U, so some variations in stain rate should be covered by the data volume. On the other hand, it can be argued that this uniform interface thickness might be an artifact of the missing component in the scalar gradient vector. Resolution of this issue can be addressed by data spanning all three space dimensions to allow determination of the three components of the scalar gradient field.

Conclusions

We have obtained 256^3 volumes of three-dimensional spatio-temporal measurements for the conserved scalar field in a turbulent shear flow, with spatial and temporal resolution sufficient to resolve scales below the local strain-limited molecular diffusion scale in the flow. The high resolution and signal quality afforded by these measurements allow accurate determination of spatial derivatives involved in the corresponding scalar dissipation field $(\mathrm{ReSc})^{-1}\nabla\zeta\cdot\nabla\zeta(\mathbf{x},t)$ giving the local instantaneous rate of molecular mixing in the flow. These scalar gradient components are obtained through direct differentiation of the raw data. No explicit smoothing or filtering is found to be necessary.

Results directly show maps of the local instantaneous molecular mixing rate in the flow. Throughout most of these data planes, the scalar dissipation rate is low. Large values of the scalar dissipation occupy only a small fraction of the data volume. All values of the scalar dissipation are essentially confined along thin interfaces in the flow, giving direct evidence of molecular mixing occurring in thin strained laminar diffusion layers in a turbulent flow. Preliminary numerical experiments performed on these data bases appear to suggest good agreement with asymptotic solutions for internal structure within these strained diffusion layers.

Acknowledgements: The work reported here was supported, in part, by the Gas Research Institute (GRI) under Contract No. 5087-260-1443 and by The University of Michigan.

References

Biringen, S. (1965): An experimental investigation of a turbulent axisymmetric jet issuing into a coflowing airstream. *VKI Tech. Note 110*

Burgers, J. M. (1948): A mathematical model illustrating the theory of turbulence. pp. 171–199, In *Adv. Applied Mech.* (R. von Mises, T. von Kármán, eds), Academic Press, New York

Corcos, G. M., Sherman, F. S. (1984): The mixing layer: deterministic models of a turbulent flow. Part 1. Introduction and the two-dimensional flow. J Fluid Mech. **139**, 29–65

Dahm, W. J. A. (1985): Experiments on entrainment, mixing and chemical reactions in turbulent jets at large Schmidt number. *Ph.D. thesis*, Caltech

Dahm, W. J. A., Dimotakis, P. E. (1986): Measurements of entrainment and mixing in turbulent jets. AIAA J **25**, 1216–1223

Dowling, D. R. (1988): Mixing in gas phase turbulent jets. *Ph.D. thesis*, Caltech

Dowling, D. R., Dimotakis, P. E. (1988): On mixing and structure of the concentration field of turbulent jets. *Proc. 1st Nat. Fluid Dyn. Congr.*, AIAA, Washington

Hanson, R. K. (1986): Combustion diagnostics: planar imaging techniques. *Proc. 21st Int'l. Symp. on Comb.*, The Combustion Institute, 1677–1691

Maczynski, J. F. J. (1962): A round jet in an ambient co-axial stream. J. Fluid Mech. **13**, 597–608

Reichardt, H. (1975): Turbulente Strahlausbreitung in gleich-gerichteter Grundströmung. Forsch. Ingenieurwesen **30**, 133–139

Direct Simulation of Homogeneous Turbulence and Gravity Waves in Sheared and Unsheared Stratified Flows

Thomas Gerz and Ulrich Schumann

DLR, Institute of Atmospheric Physics, W-8031 Oberpfaffenhofen, Germany

Abstract

Differences in the evolution of stably stratified turbulence with and without mean shear are investigated by means of direct numerical simulations for moderate Reynolds number ($Re_\ell = 42.7$) and seven values of the Froude or Richardson number. The molecular Prandtl number is unity. In stratified flows without shear, energy decays quickly initially but at reduced rate later when gravity waves dominate the flow pattern. Gravity waves occur when the Ellison length scale reaches about 0.3 to 0.8 times the Ozmidov length scale. The flow becomes anisotropic before the first waves arise. The developed flow is dominated by gravity waves, and turbulent mixing is considerably suppressed, when the Ellison length scale is about six times the Kolmogorov length scale. For sheared turbulence the importance of buoyancy relative to shear forcing depends on the Richardson number Ri. For an initial shear number $Sh_0 = 3$, we find a critical Richardson number of 0.13 which is smaller than the value 0.25 predicted by linear inviscid theory because of the rather strong dissipation in the present simulations. In subcritical flows ($Ri < Ri_{crit}$), turbulence is dominated by shear. If the Richardson number is supercritical ($Ri > Ri_{crit}$), the turbulence is controlled by gravity and behaves at large scales as if no shear would be present. But shear causes small-scale turbulence (possibly by wave breaking) and hence the dissipation is larger than without shear. The degree of anisotropy increases with increasing Richardson number but gets limited when counter-gradient fluxes (CGF) of heat and momentum appear in the vertical direction. Temporally oscillating and sign-changing vertical fluxes at large scales have to be distinguished from persistently positive fluxes (p-CGF) at small scales. Both types develop in sheared as well as in unsheared stratified flows. The oscillating flux exchanges energy between kinetic and potential energy reservoirs and can be described by rapid-distortion calculations. The p-CGF is due to an imbalance between kinetic and potential energy sources and sinks at small scales.

Introduction

By means of direct numerical simulation we investigate homogeneous initially isotropic turbulence under the influence of four different forces, i.e. inertia, viscous forces, buoyancy and shear forces. Turbulence in stably stratified fluids decays quite differently from unstratified turbulence since gravity causes anisotropy in the horizontal and vertical velocity variances as well as in the horizontal and vertical length scales. Due to stable stratification, gravity waves are formed in the flow which contribute to the velocity variance with little vertical mixing. Counter-gradient fluxes develop in order to keep the energy budget in equilibrium. Ultimately, turbulence collapses and might get converted into horizontal motion patterns (Riley et al. 1981; Stillinger et al. 1983; Itsweire et al. 1986; Métais and Herring 1989).

Turbulent Shear Flows 7

Here, results of seven unsheared experiments will be compared with corresponding cases with shear in order to investigate differences between both types of flow. We investigate the appearance of gravity waves as a function of the Ellison length scale L_E and the Ozmidov length scale L_O, which are defined as

$$L_E = \frac{T'}{dT_R/dz} \quad \text{and} \quad L_O = \left(\frac{\varepsilon}{N^3}\right)^{1/2}, \tag{1}$$

where z, T_R, ε, and N describe the vertical space coordinate, the reference (mean) temperature, the mechanical dissipation rate, and the Brunt–Väisälä frequency, respectively. Primes denote root-mean-square (rms) values. L_E is a measure for a typical vertical size of an eddy in a buoyant medium and L_O is the length where, at large Reynolds numbers, buoyancy is of the same order of magnitude as inertia. For stratified turbulent flows, Gibson's theory (Gibson 1981) predicts that three-dimensional turbulence is active (overturning) when L_E is in the range

$$0.6L_O \geqq L_E \geqq 7.5L_K \text{ (a)} \quad \text{where} \quad L_K = \left(\frac{\nu^3}{\varepsilon}\right)^{1/4} \text{ (b)} \tag{2}$$

is the Kolmogorov length scale and ν the kinematic viscosity. According to Gibson's theory, the limit $0.6L_O$ marks the vertical size of turbulent eddies which are first influenced by the gravity force such that they induce gravity waves. The limit $7.5L_K$ is the vertical length scale of an eddy when the mean vertical heat flux first goes to zero, indicating that buoyancy acts at all scales. Itsweire and Helland (1989) analysed direct measurements of the buoyancy flux in a density-stratified turbulent flow and found that at this stage vertical mixing gets strongly suppressed at all scales (extinction of turbulence). Measurements of Stillinger et al. (1983) and Itsweire et al. (1986) yielded a range of active turbulence in unsheared stratified flows of $(0.70 \cdots 0.85)L_O > L_E > (7.6 \cdots 9.9)L_K$, which confirms Gibson's predictions fairly well.

If stably stratified turbulence is forced by shear, the gradient-Richardson number Ri controls whether turbulence grows or decays in time so that the flow is either subcritical or supercritical. Linear inviscid theories predict a critical Richardson number $Ri_{crit} = 0.25$ (Miles 1961). This critical value has been corroborated by observations in the atmosphere where the Reynolds numbers are large (Businger et al. 1971). At lower Reynolds numbers, to which laboratory flow experiments (Rohr et al. 1988) and direct simulations (Holt et al. 1989) are limited, also lower values of Ri_{crit} are observed. Holt et al. (1989) found a dependence of Ri_{crit} on the integral-scale Reynolds number Re_ℓ. We will determine the critical value from a series of seven experiments with different Richardson numbers where the Reynolds numbers are rather small ($Re_\ell = 42.7$, $Re_\lambda = 26.4$).

Some recent discussions are addressed on the degree of anisotropy of stably stratified turbulence. Based on observations of isotropic turbulence in stably stratified layers in the ocean, Gargett (1988) argues that it is reasonable that the observed anisotropy belongs entirely to gravity waves and hence turbulence itself remains isotropic in stably stratified environments. Gargett further asserts that the energy-containing eddies of the stably stratified salt-water flows studied by Stillinger et al. (1983) and Itsweire et al. (1986) were affected by buoyancy already initially due to grid-generated gravity waves and that the turbulence approached an isotropic state further downstream depending on the thermal stability. This explanation of a return to isotropy under stable stratification is quite the opposite to the interpretations advanced by Stillinger et al. (1983) and Itsweire et al. (1986). They assumed that the grid-generated turbulence was fully developed, three-dimensional and isotropic a short distance behind the grid and that it became anisotropic further downstream due to gravity.

Discrepancies with respect to anisotropy in shear flows have been discussed in the literature. Webster (1964) found a steady increase of anisotropy with increasing Richardson number as does Launder's (1975) model. On the other hand, Komori et al. (1983) found reduced anisotropy at high Richardson numbers. This is corroborated by direct numerical simulations which we performed for $Pr = 5$ (Gerz et al. 1989) and by eigen-analysis of a second-order closure model for comparable conditions (Schumann 1987). Schumann (1987) and Gerz et al. (1989) explain this inverted trend by the effect of a temporally persistent counter-gradient heat-flux.

Temporally persistent counter-gradient heat-fluxes (p-CGHF) arise to compensate an imbalance between the dissipation rates of kinetic energy and available potential energy (for high Prandtl numbers). A second mechanism, mentioned by Holloway (1988), results from differences in the efficiency of nonlinear energy transfer from large to small scales for both kinetic and potential energy. The present study will discuss both explanations.

Method and Parameters

The three-dimensional Navier–Stokes and temperature equations for perturbation velocities (u, v, w) or (u_1, u_2, u_3) and temperature T are integrated in a cubic domain with coordinates (x, y, z) or (x_1, x_2, x_3) pointing in downstream, spanwise and vertical directions, respectively. The Boussinesq approximation is used. The vertical profiles of reference temperature $T_R(z)$ and mean velocity $U(z)$ are superimposed on the fluctuation fields. The profiles are linear so that the turbulence statistics remain spatially homogeneous. Periodic and shear periodic boundaries are applied in the horizontal and vertical directions, respectively. Details of the numerical method have been documented and verified by comparison to various experiments in Gerz et al. (1989). A spatial resolution of 64^3 grid points provides sufficient accuracy for cases with $Re_\lambda \approx 25$, $Pr = 1$, and narrow initial spectra. Simulations with a much more refined grid of 160^3 gridpoints have been performed by Gerz (1990) to reveal coherent vortex structures in stratified and sheared turbulence.

In order to normalize the results, we define the reference scales ρ_0 for density and ℓ_0, v_0 and $\ell_0 dT_R/dz$ for the initial values of the integral length scale, the rms velocity and the temperature fluctuation related to the mean temperature gradient, respectively. The box length L is chosen such that $L/\ell_0 = 4\pi$. For sheared turbulence the initial value of the shear number $Sh = (dU/dz)(\ell/v)$ is chosen to be $Sh_0 = 3$ as in Gerz et al. (1989). Buoyancy effects grow with Fr^{-1} and $Ri^{1/2}$ where $Fr = v/(N\ell)$ and $Ri = (Sh\ Fr)^{-2}$ are the Froude number and the Richardson number, respectively. The values of Fr and Ri in cases A to G are selected such that comparable sheared and unsheared flows have the same values of Fr^{-1} and $Sh_0 Ri^{1/2}$. The Brunt–Väisälä

Table 1. Initialization parameters of stratified unsheared and sheared turbulence. For both types of turbulence seven cases A to G are considered. In case A the temperature is passive. The symbols ν, γ, $N = (\alpha g dT_R/dz)^{1/2}$, α, and g are constant and stand for kinematic viscosity, temperature conductivity, Brunt–Väisälä frequency, isobaric volumetric expansion coefficient, and gravitational acceleration, respectively. The bar denotes the average over the entire domain. In the text the initial values are denoted by subscript zero

Grid points		M^3					64^3
Box length		L/ℓ					4π
Rms velocity		$v = (\overline{u_i u_i}/3)^{1/2}$					0.8226
Rms temperature		$T' = (\overline{TT})^{1/2}$					0
Shear number		$Sh = (dU/dz)(\ell/v)$					0, 3.0
Reynolds numbers		$Re_\lambda = v\lambda/\nu$, $Re_\ell = v\ell/\nu$					26.4, 42.7
Prandtl number		$Pr = \nu/\gamma$					1
Froude number		$Fr = v/(N\ell)$					
Richardson number		$Ri = (ShFr)^{-2}$					
Cases	A	B	C	D	E	F	G
$Fr =$	∞	1.42	1.16	0.92	0.58	0.41	0.29
$Ri =$	0	0.055	0.0825	0.13	0.33	0.66	1.32
$Fr^{-1} \propto ShRi^{1/2} =$	0	0.70	0.86	1.08	1.72	2.44	3.45
Integral length		ℓ					0.9044
Taylor micro-length		λ					0.5581
Kolmogorov length		L_K					0.05524
Mean velocity-derivative skewness coefficient							−0.45

frequency N is the same in both types of flow so that times are comparable directly relative to gravity oscillations. The isotropic velocity field is initialized with the energy-density spectrum $\hat{E}(k) = 16\sqrt{2/\pi}(2\pi)^{-1}v_0^2k^4k_p^{-5}\exp(-2(k/k_p)^2)$. The transfer spectrum is zero initially. To establish nonlinear turbulent interactions, the velocity components are integrated without external forces until the mean velocity-derivative skewness coefficient has reached the typical value of -0.45 in isotropic turbulence. This occurs at time $(v_0/\ell_0)t = 1$ indicating that the transfer spectrum is fully developed. The data at this time are taken as the initial parameters for all simulations and are summarized in Table 1.

We will discuss the exchange of energy between kinetic and available potential energy reservoirs. Therefore, the balance equations for the volume-averaged fluxes $\overline{u_i u_j}$, $\overline{u_i T}$ and kinetic energy $E_{kin} = \overline{u_i u_i}/2 = 3v^2/2$, available potential energy $E_{pot} = Fr_0^{-2}\overline{TT}/2$ and total energy $E_{tot} = E_{kin} + E_{pot}$ are listed here. They read for homogeneous stratified and sheared turbulence

$$\frac{d\overline{uu}}{dt} = -2Sh_0\overline{uw} + \phi_{11} - \varepsilon_{11} \tag{3}$$

$$\frac{d\overline{vv}}{dt} = + \phi_{22} - \varepsilon_{22} \tag{4}$$

$$\frac{d\overline{ww}}{dt} = 2Fr_0^{-2}\overline{wT} + \phi_{33} - \varepsilon_{33} \tag{5}$$

$$\frac{d\overline{uw}}{dt} = -Sh_0\overline{ww} + Fr_0^{-2}\overline{uT} + \phi_{13} - \varepsilon_{13} \tag{6}$$

$$\frac{d\overline{uT}}{dt} = -\overline{uw} \quad - Sh_0\overline{wT} \quad + \phi_{1T} - \varepsilon_{1T} \tag{7}$$

$$\frac{d\overline{wT}}{dt} = -\overline{ww} \quad + Fr_0^{-2}\overline{TT} \quad + \phi_{3T} - \varepsilon_{3T} \tag{8}$$

$$\frac{d\overline{TT}}{dt} \quad - 2\overline{wT} \quad - \varepsilon_T \tag{9}$$

$$\frac{dE_{\text{kin}}}{dt} = -Sh_0\overline{uw} + Fr_0^{-2}\overline{wT} \quad - \varepsilon \tag{10}$$

$$\frac{dE_{\text{pot}}}{dt} = \quad - Fr_0^{-2}\overline{wT} \quad - \chi \tag{11}$$

$$\frac{dE_{\text{tot}}}{dt} = -Sh_0\overline{uw} \quad - \varepsilon - \chi \tag{12}$$

where

$$\varepsilon_{ij} = \frac{2}{Re_{\ell 0}}\overline{\frac{\partial u_i}{\partial x_k}\frac{\partial u_j}{\partial x_k}}, \quad \varepsilon_{iT} = \frac{1+Pr}{Re_{\ell 0}Pr}\overline{\frac{\partial T}{\partial x_k}\frac{\partial u_i}{\partial x_k}}, \quad \varepsilon_T = \frac{2}{Re_{\ell 0}Pr}\overline{\frac{\partial T}{\partial x_k}\frac{\partial T}{\partial x_k}}$$

and

$$\varepsilon = \frac{1}{2}\varepsilon_{ii}, \quad \chi = \frac{1}{2Fr^2}\varepsilon_T \tag{13}$$

are the respective dissipation rates, and

$$\phi_{ij} = \overline{p\left(\frac{\partial u_i}{\partial x_j} + \frac{\partial u_j}{\partial x_i}\right)} \quad \text{and} \quad \phi_{iT} = \overline{p\frac{\partial T}{\partial x_i}} \tag{14}$$

denote the mean pressure–strain and the pressure–temperature gradients. The Eqs. (3)–(12) display the terms of the total time change, the changes due to shear production, buoyancy, pressure interaction and dissipation from left to right, respectively.

Unsheared Stratified Turbulence

The influence of increasing stable stratification on an initially isotropic turbulent flow with zero initial temperature fluctuations is illustrated in Fig. 1 by plots of E_{tot}, E_{kin}, E_{pot}, normalized rms velocities u'/u'_0 and w'/w'_0, and correlation coefficient of the negative vertical heat flux versus time for cases A to G. While the decay of the total energy is quite uninfluenced for weak stratification with small values of the inverse Froude number Fr^{-1} (cases A to D), it decays at a lower rate for $Fr^{-1} \geqq 1.72$ (cases E to F). Energy is exchanged periodically between the kinetic and potential energy reservoirs for positive values of Fr^{-1} as indicated by the phase angle of π between

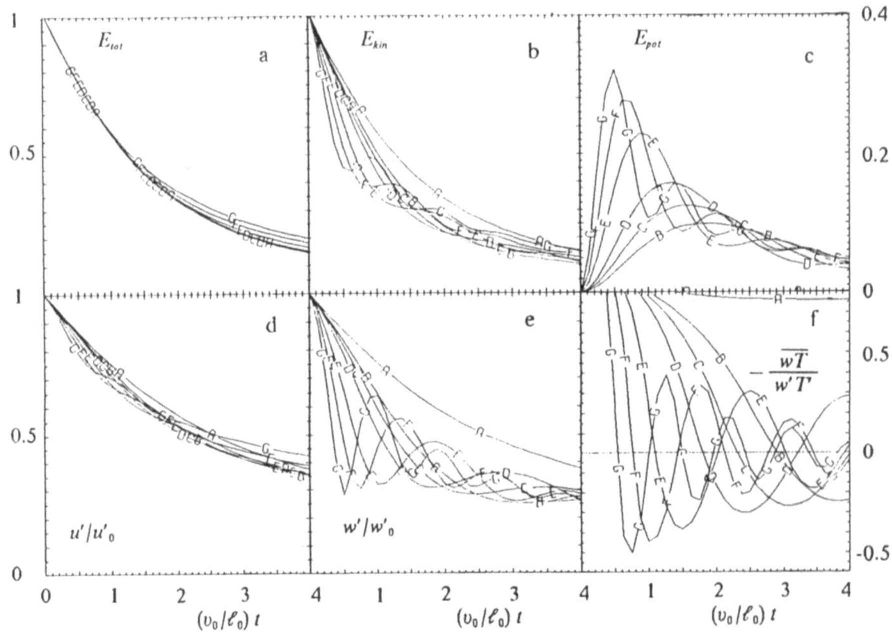

Fig. 1 a–f. Energies (**a–c**), normalized rms velocities (**d, e**) and the correlation coefficient of the vertical heat flux (**f**) versus time in stratified unsheared turbulence for cases A to G according to Table 1

E_{kin} and E_{pot}. These oscillations are called forth by the initial imbalance between kinetic and available potential energy since the thermally equilibrated flow is suddenly exposed to a mean temperature gradient at $t = 0$. This explains also why the amplitudes increase with increasing Fr^{-1}.

Since $Sh = 0$, the vertical heat flux $\overline{wT}$ is the only non-vanishing cross-correlation in unsheared stratified homogeneous flows, see Eqs. (6)–(8). This flux accomplishes the transfer of energy between the potential and the kinetic reservoir (Eqs. (10), (11)). It oscillates periodically between negative and positive values. We found for all values of Fr and at early times that the oscillation period scaled by N, $N\tau = (v_0/\ell_0)\tau/Fr$, was 3.4 which is little larger than half the Brunt–Väisälä period $N\tau_B = 2\pi$. The period decreases to $N\tau = 3.2$ at later times approaching $N\tau_B/2$. In a uniformly stratified flow the Brunt–Väisälä period is the only mode of linear oscillations. The variances and covariances of energy and heat flux oscillate with half this period. Thus, this value was also found by rapid distortion calculations as performed by Hunt et al. (1988), who considered a linear and inviscid flow. We conclude that in our simulations the initially larger oscillation period of $\overline{wT}$ compared to linear theory is due to nonlinear inertia forces in the flow which are neglected in rapid distortion theory.

The evolution of nonlinear energy transfer from large to small scales is reflected in the evolution of the skewness coefficients of velocity and temperature derivatives

$$S_{f_i} = \frac{\overline{(\partial f/\partial x_i)^3}}{[\overline{(\partial f/\partial x_i)^2}]^{3/2}} \tag{15}$$

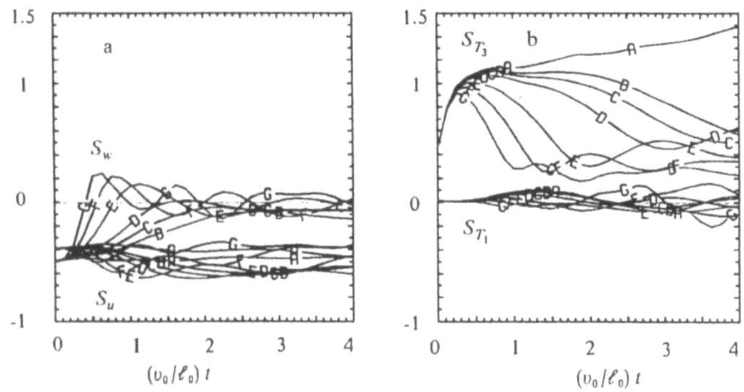

Fig. 2 a, b. Skewness coefficients of velocity gradients (**a**) and temperature gradients (**b**) versus time in stratified unsheared turbulence

(Batchelor 1953). The variable f stands for u_i (no summation on i) or T, respectively. Figure 2(a) illustrates by means of the skewness coefficients for $\partial u/\partial x$ and $\partial w/\partial z$ that nonlinearity is maintained in the downstream velocity component for small inverse Froude numbers and decreases slightly for large values of Fr^{-1}. The same behaviour was found for the lateral coefficient S_v (not shown). Depending on Fr, the skewness of the vertical component is reduced more rapidly and approaches zero at late times. Horizontal temperature-derivative skewnesses (see e.g. S_{T_1} in Fig. 2(b)) are found to be negligible due to the horizontally almost isotropic temperature field whereas S_{T_3} is large for $Fr^{-1} = 0$ (case A) due to the mean temperature gradient and diminishes with increasing inverse Froude number to a minimum level of 0.3. Thus, for small inverse Froude number, the nonlinear interactions measured by the skewness coefficient stay large and close to the value in unstratified isotropic flow ($S_{u_i} \approx -0.45$). They get reduced if the flow is exposed to stable stratification when gravity oscillations have developed. Since stable stratification reduces vertical mixing, this reduction is largest in the skewness of vertical derivatives.

In agreement with laboratory experiment reported by Britter (1988), we observe a reduced dissipation of total energy as seen in the reduced decay rate of E_{tot} for $Fr^{-1} \geqq 1.72$ (Fig. 1a). This stems from the reduced dissipation rate of temperature fluctuations as illustrated in Fig. 3. We see that $\varepsilon(t)$ behaves very similar as $E_{kin}(t)$ initially, but for late times it decreases at the same rate independent of the degree of stratification. On the other hand, ε_T reflects the behaviour of the temperature variance and diminishes monotonously when $Fr^{-1} \to \infty$.

All rms-velocity components decay at the same rate in case A with $Fr^{-1} = 0$. They decay faster for $Fr^{-1} > 0$ because part of kinetic energy is stored in the potential energy reservoir (see Fig. 1d, e for the downstream and vertical components). However, there are remarkable differences between the evolution of u' and w'. The latter receives and delivers energy from and to the potential energy reservoir periodically (see Eq. (5)) reaching extreme values at times when $\overline{wT}$ changes sign. This is a linear effect as can be seen from the respective balance equations and has been described by rapid distortion theory (Hunt et al. 1988). The horizontal fluctuations u' experience only weak oscillations and decay monotonously but at reduced rates when stratification is increased. Note that the decay rate

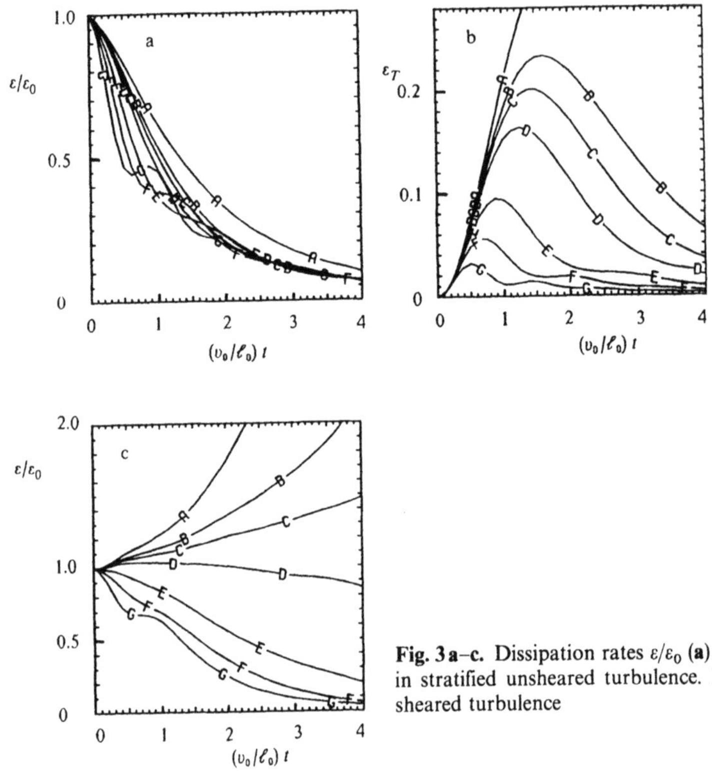

Fig. 3 a–c. Dissipation rates $\varepsilon/\varepsilon_0$ (**a**) and ε_T (**b**) versus time in stratified unsheared turbulence. Plot (**c**) shows $\varepsilon/\varepsilon_0$ in sheared turbulence

of u' is reduced compared to that of w' for this case. This corroborates results from corresponding laboratory flows (Britter 1988) and reflects the change of flow character from decaying isotropic turbulence to decaying anisotropic buoyancy influenced motions. For strong stratification ($Fr^{-1} > 1.72$), the decay rate of u' is even weaker than in the isotropic case A. The energy transfer among the velocity components is effected by nonlinear interactions between pressure and local strain (Eqs. (3)–(5)). From the plots it is obvious that these forces become more significant when stratification is increased which underlines the importance of buoyancy effects on pressure-strain terms and on its proper modelling (Launder 1975 and Gerz et al. 1989).

Sheared and Stratified Turbulence

The development of sheared and stratified flows at different Richardson numbers (cases A to G) is depicted in Fig. 4 by means of total, kinetic and available potential energy and the correlation coefficients of $-\overline{uw}$, $\overline{uT}$ and $-\overline{wT}$ versus time. These cases differ from the unsheared cases solely by the addition of a mean shear rate to the flow. The values of the Richardson number $Ri = (Sh_0 Fr_0)^{-2}$ are given in Table 1. As reflected by the close to steady-state behaviour of case D, the critical value of the

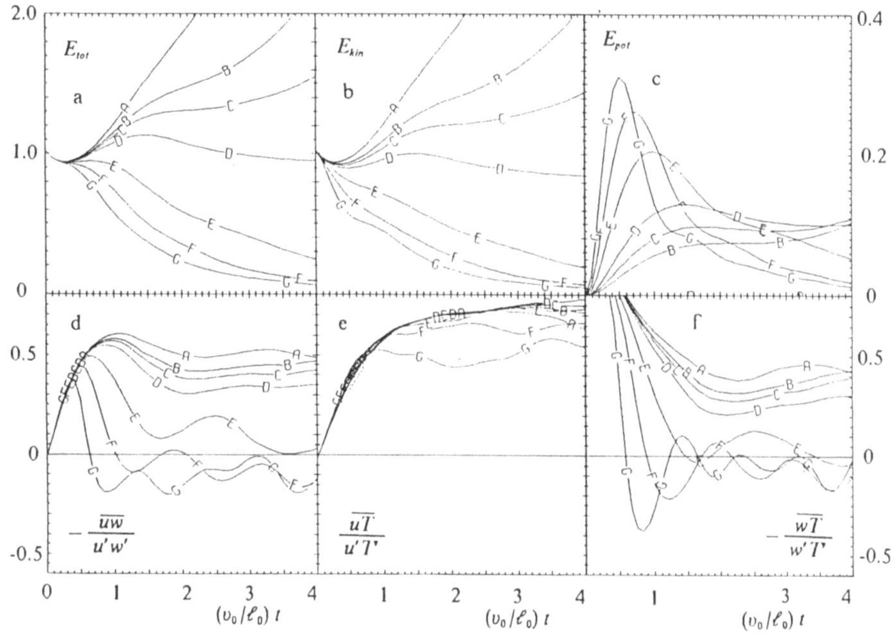

Fig. 4 a–f. Energies (**a–c**) and flux-correlation coefficients (**d–f**) versus time in sheared and stratified turbulence for cases A to G according to Table 1

Richardson number is found to be $Ri_{\text{crit}} \approx 0.13$. The critical or stationary Richardson number of a sheared and stratified flow is defined as the value of Ri where the kinetic energy E_{kin} approaches a constant level in time. For an inviscid, laminar, continuously stratified shear flow, Miles (1961) derived a stability criterion requiring a gradient Richardson number larger than $Ri_{\text{crit}} = 0.25$. In viscous homogeneous flows we yield from Eq. (10) for $dE/dt = 0$ the criterion

$$Ri_{\text{crit}} = \frac{\overline{uw}}{\overline{wT}\, Sh_0}\left(1 + \frac{\varepsilon}{Sh_0 \overline{uw}}\right) = \sigma_t \left(1 + \frac{\varepsilon}{Sh_0 \overline{uw}}\right) \tag{16}$$

where σ_t is the turbulent Prandtl number. If dissipation ε equals production $Sh_0 uw$ (< 0) it follows $Ri_{\text{crit}} = 0$. Hence, any small but positive temperature gradient will cause damping of the flow in this case. If viscous forces are much smaller than production, Ri_{crit} may approach the corresponding turbulent Prandtl number. Gerz et al. (1989) suggested that the critical Richardson number is a function of the shear number $Sh = (\ell/v)(dU/dz)$ when dissipation is important. For viscous flows Holt et al. (1989) found increasing values of Ri_{crit} with increasing Reynolds number Re_ℓ at fixed shear number. Rohr et al. (1988) found a critical value of $Ri_{\text{crit}} \approx 0.21$ and derived an approximate solution for $E_{\text{kin}}(t)$ suggesting a weak dependency of the terms of the right-hand side of Eq. (16) on the gradient-Richardson number.

The flows presented here are characterized by $Re_{\ell 0} = 42.7$, $Pr = 1$ and $Sh_0 = 3$. For this parameter configuration we found a critical or stationary Richardson number of $Ri_{\text{crit}} = 0.13$. This was corroborated by extending the simulation with

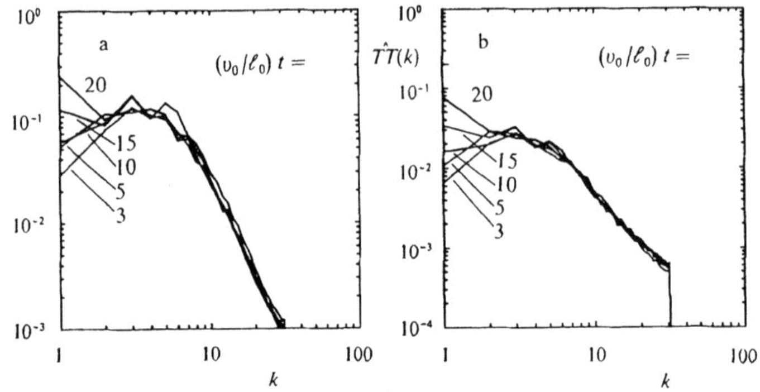

Fig. 5 a, b. Power spectrum of kinetic energy density (**a**) and temperature-variance density (**b**) versus integer wavenumber for $Ri = 0.13$ (case D) and various times

$Ri = 0.13$ (case D) up to time $(v_0/\ell_0)t = 20$ for which the kinetic and potential energy stayed close to constant until the final time. For this run we yield $\sigma_t = 0.92 \pm 0.19$ and $\varepsilon/(Sh_0\overline{uw}) = -0.89 \pm 0.19$ in a temporal mean which results in $Ri_{crit} = 0.10 \pm 0.04$ according to (16).

A slight change with time is observed in the power spectra of kinetic energy $\hat{E}$ and temperature variance $\widehat{TT}$ versus integer wavenumber k as shown in Fig. 5. $\hat{E}(k)$ and $\widehat{TT}\,(k)$ stay fairly constant within the entire wavenumber range and for all times, except for $k = 1$ where a continuous growth of both $\hat{E}$ and $\widehat{TT}$ is observed. The power spectra further show that the simulations sufficiently resolve the small scale dissipation of both E_{kin} and $\overline{TT}$. Insufficient resolution would cause an increase of intensity at the highest wavenumbers.

For subcritical cases A, B and C with $Ri < Ri_{crit}$, both the kinetic energy E_{kin} and the available potential energy $E_{pot} = Ri\overline{TT}/2$ and, thus, the total energy E_{tot} grow in time according to the source terms in Eqs. (3), (10) and (12), see Fig. 4. In the supercritical cases E, F and G, E_{pot} first increases strongly due to turbulent motions in the vertical direction building up the potential energy reservoir. Later E_{pot} decays in time. This decay starts at time $Nt = (v_0/\ell_0)t/Fr > 1.72$ for all three cases E, F and G. In contrast to the unsheared cases, the evolution of $-\overline{wT}/(w'T')$ depends strongly on the degree of stratification if also shear is present (Fig. 4f). For supercritical Richardson numbers (cases E, F and G) oscillations of both vertical fluxes of momentum and heat occur. Oscillations also can be detected in the graphs of E_{kin} and E_{pot} in cases F and G. Such oscillations do not appear in $E_{tot}(t)$. Analogously to the unsheared cases, we conclude from these facts that a large part of energy is stored in gravity waves for large Richardson numbers. For $Ri = 0.66$ and 1.32 we further observe that $\overline{uw}$ and $\overline{wT}$ remain *persistently* positive in the temporal mean (persistent counter-gradient fluxes of momentum and heat, p-CGMF, p-CGHF).

The oscillating exchange between kinetic and potential energy occurs only for approximately one time period π/N. For $Nt > 5.4$, both E_{kin} and E_{pot} decay monotonously. Note that they decay at a *stronger* rate than they do in the unsheared cases. We offer two mechanisms to explain this rather unexpected behaviour. The first possible mechanism is wave breaking. Gravity waves get unstable and break into turbulence when the effective Richardson number Ri_{eff} gets subcritical. Ri_{eff} depends on local stratification and local shear,

$$Ri_{\text{eff}} = \alpha g \frac{\partial(T + T_{\text{R}})/\partial z}{(\partial(u + U)/\partial z)^2}. \tag{17}$$

Because of the fact that $\partial u/\partial z$ is larger at small scales than at large scales, Ri_{eff} is smaller and may first get subcritical for small-scale motions. Hence, small-scale waves break more easily than large-scale waves and cause additional turbulence at small scales which increases dissipation. This explanation is corroborated by the simulation data which show such an increase of dissipation in the sheared case relative to the unsheared case (Fig. 3c, a). The second possible mechanism is a consequence of CGMF which also has been observed in stably stratified laboratory flows (Britter 1988). Such buoyancy induced CGMF causes *negative* production of kinetic energy, see Eq. (10).

A downstream heat flux $\overline{uT}$ occurs in a vertically sheared and stratified flow although no mean temperature gradient exists in the x-direction, see Eq. (7). The flux is positive and its correlation coefficient in Fig. 4e reaches large values of 0.5 to 0.8. Hence, cold (hot) fluid is strongly correlated with slow (fast) downstream velocity. In the temporal mean of cases A to E we observe that according to shear and weak-to-moderate buoyancy fast (slow) downstream and hot (cold) fluid moves down (up) ($\overline{uw}, \overline{wT} < 0$) and, thus, $\overline{uT} > 0$. The correlation coefficient $\overline{uT}/(u'T') \approx 0.7$ is unaffected by Ri for these cases but drops to little lower values 0.65 and 0.5 in cases F and G when buoyancy dominates shear. Then slow (fast) and cold (hot) fluid moves down (up) in the temporal mean ($\overline{uw}, \overline{wT} > 0$), again resulting in $\overline{uT} > 0$.

The large amplitudes of $\overline{wT}$ in the first oscillation periods indicate an exchange of energy between the potential and kinetic energy reservoir due to gravity waves as in the unsheared cases. The time scale of these oscillations, however, increases with decreasing Ri and amount to $N\tau = 3.5, 3.4, 3.7$ and ≈ 4 for $Ri = 1.32, 0.66, 0.33$ and 0.13, respectively. Qualitatively the same fact is displayed in the oscillations of $\overline{uw}$. Hence, as in the situation without shear, we conclude that the deviation of the

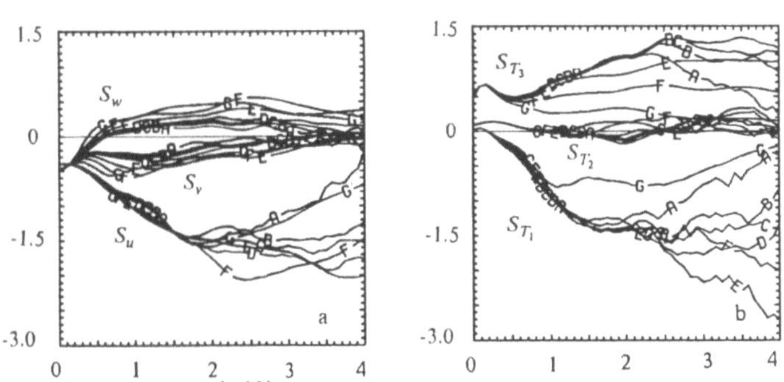

Fig. 6 a, b. Skewness coefficients of velocity gradients (**a**) and temperature gradients (**b**) versus time in stratified and sheared turbulence

oscillation periods from half the Brunt–Väisälä period is due to the nonlinear inertia forces in the flow which are much stronger in sheared than in unsheared flows. In Fig. 6 skewness coefficients for velocity and temperature derivatives (Eq. (15)) are plotted. Due to the mean shear force we observe strongly enhanced values of S_u and S_{T_1} for all Ri. The coefficient S_w changes sign as in the unsheared case but then grows with increasing Ri. S_{T_3} behaves as in the unsheared case and the lateral coefficients vanish at large times for all Ri. In general it is clear from the picture that the derivative skewnesses (and hence the nonlinearities) are the larger the stronger the influence of shear is compared to buoyancy. The derivative skewness coefficients are smallest for cases F and G with $Ri = 0.66$ and 1.32 where we observed an oscillation period of the fluxes closest to $N\tau_B$, which is consistent with the previous results.

Persistent Counter-Gradient Fluxes

We now discuss the counter-gradient heat-fluxes (CGHF) in stratified turbulence. We have to distinguish between temporally oscillating fluxes, as can be described by linear analysis as in rapid distortion calculations, and temporally persistent positive heat fluxes (p-CGHF). Figure 7 depicts three cospectra of the vertical heat flux $\widehat{wT}$ for the unsheared case D and the two sheared flows D and G at several times. Considering pictures (a) and (c) we observe that $\widehat{wT}$ oscillates between positive and negative values at large scales (small wavenumbers k) whereas it has small but persistently positive values at small scales, regardless if shear is present or not. For $Ri = 0.13$, we observe positive values of $\widehat{wT}$ at large wavenumbers k whereas it stays strongly negative at small values of k. Such negative fluxes indicate strong turbulent mixing. All pictures are consistent with the curves of $\overline{wT}(t)$ in Fig. 1f and 4f: A p-CGHF can be recognized in the integral over all scales only when the large and sign-changing contributions at small k cancel such that the small positive contributions at large wavenumbers remain.

The small-scale motion related to the p-CGHF is buoyancy driven. It is expected that this kind of motion creates the corresponding momentum flux, p-CGMF

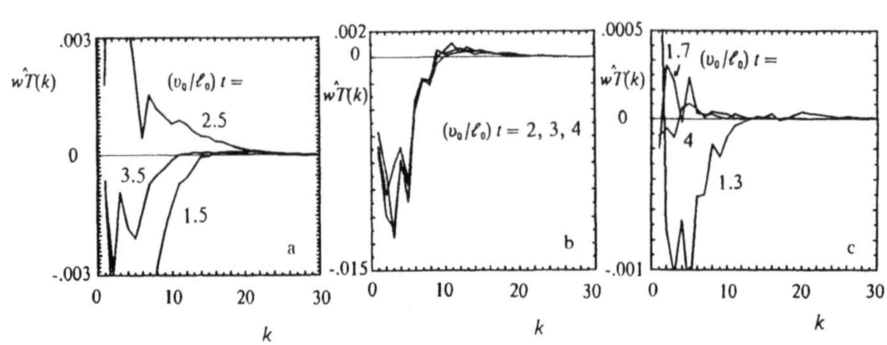

Fig. 7 a–c. Cospectra of the vertical heat-flux density for (**a**) unsheared flow with $Fr = 0.92$ (case D) and for sheared flows with (**b**) $Ri = 0.13$ (case D) and (**c**) $Ri = 1.32$ (case G) at several times

(Fig. 4d), since the cospectrum of uw (not shown) behaves similar as $\widehat{wT}$. It has strong oscillating contributions at small wavenumbers and persistently positive values at large k.

Based on these findings and previous explanations as given by Schumann (1987), Holloway (1988) and Gerz et al. (1989), we interpret the p-CGHF as follows. The p-CGHF is a property of small-scale motion in quasi-steady state which requires a balance between production due to shear, buoyancy effects and molecular dissipation for both kinetic and potential energy. An imbalance arises if the molecular dissipations of kinetic and potential energy differ from each other. If the molecular Prandtl number is about unity or larger, potential energy is dissipated more slowly than kinetic energy. Therefore, a p-CGHF is required to convert potential energy into kinetic energy to maintain a steady state. A second reason for an imbalance originates from differences in the efficiency of nonlinear transfer of kinetic and potential energy to small scales (Holloway 1988). It is interesting to note that the effective turbulent Prandtl number of locally isotropic turbulence is 0.4 (Townsend 1976). This indicates that the turbulent energy cascade from large to small scales is more efficient for potential than for kinetic energy. For this reason, steady state requires a p-CGHF, even for $Pr < 1$, as air, to convert some of the excessive potential energy into kinetic energy. At moderate Reynolds numbers, a sheared and stratified flow is far from isotropy even at small scales. Hence, for such flows the first reason for an imbalance suits better: Gerz et al. (1989) simulated flows with $Re_\ell \approx 47$ and $Pr = 5$ and found a strong p-CGHF; the simulations presented here with $Re_\ell \approx 43$ and $Pr = 1$ reveal a weak but still positive p-CGHF. For large Reynolds numbers, the second explanation is more appropriate: Sidi and Dalaudier (1989) have observed an increase of intensity in the temperature variance spectra at high wavenumbers obtained from the lower stably stratified stratosphere ($Re_\ell \approx 10^7 \cdots 10^8$, $Pr = 0.7$). They also found indications of a corresponding CGHF.

Length Scales

In Fig. 8 the integral length scale

$$\ell = \frac{1}{2v^2} \int \frac{1}{k} \hat{E}(k) dk \tag{18}$$

and the Kolmogorov length scale L_K (Eq. (2)b) of both the unsheared and sheared experiments are plotted versus time. In decaying turbulence ℓ grows due to energy loss by dissipation at large wavenumbers and due to reduced energy transfer from small to large wavenumbers when the flow is stably stratified. L_K grows quite uninfluenced by the degree of stratification according to similar decay rates of dissipation for the cases B to G, see Fig. 3a. If also shear is present, we observe a strong increase of the integral length for all cases A to G initially, followed by a decrease and a second increase at late times. In the initial period, shear is suddenly imposed to the preexisting isotropic field. This augments the large scales resulting in a strong increase of ℓ. Later, in the period $(v_0/\ell_0)t = 1.5 \dots 2$ when the skewness

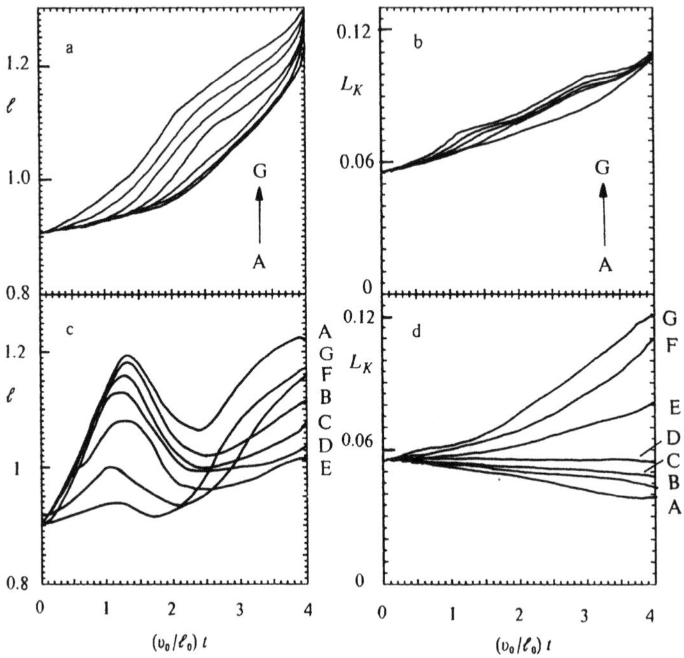

Fig. 8 a–d. Integral scale ℓ and Kolmogorov scale L_k of unsheared (**a, b**) and sheared (**c, d**) stably stratified turbulence versus time for cases A to G

coefficients approach their maximum values (see Fig. 6), ℓ is reduced since the large scales start to lose energy by feeding the small scales. The flow is now at a quasisteady state, see also Fig. 4. Once the transfer spectrum is established, energy at large scales can be rebuild by the mean shear flow resulting in a second increase of the integral length. According to the growing influence of buoyancy from case A to case G, the second increase of ℓ is lowered for subcritical values of Ri and is enhanced again if Ri is supercritical. The increase in the latter cases is equivalent to the increase of ℓ found in unsheared flows. Due to increased dissipation in shear flows, the Kolmogorov length (Fig. 8d) decreases for subcritical flows, stays constant when $Ri = Ri_{crit}$ (case D) and increases when buoyancy forces dominate.

In Fig. 9 the Ellison length scale L_E and the Ozmidov scale L_O of the present experiments with and without shear are plotted versus time. The time when the value of L_E of actively stratified flows first deviates notably (i.e. by 2%) from L_E of the passive flow (case A) marks the onset of gravity waves generated by the largest eddies. As listed in Table 2, this event occurs for all cases at the same frequency-normalized time Nt_{on}. But—in contrast to the findings of Itsweire et al. (1986)—we observe that the ratio L_E/L_O grows with increasing stability. According to the theory (Gibson 1981), the extinction of turbulence at all scales is reached when the mean heat flux is zero for the first time. This has been confirmed by measured density-flux cospectra (Itsweire and Helland 1989) which vanished almost entirely at all scales at this time. Our data of the unsheared flow simulations show that at the time when $\overline{wT}$ is zero first (see Fig. 1f), the corresponding ratio L_E/L_K is not constant but increases

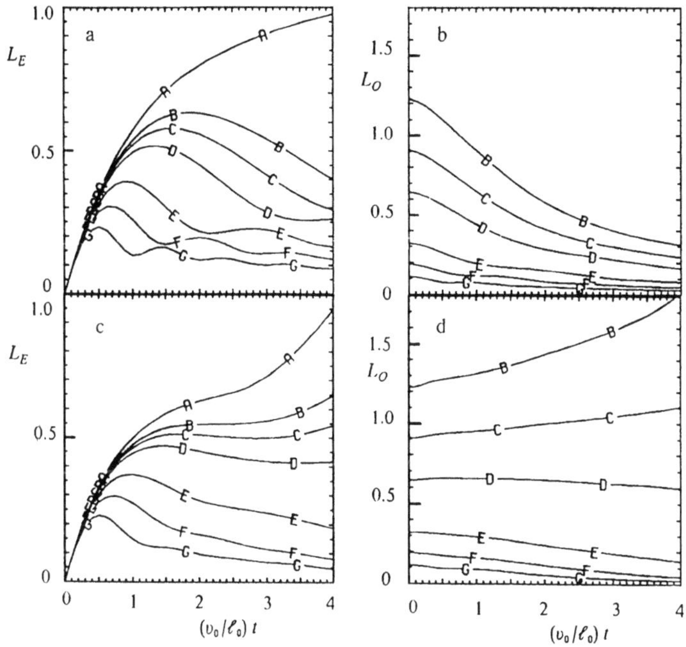

Fig. 9 a–d. Ellison scale L_E and Ozmidov scale L_0 of unsheared (**a, b**) and sheared (**c, d**) stably stratified turbulence versus time for cases A to G

Table 2. Length-scale ratios for unsheared and sheared stratified turbulence. Nt_{on} is the time at onset of buoyancy effects; Nt_{ex} is the time at extinction of turbulence according to Gibson (1981), see Eq. (2)a

Cases	L_O/L_K at $t=0$	L_E/L_O	at Nt_{on}	L_E/L_K	at Nt_{ex}
B: $Fr = 1.42$	22.2	0.34	0.40	6.1	2.04
$Ri = 0.055$	22.2	0.27	0.41		
C: $Fr = 1.16$	16.5	0.38	0.40	6.0	1.98
$Ri = 0.083$	16.5	0.34	0.41		
D: $Fr = 0.92$	11.6	0.48	0.43	6.0	1.95
$Ri = 0.13$	11.6	0.42	0.43		
E: $Fr = 0.58$	5.9	0.59	0.40	5.3	1.90
$Ri = 0.33$	5.9	0.54	0.40	5.7	2.53
F: $Fr = 0.41$	3.5	0.67	0.39	4.5	1.93
$Ri = 0.66$	3.5	0.68	0.41	4.9	2.11
G: $Fr = 0.29$	2.0	0.75	0.38	3.5	1.90
$Ri = 1.32$	2.0	0.82	0.40	3.6	2.01

with decreasing stratification reaching an asymptotic level of six for cases B, C and D (Table 2). This result is reasonable because Gibson's theory presumes that $L_O \gg L_K$ initially, a condition which is satisfied only in cases B, C and D. Hence, the theoretical estimate Eq. (2)a for active turbulence according to Gibson (1981) is corroborated within reasonable limits.

The temporal evolution of the length scales L_K, L_E and L_O in unsheared flows is in very good agreement with corresponding data from measurements (Itsweire et al.

1986) and direct numerical simulations (Métais and Herring 1989, see their Fig. 4). The evolution of the Ozmidov scale starts at different levels depending on N. In all cases without shear L_O decreases in time according to the decreasing dissipation rate. In cases A to C with shear ($Ri < Ri_{crit}$), both L_E and L_O grow continuously as does the total energy of turbulence. The correlation coefficient of the vertical heat flux never approaches zero in these cases (Fig. 4d). Hence, the flow remains actively turbulent although gravity waves have been generated by the largest eddies. This result confirms the measurements of Rohr et al. (1988) who also observed a continuous growth of all length scales as long as $Ri < Ri_{crit}$. For the supercritical cases E to G, L_E and L_O decay and behave very similar as in the unsheared flows (Rohr et al. 1988).

Anisotropy

We have computed the components of the anisotropy tensor

$$b_{ij} = \frac{\overline{u_i u_j}}{2E_{kin}} - \frac{1}{3}\delta_{ij} \tag{19}$$

from our simulation data and plotted the trace components b_{ii} versus time in Fig. 10. Unsheared cases will be discussed first. Since stable stratification reduces the vertical motion of fluid parcels compared to isotropic motion, buoyancy-induced anisotropy manifests in a negative vertical component b_{33} which is compensated by the two equally large horizontal components, i.e. $b_{11} \approx b_{22} \approx -b_{33}/2$. The temporal mean value of b_{33} during integration time is ≈ -0.14. The oscillations of the trace components have a time period of about π/N which is also observed for oscillations of the energies and the fluxes, see Fig. 1. This suggests that the anisotropy is correlated with gravity waves in the turbulent flow as one would expect. It is remarkable, however, that in all cases the onset of anisotropy occurs significantly *before* the value of L_E departs from that in the passive scalar case at $Nt \approx 0.4$ (see

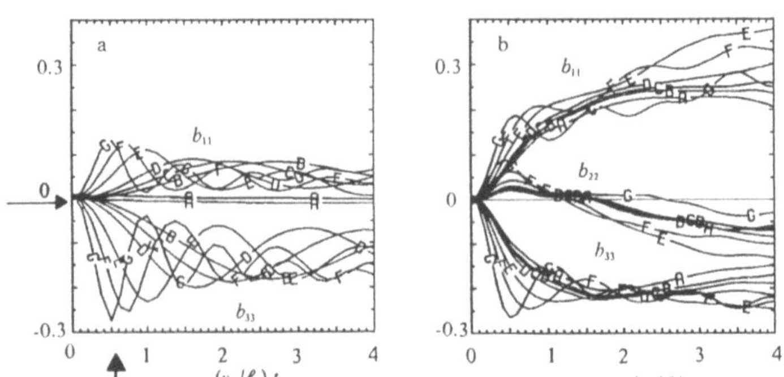

Fig. 10. a, b. Trace components of the anisotropy tensor b_{ij} of (**a**) unsheared and (**b**) sheared stably strafied turbulence versus time

Table 2) and, hence, *before* the first gravity wave is generated by the turbulent eddy with the largest vertical extension. For the unsheared case B e.g., $Nt = 0.4$ corresponds to $(v_0/\ell_0)t = 0.57$. At this time, b_{33} is already significantly smaller than zero, see arrows in Fig. 10a. In fact, at $Nt \approx 0.4$ all simulated flows have reached the same level of anisotropy, $b_{33} \approx -0.028$, which is already about 20% of the temporal mean of b_{33}. Hence, our results for unsheared flows support the results and interpretations of Stillinger et al. (1983) and Itsweire et al. (1986): Stratification causes anisotropy in a formerly isotropic flow. The anisotropy manifests in gravity waves *and* in turbulence.

The influence of shear and stratification on the degree of anisotropy is illustrated in the plots of the trace components of the anisotropy tensor $b_{ij}(t)$ in Fig. 10b. We observe a much stronger degree of anisotropy of the downstream velocity component b_{11} compensated by unequal shares of b_{22} and b_{33} when shear is present. The degree of anisotropy in shear flows grows with increasing Richardson number as predicted by Launder (1975). For $Ri < Ri_{crit}$, the anisotropy of the flow is only weakly affected by buoyancy whereas for supercritical flows the oscillations of b_{11} and b_{33} indicate dominating stratification effects. Further we observe that for cases F and G the degree of anisotropy is slightly reduced which coincides with the occurrence of CGMF and CGHF, see Fig. 4d, f. Similar findings were reported by Gerz et al. (1989) who observed in flows with $Pr = 5$ that the anisotropy is considerably reduced for supercritical Richardson numbers where a CGHF occurred. Both results are likely to sustain the findings of Gargett (1988) insofar as shear-induced anisotropy is diminished when $Ri > Ri_{crit}$.

Analogous to b_{ij}, the anisotropic dissipation rates d_{ij} are defined as

$$d_{ij} = \frac{\varepsilon_{ij}}{2\varepsilon} - \frac{1}{3}\delta_{ij}. \tag{20}$$

In flows with low-to-moderate Reynolds number we expect that the anisotropy of the flow is compensated by both pressure-strain interactions and anisotropic dissipation rates (Gerz et al. 1989). In all flows presented here we find considerably large values of d_{ij} reaching between 50% and 80% of the magnitude of b_{ij}. The largest contributions are observed in supercritical shear flows which may indicate breaking gravity waves.

Conclusions

The method of Gerz et al. (1989) has been successfully applied to homogeneous turbulence with and without shear for moderate Reynolds number. In unsheared stratified flows, initially isotropic turbulence becomes anisotropic even before gravity waves are generated by the eddies with the largest vertical extent. This is corroborated by Riley et al. (1981). Gravity waves occur when the Ellison length reaches about the magnitude of the Ozmidov length scale. At the extinction of turbulence, L_E is about six times the Kolmogorov length scale. This result confirms the estimate, Eq. (2)a, given by Gibson (1981) within reasonable limits. Further, it supports the findings of Stillinger et al. (1983) and Itsweire et al. (1986) but contradicts the arguments of Gargett (1988).

An important phenomenon of strongly stratified turbulence is the appearance of temporally persistent (non-zero) counter-gradient fluxes of momentum and heat, p-CGF. The p-CGHF and the related p-CGMF belong to buoyancy driven motions at small scales. A p-CGHF reduces excessive potential energy. The excessive potential energy arises at small scales either due to the more efficient energy cascade of potential energy than that of kinetic energy (Holloway 1988) or due to differences in the molecular dissipation rates of E_{kin} and E_{pot} (Schumann 1987; Gerz et al. 1989). For large Reynolds number, the former effect dominates even if $Pr < 1$. For example, Sidi and Dalaudier (1989) have measured an increase of intensity in the temperature-variance spectra at high wavenumbers and found indications of a corresponding CGHF in the lower stably stratified stratosphere. For moderate Reynolds numbers the second effect is more important, in particular for flows with large Prandtl numbers. Gerz et al. (1989) found a strong p-CGHF for $Pr = 5$ and the present simulations show the appearance of a rather weak p-CGHF for $Pr = 1$.

The CGHF and the related CGMF have important effects on the degree of anisotropy. If shear is present, the degree of anisotropy grows with increasing Richardson number as predicted by Launder (1975) but gets limited when counter-gradient fluxes appear. Since the p-CGHF depends on the molecular Prandtl number, so does the degree of anisotropy.

The critical Richardson number 0.13 of the present flow simulations with $Re_{\ell 0} = 42.7$ and $Sh_0 = 3$ is smaller than the linear and inviscid results 0.25 (Miles 1961). It depends on the ratio of dissipation relative to shear production and may become zero if both have equal magnitude. For high Reynolds numbers this ratio may vanish so that the inviscid theory applies. If $Ri < Ri_{\text{crit}}$, turbulence is dominated by shear. Supercritical flows behave at large scales as if no shear would be present. These results corroborate the experimental findings of Rohr et al. (1988). However, shear causes wave breaking and, thus, converts gravity waves into small-scale turbulence which increases dynamical and thermal dissipation compared to the unsheared flows.

Acknowledgement: We wish to thank Steve Holt and Prof. Koseff for discussions on sheared and stratified turbulence during the TSF 7 conference at Stanford.

References

Batchelor, G. K. (1953): The Theory of Homogeneous Turbulence. Cambridge University Press, Cambridge 1982

Britter, R. E. (1988): Laboratory experiments on turbulence in density-stratified fluids. Proc. 8th AMS-Conf. on Turbulence and Diffusion, April 26–29, 1988, San Diego

Businger, J. A., Wyngaard, J. C., Izumi, Y., Bradley, E. F. (1971): Flux-profile relationships in the atmospheric surface layer. J. Atmos. Sci. **28**, 181–189

Gargett, A. (1988): The scaling of turbulence in the presence of stable stratification. J. Geophys. Res. **93**, 5021–5036

Gerz, T. (1990): Coherent structures in stratified turbulent shear flows deduced from direct simulations. In *Turbulence and Coherent Structures* (O. Métais, M. Lesieur, eds.), Kluwer Academic Publishers, 449–468

Gerz, T., Schumann, U., Elghobashi, S. (1989): Direct numerical simulation of statified homogeneous turbulent shear flows. J. Fluid Mech. **200**, 563–594

Gibson, C. H. (1981): Fossil turbulence and internal waves. Nonlinear Properties of Internal Waves (B. J. West, ed.) AIP Conf. **76**, pp. 159–179

Holloway, G. (1988): The buoyancy flux from internal gravity wave breaking. Dyn. Atmos. Oceans **12**, 107–125

Holt, S. E., Koseff, J. R., Ferziger, J. H. (1989): The evolution of turbulence in the presence of mean shear and stable stratification. Proc. 7th Symp. on Turbulent Shear Flows, 21–23, Aug. 1989, Stanford Univ. 12–2

Hunt, J. C. R., Stretch, D. D., Britter, R. E. (1988): Length scales in stably stratified turbulent flows and their use in turbulence models. in: Stably Stratified Flows and Dense Gas Dispersion (J. S. Puttock, ed.), Clarendon Press, Oxford, pp. 285–321

Itsweire, E. C., Helland, K. N., van Atta, C. W. (1986): The evolution of grid-generated turbulence in a stably stratified fluid. J. Fluid Mech. **162**, 299–338

Itsweire, E. C., Helland, K. N. (1989): Spectra and energy transfer in stably stratified turbulence. J. Fluid Mech. **207**, 419–452

Komori, S., Ueda, H., Ogino, F., Mizushina, T. (1983): Turbulence structure in stably stratified open-channel flow. J. Fluid Mech. **130**, 13–26

Launder, B. E. (1975): On the effects of a gravitational field on the turbulent transport of heat and momentum. J. Fluid Mech. **67**, 569–581

Métais, O., Herring, J. R. (1989): Numerical simulation of freely evolving turbulence in stably stratified fluids. J. Fluid Mech. **202**, 117–148

Miles, J. W. (1961): On the stability of heterogeneous shear flows. J. Fluid Mech. **10**, 496–508

Riley, J. J., Metcalfe, R. W., Weissman, M. A. (1981): Direct numerical simulations of homogeneous turbulence in density-stratified fluids. Nonlinear Properties of Internal Waves (B. J. West, ed.), AIP Conf. **76**, pp. 79–112

Rohr, J. J., Itsweire, E. C., Helland, K. N., van Atta, C. W. (1988): An investigation of the growth of turbulence in a uniform-mean-shear flow. J. Fluid Mech. **187**, 1–33

Schumann, U. (1987): The counter-gradient heat-flux in turbulent stratified flows. Nucl. Engrg. Desg. **100**, 255–262

Sidi, C., Dalaudier, F. (1989): Temperature and heat flux spectra in the turbulent buoyancy subrange. PAGEOPH. **130**, 547–569

Stillinger, D. C., Helland, K. N., and van Atta, C. W. (1983): Experiments on the transition of homogeneous turbulence to internal waves in a stratified fluid. J. Fluid Mech. **131**, 91–122

Townsend, A. A. (1976): The structure of turbulent shear flow. Cambridge University Press, London, 2nd ed., p. 359

Webster, C. A. G. (1964): An experimental study of turbulence in a density stratified shear flow. J. Fluid Mech. **19**, 221–245

Turbulence Model for Triple Velocity and Scalar Correlations

Y. Nagano and M. Tagawa

Department of Mechanical Engineering, Nagoya Institute of Technology, Gokiso-cho, Showa-ku, Nagoya 466, Japan

Abstract

The conventional assumption for triple products, i.e., gradient-type diffusion, is not correct in near-wall turbulence and thus the use of this assumption in Reynolds-stress and scalar-flux equation modeling may introduce serious errors into the calculations. The present paper is centered on the development of a new model for triple correlations of velocity and scalar fluctuations. The model is based on the structural characteristics of shear-generated turbulence and the transport equations for velocity and scalar fields. It is shown that the proposed model works quite well in various flow fields.

Nomenclature

a	thermal diffusivity
C	universal constant of structural model for triple correlation
C_{u1}, C_{u2}	turbulence model constants for $\overline{u_i^3}$
$C_{\theta 1}, C_{\theta 2}$	turbulence model constants for $\overline{\theta^3}$
k	turbulent kinetic energy, $= \overline{u_m u_m}/2$
p	pressure fluctuation
$S(\chi)$	skewness factor of χ, $= \overline{\hat{\chi}^3}$
$\bar{U}_i$, u_i	components of mean and fluctuating velocities
u, v	streamwise and normal velocity fluctuations
u_τ	friction velocity
x_i	coordinate in tensor notation
y^+	dimensionless distance from wall, $= u_\tau y/\nu$
$\varepsilon, \varepsilon_i$	dissipation rates of k and $\overline{u_i^3}$
$\varepsilon_\theta, \varepsilon_{\theta^3}$	dissipation rates of $\overline{\theta^2}/2$ and $\overline{\theta^3}$
$\bar{\Theta}, \theta$	mean and fluctuating scalars (temperatures)
θ_τ	friction temperature
ν	kinematic viscosity
ρ	density
σ_x	sign function
τ	time
τ_u, τ_θ	time scales of velocity and scalar fields, $= k/\varepsilon$, $(\overline{\theta^2}/2)/\varepsilon_\theta$
χ	stochastic variable
$\overline{(\ )}$	time mean value

Turbulent Shear Flows 7

$(\hat{\ })$	normalization by r.m.s. value
$(\ )'$	normalization by u_τ or θ_τ

subscripts

$,i$	differentiation by x_i
m	tensor notation with summation convention

Introduction

As the most advanced numerical strategy, the Reynolds-stress and scalar-flux equation models have been increasingly applied to various flow-field problems. Results, however, are not as satisfactory as initially expected. Nagano and Tagawa (1988) pointed out that this is in part due to the inadequate "gradient-type diffusion" assumption used in the triple correlation modeling for closing the transport equations of Reynolds stresses and turbulent scalar fluxes. Since the statistical characteristics of triple correlations such as probability density distributions and their internal structures are dominated completely by the dynamic flow structures (Nagano and Tagawa 1988), we need to develop an appropriate new model based not on the static gradient-type diffusion hypothesis but on the physical behavior of triple correlations. In another paper (Nagano and Tagawa 1990), we have developed an entirely new "structural" turbulence model for triple velocity and scalar correlations considering their physical behavior and verified the effectiveness by application to various types of flows.

In the present study, by combining this structural model with the transport equations of triple correlations, we construct a set of simple models to predict accurately the turbulent transport of Reynolds shear stress $\overline{uv}$ and scalar flux $\overline{v\theta}$ in turbulent shear flows, which have been often predicted erroneously by the existing models.

Outline of The Structural Models for Triple Correlations

In this section, we outline the structural models for triple velocity and scalar correlations, which have been developed in the previous study (Nagano and Tagawa 1990).

The triple correlations of velocity-fluctuation components, u (streamwise) and v (normal to the wall), and scalar (temperature) fluctuations, θ, in wall turbulence are determined almost completely by the organized fluid motions such as the ejections and sweeps (Nagano and Tagawa 1988) classified in the second ($i = 2$) and fourth ($i = 4$) quadrants on the (u, v)-plane. As an example, the measured fractional contributions of the classified motions to the turbulent transport of Reynolds shear stress, $(\overline{vuv})_i$, in pipe-flow turbulence are presented in Fig. 1. In what follows, a prime denotes the normalization of velocity and scalar by the friction velocity and friction temperature, respectively. As is apparent from Fig. 1, the turbulent transport of $\overline{uv}$, i.e., $\overline{vuv}$, is dominated almost completely by the $Q2$- ($i = 2$) and $Q4$-motions ($i = 4$), and hence the net value is determined by the disparity in contributions between these two types of motions. The interactive fluid motions classified into the first- and third-quadrants in the (u, v)-plane, i.e., $Q1$- and $Q3$-motions, contribute very little to the net value because both contributions are quite small in absolute value and have the signs to cancel out. In short, the triple products of velocity and scalar are determined by the dynamic fluid motions such as the ejections ($Q2$-motions) and sweeps ($Q4$-motions). And this may explain why the triple products could not be described adequately by a conventional static model such as gradient-type diffusion of second-order moments.

Lines in Fig. 1 represent the theoretical predictions from the following equation developed in our earlier study (Nagano and Tagawa 1988).

$$\overline{(\hat{u}^l\hat{v}^m\hat{\theta}^n)}_i = \sigma^l_{u,i}\sigma^m_{v,i}\int_0^{\infty}\left[\int_0^{\infty}\left\{\int_{-\infty}^{\infty}\hat{u}^l\hat{v}^m\hat{\theta}^n P(\sigma_{u,i}\hat{u}, \sigma_{v,i}\hat{v}, \hat{\theta})d\hat{\theta}\right\}d\hat{v}\right]d\hat{u}, \tag{1}$$

where $P(\hat{u}, \hat{v}, \hat{\theta})$ is a three-dimensional joint probability density function (p.d.f.), a circumflex "∧" denotes the normalization by the respective r.m.s. value, and $\sigma_{u,i}$ and $\sigma_{v,i}$ are sign functions which represent the signs of u and v of the ith quadrant in the (u, v)-plane:

$$\sigma_{u,i} = (1, -1, -1, 1), \quad \sigma_{v,i} = (1, 1, -1, -1). \tag{2}$$

For example, in the case of $i = 2$, we have $(\sigma_{u,i}, \sigma_{v,i}) = (-1, 1)$. Using cumulants and Hermite polynomials, the joint p.d.f. $P(\hat{u}, \hat{v}, \hat{\theta})$ can be written (Nagano and Tagawa 1988) as

$$P(\hat{u}, \hat{v}, \hat{\theta}) = \frac{1}{(2\pi)^{3/2}}\sum_{p,q,r=0}^{K\leqq 4} C_{pqr}H_p(\hat{u})H_q(\hat{v})H_r(\hat{\theta})\exp\{-\tfrac{1}{2}(\hat{u}^2 + \hat{v}^2 + \hat{\theta}^2)\}, \tag{3}$$

where K is an integer which satisfies the relation $K = p + q + r \leqq 4$. The expansion coefficient C_{pqr} represented by cumulants k_{pqr} and an Hermite polynomial $H_n(\chi)$ are given by

$$\begin{aligned}
&C_{000} = 1, C_{100} = C_{010} = C_{001} = 0, C_{200} = C_{020} = C_{002} = 0,\\
&C_{110} = k_{110} = \overline{\hat{u}\hat{v}} = R_{uv}, C_{101} = k_{101} = \overline{\hat{u}\hat{\theta}} = R_{u\theta}, C_{011} = k_{011} = \overline{\hat{v}\hat{\theta}} = R_{v\theta},\\
&C_{300} = \tfrac{1}{6}k_{300} = \tfrac{1}{6}\overline{\hat{u}^3} = \tfrac{1}{6}S(u), C_{030} = \tfrac{1}{6}S(v), C_{003} = \tfrac{1}{6}S(\theta),\\
&C_{210} = \tfrac{1}{2}k_{210} = \tfrac{1}{2}\overline{\hat{u}^2\hat{v}}, \ldots, C_{012} = \tfrac{1}{2}k_{012} = \tfrac{1}{2}\overline{\hat{v}\hat{\theta}^2}, C_{111} = k_{111} = \overline{\hat{u}\hat{v}\hat{\theta}},\\
&C_{400} = \tfrac{1}{24}k_{400}, \ldots, C_{310} = \tfrac{1}{6}k_{310}, \ldots, C_{013} = \tfrac{1}{6}k_{013},\\
&C_{220} = \tfrac{1}{4}(k_{220} + 2k_{110}^2), \ldots, C_{112} = \tfrac{1}{2}(k_{112} + 2k_{101}k_{011}),
\end{aligned} \tag{4}$$

and

$$H_n(\chi) = (-1)^n \exp\left(\frac{\chi^2}{2}\right)\frac{d^n}{d\chi^n}\exp\left(-\frac{\chi^2}{2}\right). \tag{5}$$

Thus, substitution of (2)–(5) into (1) yields

$$\overline{(\hat{u}^l\hat{v}^m\hat{\theta}^n)}_i = \frac{2}{(2\pi)^{3/2}} \sum_{\substack{p,q,r=0 \\ n+r=\text{even}}}^{K\leq 4} \sigma_{u,i}^{l+p}\sigma_{v,i}^{m+q} C_{pqr} B_{l,p} B_{m,q} B_{n,r}, \tag{6}$$

where $B_{j,k}$ is defined as

$$B_{j,k} = \int_0^\infty \chi^j H_k(\chi)\exp(-\chi^2/2)d\chi. \tag{7}$$

For $\overline{(vuv)}_i$ (with $l=1$, $m=2$, $n=0$), Eq. (1) is reduced to

$$\overline{(\hat{v}\hat{u}\hat{v})}_i = \tfrac{1}{4}\overline{\hat{v}\hat{u}\hat{v}} + \sigma_{u,i}\frac{1}{2\sqrt{2\pi}}(1 + 2C_{220} - C_{400})$$
$$+ \sigma_{v,i}\frac{1}{\sqrt{2\pi}}(C_{110} + C_{130}) + \sigma_{u,i}\sigma_{v,i}\frac{1}{\pi}(C_{210} + C_{030}). \tag{8}$$

As seen in Fig. 1, Eq. (8) reproduces the above mentioned dynamic behavior of the third-order moment very well.

As suggested from these analyses, the turbulent transport process of second-order moments is governed almost completely by the coherent structures in a turbulent shear flow. Hence, we may utilize this experimental fact for modeling the triple correlations. Also, considering the structural similarity which is common to shear-generated turbulence (Lee et al. 1987; Moin 1990), we can expect a wide applicability of the foregoing quadrant analysis. The $Q2$- and $Q4$-motions, which correspond to the ejections and sweeps, are found to play a key role in determining the internal structures of turbulence statistics in wall turbulence. On the other hand, the $Q1$- and $Q3$-motions are important in pattern recognition of the coherent structures in wall turbulence (Nagano and Hishida 1990). However, as shown in Fig. 1, we may discard the contributions of $Q1$- and $Q3$-motions to the net values of the triple products. This fact can be described by the following equation for $\overline{vuv}$.

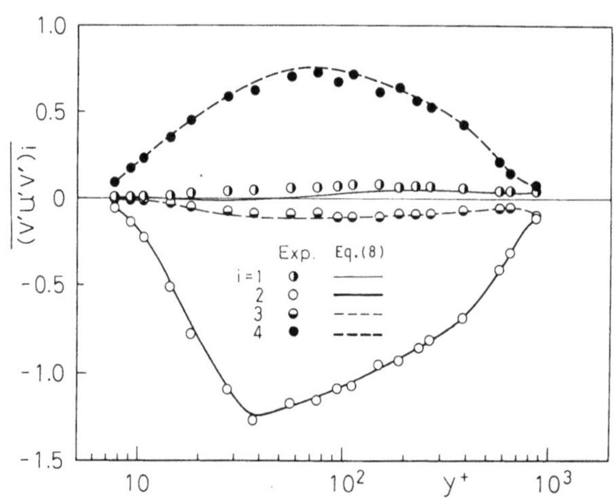

Fig. 1. Contributions of organized motions to the triple correlation $\overline{vuv}$

$$\overline{vuv} \simeq \overline{(vuv)}_{i=2} + \overline{(vuv)}_{i=4}. \tag{9}$$

The similar equation also holds for $\overline{vu^2}$, $\overline{vv\theta}$, $\overline{v\theta^2}$, etc. (Nagano and Tagawa 1990).

The calculations from (8), as shown in Fig. 1, agree well with the experiment. Hence, substituting (8) into (9), we may model the triple product which is based on the physical behavior. The result is

$$\begin{aligned}\overline{\hat{v}\hat{u}\hat{v}} &\simeq \tfrac{1}{2}\overline{\hat{v}\hat{u}\hat{v}} - \frac{2}{\pi}(C_{210} + C_{030}) \\ &= \tfrac{1}{2}\overline{\hat{v}\hat{u}\hat{v}} - \frac{1}{\pi}[\overline{\hat{v}\hat{u}^2} + \tfrac{1}{3}S(v)].\end{aligned} \tag{10}$$

A similar calculation for $\overline{vu^2}$ yields

$$\overline{\hat{v}\hat{u}^2} \simeq \tfrac{1}{2}\overline{\hat{v}\hat{u}^2} - \frac{1}{\pi}[\overline{\hat{v}\hat{u}\hat{v}} + \tfrac{1}{3}S(u)], \tag{11}$$

where $S(\chi)$ is the skewness factor of a stochastic variable χ. From (10) and (11), $\overline{vuv}$ and $\overline{vu^2}$ can be represented with $S(u)$ and $S(v)$ as

$$\overline{\hat{v}\hat{u}\hat{v}} = C\left[S(u) + \sigma_{\overline{uv}}\frac{\pi}{2}S(v) \right], \tag{12}$$

$$\overline{\hat{v}\hat{u}^2} = C\left[\sigma_{\overline{uv}}\frac{\pi}{2}S(u) + S(v) \right], \tag{13}$$

$$C = \frac{1}{3[(\pi/2)^2 - 1]}, \tag{14}$$

where $\sigma_{\overline{uv}}$ represents a sign function introduced to make a model independent of a coordinate system:

$$\sigma_x = \begin{cases} 1, & x \geqq 0 \\ -1, & x < 0 \end{cases}. \tag{15}$$

For $\overline{vuv}$, $\sigma_{\overline{uv}}$ is equal to -1 because $x = \overline{uv} < 0$. We term the present model such as (12) or (13) a "structural" model for a triple product, since it is constructed on the basis of the structure of turbulence.

We can also apply the above theoretical approach to the modeling of triple products in scalar field, $\overline{vv\theta}$, $\overline{v\theta^2}$, etc. The resultant models for $\overline{vv\theta}$ and $\overline{v\theta^2}$ can be written as

$$\overline{\hat{v}\hat{v}\hat{\theta}} \simeq C\left[\sigma_{\overline{v\theta}}\frac{\pi}{2}S(v) + S(\theta) \right], \tag{16}$$

$$\overline{\hat{v}\hat{\theta}^2} \simeq C\left[S(v) + \sigma_{\overline{v\theta}}\frac{\pi}{2}S(\theta) \right], \tag{17}$$

where C is identical to (14) and $\sigma_{\overline{v\theta}}$ is defined by (15).

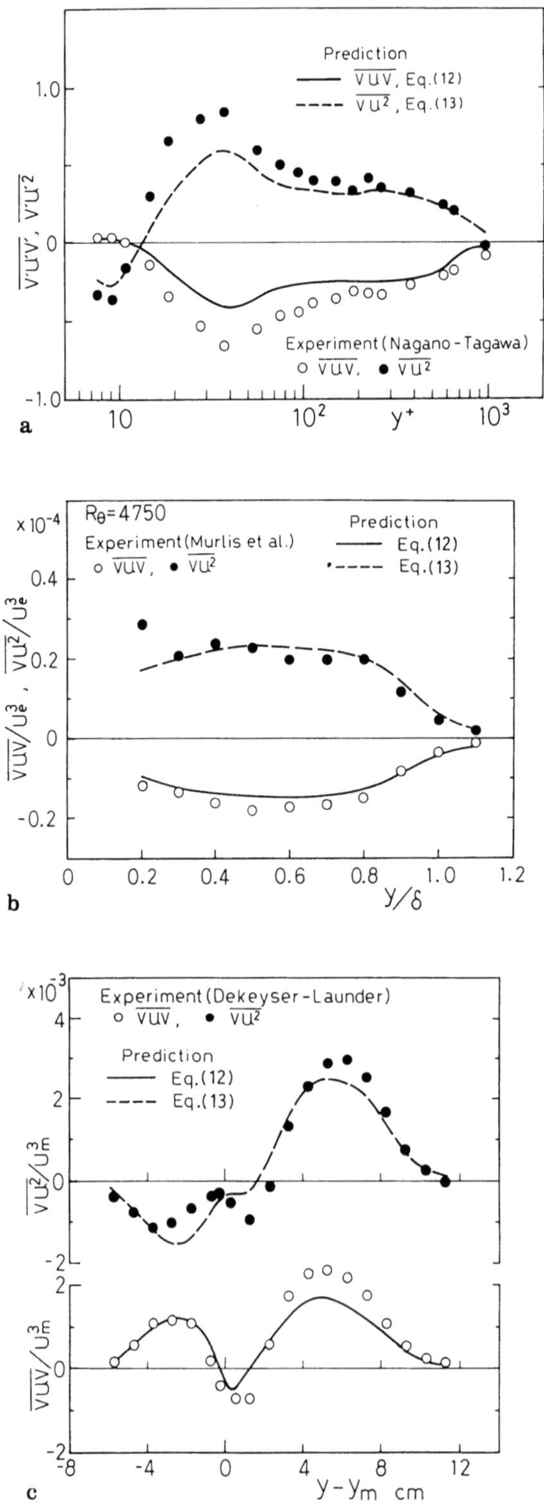

Fig. 2. (*Continued*)

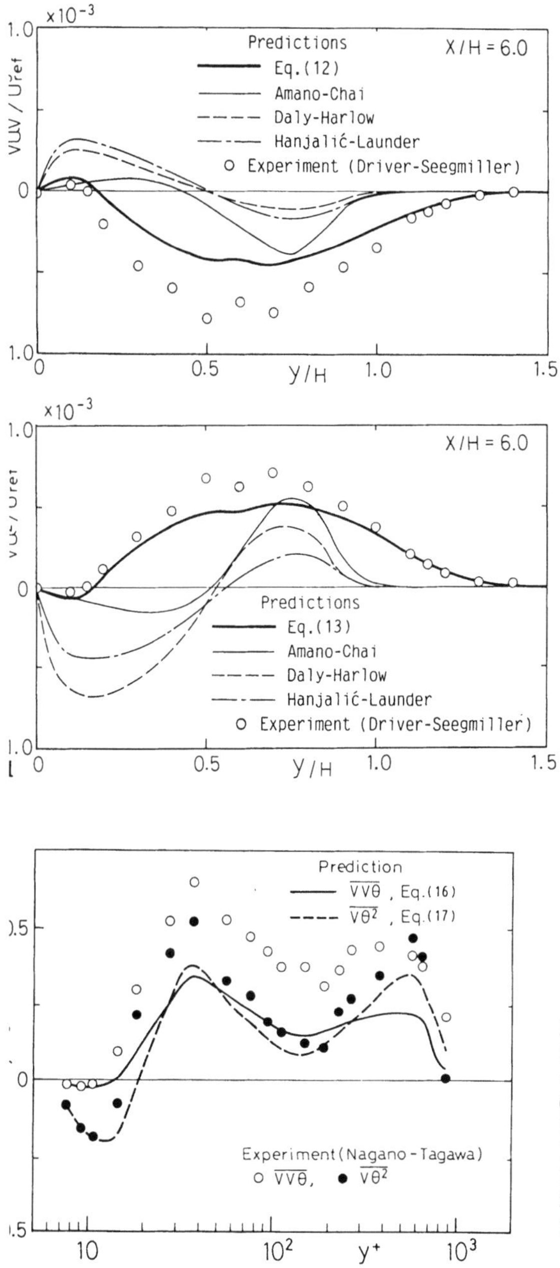

Fig. 2 a–e. Sample results predicted from the structural turbulence models for triple velocity and scalar correlations. **a** Pipe flow (Nagano and Tagawa 1988); **b** Flat plate boundary layer flow (Murlis, Tsai and Bradshaw 1982); **c** Jet (Dekeyser and Launder 1985); **d** Backward-facing step flow (Driver and Seegmiller 1985); **e** Heated pipe flow (Nagano and Tagawa 1988)

These structural models are very useful for the following reasons (Nagano and Tagawa 1990): (i) the number of the triple correlations to be modeled can be reduced substantially; (ii) modeling is required only for the transport equation of a single component $\overline{\chi^3}$ and thus becomes much simpler than for a general triple correlation

$\overline{\chi_1\chi_2\chi_3}$; (iii) parameters characterizing the triple correlations can be made clearer (for example, since the skewnesses $S(u)$ and $S(\theta)$ are nearly equal to zero in the log-law region of wall turbulence (Nagano and Tagawa 1988), $S(v)$ dominates the behavior of other triple correlations).

The effectiveness and universality of the structural models have been verified by application to twelve structurally different flow cases (Nagano and Tagawa 1990). Here, we cite five of them. The predictions of $\overline{vuv}$ and $\overline{vu^2}$ from (12) and (13) are shown in Figs. 2a–d for a pipe flow (Nagano and Tagawa 1988), a flat plate boundary layer flow (Murlis et al. 1982), a jet (Dekeyser and Launder 1985) and a backward-facing step flow (Driver and Seegmiller 1985), respectively. In Fig. 2d, also included are the results predicted from gradient-type diffusion models (Daly and Harlow 1970; Hanjalić and Launder 1972) and those from a transport equation model (Amano and Chai 1988b). The predictions of $\overline{vv\theta}$ and $\overline{v\theta^2}$ in a heated pipe-flow (Nagano and Tagawa 1988) from (16) and (17) are shown in Fig. 2e. As is obvious from these figures, the structural models for triple correlations given by (12), (13), (16) and (17) work quite well regardless of the flow types.

Modeling of Turbulent Transport of Second-Order Moments

Transport Equations and Their Previous Modeling

The Reynolds-stress and scalar-flux equation models need all the terms in the transport equations to be closed within the second-order moments of velocity and scalar. The existing models for triple correlations based on the gradient-type diffusion of second-order moments are those proposed by Daly and Harlow (1970), Hanjalić and Launder (1972) and Cormack, Leal and Seinfeld (1978) for velocity fields, and by Donaldson et al. (1972), Deardorff (1973), Owen (1973), Wyngaard and Coté (1974) and Wyngaard (1975) for scalar fields. Lumley (1978) proposed alternative models for triple velocity and scalar correlations based on a type of "realizability" that turbulence will relax to Gaussian behavior in the absence of inhomogeneity. On the other hand, the modeling of triple correlations without an assumption of the gradient-type diffusion has been done by Amano et al. (1988) and Amano and Chai (1988a) for triple velocity correlations and by André et al. (1979), Dekeyser and Launder (1985) and Amano and Chai (1988b) for triple velocity and scalar correlations on the basis of the relevant transport equations. In the following, we review the most recent models for triple velocity correlations.

The exact transport equation of a general triple correlation $\overline{u_i u_j u_k}$ may be written as

$$\frac{D\overline{u_i u_j u_k}}{D\tau} = P_{ijk1} + P_{ijk2} + d_{ijk} + \phi_{ijk} - \varepsilon_{ijk}, \tag{18}$$

where

$$\left.\begin{aligned}
P_{\mathrm{ijk1}} &= \overline{u_i u_j}(\overline{u_k u_m})_{,m} + \overline{u_j u_k}(\overline{u_i u_m})_{,m} + \overline{u_k u_i}(\overline{u_j u_m})_{,m} \\
P_{\mathrm{ijk2}} &= -(\overline{u_i u_j u_m}\bar{U}_{k,m} + \overline{u_k u_i u_m}\bar{U}_{j,m} + \overline{u_j u_k u_m}\bar{U}_{i,m}) \\
d_{\mathrm{ijk}} &= -[\overline{u_i u_j u_k u_m} + \overline{(p/\rho) u_j u_k}\delta_{im} + \overline{(p/\rho) u_k u_i}\delta_{jm} \\
&\quad + \overline{(p/\rho) u_i u_j}\delta_{km} - \nu(\overline{u_i u_j u_k})_{,m}]_{,m} \\
\phi_{\mathrm{ijk}} &= \overline{(p/\rho)[(u_i u_j)_{,k} + (u_k u_i)_{,j} + (u_j u_k)_{,i}]} \\
\varepsilon_{\mathrm{ijk}} &= 2\nu(\overline{u_k u_{i,m} u_{j,m}} + \overline{u_j u_{k,m} u_{i,m}} + \overline{u_i u_{j,m} u_{k,m}})
\end{aligned}\right\} \tag{19}$$

The notations P_{ijk1} and P_{ijk2} represent generation due to Reynolds stresses interacting with their gradients and that due to mean strains, and d_{ijk}, ϕ_{ijk} and $\varepsilon_{\mathrm{ijk}}$ represent the terms of diffusion, pressure interaction and dissipation, respectively. The notation $D/D\tau$ represents the mean substantial derivative, and $f_{,m}$ denotes the differential operator defined by $\partial f/\partial x_m$ (x_m: coordinates).

Dekeyser and Launder (1985) modeled (19), applying the idea used widely in the second-order closure modeling (e.g., Launder, Reece and Rodi 1975), as

$$d_{\mathrm{ijk}} = -[(\overline{u_i u_j})(\overline{u_k u_m}) + (\overline{u_j u_k})(\overline{u_i u_m}) + (\overline{u_k u_i})(\overline{u_j u_m})]_{,m} \tag{20}$$

$$\phi_{\mathrm{ijk}} = -\overline{u_i u_j u_k}/c\tau_u - 0.5P_{\mathrm{ijk1}} - 0.5P_{\mathrm{ijk2}} \tag{21}$$

$$\varepsilon_{\mathrm{ijk}} = -c_\varepsilon \tau_u \varepsilon_{,m}(\overline{u_i u_m}\delta_{jk} + \overline{u_j u_m}\delta_{ki} + \overline{u_k u_m}\delta_{ij}) \tag{22}$$

Here τ_u is the time-scale of velocity field ($= k/\varepsilon$), and the recommended values for the model constants c and c_ε are 0.075 and 0.2, respectively. The modeled diffusion term d_{ijk}, (20), was derived using the quasi-Gaussian assumption that any fourth-order cumulants are zero. The first term in the right side of (21) was obtained applying the extended "return to isotropy" idea used to model the "slow term" of the pressure-strain correlation in the $\overline{u_i u_j}$ transport equation, while the second and third ones in (21) emerge from the "isotropization-of-production" idea for the "rapid term". If we discard P_{ijk2}, $\varepsilon_{\mathrm{ijk}}$, and the second and third terms in ϕ_{ijk}, the modeled Eqs. (20)–(22) reduce to the model of Hanjalić and Launder (1972) (with $c = 0.11$).

On the other hand, Amano and Chai (1988a) modeled (19), including the low Reynolds number effect on ϕ_{ijk}, as

$$\phi_{\mathrm{ijk}} = -C_\gamma(\overline{u_i u_j u_k}/k)\{\varepsilon + \max[C_{\gamma w}k^{3/2}/C_\ell y, 2\nu(\partial k^{1/2}/\partial y)^2]\} \tag{23}$$

$$\varepsilon_{\mathrm{ijk}} = C_{\varepsilon\gamma}\varepsilon k^{1/2} \tag{24}$$

where C_γ, $C_{\gamma w}$, C_ℓ and $C_{\varepsilon\gamma}$ are model constants with the recommended values as 3.0, 8.0, 2.55 and 0.20, respectively.

Similarly, modeling triple correlations in scalar field is also possible on the basis of the relevant transport equations (André et al. 1979; Dekeyser and Launder 1985; Amano and Chai 1988b). However, none of these models for triple velocity and scalar correlations yield satisfactory results in the near-wall region where the turbulent transport takes its maximum and accurate prediction is required (Moin and Kim 1982; Nagano and Hishida 1985; Krishnamoorthy and Antonia 1987), although the recent models have made a marked step forward. Therefore, these models may not be sufficiently suitable to be incorporated into turbulence models which allow calculations right up to a wall boundary. Moreover, since André et al. (1979), Dekeyser and Launder (1985) and Amano and Chai (1988a, b) modeled all the triple correlations which appear in the second-order closure models, their systems of the modeled equations for triple correlations are more complex and numerous than the main systems of the Reynolds-stress and scalar-flux equation models.

Present Model

To overcome the above-mentioned problems, we propose a new model with the aid of the structural models for the triple correlations described previously. The present structural models for triple correlations indicate that additional modeling is not necessary if the skewness factors of u, v and θ are modeled correctly.

For triple correlations of single turbulent quantities, $\overline{u_i^3}$ and $\overline{\theta^3}$, the transport equations may be written as

$$\left.\begin{aligned} \frac{D\overline{u_i^3}}{D\tau} &= P_{i1} + P_{i2} + d_i + \phi_i - \varepsilon_i \\ P_{i1} &= 3\overline{u_i^2}(\overline{u_i u_m})_{,m}, \quad P_{i2} = -3\overline{u_i^2 u_m}\bar{U}_{i,m} \\ d_i &= -[\overline{u_i^3 u_m} + 3\overline{(p/\rho)u_i^2}\delta_{im} - \nu(\overline{u_i^3})_{,m}]_{,m} \\ \phi_i &= 6\overline{(p/\rho)u_i u_{i,i}}, \quad \varepsilon_i = 6\nu\overline{u_i u_{i,m} u_{i,m}} \end{aligned}\right\} \tag{25}$$

$$\left.\begin{aligned} \frac{D\overline{\theta^3}}{D\tau} &= P_{\theta 1} + P_{\theta 2} + d_\theta - \varepsilon_{\theta^3} \\ P_{\theta 1} &= 3\overline{\theta^2}(\overline{u_m\theta})_{,m}, \quad P_{\theta 2} = -3\overline{u_m\theta^2}\bar{\Theta}_{,m} \\ d_\theta &= -[\overline{\theta^3 u_m} - a(\overline{\theta^3})_{,m}]_{,m}, \quad \varepsilon_{\theta^3} = 6a\overline{\theta\theta_{,m}\theta_{,m}} \end{aligned}\right\} \tag{26}$$

where the summation convention in the tensor notation is applied only to the subscript m.

Now, we represent the pressure interaction term ϕ_i in (25) as

$$\phi_i = -\overline{u_i^3}/C_{u1}\tau_u + C_{u2}\tau_u\overline{u_i u_m}[(\varepsilon/k)(\overline{u_i^2} - 2k/3)]_{,m} \tag{27}$$

where C_{u1} and C_{u2} are model constants, and k, ε and τ_u are the turbulent kinetic energy ($= \overline{u_m u_m}/2$), the dissipation rate of k, and the time-scale of a velocity field ($= k/\varepsilon$), respectively. The first term in the right side of (27) is identical to the Hanjalić and Launder (1972) model for ϕ_i. The second term in (27) is derived from an alternative representation for ϕ_i. We rewrite ϕ_i as $6\overline{u_i[(p/\rho)u_{i,i}]}$ and then by applying the gradient-type diffusion assumption to this term with the Rotta's model for the pressure-strain correlation, we obtain the second formulation. As is obvious from (27), the second term becomes more important as an anisotropy in turbulence components increases. On the other hand, Chou (1987) recently made another approach to the modeling of ϕ_i based on a Poisson equation for pressure fluctuations. However, this model is too complicated to be readily incorporated in the Reynolds stress model.

The diffusion terms, d_i and d_θ, in (25) and (26) are modeled on the assumption that the fourth-order cumulants are consistently zero, i.e., the quasi-Gaussian assumption (Hanjalić and Launder 1972; Dekeyser and Launder 1985).

$$d_i = -3[(\overline{u_i^2})(\overline{u_i u_m})]_{,m} \tag{28}$$

$$d_\theta = -3[(\overline{\theta^2})(\overline{u_m \theta})]_{,m} \tag{29}$$

Modeling of the dissipation term ε_i in (25) is easily made if we use (22) with $i = j = k$. However, if we make the common assumption that the fine-scale motion is isotropic, ε_i can be discarded (André et al. 1982), so that

$$\varepsilon_i = 0. \tag{30}$$

For the sake of brevity, we adopt (30) as a model for ε_i. On the other hand, ε_{θ^3} in (26) is not zero even at high Reynolds numbers and may be modeled as

$$\varepsilon_{\theta^3} = \overline{\theta^3}/C_{\theta 1}\tau_\theta - C_{\theta 2}\tau_\theta \overline{u_m \theta}\,\varepsilon_{\theta,m} \tag{31}$$

where $C_{\theta 1}$ and $C_{\theta 2}$ are model constants, and τ_θ is the time-scale of a scalar field $(= (\overline{\theta^2}/2)/\varepsilon_\theta$ and ε_θ is the dissipation rate of $\overline{\theta^2}/2)$. The first term in the right side of (31) corresponds to the model of André et al. (1979) given by

$$\varepsilon_{\theta^3} = \overline{\theta^3}/C'_\theta \tau_u \tag{32}$$

except for a time scale (they used τ_u instead of τ_θ). The value of the model constant C'_θ is approximately equal to 1/6–1/4 (André et al. 1979; 1982). The second term in (31) is a model similar to (22), which represents the effect of anisotropy on ε_{θ^3} and improves the predictions near the wall.

The structural models given by (12), (13), (16) and (17) are applied to the triple correlations of the generation terms in (25) and (26), $\overline{u_i^2 u_m}$ and $\overline{u_m \theta^2}$, for closing these terms. The model constants used in the present study are:

$$\left.\begin{array}{ll} C_{u1} = 0.12, & C_{u2} = 1.4 \\ C_{\theta 1} = 0.26, & C_{\theta 2} = 3.4 \end{array}\right\} \tag{33}$$

The constant $C_{u1} = 0.12$ is almost equal to that of the Hanjalić and Launder model (Launder et al. 1975). And a value of $C_{\theta 1}$ corresponds to that of André et al. (1979) with $\tau_\theta/\tau_u \simeq 0.5$–1.0 (Nagano and Hishida 1985; Nagano and Kim 1988).

Results

The predicted results of $S(u)$, $S(\theta)$ and $\overline{v'^3}$ in a heated pipe flow by the present models (25) and (26) are respectively demonstrated in Figs. 3 and 4, compared with the experimental values (Nagano and Tagawa 1988). Obviously, the present models can predict the experimental values with sufficient accuracy. When using the structural models (12), (13), (16) and (17) to predict the triple correlations $\overline{vuv}$, $\overline{vu^2}$, $\overline{vv\theta}$ and $\overline{v\theta^2}$, accurate prediction for the triple products of single turbulent quantities is crucial. Especially in wall turbulence, as mentioned previously, the $S(v)$ distribution determines the behavior of other triple correlations, and so the highest accuracy is required for the $\overline{v^3}$ prediction. The values of $\overline{v'^3}$ predicted by the existing models are also included in Fig. 4, but all of these predictions are too unsatisfactory to be used near the wall. (For full details of the existing models, refer to Nagano and Tagawa (1988).)

The present predictions of $\overline{vuv}$, $\overline{vu^2}$, $\overline{vv\theta}$ and $\overline{v\theta^2}$ in wall turbulence are shown in Figs. 5 and 6, compared with those given by the existing models. The greatest

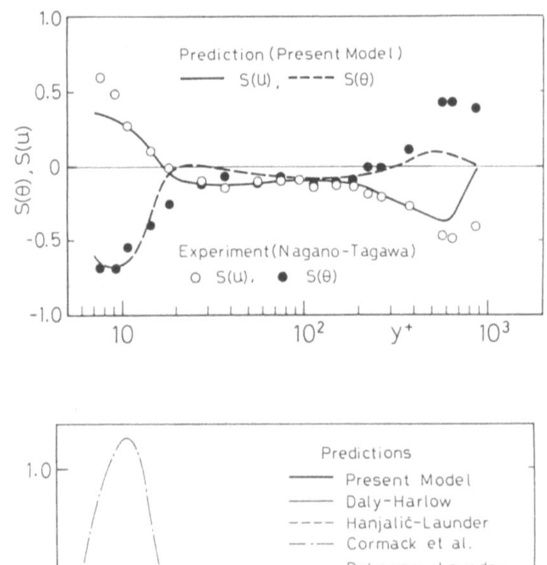

Fig. 3. Predictions of $S(u)$ and $S(\theta)$ in heated wall turbulence

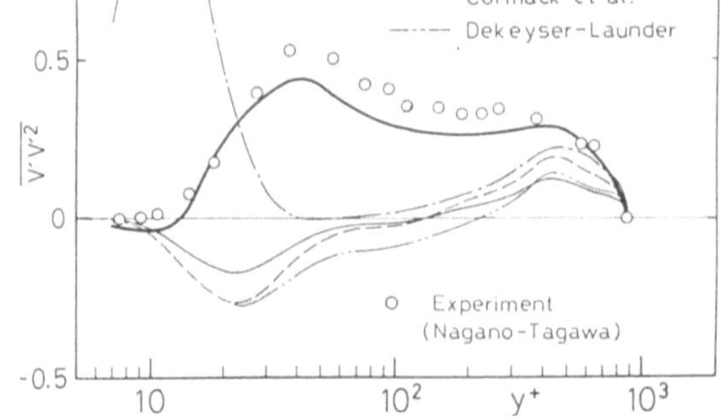

Fig. 4. Prediction of $\overline{v'^3}$ in wall turbulence

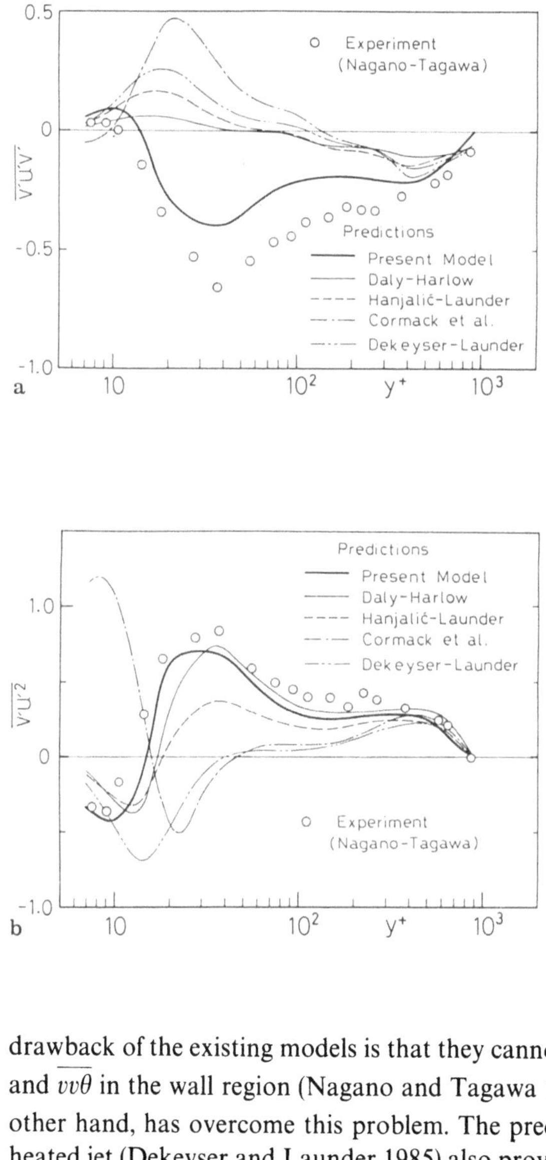

Fig. 5. a,b. Comparison of model results for triple velocity correlations in wall turbulence. **a** $\overline{v'u'v'}$; **b** $\overline{v'u'^2}$

drawback of the existing models is that they cannot predict the distributions of $\overline{vuv}$ and $\overline{vv\theta}$ in the wall region (Nagano and Tagawa 1988). The present model, on the other hand, has overcome this problem. The predictions of $\overline{vuv}$ in an asymmetric heated jet (Dekeyser and Launder 1985) also prove that the present model predicts the values closest to the experiments in all available models (Fig. 7).

Conclusions

The gradient-type diffusion hypothesis for triple correlations is not correct in near-wall turbulence, and thus the use of this hypothesis in Reynolds-stress and scalar-flux equation models may cause erroneous predictions. In the present study, to resolve this problem, we have combined the previously developed structural models

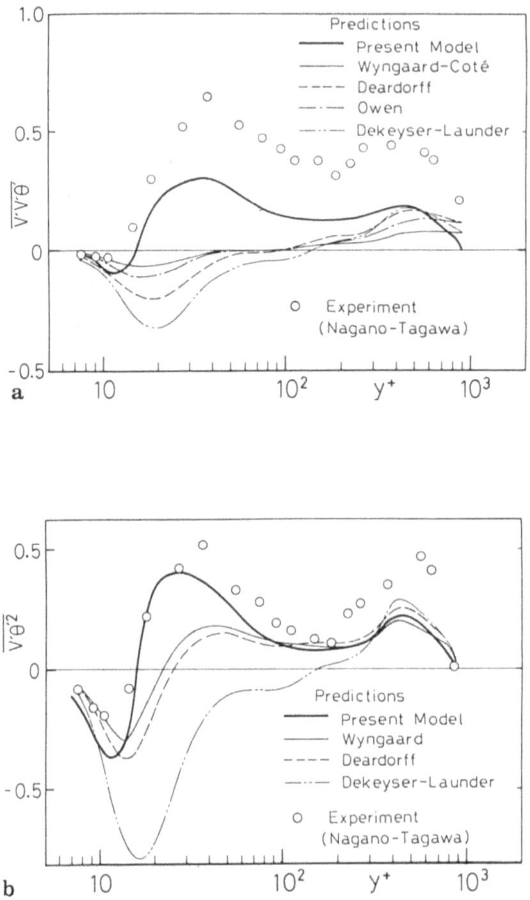

Fig. 6. Comparison of model results for triple velocity and scalar correlations in heated wall turbulence. **a** $\overline{v'v'\theta'}$; **b** $\overline{v'\theta'^2}$

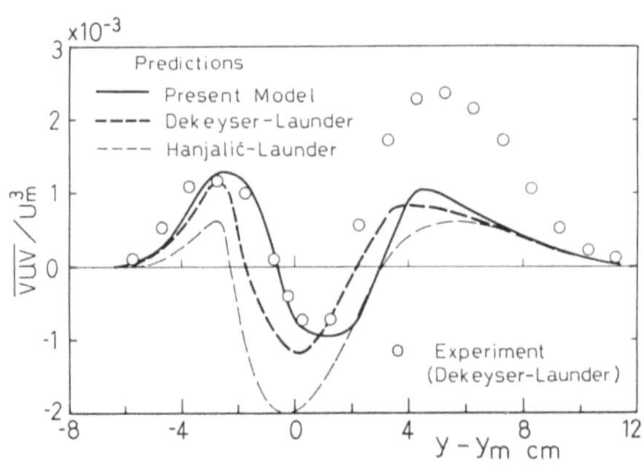

Fig. 7. Comparison of model results for $\overline{vuv}$ in a jet

for triple velocity and scalar correlations [(12), (13), (16), (17)] with their transport equations, and constructed a set of simple models for the turbulent transport terms [(25)–(31), (33)] in the Reynolds-stress and scalar-flux equation models. Comparison with the experiment has proven that the present models for triple correlations work quite well in various flow fields.

Acknowledgement: This research was partially supported by a Grant-in-Aid for Scientific Research on Priority Areas from the Ministry of Education, Science and Culture of Japan (No. 01613002).

References

Amano, R. S., Chai, J. C. (1988a): "A closure model of diffusion transport of the Reynolds-stress equations and its application to a turbulent step flow," in *Transport Phenomena in Turbulent Flows: Theory, Experiment, and Numerical Simulation* (M. Hirata and N. Kasagi, ed.) (Hemisphere), pp. 649–660

Amano, R. S., Chai, J. C. (1988b): Transport models of the turbulent velocity-temperature products for computations of recirculating flows. Numerical Heat Transfer **14**, 75–95

Amano, R. S., Goel, P., Chai, J. C. (1988): Turbulence energy and diffusion transport of third-moments in a separating and reattaching flow. AIAA J. **26**, 273–282

André, J. C., De Moor, G., Lacarrère, P., Therry, G., du Vachat, R. (1979): "The clipping approximation and inhomogeneous turbulence simulations," in *Turbulent Shear Flows 1*, (F. Durst, B. E. Launder, F. W. Schmidt, J. H. Whitelaw, eds) (Springer, Berlin, Heidelberg, New York), pp. 307–318

André, J. C., Lacarrère, P., Traoré, K. (1982): "Pressure effects on triple correlations in turbulent convective flows," in *Turbulent Shear Flows 3* (L. J. S. Bradbury, F. Durst, B. E. Launder, F. W. Schmidt, J. H. Whitelaw, eds) (Springer, Berlin, Heidelberg, New York), pp. 243–252

Chou, P. Y. (1987): "On the theory of turbulence for incompressible fluids," in *Perspectives in Turbulence Studies* (H. U. Meier and P. Bradshaw, eds) (Springer, Berlin, Heidelberg, New York), pp. 351–381

Cormack, D. E., Leal, L. G., Seinfeld, J. H. (1978): An evaluation of mean Reynolds stress turbulence models: The triple velocity correlation. Trans. ASME, J. Fluids Engng. **100**, 47–54

Daly, B. J., Harlow, F. H. (1970): Transport equations in turbulence. Phys. Fluids **13**, 2634–2649

Deardorff, J. W. (1973): "Three-dimensional numerical modeling of the planetary boundary layer," in *Proceedings of Workshop on Micrometeorology*, (American Meteorological Society), pp. 271–311

Dekeyser, I., Launder, B. E. (1985): "A comparison of triple-moment temperature-velocity correlations in the asymmetric heated jet with alternative closure models," in *Turbulent Shear Flows 4* (L. J. S. Bradbury, F. Durst, B. E. Launder, F. W. Schmidt, J. H. Whitelaw, eds) (Springer, Berlin, Heidelberg, New York), pp. 102–117

Donaldson, C. duP., Sullivan, R. D., Rosenbaum, H. (1972): A theoretical study of the generation of atmospheric-clear air turbulence. AIAA J. **10**, 162–170

Driver, D. M., Seegmiller, H. L. (1985): Features of a reattaching turbulent shear layer in divergent channel flow. AIAA J. **23**, 163–171

Hanjalić, K., Launder, B. E. (1972): A Reynolds stress model of turbulence and its application to thin shear flows. J. Fluid Mech. **52**, 609–638

Krishnamoorthy, L. V., Antonia, R. A. (1987): Temperature-dissipation measurements in a turbulent boundary layer. J. Fluid Mech. **176**, 265–281

Launder, B. E., Reece, G. J., Rodi, W. (1975): Progress in the development of a Reynolds-stress turbulence closure. J. Fluid Mech. **68**, 537–566

Lee, M. J., Kim, J., Moin, P. (1987): "Turbulence structure at high shear rate," in *Proceedings of 6th Symposium on Turbulent Shear Flows*, Toulouse, France, pp. 22.6.1–22.6.6

Lumley, J. L. (1978): "Computational modeling of turbulent flows," in *Advances in Applied Mechanics* **18**, (C.-S. Yih, ed.) (Academic Press), pp. 123–176

Moin, P. (1990): "Similarity of organized structures in turbulent shear flows," in *Near-Wall Turbulence*, (S. J. Kline, N. H. Afgan, eds) (Hemisphere), pp. 2–6

Moin, P., Kim, J. (1982): Numerical investigation of turbulent channel flow. J. Fluid Mech. **118**, 341–377

Murlis, J., Tsai, H. M., Bradshaw, P. (1982): The structure of turbulent boundary layers at low Reynolds numbers. J. Fluid Mech. **122**, 13–56

Nagano, Y., Hishida, M. (1985): "Production and dissipation of turbulent velocity and temperature fluctuations in fully developed pipe flow," in *Proceedings of 5th Symposium on Turbulent Shear Flows*, Ithaca, USA, pp. 14.19–14.24

Nagano, Y., Hishida, M. (1990): "Turbulent heat transfer associated with coherent structures near the wall," in *Near-Wall Turbulence* (S. J. Kline, N. H. Afgan, eds) (Hemisphere), pp. 568–581

Nagano, Y., Kim, C. (1988): A two-equation model for heat transport in wall turbulent shear flows. Trans. ASME, J. Heat Transfer **110**, 583–589

Nagano, Y., Tagawa, M. (1988): Statistical characteristics of wall turbulence with a passive scalar. J. Fluid Mech. **196**, 157–185

Nagano, Y., Tagawa, M. (1990): A structural turbulence model for triple products of velocity and scalar. J. Fluid Mech. **215**, 639–657

Owen, R. G. (1973): "An analytical turbulent transport model applied to nonisothermal fully-developed duct flows," *Ph.D. Thesis*, Pennsylvania State University

Wyngaard, J. C. (1975): Modeling the planetary boundary layer: Extension to the stable case. Boundary-Layer Meteorology **9**, 441–460

Wyngaard, J. C., Coté, O. R. (1974): The evolution of a convective planetary boundary layer: A higher-order-closure model study. Boundary-Layer Meteorology **7**, 289–308

Fossil Two-Dimensional Turbulence in the Ocean

Carl H. Gibson

Departments of Applied Mechanics and Engineering Sciences and Scripps Institution of Oceanography, University of California at San Diego, La Jolla, CA 92093, USA

Abstract

Turbulence in the stratified ocean is constrained at large vertical scales by buoyancy forces, leaving remnant internal waves and temperature and salinity fluctuations, termed *fossil turbulence*, that persist after the flow is no longer actively turbulent (with inertial-force dominated eddies) at the scales of the fluctuations. Most patches of ocean temperature microstructure are fossil turbulence (no longer turbulent) at the largest scales, and moving as buoyancy-dominated internal waves with embedded active turbulent motions (if any) only at smaller scales. Larger scale two-dimensional turbulent motions may grow in the horizontal until contrained by Coriolis forces to form eddy-like Coriolis-inertial "waves" with embedded, smaller scale, active 2-D and 3-D turbulence. The remnant horizontal fluctuations in the scalar and vorticity fields that persist after the fluid motions are no longer active 2-D turbulence at the scale of the fluctuations are termed *fossil 2-D turbulence*. By preserving the largest scales L_{W0} of the previous active 2-D turbulence the 2-D fossils preserve information about the dissipation rate ε_0 that existed when fossilization of the 2-D turbulence began assuming $L_{W0} = C_\Omega L_{\Omega 0}$, where $L_{\Omega 0} \equiv (\varepsilon_0/f^3)^{1/2}$ is the Coriolis-inertial (or Hopfinger) length scale at fossilization, f is the Coriolis parameter $2\Omega \sin\phi$, Ω is the rotation rate, ϕ is the latitude and C_Ω is a universal constant of order 1.

1. Introduction

Small scale internal waves in the ocean are produced when turbulence patches are damped by buoyancy forces. These internal waves, and the associated temperature and salinity microstructure remnants, persist as fossil turbulence footprints of the largest scales of the previous active turbulence. The highest wavenumber portions of composite deep ocean velocity gradient and temperature gradient spectra formed by averaging a few scattered data samples are dominated by such "fossil vorticity turbulence" internal waves and "fossil temperature turbulence" temperature microstructure, according to Gibson (1986). This suggests the turbulence and mixing processes have been undersampled, and that larger data samples would include the more active regions implied by the fossils and indicate larger mean squared velocity and temperature gradient values. Similarly, two-dimensional turbulence may be constrained by Coriolis forces to produce Coriolis-inertial waves (eddies) and associated partially mixed scalar fields (mesostructure) that will persist as fossil 2-D turbulence remnants and may also preserve information about previous active turbulence events.

Properties of stratified 3-D turbulence and stratified 3-D fossil turbulence (turbulence fossilized in various fields by stable stratification) are explored in a series

Turbulent Shear Flows 7

of papers Gibson (1980–1991) and are summarized by Gibson (1986, 1987b, 1990a). Hydrodynamic phase diagrams are developed to determine the *hydrodynamic state* (active, active-fossil, fossil) of a measured patch of microstructure, where the state is determined by a comparison of length scales of the microstructure with length scales of transition between fluid forces. If the inertial-buoyancy (Ozmidov) scale $L_R \equiv (\varepsilon/N^3)^{1/2}$ is larger than the energy (or integral, or largest) scale L_O of the turbulence then the microstructure is actively turbulent at all scales; if $L_{RO} > L_O \approx L_R > L_K$ the large scales are fossil and the small scales active; and all scales are fossil if $L_R < L_{KF}$, where L_{RO} is the Ozmidov scale of beginning fossilization, L_K is the inertial-viscous (Kolmogorov) scale and $L_{KF} \equiv (\nu/N)^{1/2}$ is the buoyancy-inertial-viscous (fossil Kolmogorov) scale of complete fossilization. Universal proportionality constants of transition between states have been omitted. Laboratory tests confirm that various maximum overturn scales [Thorpe et al.: see Gibson (1990a, b)] of temperature microstructure fields produced by turbulence collapse more slowly in stratified flows than the turbulence energy scales $L_O \approx L_R$, and therefore preserve information about L_{RO} and ε_0 at fossilization.

The persistence of large scale scalar and flow field perturbations after the turbulence which produced them has been damped is the distinguishing property of fossil turbulence. Properties of 2-D turbulence are not nearly as well established as those of 3-D turbulence, and 2-D fossil turbulence has apparently not even been previously suggested. The purpose of the present paper is to explore the possibility of, and evidence for, rotationally fossilized turbulence with 2-D fossil turbulence as a special case. It is suggested that some oceanic mesoscale eddy motions identified as 2-D turbulence may actually be Coriolis-inertial wave fossils of turbulence.

Just as most internal wave motions in the ocean at scales larger than about ten meters were probably not produced by damping turbulence, and are therefore neither active turbulence nor fossil vorticity turbulence, many oceanic Coriolis-inertial wave motions at scales of tens or hundreds of kilometers are probably neither active 2-D turbulence nor fossil 2-D vorticity turbulence. However, many smaller scale eddies in the ocean that are presently identified as 2-D turbulence, sometimes indicated by mushroom shaped temperature or chlorophyll concentration patterns at the sea surface (Federov eddies) observed from satellites, may not be active 2-D turbulence at the time of observation but fossil 2-D vorticity turbulence. Valuable information about previous active 2-D turbulence may be preserved by these horizontal scalar field patterns, which should be classified as fossil 2-D scalar turbulence fluctuations. Methods to classify the hydrodynamic state of such horizontal "mesostructures" similar to hydrodynamic phase diagrams developed to classify the hydrodynamic state of 3-D active and 3-D fossil turbulence microstructure fluctuations are discussed in the Sects. 4 and 5 following a review of fundamental principles in Sect. 3 and definitions in Sect. 2. Laboratory experiments of rotational, non-stratified turbulence show the formation of Coriolis-inertial waves that are a form of rotational fossil turbulence, and are discussed in Sect. 4.

Definitions of terms are given in the following Sect. 2. They must be read carefully to avoid confusion. Fossil turbulence is not turbulence just as fossil dinosaurs are not dinosaurs. A narrow definition of active turbulence is given that requires that the inertial-vortex forces of the eddy motions dominate all other forces that may damp the eddies. More details are given by Gibson (1990ab, 1991). It is not enough that the flow be rotational, or random, or both, for it to be considered turbulent. Critical Reynolds et al. numbers must be exceeded by the eddy motions. All that wiggles is not turbulence.

2. Definitions

Turbulence is an eddy-like state of fluid motion that arises when inertial-vortex forces of a shear flow are larger than viscous, buoyancy, Coriolis, or other forces opposing the growth of the eddies. Because shear layers are unstable at all scales, and because overturn time scales generally decrease with increasing eddy size, turbulent motions form first at the smallest possible length scale, and grow to larger scales by processes of entrainment and vortex interaction. The smallest eddies are at the inertial-viscous, or Kolmogorov, length scale, $L_K \equiv (\nu^3/\varepsilon)^{1/4}$, where ν is the kinematic viscosity and ε is the viscous dissipation rate. They move with a universal critical Reynolds number of about 100. If motions are constrained in one (Cartesian) direction by, for example, the presence of walls, surface tension in soap bubbles, or stable stratification, the turbulence will grow in the other two directions, and is termed "two-dimensional turbulence", or simply, *2-D turbulence.* In the absence of any such constraints turbulence loses orientation at small scales and becomes locally isotropic and homogeneous in space and time, and is termed *3-D turbulence* or simply *turbulence.* As a turbulent patch grows in a stably stratified fluid such as the ocean, buoyancy forces constrain the vertical overturning of the eddies at the buoyancy-inertial, or Ozmidov length scale $L_R \equiv (\varepsilon/N^3)^{1/2}$, where the intrinsic stratification frequency $N \equiv (-g\rho_{,z}/\rho)^{1/2}$, ρ is density, g is gravity, and the, z subscript indicates the gradient in the vertical direction z.

Horizontal eddy motions of thick ocean layers are not constrained by buoyancy or viscous forces, but may continue to grow as *2-D active turbulence* until constrained by Coriolis forces at the Coriolis-inertial length scale $L_\Omega \equiv (\varepsilon/f^3)^{1/2}$, where f is $2\Omega \sin\phi$, Ω is the earth's rotation rate and ϕ is the latitude. [After writing this paper it was discovered that the length scales $L_\Omega \equiv (\varepsilon/f^3)^{1/2}$ had been independently, and previously, discovered by Hopfinger in Mory and Hopfinger (1988) by means discussed in Section 4, but given a different physical interpretation. It is hereby suggested that it should be termed the Hopfinger length scale with symbol L_H.]

As proposed in the present paper, the kinetic energy of these Coriolis force constrained, 2-D turbulent motions is converted to persistent, Coriolis-inertial wave motions termed *2-D fossil vorticity turbulence.* Active 2-D turbulence and fossil 2-D vorticity turbulence motions of the present paper are a special case of the *vortical mode waves* defined by Müller (1988) as wave modes for which vorticity is frozen in the fluid and does not propagate. They may also be a subspecies of various other oceanic motions termed *mode eddies, modons, quasi-geostrophic eddies, inertial waves* and *Coriolis-inertial waves.* Scalar fields such as temperature and salinity stirred but not mixed by the turbulence before its damping may also persist, and are termed *2-D fossil temperature turbulence* and *2-D fossil salinity turbulence*, respectively. Turbulence damped by buoyancy forces in the ocean is three-dimensional, and produces *3-D fossil (vorticity, temperature, salinity, etc.) turbulence. Oceanic* turbulence constrained (damped) by Coriolis forces will generally be 2-D turbulence, and it is proposed in the present paper that its damping produces *2-D fossil (vorticity, temperature, salinity, etc.) turbulence.* When the turbulent flow is rotating but not stratified or otherwise constrained, Coriolis forces may produce *rotational fossil (vorticity, temperature, salinity, etc.) turbulence.* Experimental examples of rotationally constrained turbulence are given in Sect. 4.

3. The Formation of Turbulent Flows

Figure 1 shows a shear layer, the source of both 3-D and 2-D turbulence. The origin of shear instability, and therefore turbulence, is the *inertial-vortex force per unit mass* $\vec{\mathbf{v}} \times \vec{\boldsymbol{\omega}}$ in the momentum equation,

$$\frac{\partial \vec{\mathbf{v}}}{\partial t} = -\nabla B + \vec{\mathbf{v}} \times \vec{\boldsymbol{\omega}} + \vec{\mathbf{v}} \times 2\vec{\boldsymbol{\Omega}} + \vec{\mathbf{g}}' + \nabla\cdot(\overleftrightarrow{\tau}/\rho) \tag{1}$$

where $\vec{\mathbf{v}}$ is the velocity field, $\vec{\boldsymbol{\omega}}$ is the vorticity, $\vec{\boldsymbol{\Omega}}$ is the angular velocity of the coordinate system, $\vec{\mathbf{g}}'$ is the reduced gravitational force per unit mass $\vec{\mathbf{g}}' \equiv (\rho'/\bar{\rho})\vec{\mathbf{g}}$, ρ' is the fluctuation of density ρ about the mean $\bar{\rho}$, $\overleftrightarrow{\tau}$ is the viscous stress tensor and $B \equiv (p/\rho + v^2/2 + gz)$ is the Bernoulli group of mechanical energy terms. Perturbations of the shear layer result in amplifying inertial-vortex forces in the directions of the perturbation, as shown for time $t_1 > t_0$, causing a roll-up of the vortex sheet into eddies if the other terms in the equation permit this to happen. Ratios of the inertial-vortex force term to the other terms in (1) form the Reynolds et al. and inverse Euler numbers, from right to left. Viscous, Coriolis and, for stable density stratification, gravitational forces develop in directions that tend to suppress the eddies.

Neglecting gravity, the first eddies to form will be on scales of order $L_K \equiv (\nu^3/\varepsilon)^{1/4}$, the Kolmogorov length scale, in times $t_2 \approx (\varepsilon/\nu)^{1/2} \equiv T_K$, the Kolmogorov time scale, where ε is the viscous dissipation rate per unit mass characteristic of the shear layer. The shear layer thickness will also generally be of order L_K. Perturbations on length scale L smaller than L_K will be damped by viscous forces produced by the perturbation since the magnitude of such forces are proportional to L^{-1}. The time for turnover should be of order $L/v(L)$, which increases with L unless the velocity $v(L)$ increases as $v(L) \approx L^n$ where the exponent $n \geqq 1$. For perturbations induced by turbulence, the exponent $n = 1/3$ so that the smallest scale eddies permitted by viscous damping will grow first. This will be on wavelengths of about 10–60 L_K. Six eddies are shown on the shear layer, numbered for identification, in the middle sketch of Fig. 1, after a time interval of order T_K after the initial condition of a nonturbulent shear layer. The vorticity magnitude should be maximum near the

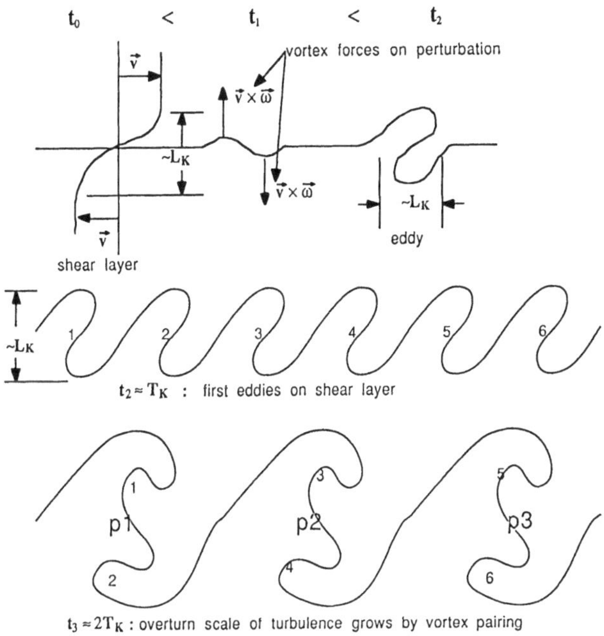

Fig. 1. Schematic development of eddies on a shear layer due to development of inertial-vortex forces which amplify a perturbation. Starting from an unperturbed state at t_0, the first eddies to develop are on the inertial-viscous, or Kolmogorov, scale L_K in a Kolmogorov time $t_2 \approx T_K$. The numbered eddies have vorticity maxima, which pair in times of order $t_3 \approx 2T_K$ to form secondary eddy pairs with centers indicated by $p1$, $p2$ and $p3$ below

centers of the numbered eddies, with the same sign into the paper. Any perturbation in the uniform spacing of these eddies will cause nearest neighbors to circle about each other in response to mutually induced velocities, forming patterns represented by the three adjacent pairs of larger eddies shown at the bottom of Fig. 1 labelled with pair numbers $p1$, $p2$ and $p3$. If these larger eddies encounter fluid with velocity of the same sense as the original shear layer, new vortex forces will develop to continue the turbulence cascade to larger and larger scales, energy is entrained to maintain the small scale structure, and the Reynolds number of the turbulence increases.

Note that although the energy driving this "cascade" of turbulence is extracted from velocity differences on larger and larger scales as the largest eddy scale grows, the basic direction of the *turbulence cascade* is from small to large scales (which is generally termed an *inverse cascade*). This is true whether the turbulence is 2-D or 3-D, contrary to a common assumption that 3-D turbulence differs from 2-D turbulence because the 3-D turbulence cascade is from large to small scales but the 2-D turbulence cascade is from small to large scales. Actually their basic formation and growth mechanisms are nearly identical. They differ geometrically in the coupling between small and large scales, and in the physical mechanisms limiting their largest scales. The primary difference between their governing equations is the vortex line stretching term $\vec{\omega}\cdot\vec{e}$ in the conservation of vorticity equation

$$\frac{D\vec{\omega}}{Dt} = \vec{\omega}\cdot\vec{e} + \left[\frac{\nabla\rho \times \nabla \mathbf{p}}{\rho^2}\right] + \nu\nabla^2\vec{\omega} \tag{2}$$

which is absent for strictly 2-D flows. However, since oceanic 2-D turbulence will generally have embedded 3-D turbulence, and since vortex line stretching occurs on scales of order L_K, it is not clear that this difference is crucial. Vortex lines of 2-D turbulence can be stretched by the embedded 3-D turbulence.

The distinguishing characteristic of 2-D turbulence is that the turbulent eddy motions are constrained both in an axial direction and also in the azimuthal directions in the plane normal to the axis. In rotational or electromagnetic 2-D turbulence, the eddy motions in the axial plane are inhibited at scales larger than the Coriolis-inertial force or electromagnetic-inertial force scales, Gibson (1991), and by buoyancy forces in the axial direction.

4. Evolution of Turbulence in Stratified, Rotating Flows

The development of 3-D turbulence and fossil 3-D turbulence is illustrated schematically by Fig. 2.1, which shows, from left to right, the growth of eddy motions from Kolmogorov to Ozmidov scales, fossilization to form mixed active and fossil turbulence, and finally completely fossil 3-D turbulence. Temperature overturn scales of the fossil preserve the overturn scale of the turbulence when fossilization begins; that is L_{RO}, even though the vertical motions are buoyancy dominated at all scales, and hence are nonturbulent by definition. Figure 2.2 is a laboratory demonstration of the fossilization process for a turbulent jet in salt stratified water. The dissipation rate is estimated by the expression $\varepsilon = 48(U^3/D)(x/D)^{-4}$ for turbulent jets, from Friehe et al. (1971), where U is the exit jet velocity, D is the diameter and x is the distance downstream. Assuming the energy scale of the turbulence L_O is initially equal to the jet height L_P, the universal constant

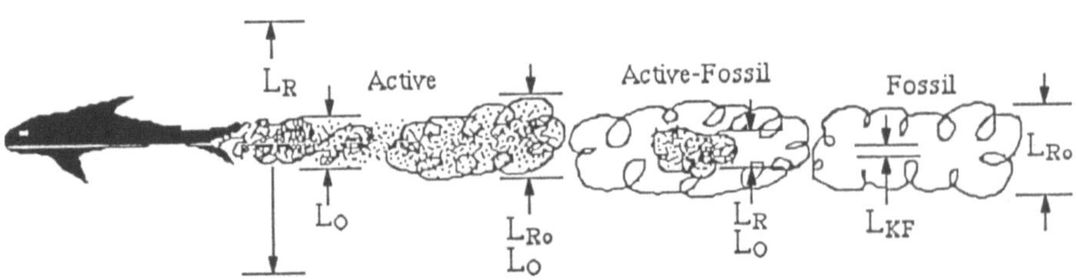

Fig. 2.1. Evolution of 3-D turbulence in stably stratified wake of a large fish. In the "Active" turbulence state, the energy scale of the turbulence L_O is less than the buoyancy-inertial, or Ozmidov, scale L_R, so the turbulence is unaffected by buoyancy. Fossilization begins where $L_O \approx L_R \equiv L_{RO}$. The "Active-Fossil" regime is large scale forced internal waves with embedded active patches. In the completely "Fossil" regime turbulence is damped at all scales, with transition at the buoyant-inertial-viscous scale $L_{KF} = (\nu/N)^{1/2}$. Density overturns and internal wave motions persist at the buoyant-inertial transition scale L_{RO}

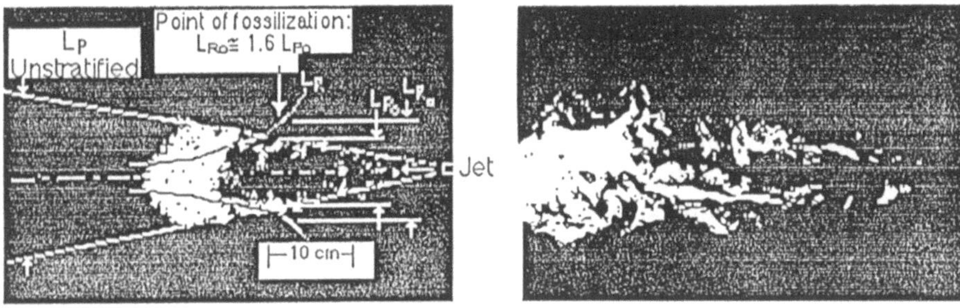

a b

Fig. 2.2. Laboratory demonstration of the evolution of turbulence in stably stratified water, with $N = 0.27$ rad s^{-1}, from Gibson (1990a, Fig. 7a, b). In a thin, vertical sheet of laser light causes fluorescence of dye in a turbulent jet as it spreads at time $tN = 1.4$. The height L_P departs from the constant spreading angle of 20° for an unstratified jet when the Ozmidov scale L_R is about $1.6\,L_P$, at the point where fossilization begins, shown by *o*-subscripts. Active turbulence at the downstream end of the jet is constrained by buoyancy to scales $L_O \approx L_R/1.6$ much smaller than L_P. In b. the fossil turbulence remnant is shown at time $tN = 81$ after all turbulence has been damped at all scales by buoyancy and viscosity

$C_S \approx 0.6$ in the expression $L_{P0} = C_S L_{R0}$, where 0-subscripts indicate the point of fossilization when the spreading rate departs from that for an unstratified jet.

The development of 2-D turbulence and fossil 2-D turbulence in a stably stratified ocean is shown schematically by the top view of the seamount wake in Fig. 3.1. The assumption is that the Coriolis parameter f is much smaller than the buoyancy frequency N and that horizontal frictional forces, both laminar and turbulent, can be neglected. The 2-D turbulence source size (seamount) is much

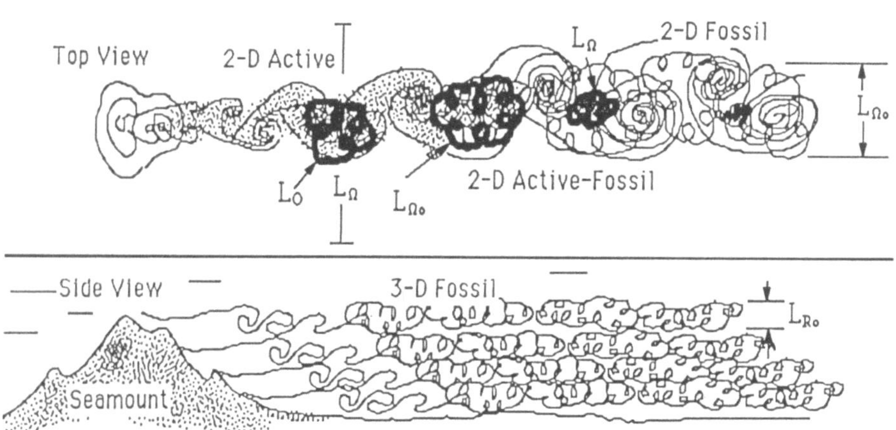

Fig. 3.1. Evolution of turbulence in stably stratified, rotating, wake of an ocean seamount ($f < N$). Above, the top view shows 2-D active turbulence eddies growing by entrainment until their energy scale reaches the Coriolis-inertial, or Coriolis, scale L_Ω, where $L_O \approx L_\Omega \equiv L_{\Omega 0}$ at the point of 2-D turbulence fossilization. Thereafter, the 2-D turbulence is damped to smaller scales, leaving fossil 2-D turbulence remnants on all scales up to $L_{\Omega 0}$. Below, the side view shows that 3-D turbulence is constrained by buoyancy to overturn scales of order L_{R0}, the Ozmidov scale at fossilization of the 3-D turbulence

larger than the 3-D turbulence source of Fig. 2.1 (fish) so that Reynolds numbers are enormous and the eddies may grow to scales for which Coriolis forces will be relevant before the turbulence is damped. We may estimate this size by comparing inertial forces and Coriolis forces for an eddy of size L and thickness H. The inertial force generated by the eddy is of order $F_I \approx \rho V^2 LH$ and the Coriolis force is $F_\Omega \approx \rho f V L^2 H$. Estimating $V \approx (\varepsilon L)^{1/3}$ and solving for the critical length $L \equiv L_\Omega$ for which the two forces are equal gives

$$L_\Omega \equiv (\varepsilon/f^3)^{1/2}, \tag{3}$$

where f is the Coriolis parameter and ε is the average dissipation rate for the eddy. Eddy motions will be deflected to the right in the northern hemisphere and to the left in the southern hemisphere by the Coriolis force, and their turbulent kinetic energy converted to anticyclonic inertial-Coriolis wave motions (clockwise in the north and counterclockwise in the south). Since fluid and kinetic energy is no longer entrained by the 2-D turbulence, but is dissipated by embedded 3-D turbulence, the Coriolis scale will monotonically decrease leaving 2-D turbulence remnants, or fossils, in all partially mixed scalar fields at all scales $L_{\Omega 0} > L > L_\Omega$. The smallest scale for 2-D turbulence will be the layer thickness or vertical amplitude of embedded internal waves, whichever is smaller. The layer thickness for this example will be set by the largest scale for 3-D turbulence, past or present, L_{R0}. In the seamount wake example given, this will be the thickness of the boundary layers shed from the seamount surface, or less, as shown in the side view of the seamount at the bottom of Fig. 3.1. When both 2-D and 3-D fossil turbulence exist, the dissipation rates at 2-D fossilization may be designated as $\varepsilon_{\Omega 0}$ and the dissipation at 3-D fossilization as ε_{S0}. Possibly they are equal. Probably they are related.

Figure 3.2 is a laboratory demonstration of the effects of rotation on a turbulent jet. Flierl et al. (1983) show that a dyed turbulent jet in a rotating flow forms dipole pairs, or *modons*. The size of the remnant monopole eddies formed when the turbulence was damped is approximately equal to that of the Coriolis scale $L_{\Omega 0} \equiv (\varepsilon_0/f^3)^{1/2}$, where ε_0 (or $\varepsilon_{\Omega 0}$) is the dissipation rate of the jet at the point when Coriolis deflection begins, as shown by Fig. 3.2, from Gibson (1989, Fig. 1 h), using the expression for ε of Friehe et al. (1971) given above. Such motions and their associated scalar fields are a form of fossil turbulence, and their size preserves the quantity $\varepsilon_{\Omega 0}$. From Fig. 3.2, the universal constant C_Ω in the expression $L_{W0} = C_\Omega L_{\Omega 0}$ appears to be in the range 0.5–1, where L_{W0} is the energy scale of the turbulence when fossilization begins. If the turbulence were 2-D turbulence when fossilization begins, then the remnants would be 2-D fossil turbulence. In this case, the remnant Coriolis-inertial waves (monopole eddies, modons) are rotational fossil vorticity turbulence using the terminology of Sect. 2.

Figure 3.3 shows the Mory and Hopfinger (1988) experiment. A grid with mesh size M oscillates vertically a distance S in a tank of water rotating on a turntable with angular velocity Ω rad/s. The flow is visualized by means of streak photographs of particles in the water illuminated by horizontal sheets of laser light. The top view shows the eddy sizes indicated by a horizontal light sheet near the top of the tank. The side view show the flow pattern inferred by Mory and Hopfinger (1988), as well as their interpretation of the hydrodynamic states of the flow at various levels. Near

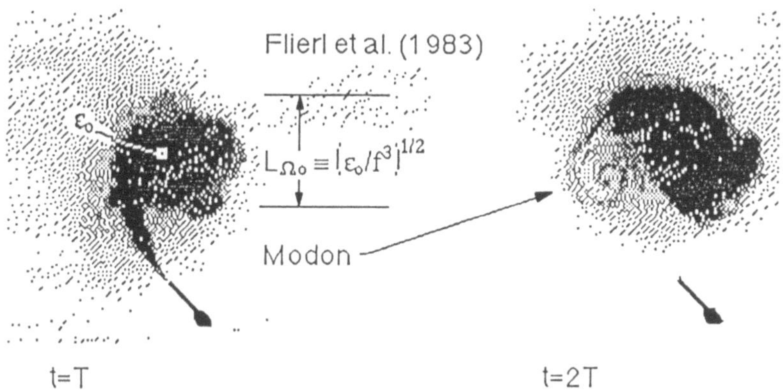

Fig. 3.2. Laboratory demonstration of the effects of rotation on turbulence, from Gibson (1989, Fig. 1h) and Flierl et al. (1983, Fig. 2b,c). Laboratory turbulent jet pulse in rotating tank. On the left, the jet of dyed fluid is "fossilized" by Coriolis forces after one counterclockwise rotation period $t = T = 15.1$ s to form a clockwise eddy with scale $L_{\Omega 0} = (\varepsilon_0/f^3)^{1/2}$, where the location of ε_0 for the turbulent jet is shown by the square. On the right, a counterclockwise eddy has been induced after another rotation period. The velocity and dye fields are rotational fossil turbulence because the eddy and dye sizes preserve information ($L_{\Omega 0}$) about turbulence at a previous time when first constrained by Coriolis forces

the grid they assume the flow is 3D turbulence unaffected by rotation. At some point above the grid, they assume the advection of energy out of the 3D turbulence layer by inertial waves becomes comparable to the cascade rate of the kinetic energy of turbulence from large to small scales, and that this energy transfer somehow blocks the turbulence cascade so that inertial waves (which they term "2D turbulent eddies") take over. By setting the dissipation rate ε equal to the advection rate $\varepsilon_I \sim V^2 C_g/L$ and assuming $C_g \sim fL$ and $V^2 \sim (fL)^2$ a critical scale $L_I \sim (\varepsilon/f^3)^{1/2}$ emerges, where C_g is the (linear) group velocity of inertial waves of scale L and V^2 is their kinetic energy (saturated).

The physical basis of these assumptions and this energy advection explanation of the formation of 2-D turbulence are somewhat obscure and questionable, and the derivation of L_Ω reduces to a dimensional analysis that assumes ε, f and L are the only relevant dimensional parameters of the rotating turbulence. The axial eddy motions formed are not 2-D turbulence according to the definitions of Sect. 2, but Coriolis-inertial waves. As shown by Fig. 1 and the discussion above, turbulence forms at small scales and propagates to large scales by entrainment. Converting the energy of the large scale turbulent eddies to inertial waves (or 2-D turbulence) cannot block the formation of smaller scale turbulence eddies because these are formed at small scales from local shear instabilities and grow by entrainment. Hopfinger et al. (1982) notes that the local Rossby number based on the velocity and size of the inertial waves at formation is a constant, $Ro_0 \approx 0.2$, independent of the grid Rossby number, which supports the view the inertial waves result from a Coriolis force constraint of inertial-vortex forces of growing turbulent eddies. From the Kolmogorov–Oboukov relation $u(L) \approx (\varepsilon L)^{1/3}$ for eddies of size L in the inertial subrange of turbulence and the assumption of a universal critical Rossby number

$Ro_{\mathrm{crit}} \equiv u(L_\Omega)/L_\Omega f$ it follows that the maximum turbulence scale must be L_Ω as in (3), where $L_\Omega = (\varepsilon/f^3)^{1/2}(Ro_{\mathrm{crit}})^{-2/3}$.

The Hopfinger energy blocking argument by inertial wave advection would apparently not predict inertial wave formation in a decaying, rotating, *homogeneous* turbulence field since then there would be nowhere for the inertial wave energy to propagate. However, the force balance argument given in the derivation of (3) does predict that inertial waves should appear as a fossil vorticity turbulence remnant as the homogeneous turbulence decays, preserving information about $\varepsilon_{\Omega 0}$. When the turbulence energy scale L_{O} increases to a size that matches the decreasing Coriolis-inertial scale $(\varepsilon/f^3)^{12}$ at $L_{\mathrm{O}} = L_{\mathrm{WO}} = C_\Omega(\varepsilon_0/f^3)^{1/2}$ then formation of eddies on larger scales will be inhibited by Coriolis forces. The formation of 2-D eddy motions in a decaying, homogeneous grid turbulence in a rotating tank has been demonstrated by Stuart B. Dalziel (1989 work in progress, Department of Applied Mathematics and Theoretical Physics, Cambridge University).

Ignoring stratification and universal proportionality constants, the length scale criteria for the various hydrodynamic states of rotational turbulence with energy scale L_{O} are therefore:

1. Active rotational turbulence: $L_\Omega > L_{\mathrm{O}} > L_{\mathrm{K}}$
2. Rotational turbulence at beginning fossilization: $L_{\mathrm{O}} = L_\Omega = L_{\Omega 0} > L_{\mathrm{K}}$
3. Active-fossil rotational turbulence: $L_{\Omega 0} > L_{\mathrm{O}} = L_\Omega > L_{\mathrm{K}}$
4. Rotational turbulence at complete fossilization: $L_{\Omega 0} > L_{\mathrm{K}} = L_\Omega = L_{\mathrm{Kf}}$
5. Completely fossil rotational turbulence: $L_{\Omega 0} > L_{\mathrm{Kf}} \equiv (\nu/f)^{1/2} > L_\Omega$.

These criteria may be compared to those given by Mory and Hopfinger (1988):

A. Three-dimensional turbulence: $L_\Omega > L_{\mathrm{O}}$
B. Inertial waves and 3D turbulence: $L_{\mathrm{K}} > L_{\mathrm{O}} > L_\Omega$
C. Inertial waves, quasi-2D turbulence: $L_{\mathrm{O}} > L_{\mathrm{K}}$.

Criterion 1. and A. are almost equivalent except that the Reynolds number requirement for turbulence is explicit in 1. and not in A. Criterion B. is impossible for 3-D or 2-D turbulence and seems to be a typographical error, and should probably be $L_{\mathrm{O}} > L_\Omega > L_{\mathrm{K}}$ instead [this is confirmed by Hopfinger, personal communication 1989]. Even this is not equivalent to 3. Criterion C. differs from 5. because it follows for 3-D turbulence and is neutral regarding the possibility of 2-D turbulence because no reference is made to rotation. Regions labelled "Present Interpretation" in Fig. 3.3 are also different.

As shown in Fig. 3.3, active 3-D turbulence forms near the grid where $\varepsilon > \varepsilon_0$. Where $\varepsilon = \varepsilon_0$, Coriolis-inertial waves (axial eddies, Taylor columns, rotational fossil vorticity turbulence) are formed and larger scale eddy formation with nonvertical axes are also inhibited. Because ε decreases at higher levels, the maximum turbulence scale decreases, but the larger 2-D eddies persist as rotational fossil vorticity turbulence with embedded active turbulence. This layer is labeled as Active-Fossil Turbulence on the left of Fig. 3.3. When the Coriolis-inertial scale decreases to the inertial-viscous, or Kolmogorov, scale, all turbulence ceases at all scales by the definition of turbulence in Sect. 2. Setting $L_\Omega \approx L_{\mathrm{K}}$ gives the viscous dissipation

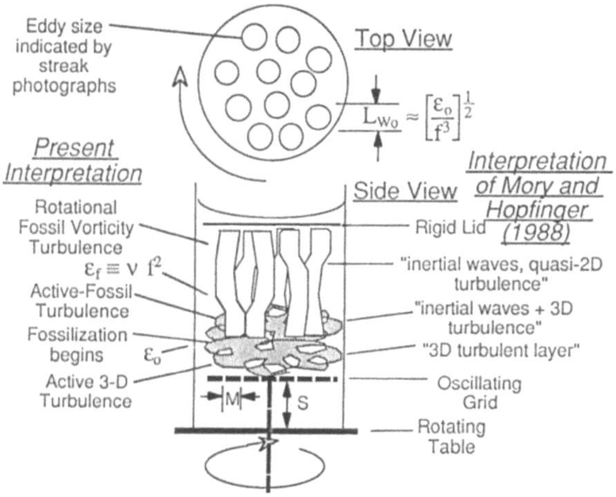

Fig. 3.3. Reinterpretation of rotating, oscillating grid experiment of Mory and Hopfinger (1988). The Coriolis-inertial waves formed when Coriolis forces damp the grid generated turbulence are interpreted as rotational fossil vorticity turbulence, and not quasi-2D turbulence or 2-D turbulent as proposed by Mory and Hopfinger (1988)

rate at the Coriolis-inertial-viscous transition point ε_f, where

$$\varepsilon_f \equiv \nu f^2 \tag{4}$$

as shown in Fig. 3.3. This occurs at a Coriolis-inertial-viscous scale L_{Kf}, or $L_{\Omega f}$,

$$L_{Kf} \equiv (\nu/f)^{1/2} \tag{5}$$

introduced by Gibson (1991), similar to the buoyant-inertial-viscous scale $L_{RF} = L_{KF} \equiv (\nu/N)^{1/2}$, from Gibson (1980). Above this layer the inertial wave eddies preserve the length scale $L_{\Omega 0}$ of the Coriolis-inertial transition at the beginning of fossilization, and are therefore a form of rotational fossil vorticity turbulence.

Figure 3.4 shows a schematic of the Dalziel experiment. At time $t = 0$ a square mesh cutout grid is towed through a $25 \times 50 \times 100$ cm tank of water rotating at 0.23 radians/sec. Pearlescent material in the water aligns with rate-of-stain tensor principal axes to permit flow visualization from photographs. The flow patterns for rotational and nonrotational grid turbulence at different times are shown below the apparatus. At 2 seconds the flows appear to be identical, so the turbulence is completely active at all scales with energy scale smaller than the Coriolis-inertial scale. At 15–30 seconds obvious eddy patterns emerge in the rotating flow, with smaller scale turbulence still persisting, indicating combined active-fossil turbulence. Coriolis-inertial wave motions have absorbed the kinetic energy of the largest turbulent eddies and preserve their largest scale at the beginning of fossilization. At 120 seconds only the 2-D eddy motions apparently remain in the rotating tank as completely fossil turbulence. A spectrum of such saturated Coriolis-inertial waves should extend from the Coriolis-inertial scale at fossilization $L_{\Omega 0}$ to the Coriolis-inertial-viscous scale $L_{\Omega f} \equiv (\nu/f)^{1/2}$. The velocity spectrum ϕ_u of saturated fossil vorticity turbulence produced by Coriolis force damping of active turbulence should be $k^2 \phi_u \approx \alpha_\Omega f^2 k^{-1}$, where α_Ω is a universal constant of order 1, for wavenumbers k in the range $L_{\Omega 0}^{-1} < k < L_{\Omega f}^{-1}$. Preliminary estimates of dissipation rates at various stages of the experiment suggest that fossilization probably begins before the eddies

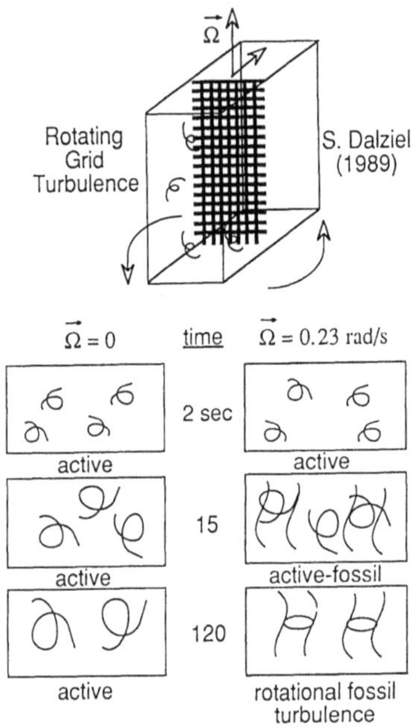

Fig. 3.4. Rotating grid turbulence experiment of Stuart Dalziel (1989). The growth of the largest grid turbulence eddy size is limited by Coriolis forces when the tank rotates, producing rotational fossil vorticity turbulence motions (inertial waves) that preserve the scale of the turbulence $L_{\Omega 0}$ when the fossilization began

can be detected by the flow visualization technique if the proportionality constant between the eddy size and Coriolis-inertial scale at the beginning of fossilization is of order one.

5. Hydrodynamic Phase Diagrams of 3-D and 2-D Turbulence

Hydrodynamic states of stratified and rotating turbulence may be determined by means of hydrodynamic phase diagrams, or HPD, based on various measurable length scales of the hydrophysical fields, as shown in Fig. 4. For the 3-D HPD on the left, L_R/L_{R0} represents $(Fr/Fr_0)^{3/2}$, the Froude number of the flow divided by the critical value at the inertial-buoyancy transition (raised to the 3/2 power), and L_R/L_{KF} represents $(Re/Re_F)^{1/2}$, the Reynolds number of the flow divided by the critical value at the inertial-buoyant-viscous transition (raised to the 1/2 power), as discussed by Gibson (1986). Most ocean microstructure patches detected fall in the *active-fossil* or completely *fossil* regimes because these are more persistent that the completely *active state.*

In the 2-D HPD on the right, $L_\Omega/L_{\Omega 0}$ compares the Rossby number of the flow to that at the Coriolis-inertial transition and L_Ω/L_{R0} compares the Burger number of the flow to the value at the inertial-buoyancy transition. The Burger number is defined as the square of the ratio of the Rossby radius of deformation u/f to a geometric scale, where u is the velocity of a flow or wave motion. If we identify the Rossby radius of the largest turbulent eddies with L_Ω setting $u = (\varepsilon L)^{1/3}$ from the Kolmogorov–Oboukov law, and the geometric scale with $L_{\Omega 0}$, the L_Ω/L_{R0} is the

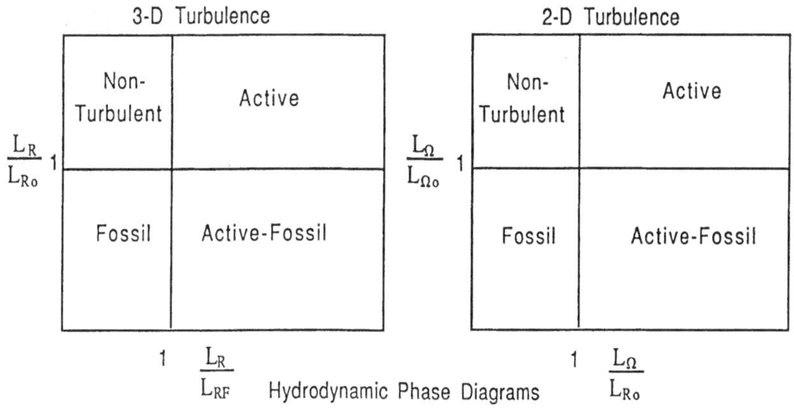

Fig. 4. Hydrodynamic phase diagrams for classifying oceanic microstructure and finestructure according to hydrodynamic state. On the left, 3-D turbulence is classified by comparing measured Ozmidov scales (from ε and N) to critical values at the buoyancy-inertial point in the ordinate (from L_P), a measure of the Froude number, and the buoyancy-inertial-viscous transition point on the abscissa (from ε, ν and N), a measure of the Reynolds number. On the right, 2-D turbulence is classified by comparing Coriolis scales (from ε and f) to the critical value at the Coriolis-inertial transition point on the ordinate (from L_W), a measure of the Rossby number, and the layer thickness on the abscissa (from L_P), a measure of the Burger number. For nonstratified, rotating turbulence such as shown in Figs. 3.3 and 3.4 the abscissa should be $L_\Omega/L_{\Omega f}$, which is the square root of Reynolds number divided by that at the Coriolis-inertial-viscous transition point, $(Re/Re_f)^{1/2}$, as shown in the text

indicated Burger number ratio raised to the 1/2 power. For stratified, rotating flows such as the ocean the vertical scale of turbulence is limited by the stable stratification to L_{R0} and the horizontal scale is limited by Coriolis forces to $L_{\Omega 0}$. In general, different values of average dissipation rates will exist at the buoyant-inertial transition ε_{S0}, and the Coriolis-inertial transition $\varepsilon_{\Omega 0}$ that will be reflected in the characteristic horizontal mesostructure and vertical finestructure scales, respectively.

For nonstratified rotating turbulent flows such as that shown in Fig. 3.3, the appropriate HPD would be a plot of $L_\Omega/L_{\Omega 0} = (Ro/Ro_0)^{3/2}$ versus $L_\Omega/L_{\Omega f} = (Re/Re_f)^{1/2}$. The Rossby number $Ro \equiv u/fL$ for turbulent motions on scale L with viscous dissipation rate ε is $\varepsilon^{1/3}/L^{2/3}f$. The Rossby number Ro_0 for the same scale motions at the beginning of 2-D turbulence formation and fossilization is $\varepsilon_0^{1/3}/L^{2/3}f$. Therefore from the ratio $Ro/Ro_0 = (\varepsilon/\varepsilon_0)^{1/3}$ it follows that $L_\Omega/L_{\Omega 0} = (Ro/Ro_0)^{3/2}$. The Reynolds number $Re \equiv uL/\nu$ for turbulent motions on scale L with viscous dissipation rate ε is $\varepsilon^{1/3}L^{4/3}/\nu$. The Reynolds number at the Coriolis-inertial-viscous transition point $Re_f = \varepsilon_f^{1/3}/L_{Kf}^{4/3}/\nu$. Therefore from the ratio $Re/Re_f = \varepsilon/\varepsilon_f$, so it follows that $L_\Omega/L_{\Omega f} = (Re/Re_f)^{1/2}$. The same arguments are used by Gibson (1986) to interpret the HPD for 3-D turbulence in stable stratification, shown on the left in Fig. 4.

Very little information is available to permit classification of ocean motions using 2-D turbulence hydrodynamic phase diagrams because of the very large scales of such flows. Typical values of $L_{\Omega 0}$ at high latitudes would likely be tens of kilometers, and at the equator hundreds of kilometers. On the other hand, copious evidence exists of horizontal variability, fronts, and horizontal eddy motions as discussed by Fedorov (1986). Surface temperature signatures of 2-D turbulence are clear from satellites in regions of strong forcing, such as coastal zones, or low Coriolis forces, such as equatorial regions, and give qualitative support for the expected decrease in 2-D eddy sizes with increasing latitude, although no direct measurements of ε in the observed eddies are available at this time.

Baker and Gibson (1987) show that oceanic temperature and viscous dissipation rates are enormously intermittent in most layers other than the surface wind mixed layer, with variances σ^2 (intermittency factors) of the natural logarithm of ε and χ in the range 3–8, compared to about 4 for the atmospheric boundary layer. Such huge intermittency factors imply cascades over a wide range of scales according to the Kolmogorov third universal similarity hypothesis, and imply that the intermittency of 3-D turbulence observed in the ocean is probably due to a 2-D turbulence cascade in such layers. The largest intermittency factors are found at equatorial latitudes, as expected if the cascade includes 2-D turbulence and the energy scale L_W is maximum.

6. Summary

A model for 2-D turbulence and fossil 2-D turbulence in rotating flows has been presented which closely parallels that for 3-D turbulence and fossil 3-D turbulence in stratified flows, except that Coriolis forces limit the largest active 2-D and 3-D turbulent eddies, and buoyancy forces limit the largest active 3-D turbulent eddies. Most oceanic microstructure is found to be fossil turbulence remnants of 3-D turbulence. It seems quite possible that many mesoscale eddy structures observed in the ocean by satellites as sea surface temperature or color patterns, or similar patterns in planetary atmospheres observed from spacecraft, are 2-D fossil turbulence. The largest scale turbulent eddy permitted by Coriolis forces is found to be at the Coriolis-inertial force scale $L_\Omega \equiv (\varepsilon/f^3)^{1/2}$. This scale was previously, and independently, proposed as a turbulence energy cascade blocking scale by Hopfinger in Mory and Hopfinger (1988) from their observations of inertial waves produced in a rotating laboratory tank. The smallest turbulence scale at complete rotational turbulence fossilization in such a nonstratified flow should be the Coriolis-inertial-viscous scale $L_{\Omega f} \equiv (\nu/f)^{1/2}$.

Laboratory and field studies are needed to see whether the fossils preserve information about previous turbulence states, such as the dissipation rates $\varepsilon_{\Omega 0}$ and ε_{S0} existing when active 2-D and 3-D turbulence were first damped by Coriolis and buoyancy forces, respectively, and to see whether these dissipation rates differ. Laboratory studies of Flier et al. (1983) and Mory and Hopfinger (1988) show the formation of rotational fossil turbulence and suggest the universal proportionality constant C_Ω is or order 1.

Turbulence plays a central role in planetary heat, mass and momentum transfer processes. However, sampling average turbulence parameters such as dissipation rates and diffusivities is complicated by the enormous range of length and time scales involved, so that undersampling errors are likely if not inevitable. Fossil turbulence parameters are presently useful as a means of evaluating the completeness of a given data set to minimize such errors by revealing possible evidence of previous turbulence activity which may not be represented by the available data. Further laboratory and field studies of fossil turbulence may permit more efficient inferences of space-time average turbulence properties in rotating, stratified flows such as the ocean and atmosphere; that is, the development of hydropaleontology.

Acknowledgements: This work was supported by ONR and NSF. Several valuable comments were provided by Emil Hopfinger on an early version of this manuscript.

References

Baker, M. A., Gibson, C. H. (1987): Sampling turbulence in the stratified ocean: Statistical consequences of strong intermittency. *J. of Phys. Oceanogr.* **17**, 1817–1837

Fedorov, K. N. (1986): *Lecture notes on Coastal and Estuarine Studies, Vol. 19: The physical Nature and Structure of Oceanic Fronts*, Springer-Verlag, Berlin

Flierl, G. R., Stern, M. E., Whitehead, J. A. (1983): The physical significance of modons: Laboratory experiments and general integral constraints. *Dynamics of Atmospheres and Oceans* **7**, 233–263

Friehe, C. A., Van Atta, C. W., Gibson, C. H. (1971): Jet turbulence: dissipation rate measurements and correlations. in *AGARD Conf. Proc., CP 93, no. 18*, 18.1–18.7

Gibson, C. H. (1980): Fossil temperature, salinity, and vorticity turbulence in the ocean. in *Marine Turbulence*, J. Nihoul (Ed.), Elsevier Publishing Co., Amsterdam, 221–257

Gibson, C. H. (1981): Fossil turbulence and internal waves. in *American Institute of Physics Conference Proceedings No 76: Nonlinear Properties of Internal Waves*, Bruce West (Ed.), American Institute of Physics, 159–179

Gibson, C. H. (1982a): Alternative interpretations for microstructure patches in the thermocline. J. Phys. Oceanogr. **12**, 374–383

Gibson, C. H. (1982b): On the scaling of vertical temperature gradient spectra. J. Geophys. Res. **87**, C10, 8031–8038

Gibson, C. H. (1982c): Fossil turbulence in the Denmark Strait. J. Geophys. Res. **87**, C10, 8039–8046

Gibson, C. H. (1983): Turbulence in the equatorial undercurrent core. in *Hydrodynamics of the Equatorial Ocean*, J. C. H. Nihoul (Ed.), Elsevier Publishing Company, Amsterdam, 131–154

Gibson, C. H. (1986): Internal Waves, fossil Turbulence, and composite ocean microstructure spectra. J. Fluid Mech. **168**, 89–117

Gibson, C. H. (1987a): Oceanic turbulence; big bangs and continuous creation. J. Physicochem. Hydrodyn. **8** (1), 1–22

Gibson, C. H. (1987b): Fossil turbulence and intermittency in sampling oceanic mixing processes. J. Geophys. Res. **92**, C5, 5383–5404

Gibson, C. H. (1990b): Hydrodynamic phase diagrams for microstructure in stratified flows. *Proceedings: Third International Symposium on Stratified Flows, Pasadena, Feb. 3–5, 1987*, J. List and G. H. Jiska (Eds.), American Society of Civil Engineers CP 775, 276–290

Gibson, C. H. (1988a): Evidence and consequences of fossil turbulence in the ocean. in *Small Scale Turbulence and Mixing in the Ocean*, J. C. H. Nihoul and B. M. Jamart (Eds.), Elsevier Publishing Company, Amsterdam, 319–334

Gibson, C. H. (1988b): Comment on: Reynolds number effects on turbulence in the presence of stable stratification, by A. E. Gargett. in *Small Scale Turbulence and Mixing in the Ocean*, J. C. H. Nihoul and B. M. Jamart (Eds.), Elsevier Publishing Company, Amsterdam, 529–530

Gibson, C. H. (1989): Fossil turbulence in rotating, stratified flows. in *Europhysics Conference Abstracts*, (Proceedings of the Fifth European Physical Society Liquid State Conference, On Turbulence, Moscow, October 16–21, 1989), K. Bethge and G. Thomas (eds.), European Physical Society, Geneva, 227–230

Gibson, C. H. (1990a): Turbulence, Mixing and Microstructure. in *The Sea-Vol. 9 Ocean Engineering Science*, Wiley Interscience, New York, 631–659

Gibson, C. H. (1991): Turbulence. in *Encyclopedia of Physics*, R. G. Lerner and G. L. Trigg (eds.), Addison-Wesley, New York, 1310–1314.

Hopfinger, E. J., F. K. Browand, and Y. Gagne (1982): Turbulence and waves in a rotating tank. J. Fluid Mech. **125**, 505–534

Mory, M. and E. J. Hopfinger (1988): Rotating Turbulence Evolving Freely from an Initial Quasi 2D State. in *Lecture Notes in Physics 230, 16–88, Macroscopic Modelling of Turbulent Flows* (Proceedings of a workshop held at INRIA, Sophia-Antipolis, France, December 10–14, 1984), Eds. U. Frisch, J. B. Keller, G. Papanicolaou, O. Pironneau, Springer-Verlag, Berlin, 218–236

Müller, P. (1988): Vortical motions. in *Small Scale Turbulence and Mixing in the Ocean*, J. C. H. Nihoul and B. M. Jamart (Eds.), Elsevier Oceanography Series, Elsevier Publishing Company, Amsterdam, 285–301

Part II

Wall Flows

Introductory Remarks

R. M. C. So and Y. G. Lai

Mechanical and Aerospace Engineering, Arizona State University, Tempe, AZ 85287, USA

Recently, activities on wall flows tend to concentrate on demonstrating that there are certain similarities between turbulent free and wall shear flows. Representatives of these activities are the studies of Moin and Kim (1985), Lee et al. (1987), Rogers and Moin (1987) and Moin (1988). Using the direct simulation database generated for turbulent channel flows, Moin and Kim (1985) demonstrated that hairpin vortices proposed by Theodorsen (1955) to explain turbulent activities in wall-bounded shear flows could be generated solely by mean-strain stretching and random velocity fluctuations. This conjecture was later verified by Rogers and Moin (1987) who found that hairpin vortices were also of common occurrence in homogeneous turbulent shear flows. This implies that there is a similarity between the organized structures in these two different types of shear flows. The similarity also extends to the one-point turbulent statistics calculated from channel flow and homogeneous shear flow (Moin 1988).

In addition, streaky structures were found in the near-wall region of wall shear flows and also in homogeneous free shear flows if the non-dimensional shear parameter, Sq^2/ε, of the two different flows is approximately the same (Lee et al. 1987; Moin 1988). Here $S = \partial U/\partial y$ is the mean shear, q^2 its turbulent kinetic energy and ε the dissipation rate of q^2. Therefore, this result further establishes the similarity between wall and homogeneous shear flows. The similarity is not just limited to the region far away from the wall where the mean shear is small. It also exists in the near-wall region where the mean shear is much larger. Experimental support for this is provided by the study of Uzkan and Reynolds (1967) whose experiments of turbulent flow over a moving belt showed that the streaky structures disappeared when the belt speed matched that of the free-stream velocity.

These findings are of interest to turbulent modellers for two reasons. The first is that homogeneous shear flow is the simplest and most basic of all shear flows. Therefore, homogeneous shear flow properties could be used to construct one-point turbulence closures for more complicated wall flows, whether external or internal in nature. The second is the possible dominance of the mean shear rate. If the similarity between wall and homogeneous shear flow is indeed true, then near-wall effects in one-point turbulence closures could be accounted for by devising models using database obtained solely from homogeneous shear flows with approximately the same non-dimensional shear parameter. Since these are simple experiments

Turbulent Shear Flows 7

compared to most wall-bounded flows, accurate and complete data sets could be easily generated.

Unfortunately, wall flows are not that simple. The five papers in this section provide new information to show that the similarity between wall and homogeneous shear flows is not as close as the previous studies (Moin and Kim 1985; Rogers and Moin 1987; Lee et al. 1987; Moin 1988) have demonstrated. Essentially, the papers show that the wall serves more than to create a mean shear rate in the flow. The wall also has a "blocking" effect which serves to diminish the rms normal fluctuating velocity and through the action of mean shear brings about a concomitant decrease in the other turbulent fluctuations. With this understanding, the contributions of the five papers can be summarized below.

The paper by Kim and Lee uses direct simulation data generated by Rogers et al. (1986) and Kim et al. (1987) to compare the fluctuating pressure behavior in a channel flow to that in a homogeneous shear flow. It represents an extension of an earlier study by Kim (1989), who examined the pressure behavior in a channel flow only. As before, the pressure is partitioned into a rapid and a slow part. The rapid part responds to changes in the mean field while the slow part changes through non-linear interactions of the fluctuating field. Their analysis showed that the rapid part is much smaller than the slow part in a channel flow, while the opposite is true for a homogeneous shear flow. It suggests that the rapid part is influenced by the presence of a wall and that this wall influence could penetrate far into the flow away from the wall. It is also found that the velocity pressure-gradient correlation term in the Reynolds-stress equation has a global behavior in the outer region of wall flows and is difficult to model by local variables. This finding suggests that mean shear and wall contribute differently to the fluctuating pressure field and hence to the modelling of the velocity pressure-gradient correlation term in the one-point Reynolds-stress transport equation.

The paper by Lee and Hunt used inhomogeneous rapid distortion theory to study the effect of mean shear in wall-bounded turbulent flows. They analyzed two cases; one in which the mean shear in the boundary layer is uniform and another in which the boundary layer experiences no mean shear. Thus, the "blocking" effect of the wall on the boundary-layer flow with and without the presence of a mean shear can be examined and compared. Their findings can be summarized as follows. Without mean shear, "blocking" just serves to enhance the anisotropy of the normal stresses. In light of the finding by Kim and Lee, one is led to ask whether this enhancement is due to the slow part of the pressure field. On the other hand, when mean shear is present, "blocking" serves to cause the normal component of the Reynolds normal stresses to decrease to zero at the wall. This decrease, in turn, brings about concomitant decrease in other Reynolds-stress components through the action of mean shear. This means that production of these stress components is being reduced as a wall is approached. In other words, near-wall modelling of the one-point Reynolds-stress equation is important in order to bring about this behavior. Their analysis also shows that near-wall streaky structures are produced by mean-strain and not by the presence of the wall. This result seems to support the similarity finding of Moin (1988) between homogeneous and wall shear flows, while at the same time, shows that the presence of the wall serves to create more than a mean shear effect on the flow.

The third paper by Suzuki et al. is an experimental study on the perturbed boundary layer downstream of a cylinder of diameter *d* placed at a distance *C* away from the wall. In all the cases studied, the cylinder was placed inside the boundary layer; therefore, only the boundary layer flow was disturbed. Flow properties downstream of the cylinder and within the boundary layer were measured using hot-wires. Skin friction was deduced from Preston tube measurements, while wall temperature and wall heat flux were obtained from thermocouples and power balance measurements. Two interesting and important findings can be mentioned. One is that all measured flow properties including skin friction and the wall heat transfer coefficient can be correlated by C/d. The behavior of these properties is different for different values of C/d. Another is the dissimilarity between the skin friction and the wall heat transfer coefficient. These findings imply that a local disturbance can give rise to a global effect and could affect heat and momentum transfer differently. They further suggest that modelling of this flow by one-point statistics may not be very suitable and point to the difficulty of devising models for the velocity pressure-gradient correlation term that could account for the long history effect noted in the measurements.

The next two papers deal with transition in wall-bounded flows. An experimental study by Sohn et al. is concerned mainly with the phenomenon of bypass transition. Here the term bypass is used to describe transition to turbulent flow that does not follow the normal process of the breakdown of small disturbances, followed by the formation of three-dimensional structures and their eventual development into turbulent spots. Instead, the transition was triggered by large disturbances that could lead to the formation of turbulent spots directly. The paper is primarily concerned with the study of the intermittent boundary layer. Conditional sampling techniques were used to analyses the hot-wire signals to give information on the turbulent and non-turbulent parts of the boundary layer. The measured skin friction and Stanton number were found to be well correlated by standard formulae for both the laminar and turbulent regions. However, the measured turbulence statistics, unconditional as well as conditional, tend to indicate that they are quite unlike those found in equilibrium turbulent boundary layers. These results again suggest the difficulty of modelling such flows.

Finally, the paper by Henningson and Kim is on the structure of a turbulent spot. Direct simulation data obtained in a laminar plane Poiseuille flow with a wave packet introduced into the flow was used to study the structure of the spot. The turbulent spot was found to consist of two regions, a turbulent region and a wave region. Their analysis showed that the flow structure and the associated turbulence statistics inside the turbulent region were very similar to those found in fully-developed turbulent flows. The turbulent spot formed inside the plane Poiseuille flow was essentially identical to that found in a laminar boundary layer. In addition, they found that the velocity fluctuations in the wave region were much higher than inside the turbulent region and this seemed to indicate a higher level of turbulence production in this region. The study contributes to an understanding of the final stage of transition, in particular, that of bypass transition. This study together with the one by Sohn et al. provide important information on bypass transition and could be used to help devise transition models like that proposed by Finson (1977).

One of the more important contribution to turbulent flows made by the five papers is the finding by Kim and Lee that the pressure-strain term has a global influence in the outer region of a wall flow, while its influence is essentially local in the near-wall region. This result suggests that more work has to be carried out to clarify wall influence on pressure fluctuation as well as on the pressure-strain behavior. Since pressure-strain modelling is of importance to Reynolds-stress closure, its behavior in homogeneous shear flows as well as in wall shear flows is of particular interest to modellers. The study of Kim and Lee and the work of Lee and Hunt suggest that, perhaps, mean shear and wall effects should be modelled separately since they are two distinct influences contributing to the pressure-strain term.

It has been remarked by Launder and Tselepidakis (1988) that wall-reflection terms added to the model for pressure-strain are responsible for causing the turbulent normal stress in the direction normal to the wall to decrease from a maximum near the wall to zero at the wall. This is essentially the "blocking" effect of the wall referred to by Lee and Hunt. In other words, near-wall modelling of the pressure-strain term is important if the measured behavior of the turbulent normal stresses near a wall is to be reproduced. Launder and Tselepidakis (1988) assumed the pressure-strain term to vanish at the wall and proceeded to model it by incorporating the stress invariant idea of Lumley (1978) to ensure the vanishing behavior of the pressure-strain term at the wall. However this was not enough to bring about a correct near-wall behavior for the turbulent normal stresses.

On the other hand, the direct simulation data of Moser and Moin (1987) showed that pressure-strain and pressure-transport in the normal direction are of equal importance near a wall, even though they add up to zero at the wall. Therefore, this suggests that, near a wall, the velocity pressure-gradient correlation term should be modelled rather than the pressure-strain term. This approach was taken by Lai and So (1990a) and So et al. (1991) who made use of the direct simulation data to help formulate their near-wall model for the velocity pressure-gradient term. Their model proposes a near-wall correction to the pressure-strain term. As such, the proposed correction not only compensates for the incorrect near-wall behavior of the pressure-strain term, but also provides for a correct balance between pressure-strain and pressure-transport in the normal direction near a wall. With this model, they have been able to obtain good agreement between calculated turbulent statistics and those derived from direct simulation of straight and curved channel flows (Mansour et al. 1988; Moser and Moin 1987; Kim et al. 1987). Lai and So (1990b) have also extended this approach to model near-wall heat fluxes and good correlations with heat transfer measurements in fully-developed pipe flows were obtained. In view of this, more studies like those of Kim and Lee, and Lee and Hunt are required, so that a better understanding of the role of pressure-transport and pressure-strain in both free and wall shear flows could be achieved, and used to formulate a physically more correct model for the velocity pressure-gradient term in the Reynolds-stress equation.

References

Finson, M. L. (1977): On the application of second-order closure models to boundary layer transition. AGARD-CP-24

Kim, J. (1989): On the structure of pressure fluctuations in simulated turbulent channel flow. J. Fluid Mech. **205**, 421–451

Kim, J., Moin, P. and Moser, R. D. (1987): Turbulence statistics in fully-developed channel flow at low Reynolds number. J. Fluid Mech. **177**, 133–166

Lai, Y. G. and So, R. M. C. (1990a): On near-wall turbulent flow modelling. J. Fluid Mech. **221**, 641–673

Lai, Y. G. and So, R. M. C. (1990b): Near-wall modelling of turbulent heat fluxes. Int. J. Heat Mass Transfer **33**, 1429–1440

Launder, B. E. and Tselepidakis, D. P. (1988): Contribution to the second-moment modelling of sublayer turbulent transport. Proceedings of Zoran Zaric Memorial Int. Seminar on Near-Wall Turbulence, 5/16-20/88, Dubrovnik, Yugoslavia

Lee, M. J., Kim, J. and Moin, P. (1987): Turbulence structure at high shear rate. Sixth Symp. Turbulent Shear Flows, 9/7-9/87, Toulouse, France

Lumley, J. L. (1978): Computational modelling of turbulent flows. Adv. Appl. Mech. **18**, 123–176

Mansour, N. N., Kim, J. and Moin, P. (1988): Reynolds-stress and dissipation-rate budgets in a turbulent channel flow. J. Fluid Mech. **194**, 15–44

Moin, P. (1988): Similarity of organized structures in turbulent shear flows. Proceedings of Zoran Zaric Memorial Int. Seminar on Near-Wall Turbulence, 5/16-20/88, Dubrovnik, Yugoslavia, 2–6

Moin, P. and Kim, J. (1985): The structure of the voriticity field in turbulent channel flow. Part 1. Analysis of vorticity fields and statistical correlations. J. Fluid Mech. **155**, 441–464

Moser, R. D. and Moin, P. (1987): The effects of curvature in wall-bounded turbulent flows. J. Fluid Mech. **175**, 479–510

Rogers, M. M. and Moin, P. (1987): The structure of vorticity field in homogeneous turbulent shear flows. J. Fluid Mech. **176**, 33–66

Rogers, M. M., Moin, P. and Reynolds, W. C. (1986): The structure and modelling of the hydrodynamic and passive scalar field in homogeneous turbulent shear flow. TF-25, Department of Mechanical Engineering, Stanford University, Stanford, CA 94305

So, R. M. C., Lai, Y. G. and Hwang, B. C. (1991): Near-wall turbulence closure for curved flows. AIAA Journal, to appear

Theodorsen, T. (1955): The structure of turbulence. 50 Jahre Grenzschichtforschung (ed. H. Görtler and W. Tollmien), 55–72

Uzkan, T. and Reynolds, W. C. (1967): A shear-free turbulent boundary layer. J. Fluid Mech. **28**, 803–821

The Structure of Pressure Fluctuations in Turbulent Shear Flows

J. Kim and M. J. Lee

Center for Turbulence Research, NASA Ames Research Center, Moffett Field, CA 94035, USA

Abstract

Pressure fluctuations in a turbulent channel and a homogeneous shear flows are investigated by analyzing databases obtained from direct numerical simulations. Detailed statistics associated with the pressure fluctuations in both flows are discussed. Characteristics associated with the rapid (linear) and slow (nonlinear) pressure are examined. The global dependance of the pressure fluctuations and pressure-strain correlations are also examined by evaluating the integral associated with Green's function representations of them.

1. Introduction

The pressure fluctuations in turbulent flows are of considerable interest in many engineering applications. Pressure fluctuations on aerodynamic surfaces of airplanes and turbine blades, for example, contribute to noise generation and often lead to undesirable structural vibrations. Understanding the nature of the pressure fluctuations is essential to alleviate such problems. From a turbulence modeling point of view, the pressure-stain correlations that appear in the Reynolds-averaged, Navier-Stokes equations are of significant interest because these terms are responsible for intercomponent energy transfer. Existing models for the pressure-strain terms, however, are not satisfactory and are regarded as one of the major deficiencies in the Reynolds stress closure. Refer to Mansour et al. (1988) for a recent assessment of the existing turbulence models based on databases obtained by direct numerical simulations.

A great deal of experimental and theoretical work on pressure fluctuations has been performed during the past several decades. Willmarth (1975) provides a comprehensive review on the subject, and Eckelmann (1988) gives an updated review of our knowledge on pressure fluctuations. In spite of the considerable efforts evidenced in the literature, progress in our understanding of the pressure fluctuations has been very slow because of inadequate measurements of pressure fields in turbulent flows. This is because we do not yet have a reliable measurement technique for turbulent pressure fluctuations within turbulent flow comparable to the modern measurement techniques for turbulent velocity fluctuations. Consequently, most measurements reported to date have been performed only on a surface using a surface-mounted pressure transducer. Furthermore, many of the

Turbulent Shear Flows 7

earlier surface-pressure measurements suffered from inadequate spatial resolution because of the large size of pressure transducer employed in the experiments. The errors involved turned out to be significant (Willmarth 1975; Schewe 1983).

Most theoretical work in which attempts were made to compute the pressure fluctuations and the power spectra—Kraichnan (1956), Lilley (1964), and Panton and Linebarger (1974), to name a few—also suffered from the lack of accurate information to guide their analyses. The analyses were based on the earlier measurements with the poor spatial resolution. Also, most of the theoretical work assumed that the contribution from the linear source term in the Poisson equation for pressure, representing the interaction of turbulence with the mean shear, is much larger than that from the nonlinear source terms representing interactions of the turbulence with itself.

The direct numerical simulation of turbulent flow has emerged as a powerful tool in studying the fundamental physics of turbulent flows. A number of flow fields have been computed with highly accurate numerical algorithms, and the databases generated from such computations have been successfully used in studying the basic turbulent flow phenomena (see, for example, Proceedings of the 1987 and 1988 Summer Program of the NASA/Stanford Center for Turbulence Research). The availability of all flow variables at many spatial locations makes this approach extremely attractive. The currently available databases are, however, limited to simple flows at low Reynolds numbers.

In this paper we use databases obtained from direct numerical simulations to study the turbulence statistics associated with pressure fluctuations in a turbulent channel (Kim et al. 1987, hereinafter referred to as KMM) and a homogeneous turbulent shear flow (Rogers et al. 1986). Both simulations were performed with a grid resolution sufficiently fine to resolve all the essential scales present in the flows. The physical realism and accuracy of the computed flow fields have been established by detailed comparison with existing experimental results. Most of the results presented in this paper include new information that cannot be validated directly by compairing it with experimental data, unlike most of the statistics and turbulence structures of the corresponding velocity field. However, the pressure fields considered in the present work are obtained from the same velocity fields whose validity has been thoroughly examined, and there is little doubt that the pressure information presented here is as valid as the velocity field.

In §2, a brief description of the databases used in the present paper and splitting of pressure are presented. Detailed statistics of pressure fluctuations are discussed in §3. A brief summary and short discussions are given in §4. In this paper x, y, z denote the streamwise, normal (to the wall), and the spanwise directions, respectively. The velocities are u, v, and w in x, y, and z directions respectively, which are used interchangeably with subscripted variables u_1, u_2, and u_3. The subscript w indicates the value at the wall, and the superscript + indicates a non-dimensional quantity scaled by the wall variables: e.g. $y^+ = yu_\tau/\nu$, where ν is the kinematic viscosity and $u_\tau = (\tau_w/\rho)^{1/2}$ is the wall-shear velocity. For the homogeneous flow, a similar viscous length scale can be defined as $y^+ = y(S/\nu)^{1/2}$ in analogy to the channel flow, where $S = dU/dy$ denotes the mean shear.

2. Databases and Splitting of Pressure

The channel flow simulation was performed at Reynolds number, $Re \approx 3300$, based on the centerline velocity (U_c) and the channel half-width (δ), using $128 \times 129 \times 128$ (x, y, and z, respectively) spectral modes. The grid spacings in the streamwise and spanwise directions were $\Delta x^+ \approx 17$, and $\Delta z^+ \approx 6$ in wall units. Nonuniform meshes were used in the normal direction; Δy^+ varied from 0.05 at the wall to 4.5 at the ceterline. Starting with an arbitrary initial field, the computation was carried out until a statistically steady state was reached.

The homogeneous flow simulation was performed with $128 \times 128 \times 128$ spectral modes. The grid spacings were $\Delta x^+ \approx 4$ and $\Delta y^+ = \Delta z^+ \approx 2$. The initial field for the homogeneous simulation was a result of a isotropic turbulence simulation. A computed field at total shear $St = 8$ was chosen for the present analysis because by this time the turbulence statistics attained an asymptotic state. The Reynolds number based on the Taylor microscale was about 70 at this time.

The pressure field in an incompressible flow satisfies the Poisson equation

$$\nabla^2 p = -u_{i,j}u_{j,i} \tag{1}$$

and appropriate boundary conditions. The pressure may be split into rapid and slow parts by dividing the source terms in Eq. (1) into two parts, one containing the mean velocity gradient and the other without it. By decomposing the velocity field into mean and fluctuations as $u_i = U_i + u_i'$ and noting that the only nonzero mean velocity gradient for the present flows is dU/dy, Eq. (1) can be split into two parts as follows

$$\nabla^2 p_r = -2\frac{dU}{dy}\frac{\partial v'}{\partial x}$$

$$\nabla^2 p_s = -u_{i,j}'u_{j,i}' \tag{2}$$

where the subscripts r and s denote the rapid and slow parts, respectively. The terms "rapid" and "slow" are derived from the fact that only the rapid part responds immediately to a change imposed on the mean field, and the slow part will feel the change through nonlinear interactions. Sometimes, the rapid and slow pressure are also referred to as linear and nonlinear pressure, respectively, because the corresponding source terms in Eq. (2) are linear in the fluctuating quantities for the rapid part and nonlinear for the slow part.

For the homogeneous shear flow, periodic boundary conditions are used in all three directions. For the channel flow, periodic boundary conditions are used in the streamwise and spanwise directions. The boundary condition in the normal direction is obtained by evaluating the normal component equation at each wall

$$\frac{\partial p}{\partial y}(y = \pm 1) = \frac{1}{Re}\frac{\partial^2 v}{\partial y^2}$$

where $y = \pm 1$ corresponds to the upper and lower walls. Decomposition of the boundary condition is not unique (Mansour et al. 1988; Kim 1989). We use

$$\frac{\partial p_r}{\partial y}(y = \pm 1) = 0 \quad \text{and} \quad \frac{\partial p_s}{\partial y}(y = \pm 1) = \frac{1}{Re}\frac{\partial^2 v'}{\partial y^2}.$$

3. Results

3.1 Root-Mean-Square Pressure Fluctuations

The profiles of the r.m.s. fluctuation of the rapid, slow, and total pressure in the channel are shown in Fig. 1. The pressure fluctuation are normalized by the wall-shear velocity, u_τ, in Fig. 1a, and by the local turbulent kinetic energy, q^2, in Fig. 1b. We note that the slow part is substantially larger than the rapid part except very close to the wall ($y^+ < 15$), contrary to the common belief that the rapid part would be the dominant one. Both the rapid and slow parts contribute approximately 75% of the total r.m.s. value at the wall. The two parts are negatively correlated, with the correlation coefficient about -0.1 at the wall. The correlation between the rapid and slow pressure is negligible, with the correlation coefficient being less than 0.1 throughout the channel.

The pressure fluctuations in the homogneous shear flow, on the other hand, are dominated by the rapid pressure: $p_r'/q^2 = 0.49$, $p_s'/q^2 = 0.32$, and $p'/q^2 = 0.60$, where

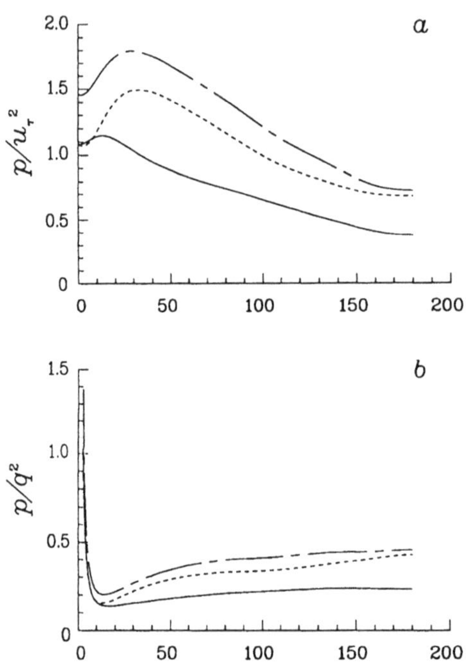

Fig. 1.1. R.m.s. pressure fluctuations in the channel (**a**) normalized by u_τ^2, (**b**) normalized by q^2: ——, rapid; -----, slow; —·—, total pressure

the prime denotes the r.m.s. fluctuations. Lee et al. (1987) showed that a strong similarity exists between a homogeneous flow and a wall-bounded flow when the nondimensional shear-rate parameter, $S^* = Sq^2/\varepsilon$, was matched, where ε denotes the dissipation rate of the mean kinetic energy. S^* in the homogeneous shear flow is about 10, comparable to that for $y^+ > 50$ in the channel flow (Lee et al. 1987). We note that p'_s/q^2 in the channel for $50 < y^+ < 180$ is comparable to that in the homogeneous flow, but p'_r/q^2 in the channel is much smaller than that in the homogeneous flow.

To obtain further insight into the pressure splitting, the source terms that appear in Eq. (2) are examined. The profiles of the mean-square rapid, slow, and total source terms are shown in Fig. 2. Figure 2a shows that the slow source term is much larger than the rapid source term throughout the channel. This is because even though the mean shear is large in the wall region, $\partial v'/\partial x$ is rather small (zero at the wall where the shear is the highest, and the mean shear drops significantly where $\partial v'/\partial x$ becomes finite away from the wall). Further breakdown of the slow source terms—there are six different terms—is shown in Fig. 2b, and it is found that $(\partial v'/\partial z)$ $(\partial w'/\partial y)$ has the largest mean-square value. Note that the magnitude of $(\partial v'/\partial z)(\partial w'/\partial y)$ would be large at the core of streamwise vortices, and indeed it has a local peak at about $y^+ = 20$, where the centers of streamwise vortices are located on the average (KMM). Note also that the locations of the maximum for each source term approximately coincide with the location of the maximum r.m.s. pressure fluctuation. The source terms for the homogeneous flow indicated the same trend: that is, the mean-square slow source term are much larger than the rapid source

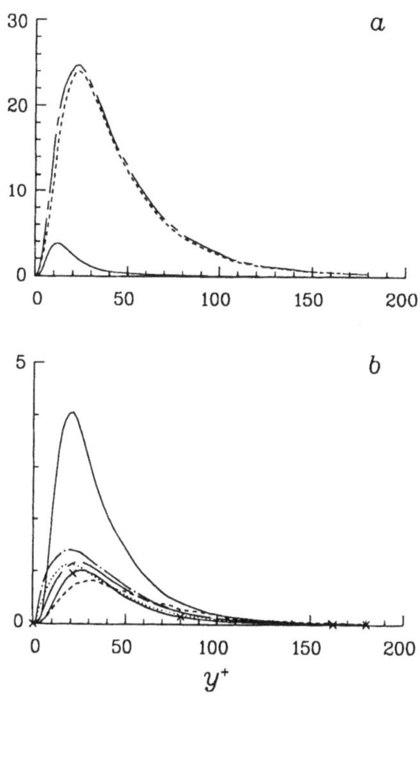

Fig. 2. (**a**) Profiles of the mean-square source terms: ——, rapid; ----, slow; —-—, total. (**b**) Profiles of the mean-square slow terms: —×—, $(\partial u'/\partial x)^2$, ----, $(\partial u'/\partial y)$ $(\partial v'/\partial x)$; —·—, $(\partial u'/\partial z)$ $(\partial w'/\partial x)$; ·······, $(\partial v'/\partial y)^2$; ———, $(\partial v'/\partial z)$ $(\partial w'/\partial y)$; —·—, $(\partial w'/\partial z)^2$

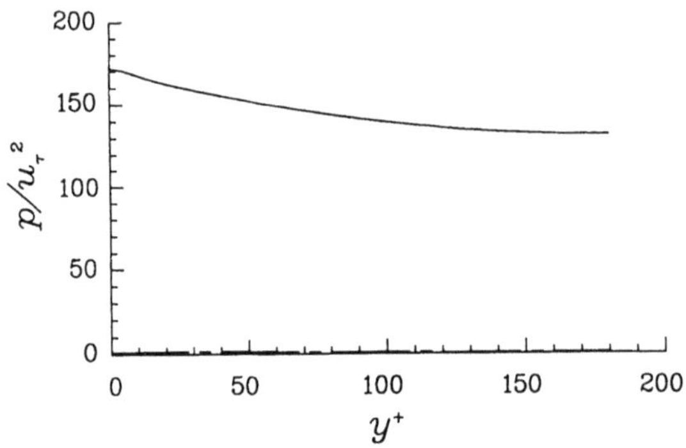

Fig. 3. R.m.s. fluctuation of the spin, splat, and total pressure: ——, p_Ω; - - - - -, p_s; —·—, p

term, and among the slow source terms, $(\partial v'/\partial z)(\partial w'/\partial y)$ has the largest mean-square value. In the homogeneous flow, the r.m.s. value of the rapid pressure is much larger than that of the slow pressure even though the slow source term is larger than the rapid term, indicating that most contribution to the rapid pressure comes from the low wavenumber content of the source term.

Bradshaw and Koh (1981) proposed a different splitting of the pressure by decomposing the source terms into rotational and irrotational components as follows:

$$\nabla^2 p = \Omega^2 - S^2, \tag{3}$$

where $\Omega^2 = \omega_{ij}\omega_{ij} = \frac{1}{4}(u_{i,j} - u_{j,i})(u_{i,j} - u_{j,i})$ and $S^2 = s_{ij}s_{ij} = \frac{1}{4}(u_{i,j} + u_{j,i})(u_{i,j} + u_{j,i})$. Although this splitting is not unique (Adrian 1982), it represents the result of two distinct physical mechanisms to which Bradshaw and Koh refer as "spin" and "splat". The spin contribution results from eddy rotation, and the splat contribution results from eddy collision. The r.m.s. values of p_Ω, p_S and p are shown in Fig. 3. Note that the r.m.s. values of p_Ω and p_S are much larger than that of p, and they are indistinguishable. In fact, instantaneous p_Ω and p_S are almost equal to each other with opposite sign (correlation coefficient being approximately -1 throughout the entire channel). It was found that $\Omega^2 \approx S^2$ (the small difference between Ω^2 and S^2 is the total source term), and hence, $p_\Omega \approx -p_S$. The high correlation indicates that splitting pressure into the rotational (spin) and irrotational (splat) parts does not appear too appealing.

3.2 Probability Density Distribution Function

Computed probability density distribution function (pdf) of the wall-pressure fluctuations in Fig. 4 shows an excellent agreement with that measured by Schewe (1983) in a turbulent boundary layer at Reynolds number, $Re_\theta = 1400$ ($Re_\theta = 280$ in a channel, for the present result). A Gaussian distribution is shown as a reference. The experimental data were obtained from the smallest transducer with its diameter equal to 19 viscous units. Other results obtained from larger transducers by Schewe (not shown here) had a considerable dependence of pdf on the transducer size. The pdfs of the rapid, slow and total pressure fluctuations at $y^+ \approx 50$ in the channel are

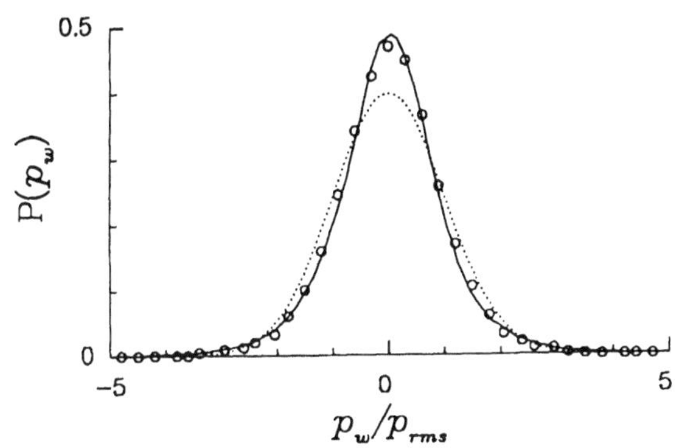

Fig. 4. Pdf of the wall pressure fluctuations in the channel: ———, computational; ----, Gaussian distribution; o, Schewe (1983). The abscissa is normalized by the respective r.m.s. values and the ordinate is normalized such that the area under the curve is equal to one

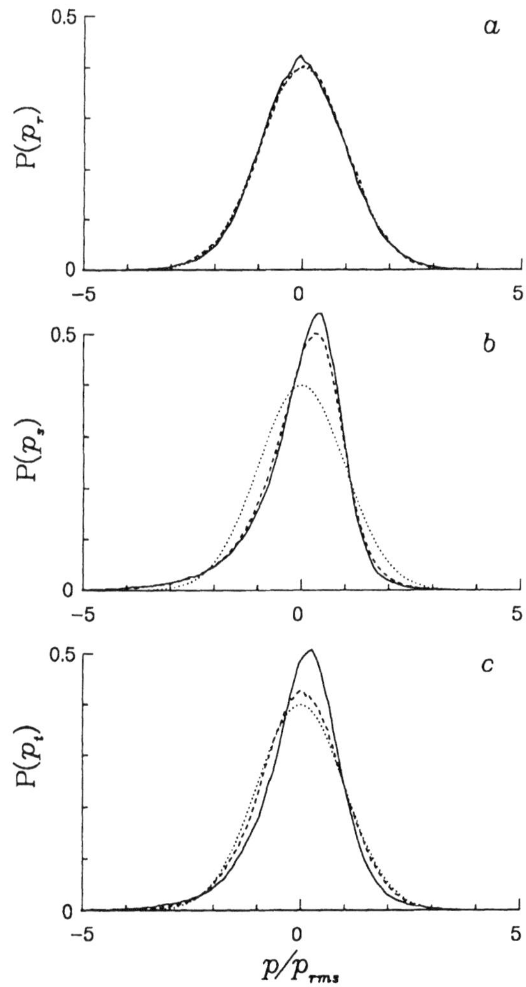

Fig. 5 a–c. Pdf of pressure fluctuations in the channel and homogeneous flows. **a** rapid, **b** slow, and **c** total pressure: ——, channel; ----, homogeneous; ·······, Gaussian.

shown in Fig. 5, and compared with those in the homogeneous shear flow. Recall that at this y-location S^* is about the same as that of the homogeneous flow. The pdfs of the rapid pressure for both flows are close to the Gaussian distribution, and those of the slow pressure show negatively skewed distributions. There exists a strong similarity in the pdfs of each pressure, although the pdfs of the total pressure show a marked difference. This is because in the channel flow the slow part is dominant, and hence, the pdf of the total pressure is similar to that of the slow pressure, while the opposite is true for the homogeneous flow.

By comparing these pdfs with those of velocity fluctuations, it is found that relatively large contributions to the r.m.s. fluctuation are from high-intensity fluctuations, especially from the negative fluctuations. For instance, at $y^+ = 12$, only 1% of the data points exceeds $\pm 3p_{\text{rms}}$, but they contribute to about 42% of the total p_{rms}. Corresponding contribution to the streamwise velocity fluctuations, on the other hand, is negligible.

Robinson et al. (1988) observed a strong correspondence between low-pressure regions and vortical structures. They reported that it was easier to identify a vortical structure using low pressure as a detecting criterion than the velocity or vorticity itself. The joint probability density distributions between pressure fluctuations and vorticity fluctuations are examined to investigate the correlations. The joint pdfs at $y^+ \approx 30$ (Fig. 6) indicate that large streamwise vorticity fluctuations are indeed more

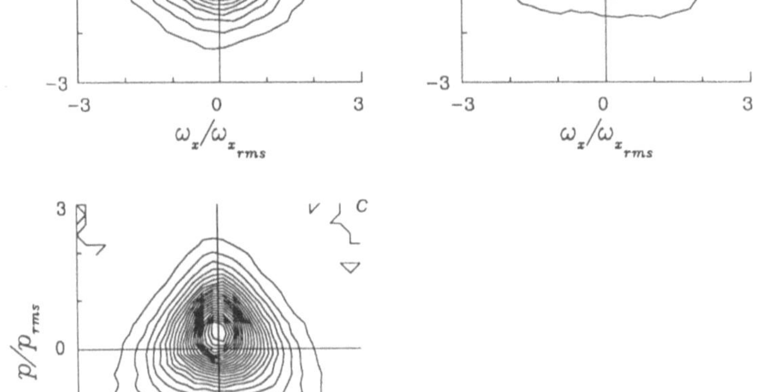

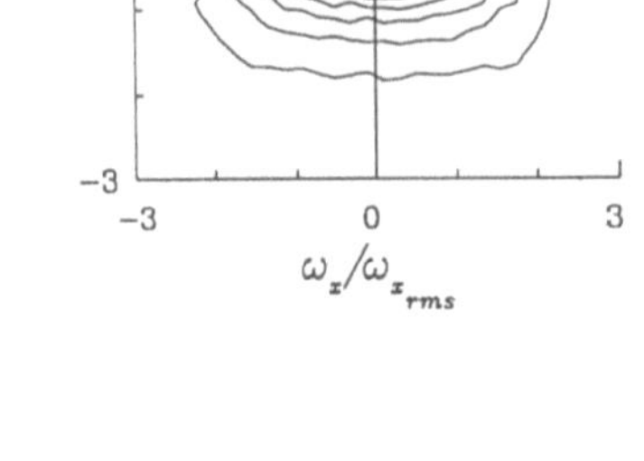

Fig. 6 a–c. Joint pdf between w_x and **a** rapid, **b** slow, **c** total pressure at $y^+ \approx 30$. The axes are normalized by the respective r.m.s. values

likely associated with large negative slow pressure fluctuations. It should be noted, however, that the reverse is not true. Large negative pressure fluctuations are not necessarily associated with the large streamwise vorticity fluctuations, suggesting that additional information is required in identifying organized vortical structures. Note also that ω_x is not correlated with p_r, but large ω_x correlates with large negative p_s. This is also true for the homogeneous flow. One could use the pressure—slow part in particular—as a diagnostic tool in detecting vortical structures, but it should be used with some caution because not all low pressure is associated with strong vortical structures.

3.3 Green's Function Representation

A solution to equation (1) can be written in terms of Green's function as follow:

$$p(x, y, z) = \int_{-1}^{1} G(x, y, z) f(x, y, y', z) dy' \tag{4}$$

where G denotes the Green's function for the problem and f denotes the source terms. For a derivation of the Green's function for the channel flow and its behavior, the reader is referred to Kim (1989). This representation is useful to gain an insight into which part of the channel in the inhomogeneous direction contributes most to the local pressure fluctuations. For instance, the y'-profile of the integrand in expression (4) represents the contribution from each y'-location to a given y-location. An example is given in Fig. 7a in which contributions to instantaneous pressure at an arbitrary (x,z)-location at the wall ($y = -1$) are shown. It illustrates that the source terms near one wall contribute significantly to the total and slow pressure at the other wall; the contribution to the rapid pressure from near the other wall is negligible. The global dependance is due to the slowly decaying Green's function for low wave numbers (Kim 1989). From Eq. (4), one can also write a similar expression for the mean-square pressure fluctuation

$$\begin{aligned}\overline{p^2(y)} &= \int_{-1}^{1} \overline{p(y) G * f(y, y')}\, dy' \\ &= \int_{-1}^{1} \int_{-1}^{1} \overline{G * f(y, \eta) G * f(y, y')}\, d\eta\, dy' \\ &= \int_{-1}^{1} h(y, y')\, dy'\end{aligned} \tag{5}$$

where the overbar denotes an average over the x and z directions. Contributions to the mean-square wall-pressure fluctuation are shown in Fig. 7b, which indicates that most contributions are from the wall region. It appears that even though the instantaneous pressure at one wall is strongly affected by the source terms from the other side of the channel (Fig. 7a), their net effects are not correlated to give a significant contribution to the mean-square fluctuation. Note that the Green's functions considered here are independent of the Reynolds number, and this nonlocal dependance would be the same for high Reynolds number flows unless the

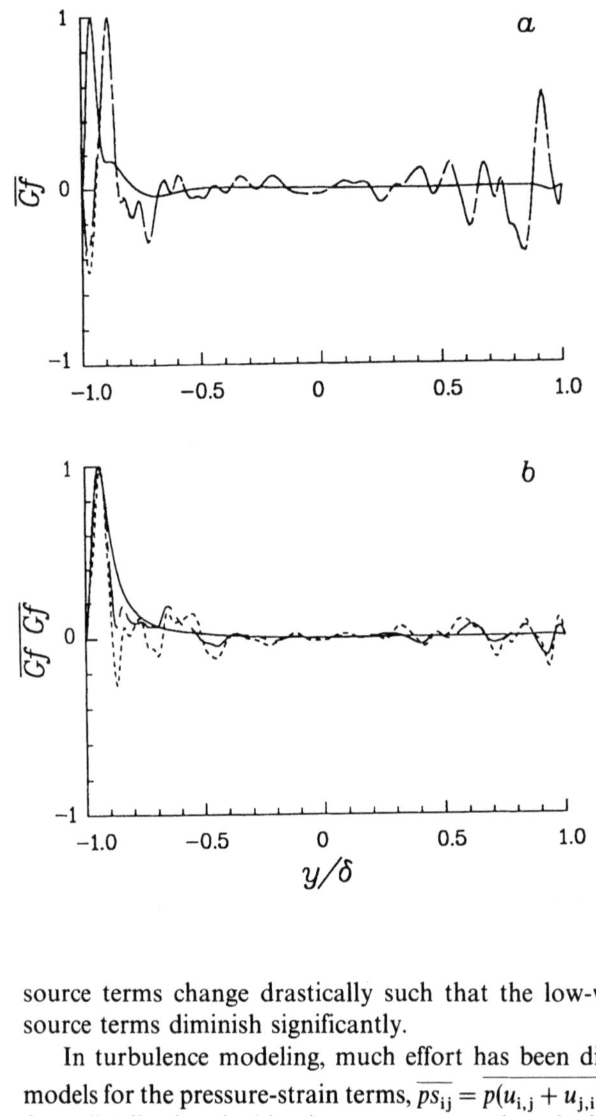

Fig. 7 a. Contribution to instantaneous wall pressure $p(x=0,\ \ y=-1,\ \ z=0)$, from across the channel, i.e., the integrand of Eq. (4). **b** Contribution to the meansquare wall-pressure fluctuation from across the channel, i.e., the integrand of Eq. (5). The ordinate is normalized such that the maximum value of each quantity is equal to one. ——, rapid; - - - -, slow; — — —, total pressure

source terms change drastically such that the low-wavenumber contents of the source terms diminish significantly.

In turbulence modeling, much effort has been directed toward in developing models for the pressure-strain terms, $\overline{ps_{ij}} = \overline{p(u_{i,j} + u_{j,i})}$. These terms are responsible for redistributing the kientic energy among the velocity components. It has been of significant interest to modelers to know how local these terms are in order to make certain approximations in developing models for them. Using Eq. (4), one can write

$$\overline{p(y)s_{ij}(y)} = \int_{-1}^{1} \overline{s_{ij}(y)\,Gf\,(y,y')}\,dy' \tag{6}$$

The pressure-strain terms that appear in the equations for $\overline{u'^2}, \overline{v'^2}, \overline{\omega'^2}$, and $\overline{u'v'}$ at two different y-locations are shown in Fig. 8. The pressure-strain correlations in the wall region (Fig. 8a) take their largest contributions from the wall-region (contributions are local), but away from the wall (Fig. 8b) they take their contributions from wide ranges of y'-locations (global). This nonlocal character of the pressure-strain terms would make them hard to model in terms of local variables.

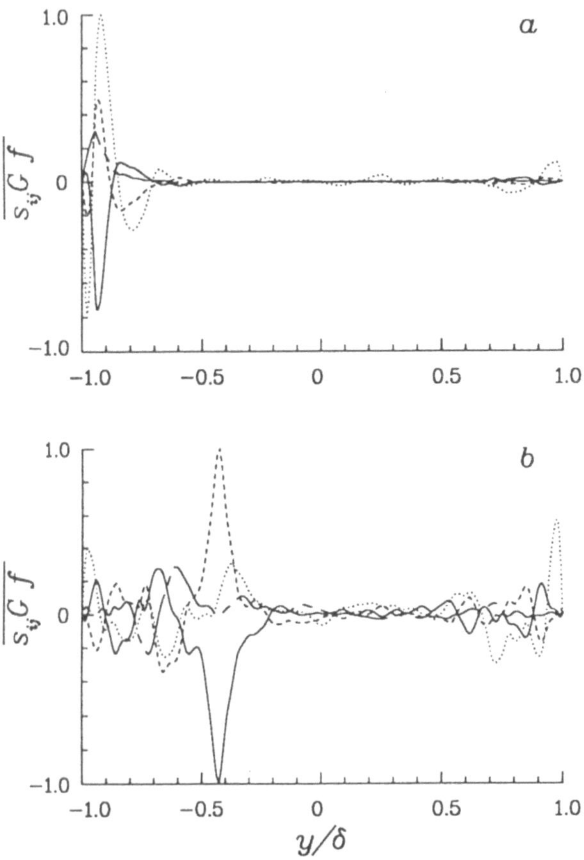

Fig. 8. Contributions to the pressure-strain correlations, i.e., the integrand of Eq. (6), at **a** $y^+ \approx 10$ **b** $y^+ \approx 100$: ——, $\overline{2p(\partial u/\partial x)}$: – – – –, $\overline{2p(\partial v/\partial y)}$; — — —, $\overline{2p(\partial w/\partial z)}$; ········, $\overline{p(\partial u/\partial y + \partial v/\partial x)}$. The ordinate is normalized such that the maximum value of the four terms is equal to one

4. Summary and Discussions

Pressure fluctuations in shear flows are investigated by analyzing databases obtained from direct numerical simulations of a turbulent channel and a homogeneous shear flow.

In the channel flow, contrary to the common belief, the slow (nonlinear) pressure is found to be dominant except near the wall, where the rapid (linear) and slow pressures are comparable. The corresponding source terms also show the same trend. Among the slow source terms, it is found that $(\partial v/\partial z)\,(\partial \omega/\partial y)$, which is large in the core of streamwise vortices, is the most dominent term. The mean-square of the slow source terms, $(\partial v/\partial z)\,(\partial \omega/\partial y)$ in particular, peak at about $y^+ \approx 20$, and the major contribution to the mean-square wall-pressure fluctuations comes from this region. The convection velocity of the wall pressure measured from the space-time correlation of the wall-pressure fluctuation is also found to be equal to the local mean velocity at about the same y-location (Kim 1989), indicating that the coherent part of the wall pressure can be regarded as a footprint connected to flow structures present at about $y^+ \approx 20$. This is consistent with Kim (1983) and Alfredsson et al. (1988), who reported that the internal shear-layer structures responsible for a major portion of turbulence production were associated with a high localized wall-pressure peak.

The probability density distribution illustrates that the pressure fluctuations are much more intermittent than velocity fluctuations, and high negative fluctuations provide a substantial contribution to the r.m.s. fluctuation. Moch of these negative fluctuations are correlated with high vorticity fluctuations. Using low pressure as a diagnostic tool in identifying vortical structures is promising, but it should be used with some caution because not all low-pressure regions are associated with organized vortical structures, although the reverse is generally true. Probability density distributions indicate a strong similarity between a turbulent channel and a homogeneous flow when the rapid and slow parts are compared separately.

Green's function representation is used to demonstrate the global dependence of pressure fluctuations. It is shown that instantaneous pressure at one wall is influenced significantly by the source terms from near the other wall. However, the same representation for the mean-square fluctuation illustrates that although the instantaneous pressure has a global dependence, most contributions to the mean-square fluctuations are local, indicating that the influence from far away is averaged out with small net effects. Contributions to the pressure-strain correlations are local near the wall, but global away from the wall. This nonlocal character of the contributions to the pressure-strain correlations raises an important question of whether one can model them in terms of local variables as is done currently.

The power spectra and two-point correlations are also examined (not presented in this paper). In the channel flow, there is little difference between the rapid and the slow pressure near the wall in the channel flow, but they differ considerably from each other away from the wall. In the homogeneous flow, the power spectra show that most contributions to the rapid pressure come from low wavenumbers, while those to the slow pressure from small wave numbers.

References

Adrian, R. J. (1982): Comment on "A note on Poisson's equation for pressure in a turbulent flow". Phys. Fluids. **25**, 577

Alfredsson, P. H., Johansson, A. V., Kim, J. (1988): Turbulence production near walls: the role of flow structures with spanwise asymmetry. *Proceedings of 1988 Summer School, CTR-S88*, Center for Turbulence Research, NASA Ames Research Center

Bradshaw, P., Koh, Y. M. (1981): A note on Poisson's equation for pressure in a turbulent flow. 24. **777**

Eckelmann, H. (1988): A review of knowledge on pressure fluctuations. *Proceedings of Zoran Zaric Memorial International Seminar on Near-Wall Turbulence*, May 16–20, 1988, Dubrovnik, Yugoslavia

Kim, J. (1983): On the structure of wall-bounded turbulent flows. Phys. Fluids. **26**, 2088

Kim, J. (1989): On the structure of pressure fluctuations in simulated turbulent channel flow. J. Fluid Mech. **205**, 421

Kim, J., Moin, P., Moser, R. D. (1987): Turbulence statistics in fully-developed channel flow at low Reynolds number. J. Fluid Mech. **177**, 133

Kraichnan, R. H. (1956): Pressure fluctuations in turbulent flow over a flat plate. J. Acoust. Soc. Am. **28**, 378

Lee, M. J., Kim, J., Moin, P. (1987): Turbulence structure at high shear rate. *Proceedings of the Sixth Symposium on Turbulent Shear Flows*, September 7–9, 1987, Toulouse, France

Lilley, G. M. (1964): A review of pressure fluctuations in turbulent boundary layers at subsonic and supersonic speeds. Arch. Mech. Stosow. **16**, 301

Mansour, N. N., Kim, J., Moin, P. (1988): Reynolds-stress and dissipation rate budgets in a turbulent channel flow. J. Fluid Mech. **194**, 15

Panton, R., Linebarger, J. H. (1974): Wall pressure spectra calculations for equilibrium boundary layers. J. Fluid Mech. **65**, 261

Robinson, S. K., Kline, S. J., Spalart, P. R. (1988): Quasi-coherent structures in the turbulent boundary layer: Part II. Verification and new information from a numerically simulated flat-plate layer. *Proc. of Zoran Zaric Memorial International Seminar on Near-Wall Turbulence*, May 16–20, 1988, Dubrovnik, Yugoslavia

Rogers, M. M., Moin, P., Reynolds, W. C. (1986): The structure and modeling of the hydrodynamic and passive scalar fields in homogeneous turbulent shear flow. *TF-25*, Department of Mechanical Engineering, Stanford University

Schewe, G. (1983): On the structure and resolution of wall-pressure fluctuations associated with turbulent boundary layer flow. J. Fluid Mech. **134**, 311

Willmarth, W. W. (1975): Pressure fluctuations beneath turbulent boundary layers. Ann. Rev. Fluid Mech. **7**, 13

The Structure of Sheared Turbulence Near a Plane Boundary

Moon Joo Lee[1] and J. C. R. Hunt[2]

[1] Center for Turbulence Research, Stanford University, Stanford, CA 94305, USA and NASA-Ames Research Center, MS 202A-1, Moffett Field, CA 94035, USA; Present address: Department of Mechanical Engineering, Pohang Institute of Science and Technology, Pohang 790–600, Korea

[2] Department of Applied Mathematics and Theoretical Physics, University of Cambridge, Silver Street, Cambridge CB3 9EW, UK

Abstract

An analysis is presented of how a plane boundary affects the structure of turbulence in a sheared free stream. A uniform-shear boundary layer (USBL) is formulated with the slip velocity condition at the surface, and inhomogeneous rapid distortion theory is applied. The effects of 'blocking' by the surface on the turbulence structure in the USBL are compared with those in the shear-free boundary layer (SFBL).

Shear produces highly anisotropic eddies elongated in the flow direction. The distinctive peaks of the spanwise spectra of the streamwise velocity, $\Theta_{11}(\kappa_3; y)$, suggest the existence of the streaky structures in the flow. The mean streak spacing estimated from the energy spectra increases with the distance from the surface, in qualitative agreement with previous measurements and computations.

The vertical velocity variance, $\overline{v^2}$, is reduced with shear at all heights, roughly in proportion to the reduction in the homogeneous value, but the shape of the profile remains unchanged only near the surface: $\overline{v^2}/\overline{v^2}^{(H)} \sim y^{2/3}$. The turbulent shear stress, $-\overline{uv}$, increases with total shear at all distances from the boundary. Scaled with the homogeneous value, the profile of the shear stress does not vary with time. The universal profile near the surface is $\overline{uv}/\overline{uv}^{(H)} \sim y^{2/3}$, similar to the vertical variance profile.

The streamwise integral length scales increase with shear, indicating elongation of the streamwise extent of eddies. At given total shear, the spanwise extent of the streaks determined from $L_{uu}^{(z)}$ widens as the boundary is approached. The smallest of the integral scales, $L_{vv}^{(z)}$, is a measure of the dissipation scale, and decreases with shear in a self-similar way: $L_{vv}^{(z)} \sim y$.

1. Introduction

1.1 Motivation and Background

In most models of sheared turbulent flows near plane boundaries, it is assumed that the mean 'shear rate,' $S = dU/dy$, controls the structure of turbulence, especially the relation between the Reynolds shear stress, $-\overline{uv}$, and the energy of the turbulence, $q^2 = \overline{u_i u_i}$. There is some support for this assumption when calculations for $\overline{u^2}$ and $-\overline{uv}$ and the cross-correlations, $R_{ij}(\mathbf{r})$, in *homogeneous* uniform-shear flows ($U = U_0 + Sy$) are compared with computations and measurements in turbulent boundary layers where S is not uniform (Lee et al. 1987; Townsend 1976). For close agreement between the two situations at a given height, y, it was found necessary that the homogeneous shear and the boundary layer should have the same dimensionless shear rate, $S^* = SL/q$ (L is an appropriate length scale), and that total shear, $\beta = St$, should be chosen appropriately. This homogeneous-shear assumption is the basis of Reynolds-stress transport modelling as explained by Lumley (1978).

Turbulent Shear Flows 7

However, it is also known that the larger scales of motion (defined as greater than y at a height y) are affected by the boundary, because vertical motions are *blocked* at $y = 0$ (Bradshaw 1967; Townsend 1961). In turbulent flows *without* shear near a *rigid* boundary or density interfaces, previous investigations (experimental, theoretical and computational) have shown that the blocking effect changes the variances of the different turbulence components ($\overline{u^2}, \overline{v^2}, \overline{w^2}$) and the length scales. The largest effect is on the variance and the streamwise integral scale, $L_{22}^{(x)}$, of the normal component, $\overline{v^2}$; near the boundary, $\overline{v^2} \sim y^{2/3}$ and $L_{22}^{(x)} \sim y$ (Hunt and Graham 1978; Hunt 1984).

Since there are many interesting and important turbulent flows near flat boundaries in which there are quite different shear rates, it is important to understand and quantify the effect of blocking by a surface in the presence of shear. There have been previous unpublished studies using rapid distortion theory by Maxey (1978) and Durbin (1979).

Hunt et al. (1987a) analysed the correlations, $\tilde{R}_{22}(y_0, y)$, between two points, y and y_0 ($y \leqq y_0$), of the normal velocity component using direct numerical simulations. It was found that the two-point correlation is, as hypothesised, a function of y/y_0, but the form differs from the linear profile in shear-free boundary layers (Hunt 1984), and is less than y/y_0 (Fig. 1).

Hunt et al. (1987b) examined the variation of the *dissipation length*, L_ε, defined as $\varepsilon/\overline{v^2}^{3/2}$. They compared computations in the boundary layer and channel flow with a simple model: $L_\varepsilon^{-1} = A_B/y + A_S(dU/dy)/\overline{v^2}^{1/2}$ (Hunt et al. 1988). The two terms represent the effect of blocking and shear, respectively. Good agreement was found except at the outer edge of the boundary layer. The *sensitivity* to shear of the *dissipation scale* is clearly much greater than for the cross-correlations, $\tilde{R}_{22}$ and $\tilde{R}_{12}$, which are dominated by blocking. An important aim of the present work is to explain this difference by looking closely at different length scales.

1.2 Uniform-Shear Boundary Layer (USBL)

It is possible to use Rapid Distortion Theory (RDT) to study how homogeneous turbulence in a uniform shear is deformed when a plate is suddenly introduced into the flow at time $t = 0$. The analysis is *inviscid* so the only effect of the plate is to block velocity fluctuations normal to itself. In addition, the *mean* velocity profile is not changed. Because the mean vorticity remains *uniform*, this leads to a great simplification in the analysis. In reality, the presence of the wall leads to Reynolds-stress gradients, which would over a long period affect the mean velocity profile; this is the effect we ignore here.

The detailed aspect of the analysis that needs considering is the difference between the blocking effect of the boundary on the sheared turbulence compared with unsheared turbulence. In a shear-free boundary layer (SFBL), there is *isotropic* turbulence in the free stream, which leads to both horizontal components, $\overline{u^2}$ and $\overline{w^2}$, increasing equally near the wall as a result of irrotational fluctuation being axisymmetric about the y-axis. However, in a uniform-shear boundary layer (USBL), the turbulence in the free stream above the surface is quite *anisotropic*. This

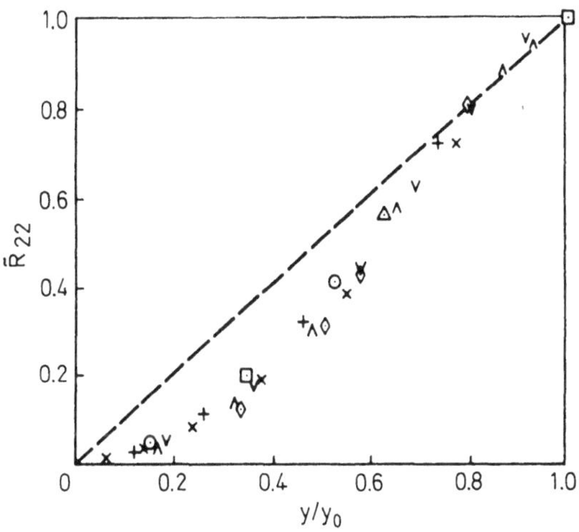

Fig. 1. Cross-correlation of v at heights y_0 and y, normalized by $\overline{v^2}(y_0)$ computed from numerical simulations (turbulent boundary layer, Spalart 1987; plane channel flow, Kim et al. 1987). Symbols, data from numerical simulation; ----, theoretical prediction and data for SFBL (Hunt 1984): $\tilde{R}_{22} \simeq y/y_0$. [From Hunt et al. (1987a).]

leads to a different redistribution of energy between u and w at the surface (Wong 1987) and a change in the interaction with the mean shear. This question has not been analysed in detail before.

The particular computations that are presented here are: (i) the variation with y of the one-dimensional spectra, $\Theta_{ij}(\kappa_1; y)$ and $\Theta_{ij}(\kappa_3; y)$, at different total shear, $\beta = St$; (ii) the Reynolds stresses, $\overline{u_i u_j}$, as a function of y; and (iii) the variation with y of the integral length scales, $L_{ij}^{(x_\kappa)}$, in the homogeneous directions (x, z) of the different velocity components. The smallest of these integral scales is usually a good indicator of the dissipation scale, L_ε, though this point has not been investigated systematically!

2. Analysis

2.1 Formulation of the Problem

We summarize here the assumptions, equations and boundary conditions governing the linearised (RDT) analysis of a turbulent velocity field, $\mathbf{u}(\mathbf{x}, t)$, in a uniform shear flow, $\mathbf{U} = (U_0 + Sy, 0, 0)$ (see Fig. 2).

The condition for ignoring the nonlinear inertial terms over the period of the distortion is that the shear is strong enough for the straining terms to dominate over nonlinear inertial terms, that is

$$S^* = SL/q \gg 1. \tag{2.1}$$

Over a period of time, the small nonlinear terms affect the redistribution of turbulence energy and momentum between different components. If $S^* \sim 1$, this relaxation time, t_r, is of order L/q, so the criterion for importance of nonlinear terms

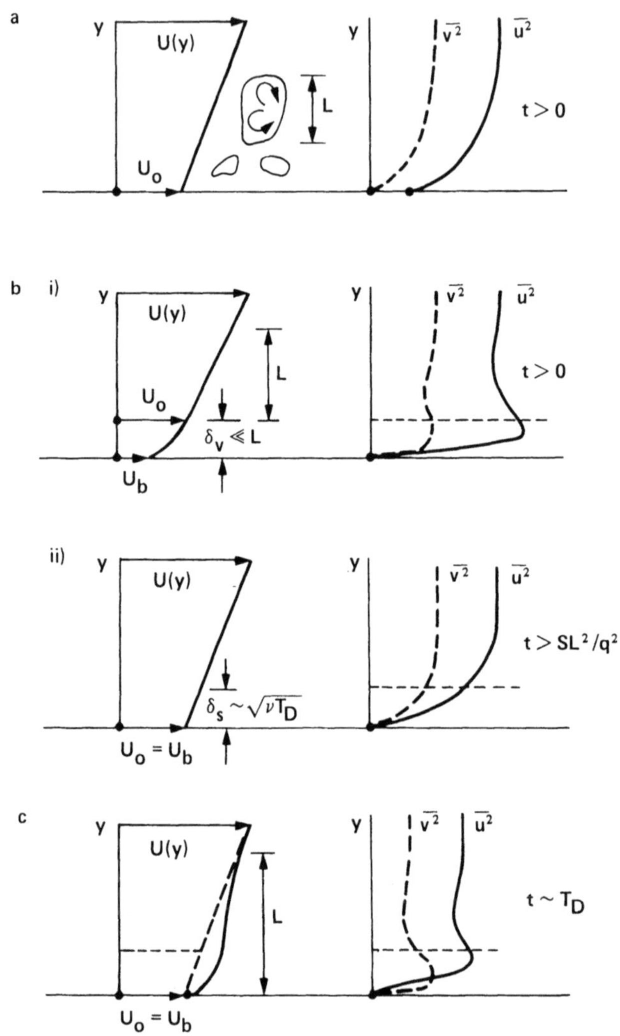

Fig. 2. Schematic diagram of a uniform-shear boundary layer (USBL), showing **a** the ideal problem; **b** the experimental reality: (i) if $U_b \neq U_0$, theory requires $\delta_v \ll L$, (ii) if $U_b = U_0$, theory requires $\delta_s \sim \sqrt{\nu T_D} \ll L$; **c** the slow distortion of the mean velocity profile ($t \sim T_D$) because of the Reynolds-shear-stress gradient

is that

$$t > t_r \sim L/q \tag{2.2}$$

(Gartshore et al. 1983; Bertoglio 1986).

But if S^* is large, the rate at which the anisotropy is growing by linear processes is so much greater than by nonlinear processes that, as Lee et al. (1987) have shown, the nonlinear redistributive processes are negligible (especially for $\overline{u^2}$ and $\overline{uv}$) even when $t \geqq t_r$.

For the linearised analysis of the USBL, it is also necessary to assume that the gradient of the Reynolds stress has a weak effect on the mean velocity profile over the time T_D of distortion:

$$T_D \left| -\frac{\partial^2 \overline{uv}}{\partial y^2} \right| \ll S \tag{2.3a}$$

or

$$S^* = SL/q \gg qT_D/L. \tag{2.3b}$$

The inviscid analysis used here can only correspond to physical experiments involving a real viscous fluid, if the viscous stresses are negligible within the flow for energy-containing eddies and if the no-slip boundary condition can be neglected (see Fig. 2). Hunt and Graham's (1978) detailed analysis of both these effects led to the following conditions for neglecting them:

$$qL/\nu \gg 1, \tag{2.4}$$

$$\sqrt{\nu T_D} \ll L \quad \text{and} \quad \delta_v \ll L, \tag{2.5a, b}$$

where δ_v is the thickness of the viscous layer on the surface, which must exist if the velocity U_0 of the free stream at $y = 0$ is not equal to the velocity of the surface U_b (Fig. 2b-i). If $U_b = U_0$, as in the moving-belt experiment of Uzkan and Reynolds (1967) and Thomas and Hancock (1977), there is no *mean* boundary layer, but the *fluctuating* velocity must still be zero. This thin surface layer affected by this no-slip fluctuating condition is of order $\sqrt{\nu T_D}$ (Fig. 2b-ii).

Given the assumptions (2.1) to (2.5), the equations for the linearised problem are

$$\frac{\bar{D}\mathbf{u}}{Dt} = -\nabla p - Sv\mathbf{i}, \tag{2.6}$$

$$\nabla \cdot \mathbf{u} = 0, \tag{2.7}$$

where $\bar{D}/Dt = \partial/\partial t + U\partial/\partial x$, $\mathbf{U} = (U_0 + Sy, 0, 0)$ and p is the kinematic pressure fluctuation. The initial and boundary conditions are given by

$$\mathbf{u}(\mathbf{x}, t) = \mathbf{u}^{(H)}(\mathbf{x}, t) \quad \text{at} \quad t = 0, \tag{2.8a}$$

where $\mathbf{u}^{(H)}$ is a prescribed random velocity field, and

$$v(x, y = 0, z, t) = 0 \quad \text{for} \quad t > 0. \tag{2.8b}$$

The solution is obtained as a sum of homogeneous flow field, $\{\mathbf{u}^{(H)}, p^{(H)}\}$ $(\mathbf{x}, t)$, subject to uniform shear, and the blocking flow field induced by the surface, $\{\mathbf{u}^{(B)}, p^{(B)}\}$ $(\mathbf{x}, t)$:

$$\mathbf{u}(\mathbf{x}, t) = \mathbf{u}^{(H)}(\mathbf{x}, t) + \mathbf{u}^{(B)}(\mathbf{x}, t), \tag{2.9a}$$

$$p(\mathbf{x}, t) = p^{(H)}(\mathbf{x}, t) + p^{(B)}(\mathbf{x}, t). \tag{2.9b}$$

Then, $\mathbf{u}^{(H)}(\mathbf{x}, t)$ satisfies (2.6) subject to (2.8a) at $t = 0$; it does *not* satisfy (2.8b). But $\mathbf{u}^{(B)}(\mathbf{x}, t)$ satisfies (2.6) subject to

$$v^{(B)}(\mathbf{x}, t) = 0 \quad \text{at} \quad t = 0. \tag{2.10a}$$

From (2.8) and (2.9), it can be shown that

$$v^{(B)}(x, y=0, z, t) = -v^{(H)}(x, y=0, z, t) \quad \text{for} \quad t>0, \tag{2.10b}$$

and

$$\mathbf{u}^{(B)}(\mathbf{x}, t) \to 0 \quad \text{as} \quad y \to \infty. \tag{2.10c}$$

By taking the curl of (2.6), it follows that

$$\frac{\bar{D}}{Dt}\nabla^2 v^{(B)} = 0. \tag{2.11}$$

So from (2.10a), one finds that this equation for $v^{(B)}(\mathbf{x}, t)$ reduces to Laplace's equation: $\nabla^2 v^{(B)} = 0$. Once $v^{(B)}(\mathbf{x}, t)$ is obtained, $p^{(B)}(\mathbf{x}, t)$ is obtained from

$$\nabla^2 p^{(B)} = -2S\frac{\partial v^{(B)}}{\partial x}. \tag{2.12}$$

Thence from (2.6), $w^{(B)}(\mathbf{x}, t)$ and $u^{(B)}(\mathbf{x}, t)$ are calculated.

The statistical and Fourier analyses are similar to the method for *inhomogeneous* RDT developed by Hunt (1973), and used for SFBL by Hunt and Graham (1978) and for a uniform-shear layer by Gartshore et al. (1983).

2.2 Physical Interpretation

The difference in their mechanics between a shear-free boundary layer (SFBL) and a uniform-shear boundary layer (USBL) can be understood from the linearised vorticity equation obtained by taking the curl of (2.6):

$$\frac{\bar{D}\boldsymbol{\omega}}{Dt} = (\boldsymbol{\omega}\cdot\nabla)\mathbf{U} + (\boldsymbol{\Omega}\cdot\nabla)\mathbf{u}. \tag{2.13}$$

In a SFBL where $\nabla\mathbf{U} = 0$ and $\boldsymbol{\Omega} = 0$, this equation reduces to

$$\frac{\bar{D}\boldsymbol{\omega}}{Dt} = 0. \tag{2.14}$$

The effect of the boundary is simply to introduce an *irrotational* velocity field, so the vorticity is not affected by the boundary, i.e.

$$\mathbf{u}^{(B)}(\mathbf{x}, t) = \nabla\phi(\mathbf{x}, t), \qquad \boldsymbol{\omega}(\mathbf{x}, t) = \boldsymbol{\omega}^{(H)}(\mathbf{x}, t). \tag{2.15}$$

This irrotational velocity field induced by image vorticity below $y = 0$ produces the *splat* effect: the vertical component $v^{(B)}$ blocks $v^{(H)}$ so that $v = 0$ at $y = 0$, while the horizontal components add to increase $\overline{u^2}$ and $\overline{w^2}$ (Fig. 3a).

But in a USBL the addition of the blocking velocity field $\mathbf{u}^{(B)}(\mathbf{x}, t)$ affects $\boldsymbol{\omega}$ because of the term $(\boldsymbol{\Omega}\cdot\nabla)\mathbf{u}$, which is the effect of velocity perturbations distorting the vorticity $\boldsymbol{\Omega} = (0, 0, \Omega_z)$ of the mean shear. The terms $\Omega_z \partial v/\partial z$ and $\Omega_z \partial w/\partial z$ primarily determine ω_y and ω_z, and u has contributions from $\Omega_z \int_0^z \omega_y dz \sim \Omega_z v$

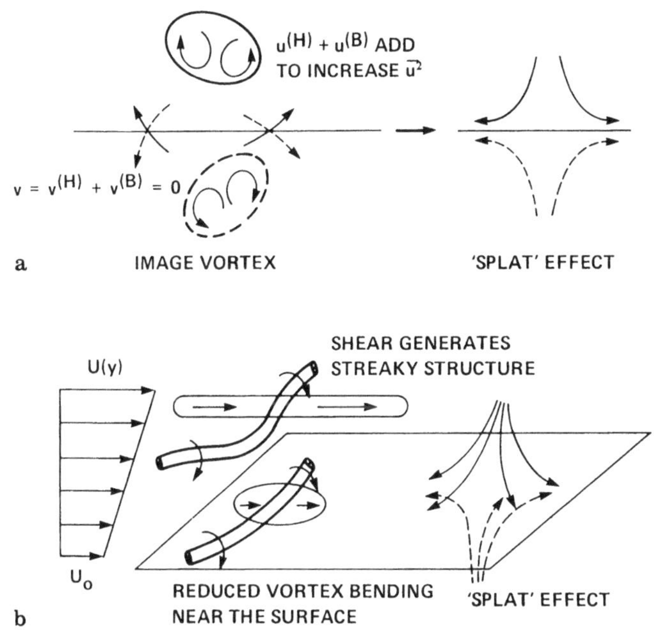

Fig. 3. Schematic diagram to show difference between the mechanisms for shear-free boundary layer (SFBL) and for uniform-shear boundary layer (USBL): **a** SFBL, image vortex below $y=0$ induces irrotational velocity $\mathbf{u}^{(\mathbf{B})}$ above $y=0$; **b** USBL, reduced vortex bending near the surface reduces $\overline{u^2}$ while the splat effect increases $\overline{u^2}$

and $\Omega_z\int_0^y \omega_z dy \sim \Omega_z\int \partial v/\partial y\, dy \sim \Omega_z v$. (Note that in a shear flow the eddies are elongated in the flow direction, so $\partial u/\partial x \ll \partial v/\partial y$.) Thus, near the boundary, as v decreases, $\overline{u^2}$ and $-\overline{uv}$ also decreases. However, in the USBL, there is also a tendency of $\overline{u^2}$ and $\overline{w^2}$ to *increase* because of the impacting (or splatting) of the normal fluctuations at the boundary. The analysis shows the different contributions by these two mechanisms.

3. Results and Discussion

We assume here that the initially homogeneous turbulence is isotropic and has the von Kármán spectrum of the dimensionless form

$$E^{(\mathrm{H})}(\kappa)=\frac{A\kappa^4}{(c_{\mathrm{K}}+\kappa^2)^{2+\mu}} \tag{3.1}$$

where $\mu=\frac{5}{6}$, $A=\frac{55}{9\pi}c_{\mathrm{K}}^{\mu}$ and $c_{\mathrm{K}}=0.558$. Note that this spectrum behaves as κ^4 in low-wavenumber region ($\kappa \ll 1$) and has the $\kappa^{-5/3}$ Kolmogorov form at high wavenumbers ($\kappa \gg 1$). All the quantities are made dimensionless by the initial r.m.s.

velocity u_0 and integral length scale L_0. Note that in USBL the turbulence quantities change with total shear $\beta = St$. In order to highlight the effects of shear on turbulence structure near a boundary, comparison is made with the results for SFBL ($\beta = 0$) (for details, see Hunt and Graham 1978; Hunt 1984).

3.1 Energy Spectra

In Fig. 4, the one-dimensional energy spectra $\Theta_{ij}(\kappa_1; y)$ are plotted for SFBL. When the turbulence is unsheared, the near-wall behavior of the spectra is such that there is no variation in $\Theta_{11}(\kappa_1 \to 0; y)$, but the increase in $\Theta_{33}(\kappa_1 \to 0; y)$ exactly balances the reduction in $\Theta_{22}(\kappa_1 \to 0; y)$, viz.

$$\Theta_{11}^{SFBL}(\kappa_1 \to 0; y) = \Theta_{11}^{(H)}(\kappa_1 \to 0) = 1/\pi, \tag{3.2}$$

$$\Theta_{22}^{SFBL}(\kappa_1 \to 0; y \to 0) + \Theta_{33}^{SFBL}(\kappa_1 \to 0; y \to 0) = 1/\pi, \tag{3.3}$$

where $\Theta_{ij}^{(H)}(\kappa_1)$ are the one-dimensional energy spectra of the initially homogeneous, isotropic turbulence, i.e. $\Theta_{ij}(\kappa_1; y \to \infty, t = 0)$. Similar results can be found by a symmetry consideration: $\Theta_{33}^{SFBL}(\kappa_3 \to 0; y) = \Theta_{33}^{(H)}(\kappa_3 \to 0) = 1/\pi$, $\Theta_{11}^{SFBL}(\kappa_3 \to 0; y \to 0) + \Theta_{22}^{SFBL}(\kappa_3 \to 0; y \to 0) = 1/\pi$. Asymptotic analysis (Hunt and Graham 1978) shows that near the surface

$$\Theta_{22}^{SFBL}(\kappa_1 \to 0; y \to 0) = \Theta_{22}^{SFBL}(\kappa_3 \to 0; y \to 0) \sim y^{5/3}. \tag{3.4}$$

In order to examine the effect of shear on the energy spectra, the results for USBL (when $\beta = 2$) are shown in Fig. 5. The most salient feature is that, as $\kappa_1 \to 0$, there is a dramatic increase with shear in the spectrum $\Theta_{11}(\kappa_1; y)$ of the streamwise velocity at distances away from the surface (cf. Figs. 4a and 5a). However, near the surface the spectrum is *not* affected by shear

$$\Theta_{11}^{USBL}(\kappa_1 \to 0; y \to 0) = \Theta_{11}^{SFBL}(\kappa_1 \to 0; y) = 1/\pi. \tag{3.5}$$

That is, very close to the surface, the streamwise motion of large-scale eddies is dominated by the induced image vortices as shown in Fig. 3(a), independent of shearing in the free stream. In contrast, shear does not change the values of $\Theta_{22}(\kappa_1 \to 0; y)$ and $\Theta_{33}(\kappa_1 \to 0; y)$ for *all* y:

$$\Theta_{22}^{USBL}(\kappa_1 \to 0; y) = \Theta_{22}^{SFBL}(\kappa_1 \to 0; y), \tag{3.6a}$$

$$\Theta_{33}^{USBL}(\kappa_1 \to 0; y) = \Theta_{33}^{SFBL}(\kappa_1 \to 0; y). \tag{3.6b}$$

At high wavenumbers, however, the spectrum $\Theta_{22}(\kappa_1; y)$ of the component normal to the boundary is significantly reduced by shear for all y (cf. Figs. 4*b* and 5*b*), which implies that the vertical motion of small-scale eddies is sensitive to shear. Interestingly enough, $\Theta_{11}(\kappa_1; y \to 0)$ is reduced much more in the high-wavenumber region ($\kappa_1 > 1$) than it is enhanced at low wavenumbers ($\kappa_1 < 1$); in fact, $\overline{u^2}$ on the surface in the USBL is lower than the SFBL value: $\overline{u^2}(y = 0) < \frac{3}{2}$.

It is of fundamental importance to observe the peaks in the spanwise spectrum $\Theta_{11}(\kappa_3; y)$ of the streamwise velocity (Fig. 6a), which implies the existence of eddies

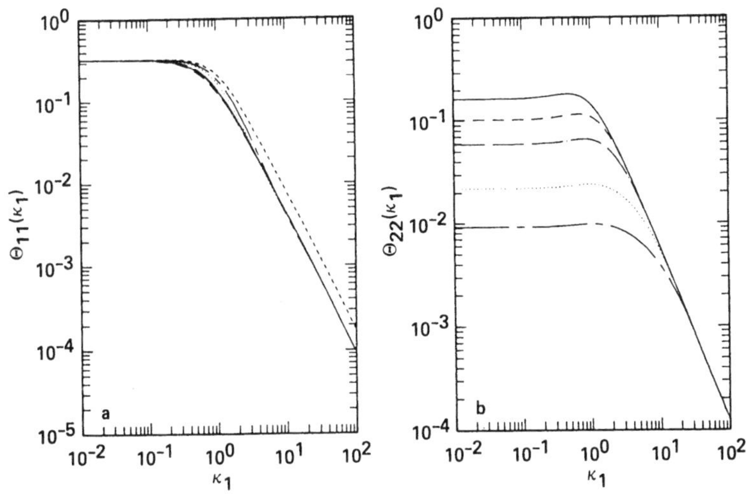

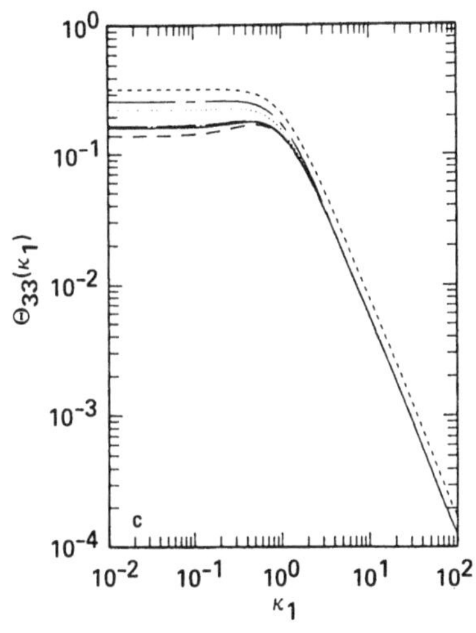

Fig. 4. The one-dimensional energy spectra in SFBL ($\beta = 0$): **a** Θ_{11} $(\kappa_1; y)$; **b** $\Theta_{22}(\kappa_1; y)$; **c** $\Theta_{33}(\kappa_1; y)$.——, $y = 0$; —-—, $y = 0.1$; ········, $y = 0.2$; —·—, $y = 0.5$; ----, $y = 1$; ——, $y = 5$

at the corresponding spanwise scales. The peaks in $\Theta_{11}(\kappa_3; y)$ at distances away from the wall and the high streamwise velocity variance (see Fig. 9) are direct evidence of the presence of *streaky structures* in the flow (see Kline et al. 1967). Notice that the peaks are much more discernible in the spectra at heights away from the wall, consistent with the earlier observation of the low- and high-speed streaks in homogeneous shear flow where there is no boundary (Lee et al. 1987). The fact that the streaks exist in sheared turbulence, independent of the presence of the boundary, but not in the SFBL strongly supports the assertion put forward by Lee et al. that the main mechanism of generating the streaks is the mean shear but not the wall blocking. This is supported also by the experiment of Uzkan and Reynolds (1967).

As shear increases, the peak in the spectrum (or streaks) is discernible at heights closer to the wall (e.g. for $\beta = 4$, see Fig. 6b). This means that, as shear is prolonged, the lateral extent of the streaks are mainly controlled by the effect of shear rather

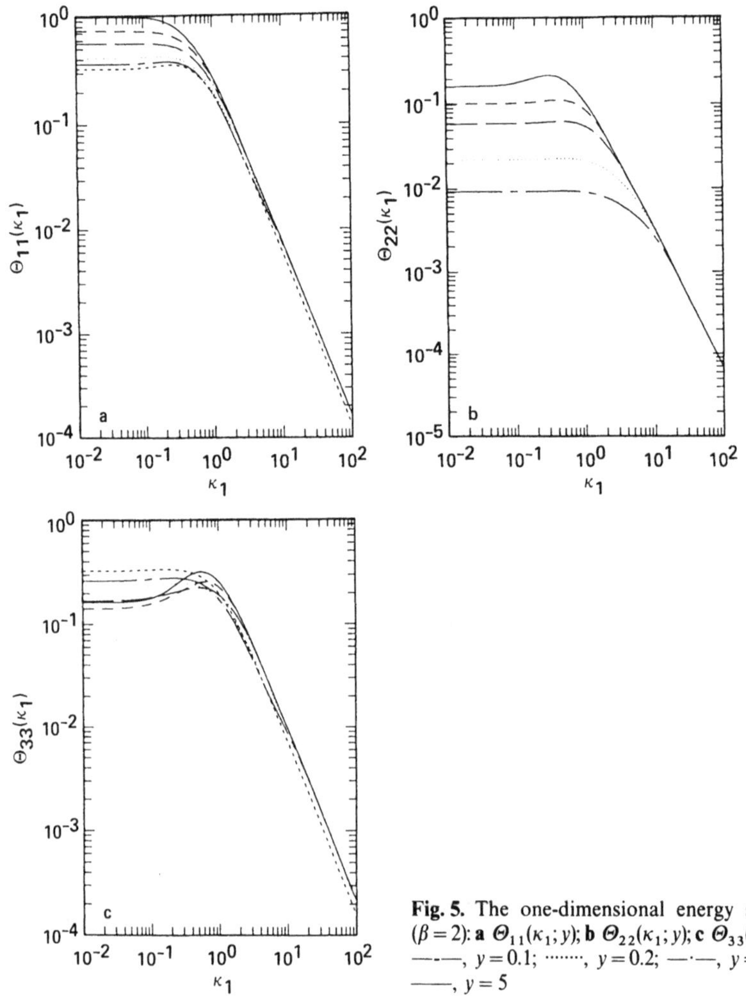

Fig. 5. The one-dimensional energy spectra in USBL ($\beta = 2$): **a** $\Theta_{11}(\kappa_1; y)$; **b** $\Theta_{22}(\kappa_1; y)$; **c** $\Theta_{33}(\kappa_1; y)$. -----, $y = 0$; —·—, $y = 0.1$; ·······, $y = 0.2$; —·—, $y = 0.5$; ——, $y = 1$: ——, $y = 5$

than by the blocking effect. Note that other aspects of the turbulence (e.g. $\overline{v^2}$ and $-\overline{uv}$) remain affected by the blocking whatever the total shear.

It is of interest to examine how the mean spacing λ_z between the streaks varies with the distance from the surface. The mean streak spacing is determined from the wavenumber $\kappa_3^{(s)}$ at which $\Theta_{11}(\kappa_3; y)$ peaks, i.e. $\lambda_z = 1/\kappa_3^{(s)}$. It is clear from Fig. 7 that λ_z increases with y, in qualitative agreement with measurements and computations in turbulent boundary layers (Kline et al. 1967; Kim et al. 1987).

The spanwise spectrum $\Theta_{33}(\kappa_3; y)$ of the spanwise velocity fluctuation as $\kappa_3 \to 0$ is unchanged for *all* y:

$$\Theta_{33}^{\mathrm{USBL}}(\kappa_3 \to 0; y) = \Theta_{33}^{\mathrm{SFBL}}(\kappa_3 \to 0; y) = \Theta_{33}^{(\mathrm{H})}(\kappa_3 \to 0) = 1/\pi. \tag{3.7}$$

Thus, (3.7) and (3.6b) indicate that in this flow the large-scale spanwise fluctuation is not affected by shear (cf. Figs. 4c and 5c). Also note that

$$\Theta_{11}^{\mathrm{USBL}}(\kappa_3 \to 0; y \to 0) = \Theta_{11}^{\mathrm{SFBL}}(\kappa_3 \to 0; y \to 0) = 1/\pi. \tag{3.8}$$

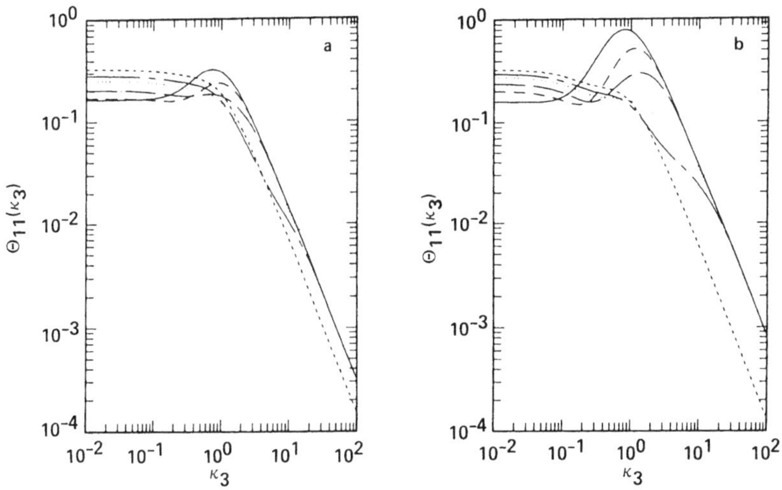

Fig. 6. The one-dimensional energy spectra $\Theta_{11}(\kappa_3; y)$ in USBL: **a** $\beta = 2$; **b** $\beta = 4$. ----, $y = 0$; —-—, $y = 0.1$; ······, $y = 0.2$; —·—, $y = 0.5$; ––––, $y = 1$; ——, $y = 5$

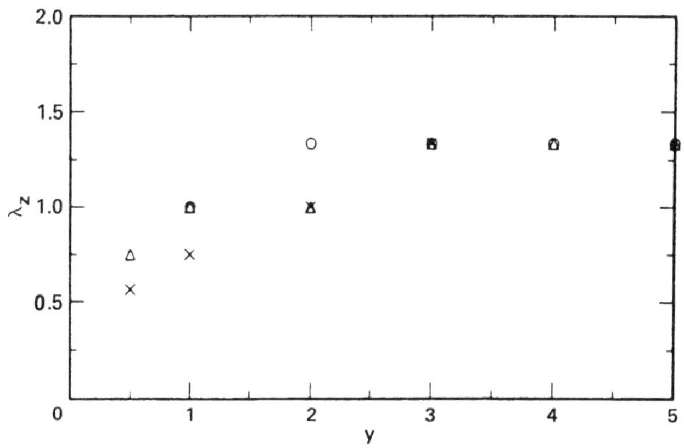

Fig. 7. Variation with the distance from the boundary of the mean streak spacing, λ_z, determined from the wavenumber at which $\Theta_{11}(\kappa_3; y)$ peaks. ○, $\beta = 2$; △, $\beta = 4$; ×, $\beta = 8$

3.2 Reynolds Stresses

In Fig. 8, the Reynolds stresses $\overline{u_i u_j}$ are shown as functions of y for SFBL. In SFBL, if the homogeneous field is isotropic, the horizontal variances are equal, $\overline{u^2} = \overline{w^2}$, for all y. Near the surface, the splat effect results in transfer of energy from the vertical component to the horizontal components:

$$\left.\begin{aligned} &\overline{u^2}(y \to 0) = \overline{w^2}(y \to 0) = \tfrac{3}{2} - 0(y^{2/3}), \\ &\overline{v^2}(y \to 0) \sim y^{2/3} \end{aligned}\right\} \quad \text{in SFBL.} \qquad (3.9\text{a, b})$$

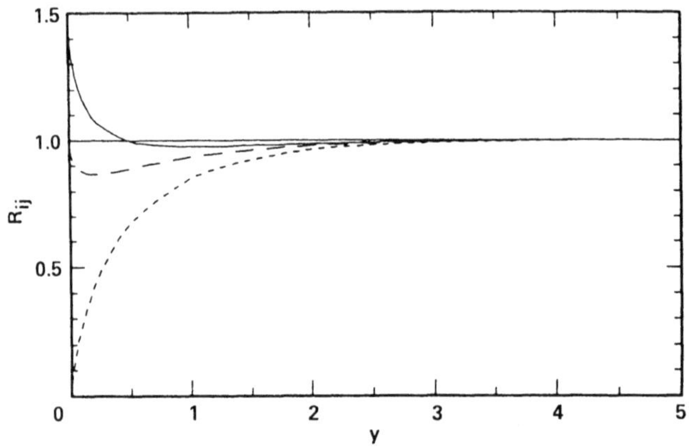

Fig. 8. Profile of the Reynolds stresses in SFBL ($\beta = 0$). ----, $\frac{1}{3}\overline{u_i u_i}$; ——, $\overline{u^2} = \overline{w^2}$; ------, $\overline{v^2}$

The horizontal variances are decreased from the wall value to the homogeneous value of unity within about $y = 0.5$ from the wall. The vertical variance $\overline{v^2}$ shows monotonic increase from zero at the wall to 95% of its homogeneous value (unity) over a distance of about 2.

Variation of the kinetic energy in SFBL is not monotonic. Except at the surface where $q^2/q_0^2 = 1$, the kinetic energy is less than its homogeneous value; the ratio q^2/q_0^2 attains its minimum value of about 0.85 at $y \simeq 0.2$.

Figure 9 shows the profile of the Reynolds stresses in USBL ($\beta = 2$). An important aspect of USBL is the increase of the turbulent kinetic energy by the transfer from the mean flow to turbulence (i.e. turbulence production). Therefore, the kinetic energy is much larger in USBL than in SFBL (except in the vicinity of the surface), e.g. 50% increase in the homogeneous value when $\beta = 2$ (cf. Figs. 8 and 9).

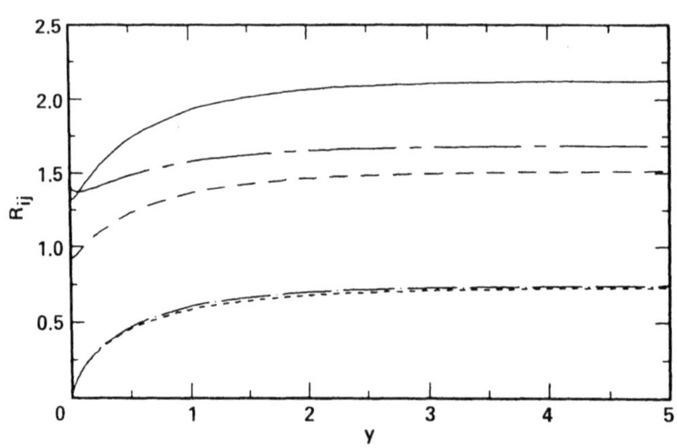

Fig. 9. Profile of the Reynolds stresses in USBL ($\beta = 2$). ----, $\frac{1}{3}\overline{u_i u_i}$; ——, $\overline{u^2}$; ----, $\overline{v^2}$; —·—, $\overline{w^2}$; —·—, $-\overline{uv}$

The dip near the surface (at $y \simeq 0.2$ when $\beta = 2$) is a direct consequence of the reduction of the horizontal variances there, as explained below. Note that in USBL the turbulent kinetic energy at the surface is not equal to the homogeneous value, unlike in SFBL.

In USBL the horizontal variances $\overline{u^2}$ and $\overline{w^2}$ are markedly enhanced by shear except very close to the wall. The reduction of these variances can be explained by the effect of the reduced bending of vortex filaments (see Fig. 3*b*); the spanwise variance $\overline{w^2}$ is less reduced than the streamwise variance $\overline{u^2}$.

Comparison of the vertical variance $\overline{v^2}$ in Figs. 8 and 9 shows that there is a substantial reduction with shear at all heights. [The apparently coincidental profiles of $\overline{v^2}$ and $-\overline{uv}$ in Fig. 9 is incidental at this particular time ($\beta = 2$).] It is interesting to consider how $\overline{v^2}$ in the vicinity of the boundary changes with shear. When scaled by the homogeneous value $\overline{v^{2(\mathrm{H})}}$, the profiles of $\overline{v^2}$ at different values of β show (Fig. 10) that as $y \to 0$

$$\overline{v^2}(y \to 0, \beta) \sim \overline{v^{2(\mathrm{H})}}(\beta) y^{2/3} \quad \text{in USBL.} \tag{3.10}$$

Another important difference in USBL is the development of the turbulent shear stress $-\overline{uv}$ with shear. Note that the shear stress is also subject to reduction by the wall blocking. The profile of the shear stress exhibits a monotonic increase from zero at the wall towards the homogeneous value that increases with shear. Thus, the same scaling used for $\overline{v^2}$ is used for the shear stress: as $y \to 0$,

$$\overline{uv}(y \to 0, \beta) \sim \overline{uv}^{(\mathrm{H})}(\beta) y^{2/3} \quad \text{in USBL.} \tag{3.11}$$

Figure 11 shows $\overline{uv}/\overline{uv}^{(\mathrm{H})}$ as a function of y at different times. This function does not change with total shear. It is within 20% of the value of $\overline{v^2}/\overline{v^{2(\mathrm{H})}}$ at $\beta = 0$.

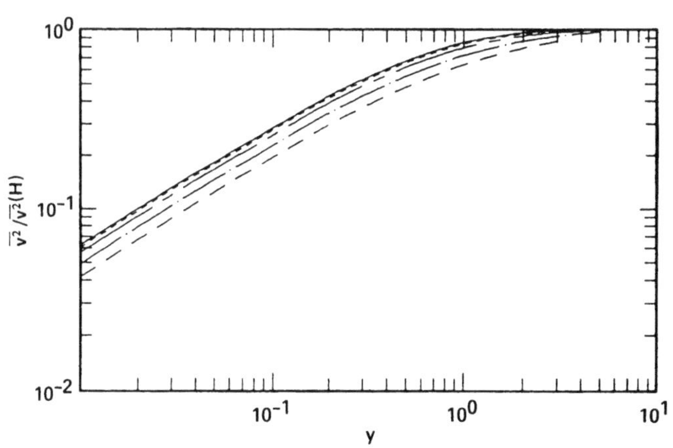

Fig. 10. Variation of the vertical variance profile with total shear: $\overline{v^2}/\overline{v^{2(\mathrm{H})}}$ *vs. y.* ——, $\beta = 0$; ----, $\beta = 1$; — — —, $\beta = 2$; —·—, $\beta = 4$; – – – –, $\beta = 8$

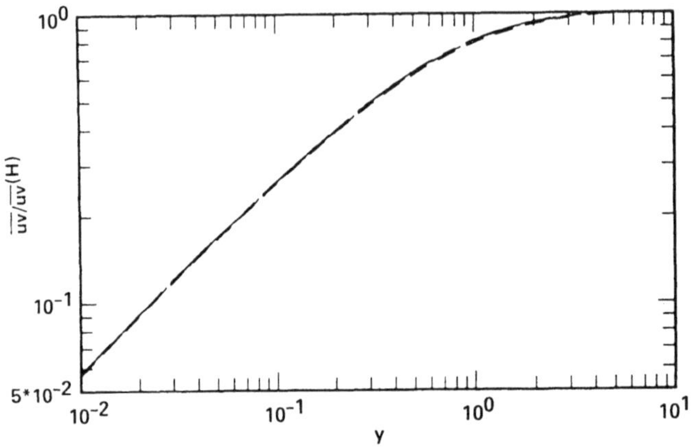

Fig. 11. Variation of the turbulent shear-stress profile with total shear: $\overline{uv}/\overline{uv}^{(H)}$ *vs.y.* ----, $\beta = 1$; —-—, $\beta = 2$; —·—, $\beta = 4$; ––––, $\beta = 8$

3.3 Integral Length Scales

Integral length scales in the x_m-direction of the correlations between the velocity components u_i and u_j at y are defined as

$$L_{ij}^{(x_m)}(y) = \frac{\pi \Theta_{ij}(\kappa_m = 0; y)}{\overline{u_i u_j}(y)} \quad \text{(no sum on } i \text{ and } j). \tag{3.12}$$

Profiles of the streamwise integral scales $L_{uu}^{(x)}$, $L_{vv}^{(x)}$ and $L_{ww}^{(x)}$ in SFBL are shown in Fig. 12. Very close to the wall, the streamwise scale of the vertical velocity grows faster than that of the streamwise velocity; from (3.2), (3.4) and (3.9a, b), one finds

$$L_{uu}^{(x)}(y \to 0) = \tfrac{2}{3} + O(y^{2/3}), \quad L_{vv}^{(x)}(y \to 0) \sim y \quad \text{in SFBL.} \tag{3.13a, b}$$

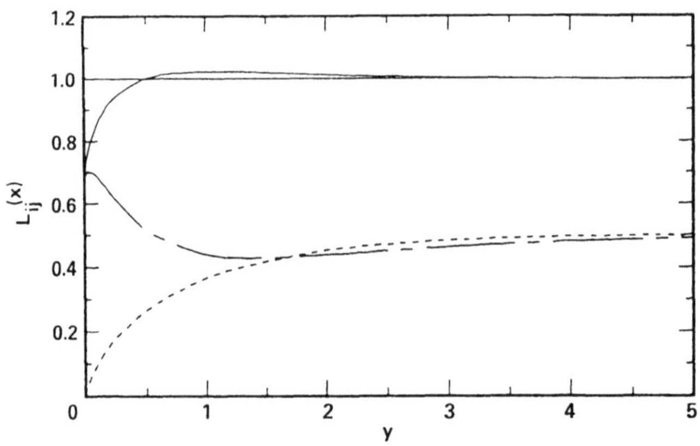

Fig. 12. Profile of the streamwise integral scales $L_{ij}^{(x)}(y)$ for SFBL ($\beta = 0$). ——, $L_{uu}^{(x)} = L_{ww}^{(z)}$; ----, $L_{vv}^{(x)} = L_{vv}^{(z)}$; —-—, $L_{ww}^{(x)} = L_{uu}^{(z)}$

However, over a distance until $L_{uu}^{(x)}$ reaches the homogeneous value of unity ($0.1 < y < 0.5$, say), the growth rate of $L_{uu}^{(x)}$ is higher. The integral scale of the spanwise velocity $L_{ww}^{(x)}$ exhibits a monotonic decrease.

Profiles of the streamwise integral length scales in USBL are plotted in Fig. 13. The effect of shear is to increase the streamwise scales of the streamwise and vertical velocity fluctuations, $L_{uu}^{(x)}$ and $L_{vv}^{(x)}$, and to reduce the scale of the spanwise fluctuation, $L_{ww}^{(x)}$, for all heights from the boundary. The increase of $L_{uu}^{(x)}$ indicates elongation with shear of the streamwise extent of the streaks. The near-wall behavior of the integral length scales can be found from (3.5)–(3.8) and (3.10):

$$\left.\begin{aligned} &L_{uu}^{(x)}(y\to 0) = L_{uu}^{(z)}(y\to 0) = 1/\overline{u^2}, \\ &L_{ww}^{(x)}(y\to 0) = L_{ww}^{(z)}(y\to 0) = 1/\overline{w^2}, \\ &L_{vv}^{(x)}(y\to 0) \sim y/\overline{v^{2(H)}} \end{aligned}\right\} \quad \text{in USBL.} \qquad (3.14a\text{–}c)$$

Near the boundary the scales of the horizontal components decrease with shear, but $L_{vv}^{(x)}$ increases with shear.

Figure 14 shows the profiles of the spanwise integral scales in USBL (for comparison with SFBL, see Fig. 12, where $L_{uu}^{(z)} = L_{ww}^{(x)}$, $L_{vv}^{(z)} = L_{vv}^{(x)}$, $L_{ww}^{(z)} = L_{uu}^{(x)}$). For heights away from the boundary, the scales of the streamwise velocity fluctuation, $L_{uu}^{(z)}$, and of the spanwise fluctuation, $L_{ww}^{(z)}$, are markedly decreased with shear, but the scale of the vertical motion, $L_{vv}^{(z)}$, is increased. Near the boundary, however, $L_{uu}^{(z)}$ (and $L_{ww}^{(z)}$) increases with shear from the unsheared value $\frac{2}{3}$. This means that since $L_{uu}^{(z)}$ may be interpreted as the characteristic scale for the spanwise extent of the streamwise eddies (streaks), at given total shear the spanwise extent of the streaky eddies widens as the boundary is approached, which is consistent with the reduced bending of vortex near the boundary (see Fig. 3b).

We note that $L_{vv}^{(z)}$ is the smallest of all the integral scales in the vicinity of the boundary, and thus it may be a good estimate for the dissipation scale L_ε. Figure 15 shows the profile of $L_{vv}^{(z)}$ at different values of total shear. The spanwise scale of the vertical fluctuation varies in a self-similar way near the boundary: $L_{vv}^{(z)} \sim A_\varepsilon y$, where $A_\varepsilon = A_\varepsilon(\beta)$ decreases with shear.

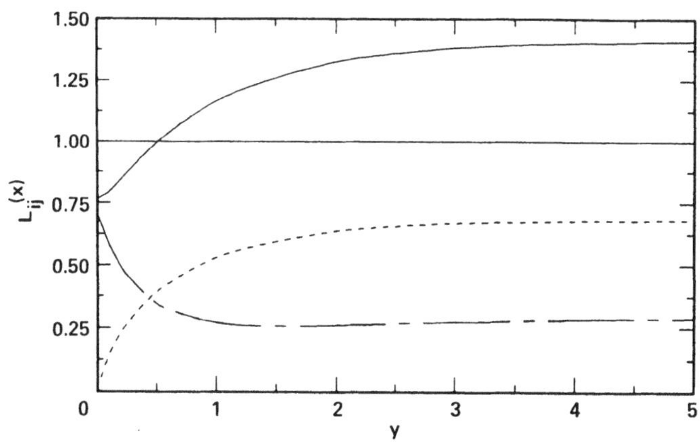

Fig. 13. Profile of the streamwise integral scales $L_{ij}^{(x)}(y)$ for USBL ($\beta = 2$). ——, $L_{uu}^{(x)}$; ----, $L_{vv}^{(x)}$; —-—, $L_{ww}^{(x)}$

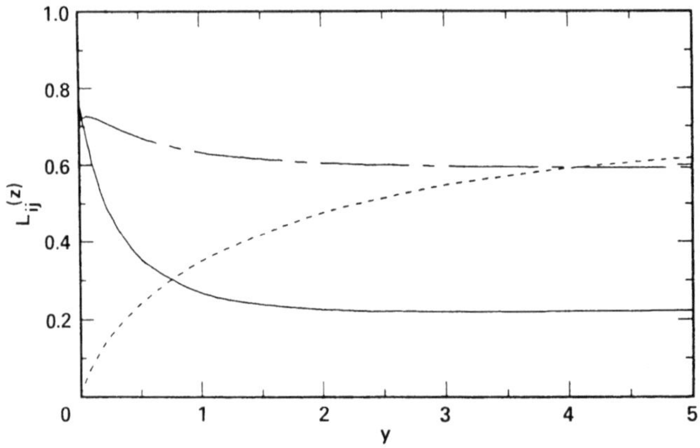

Fig. 14. Profile of the spanwise integral scales $L_{ij}^{(z)}(y)$ for USBL ($\beta = 2$). ——, $L_{uu}^{(z)}$; - - - -, $L_{vv}^{(z)}$; —-—, $L_{ww}^{(z)}$

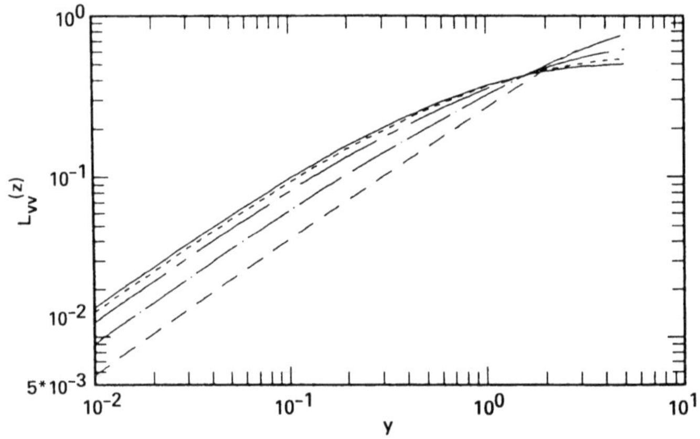

Fig. 15. Variation with total shear of the profile of $L_{vv}^{(z)}$. ——, $\beta = 0$; -----, $\beta = 1$; —·—, $\beta = 2$; ········, $\beta = 4$; ——, $\beta = 8$

4. Concluding Remarks

We have studied the effects of shear on turbulence structure near a plane boundary. It has been shown that the blocking of the vertical component of turbulence by the boundary is only slightly affected by the presence of shear (which is consistent with Fig. 1). However, the shear significantly changes the way in which the boundary affects the horizontal components $\overline{u^2}$, $\overline{w^2}$ and the Reynolds stress $-\overline{uv}$. In the SFBL the blocking (or splat) effect leads to the horizontal components being amplified. In a USBL the reduction in $\overline{v^2}$ near the boundary reduces the production of $-\overline{uv}$ and $\overline{u^2}$ by more than the splat effect increases $\overline{u^2}$. The physical explanation is given in Sect. 2.

We plan to look into other aspects of the surface blocking in sheared turbulence. In order to determine the relative effects of the wall blocking and shear on the vertical length scale, it would be of interest to examine the profiles of the two-point correlations of the vertical component of velocity at different distances from the surface and of the streamwise and vertical components.

A remarkable recent result has been the demonstration that applying the linear (RDT) distortion to an actual realisation of initially homogeneous isotropic velocity fields leads to velocity fields that contain many major large-scale coherent structures (Lee et al. 1987). These structures appear very *similar* in their scale, shape and distribution to structures that have been found in the direct simulations of homogeneous (uniform) shear flows and in the simulations and experiments of turbulent sheared boundary layers. This similarity implies that RDT is a useful way of computing the structures in shear flows. We intend to study the kinematical properties of the structures in uniform shear over a rigid surface to see how the structures are affected by the blocking effects.

References

Bertoglio, J. P. (1986): Etude d'une turbulence anisotrope, modélisation de sous-maille et approche statistique. Thése de Doctorate d'Etat, Université Claude-Bernard Lyon: Lyon, France

Bradshaw, P. (1967): 'Inactive' motion and pressure fluctuations in turbulent boundary layers. J. Fluid Mech. **30**, 241–258

Durbin, P. A. (1979): Rapid distortion theory of turbulent flows. Ph.D. Thesis, University of Cambridge: Cambridge, England

Gartshore, I. S., Durbin, P. A., Hunt, J. C. R. (1983): The production of turbulent stress in a shear flow by irrotational fluctuations. J. Fluid Mech. **137**, 307–329

Hunt, J. C. R. (1973): A theory of turbulent flow round two-dimensional bluff bodies. J. Fluid Mech. **61**, 625–706

Hunt, J. C. R. (1984): Turbulence structure in thermal convection and shear-free boundary layers. J. Fluid Mech. **138**, 161–184

Hunt, J. C. R., Graham, J. M. R. (1978): Free-stream turbulence near plane boundaries. J. Fluid Mech. **84**, 209–235

Hunt, J. C. R., Moin, P., Moser, R. D., Spalart, P. R. (1987a): Self-similarity of two-point correlations in wall-bounded turbulent flows. *Proc. of the 1987 Summer Program*, Center for Turbulence Research, Stanford University and NASA-Ames Research Center, pp. 25–36

Hunt, J. C. R., Spalart, P. R., Mansour, N. N. (1987b): A general form for the dissipation length scale in turbulent shear flows. *Proc. of the 1987 Summer Program*, Center for Turbulence Research, Stanford University and NASA-Ames Research Center, pp. 179–184

Hunt, J. C. R., Stretch, D. D., Britter, R. E. (1988): Length scales in stably stratified turbulent flows and their use in turbulence models. *Proc. IMA Conf. on Stably Stratified Flow and Dense Gas Dispersion* (J. S. Puttock, ed.), Chester, England, April 1986, pp. 285–321, Clarendon Press: Oxford, England

Kim, J., Moin, P., Moser, R. D. (1987): Turbulence statistics in fully-developed channel flow at low Reynolds number. J. Fluid Mech. **177**, 133–166

Kline, S. J., Reynolds, W. C., Schraub, F. A., Runstadler, P. W. (1967) The structure of turbulent boundary layers. J. Fluid Mech. **30**, 741–773

Lee, M. J., Kim, J., Moin, P. (1987): Turbulence structure at high shear rate. *Sixth Symp. on Turbulent Shear Flows*, Toulouse, France, Sept. 7–9, 1987 (F. Durst et al., ed.), pp. 22.6.1–22.6.6

Lumley, J. L. (1978): Computational modeling of turbulent flows. Adv. Appl. Mech. **18**, 123–176

Maxey, M. R. (1978): Aspects of unsteady turbulent shear flow, turbulent diffusion and tidal dispersion. Ph.D. Thesis, University of Cambridge: Cambridge, Engalnd

Spalart, P. R. (1988): Direct simulation of a turbulent boundary layer up to $R_\theta = 1410$. J. Fluid Mech. **187**, 61–98

Thomas, N. H., Hancock, P. E. (1977): Grid turbulence near a moving wall. J. Fluid Mech. **82**, 481–496
Townsend, A. A. (1961): Equilibrium layers and wall turbulence, J. Fluid Mech. **11**, 97–126
Townsend, A. A. (1976): *The structure of turbulent shear flow.* 2nd edn. Cambridge University Press: Cambridge, England
Uzkan, T., Reynolds, W. C. (1967): A shear-free turbulent boundary layer. J. Fluid Mech. **28**, 803–821
Wong, H. (1987): Turbulence near angled and curved surfaces. Ph.D. Dissertation, University of Cambridge: Cambridge, England

Study on a Turbulent Boundary Layer Disturbed by a Cylinder—Effect of Cylinder Size and Position

K. Suzuki, H. Suzuki, Y. Kikkawa, H. Kigawa and Y. Kawaguchi

Department of Mechanical Engineering, Kyoto University, Kyoto 606, Japan

Abstract

An experimental study was performed for a flat plate turbulent boundary layer disturbed by a cylinder. The measurement of wall heat transfer and skin friction coefficients was carried out for three different cylinder diameters and various values of the clearance between the cylinder and the flat plate. The streamwise distributions of these quantities showed that the clearance-to-diameter ratio is the most important factor. In almost all cases except for large clearance-to-diameter ratios, the existence of dissimilarity between momentum transfer and heat transfer was demonstrated. Simultaneous measurement of streamwise and normal components of velocity fluctuations was also carried out. The distributions of mean velocities, turbulent intensities, turbulent shear stresses and studied features of coherent structures were found to significantly depend on the clearance-to-diameter ratio.

Nomenclature

C	Clearance between the cylinder and the flat plate [mm]
C_f	Skin friction coefficient
d	Diameter of the cylinder [mm]
E_{11}	One-dimensional energy spectrum [m^2/s^2]
H	Distance between the cylinder axis and the flat plate [mm]
h_x	Local heat transfer coefficient [W/m^2K]
I_i	Indicator for quadrant analysis
k	Turbulence kinetic energy [m^2/s^2]
k_1	Wave number [cm^{-1}]
U	Streamwise mean velocity [m/s]
U^+	Non-dimensional U normalized with friction velocity
u	Streamwise fluctuating velocity [m/s]
u^*	Friction velocity [m/s]
$\overline{uv}$	Turbulent shear stress [m^2/s^2]
$\widetilde{uv}_i$	Fractional contribution to turbulent shear stress from the ith quadrant of u–v plane [m^2/s^2]
v	Normal fluctuating velocity [m/s]
x	Streamwise distance measured from the cylinder [mm]
x_m	Position of the second peak of h_x distribution [mm]
y	Normal distance from the flat plate [mm]

y^+ Non-dimensional y normalized with friction velocity
δ Boundary layer thickness [mm]
ν Kinematic viscosity of air [m^2/s]

Suffix

0 Undisturbed boundary layer
e Free stream
$'$ Intensity

Introduction

A turbulent boundary layer disturbed by an insertion of obstacle like a cylinder has been studied intensively by several research groups from different view points. Marumo et al. discussed the relaxation processes of averaged flow and turbulent fields [1–3]. Similar experiments were also made by Tsiolakis et al. [4]. These experimental data were used in the test of turbulence models [5, 6]. Heat transfer experiments have been made by Marumo et al. [7] and Fujita et al. [8] for the case with a single cylinder and by Kawaguchi et al. [9] for the case with multiple cylinders. Discussion has been given even on the dissimilarity existing between momentum and heat transfer processes from structural view point [10, 11]. However, there is no systematic study on the effects of cylinder size and position on the phenomenon. Fujita et al. [12] carried out heat transfer measurements but there is no data available for the discussion of the effects of cylinder size and position on momentum transfer or on turbulence structure. Moreover, their study was not made in an ideal condition. The thermal boundary layer was much thinner than the velocity boundary layer in their study so that a certain level of dissimilarity already existed at the insertion position of cylinder. This might have contaminated the dissimilarity caused by the introduction of cylinder.

In the present study, measurements were made on wall heat transfer, wall skin friction, averaged velocity and turbulent quantities by changing cylinder position variously for three cases of different cylinder diameters.

Experimental Apparatus and Procedure

The presently used experimental apparatus was basically the same as used in the previous studies [7], so that its outline is described here.

Figure 1 shows a schematic view of the wind tunnel. The boundary layer developed on the floor of the tunnel is sucked into the underneath of the flat plate by an auxiliary blower 11. This establishes the cross-sectional uniformity of the flow at the leading edge of the flat plate. Streamwise variation of the static pressure was minimized by changing the slope of the ceiling of the wind tunnel 5. The magnitude of pressure non-uniformity was kept within $\pm 1\%$ of the main stream dynamic

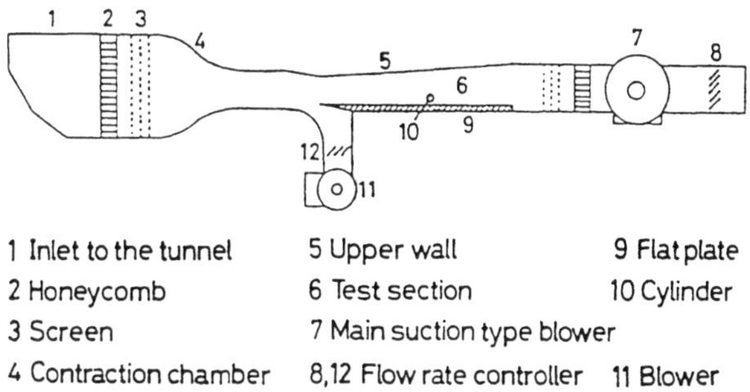

Fig. 1. Wind tunnel

pressure except for an inevitable abrupt pressure change at the streamwise position of the cylinder. The main flow velocity was 14 m/s and the level of turbulent intensity was below 1%.

The test section is shown in Fig. 2 together with the definition of the coordinates x and y, and of some geometric parameters H, C and d. The first part of the flat plate made of bakelite (from 10 to 200 mm downstream from the leading edge) was covered by a sand paper to trip the boundary layer. Heating of the surface started from a position of 213 mm downstream from the leading edge. Thin stainless steel sheets of 50 μm thickness were glued on the surface of the flat plate. These sheets were heated by passing an alternating electric current. For measuring the streamwise distribution of the surface temperature, ninety alumel-chromel thermocouples of 0.1 mm diameter were attached to the back surface of the middle one of heated sheets. Other twenty-two thermocouples were attached to the back surface of the bakelite plate. Assuming one-dimensional heat conduction inside the bakelite plate, the temperature difference across the plate was used to calculate the rate of conductive heat loss through the plate. For the calculation of radiative heat loss, the absorptivity of the metal sheet was assumed to be 0.2. The conductive and radiative heat transfer rate were subtracted from the electric heat generation rate to yield the net convective heat transfer rate. The local heat transfer coefficient measured this way for the case without an insertion of cylinder was found to agree well with an empirical equation given by Johnson and Rubesin [13] for a flat plate boundary layer. All the experiments were carried out for the heating condition of wall heat flux of about 1 kW/m^2.

A cylinder made of bakelite was located at a streamwise position of 1400 mm downstream from the leading edge of the flat plate. Both thicknesses of velocity and

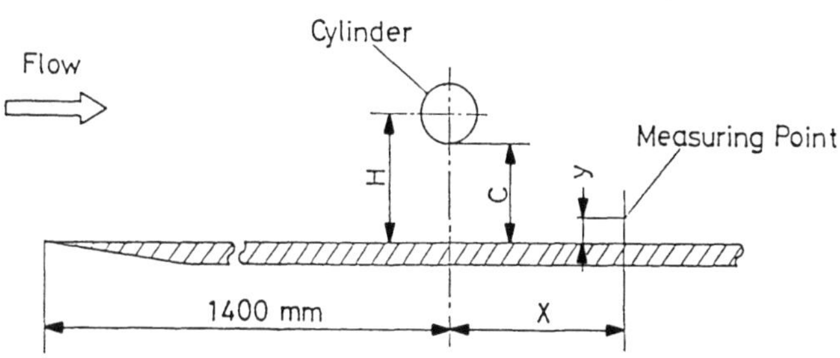

Fig. 2. Test section

temperature boundary layers at the cylinder position were about 28 mm when no cylinder was inserted. Cylinder diameter, *d*, was changed in three steps i.e., 5, 8 and 10 mm, within a range smaller than a half of thickness of undisturbed boundary layer.

The measurement of skin friction was carried out by making use of Preston tube. Relationships recommended by Patel [14] were used to transform the reading from the Preston tube into the skin frction. This was confirmed in the previous study [2] to produce no significant error except in a small region just downstream of cylinder.

For velocity measurement, a hot-wire anemometry was applied. Streamwise and normal velocity components were measured simultaneously with an X-type hot-wire probe used at constant temperature mode. Two tungsten wires of 5 μm in diameter and 1 mm in length were soldered on the prongs keeping the distance between the two wires 0.6 mm. The measurement was made at five streamwise stations of different distance from the cylinder. The output signals from the anemometers were digitized and sampled on line at the rate of 10 kHz for each signal and recorded over about 20 s on a magnetic disc with an aid of a PDP-11/23 minicomputer. Data processing was made with a computer at the Kyoto University Data Processing Center.

Measurement of heat transfer and skin friction coefficients was made [15] for nineteen cases of different combinations of cylinder diameter and clearance. Averaged velocity and turbulence measurements were made for nine cases of different combinations of cylinder diameter and clearance. All the cases for which measurements were made are tabulated in Table 1.

Table 1. Dimemsion of the diameter and position of a cylinder

d	H(mm)	C(mm)	C/d
	4.5	2.0	0.4
	5.0	2.5	0.5
	6.0	3.5	0.7
5	7.5	5.0	1.0
	12.5	10.0	2.0
	15.0	12.5	2.5
	33.5	31.0	6.2
	6.0	2.0	0.25(7)
	8.0	4.0	0.5
8	12.0	8.0	1.38(7)
	20.0	16.0	2.0
	33.5	29.5	3.69(7)
	6.0	1.0	0.1
	7.0	2.0	0.2
	10.0	5.0	0.5
10	15.0	10.0	1.0
	25.0	20.0	2.0
	33.5	28.5	2.85

Some Statical Features of the Disturbed Boundary Layer

Before entering into the main part of discussion, general features of the boundary layer disturbed by a cylinder is described in this section. For this purpose, the statistical results obtained by one of the present authors [1–3] with a cylinder of 8 mm will briefly be summarized.

Figure 3 shows several examples of semi-logarithmic plot of mean velocity profiles measured by separate type Pitot-static tubes for three cases of different normal position of cylinder [1]. y^+ and U^+ are respectively the non-dimensional normal distance from the plate and streamwise velocity normalized with the friction velocity. Velocity defect is clearly found in the velocity profiles at the stations just downstream of the cylinder. However, it is found in every case of different normal position of cylinder that the mean velocity profile obtained at $x = 837$ mm ($x/d = 104.6$) is very close to the counterpart for normal flat plate boundary layer. Therefore, the distance of one-hundred cylinder diameter is concluded to be long enough for the mean velocity field to accomplish its relaxation process.

Figure 4 shows the distribution of turbulence kinetic energy, k, in the case of $C = 2$ mm ($H = 6$ mm) [2]. The profile obtained at the first station $x = 37$ mm ($x/d = 4.5$) just downstream of the cylinder is characterized by a noticeable peak suggesting an intensive turbulence field behind the cylinder. The height of the peak becomes lower with an increase of x so that the relaxation of turbulence field also proceeds towards the downstream. However, the profile obtained at the last station $x = 837$ mm is still a little different from the profile of k for an undisturbed case, i.e. normal flat plate boundary layer. The latter shows a monotonical decrease of k towards the outer

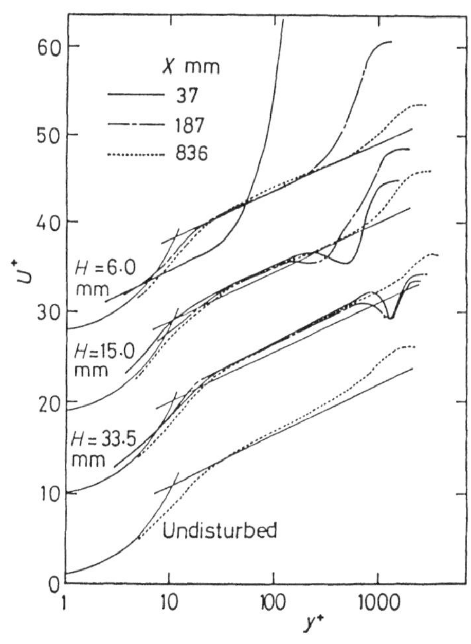

Fig. 3. Distribution of mean velocity ($d = 8$ mm) [1, 2]

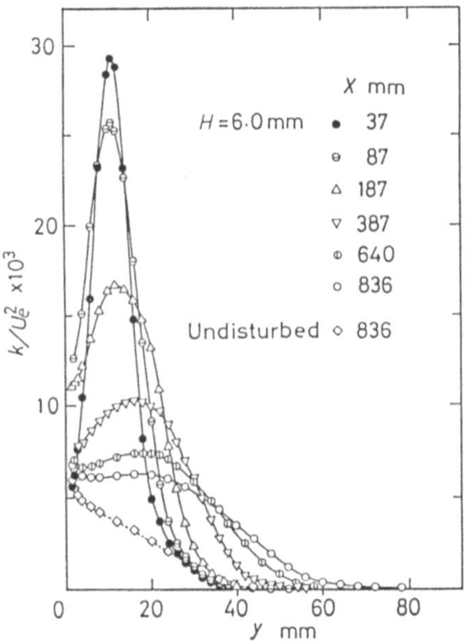

Fig. 4. Distribution of kinetic energy ($d = 8$ mm) [2]

edge of boundary layer while the profile obtained at the last station for the disturbed boundary layer still has a weak peak. In this sense, relaxation is a little slower in turbulence field than in averaged velocity field.

The cylinder Reynolds number, evaluated with the cylinder diameter and the velocity of an approaching flow, ranges from 1,500 to 8,000 depending on the cylinder size and position. When a clinder is located in a uniform flow, Karman vortex street appears in this Reynolds number range. This is also the case of the presently studied disturbed boundary layer unless the cylinder is located very close to the flat plate. Figure 5 shows some examples of one-dimensional spectrum of streamwise fluctuating velocity, u, obtained in the case of $C = 11$ mm ($H = 15$ mm) [3]. All the spectra obtained at the most upstream position $x = 37$ mm have a noticeable peak or peaks. The peak found in the spectrum is due to the existence of periodical velocity fluctuation related to the appearance of Karman vortices shedded from the cylinder. However, peak is very inconspicuous, if any, for all the spectra obtained at the next station $x = 187$ mm ($x/d = 23.4$). This indicates that the periodically fluctuating velocity field has almost vanished within the distance of about twenty cylinder diamter in this case. This suggests that the relaxation of periodically fluctuating velocity field proceeds much faster than the averaged velocity field and turbulence field do.

Results and Discussion

Now, the results of recent study are discussed. In this part of discussion, some experimental data discussed in Reference [15] will also be included. First, the streamwise distributions of skin friction coefficient and heat transfer coefficient will

		y mm					
X mm	37	1.5	8	14	20	26	32
	187	1.5	8	14	20	26	32
	387	1.5	10	15	25	—	35
	836	1.5	10	25	—	—	40
Symbols		○	⊖	△	▽	⦶	●

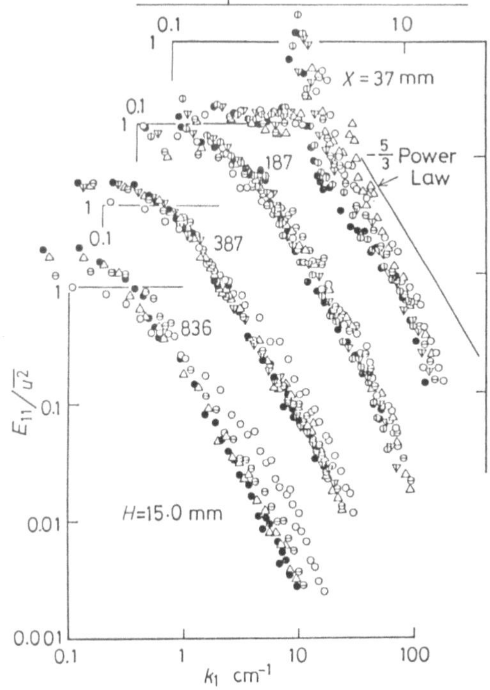

Fig. 5. Distribution of one-dimensional spectrum [3]

be discussed. Then, some additional features of turbulence quantities will be described and finally some characteristics of turbulence structure will be discussed based on the results of quadrant analysis applied to the recorded velocity signals.

Skin Friction Coefficient

The streamwise distributions of skin friction coefficient, C_f, measured for nineteen cases of different cylinder diameters and different cylinder positions were presented in reference [15]. Figure 6 shows some example of such data measured for the case of the cylinder diameter of $d = 8$ mm, and for different values of the clearance ratio, C/d. The value of C_f measured by making use of a Preston tube is not correct in and around the recirculating flow region. Since a separated flow bubble is intermittently produced behind the cylinder in the cases of small value of C/d [16], the data

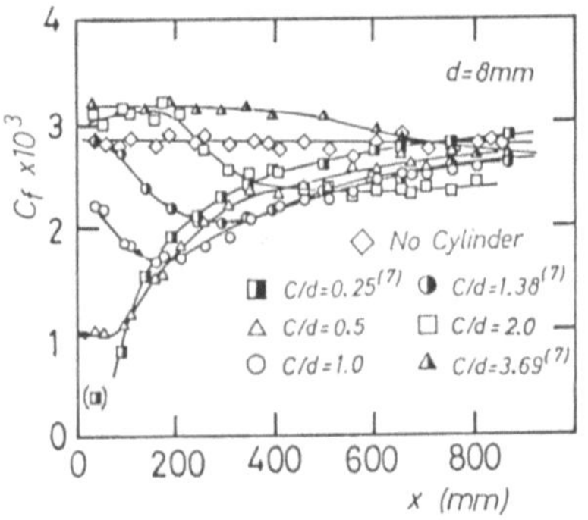

Fig. 6. Distribution of skin friction coefficient ($d = 8$ mm) [15]

obtained just downstream of the cylinder in such cases are supposed to be inaccurate. In this figure, such uncertain data are shown with the symbols in parentheses. It is found in this figure that the shape of C_f distribution changes systematically with a change of C/d.

Figure 7 shows a comparison among three sets of C_f distribution obtained with different cylinder diameters but with the same clearance ratio C/d. This figure clearly demonstrates that the cylinder diamter is a good scale available to normalize the streamwise distance in the comparisons of C_f distributions and that the clearance ratio, C/d, is a good parameter for the correlation of C_f data obtained for different cylinder positions and with different cylinder diameters. Therefore the streamwise distributions of C_f were found to be different among the cases of different cylinder diameter even if the cylinder positions C and H were the same. The thickness of undisturbed velocity boundary layer, δ_0, and the wall length scale, u_0^*/v were kept constant in the present study. Therefore, the ratio between C or H and δ_0 or u_0^*/v is not expected to significantly affect the streamwise distribution of C_f either.

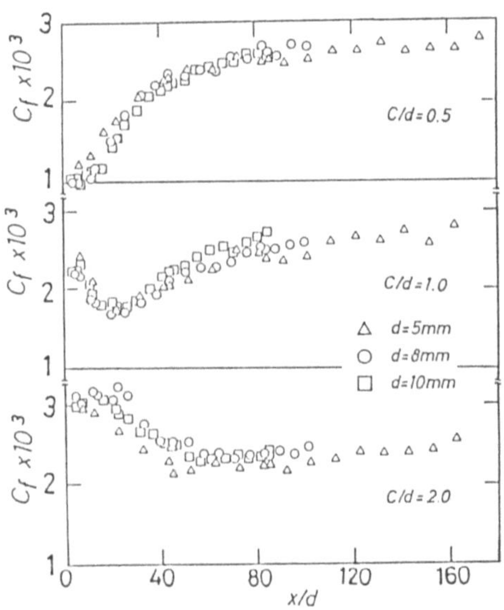

Fig. 7. Distribution of skin friction coefficient [15]

Heat Transfer Coefficient

Figure 8 shows some of the distributions of heat transfer coefficient discussed in reference [15]. These were measured for different clearance ratios but with a cylinder of the same size, i.e. 8 mm diameter. h_x and h_{0x} in the ordinates are the measured heat transfer coefficients, respectively, for the case with an insertion of cylinder and for the case without an insertion of cylinder. It is clear in this figure that heat transfer is enhanced in the downstream of cylinder in opposition to the reduction in skin friction. This opposite tendency of the change of distribution between C_f and h_x data is caused by the introduction of cylinder. It indicates the existence of dissimilarity momentum transfer and heat transfer. This kind of dissimilarity between momentum transfer and heat transfer was confirmed to exist for other cases of different cylinder diameters [15]. Previously, this dissimilarity was studied from the structural view point of turbulence [11]. Similar but more extensive discussions will be given later.

The shape of the distribution of heat transfer coefficient is found to vary systematically with a change of the clearance ratio, C/d. Therefore, the ratio C/d is expected to be a similitude parameter of the concerned heat transfer problem as well as of the wall skin friction, although dissimilarity exists between the momentum transfer and heat transfer as discussed in the above. This is confirmed in Fig. 9. The heat transfer data obtained for the same value of C/d but for three different cylinder diameters show close similarity with each other. The present data are also compared in the figure with the data obtained by Fujita et al. [12]. Their study was for the case when the thermal boundary layer was much thinner than the velocity boundary layers. From this comparison, it is concluded that the dissimilarity to be caused by the difference between the velocity and temperature fields in the thickness of approaching boundary layer does not significantly affect the dissimilarity between the heat and momentum transfer existing in the present type of disturbed boundary layer, which is caused by the introduction of cylinder [11].

Marumo et al. [7] pointed out three different mechanisms of heat transfer enhancement to occur in the turbulent boundary layer disturbed by a cylinder. One

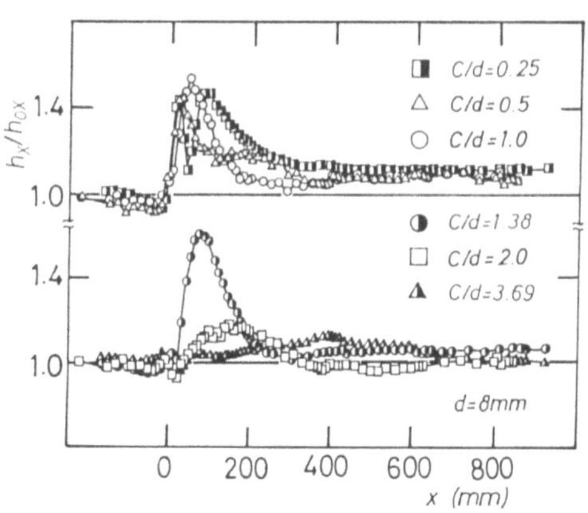

Fig. 8. Distribution of local heat transfer coefficient ($d = 8$ mm) [15]

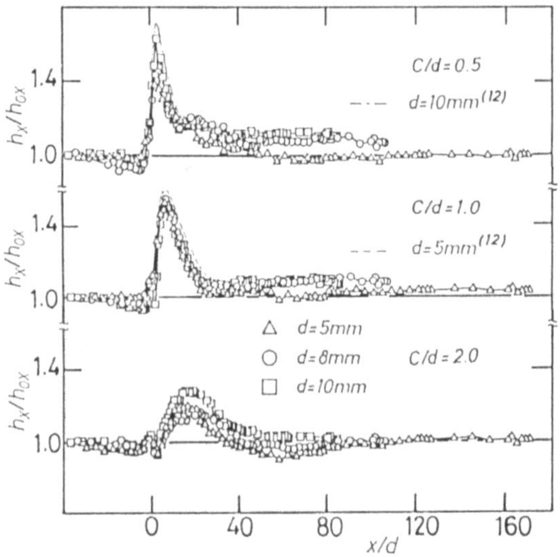

Fig. 9. Distribution of local heat transfer coefficient [15]

of them is the heat transfer enhancement resulting from the arrival of intense turbulence generated in the lower shear layer of the cylinder wake to the near wall region.

Based on their experimental data obtained with a cylinder of 8 mm diameter, they proposed an equation for a streamwise position of the peak of heat transfer coefficient, x_m, caused by such mechanism of heat transfer augmentation [7]. The proposed equation reads as follows.

$$x_m = \frac{(C + d/2)^2}{1.9^2} + 13.6\,(\text{mm}) \tag{1}$$

Paying attention to the fact that the two numerical factors in the equation should have the dimension of length and that the cylinder diameter, d, is a good scale for the

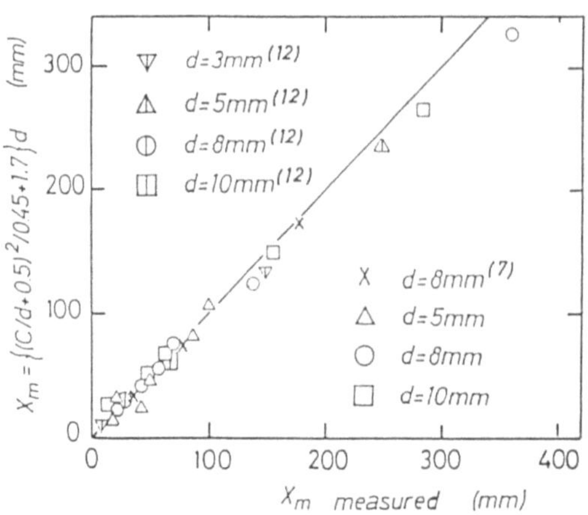

Fig. 10. Position of the peak of heat transfer coefficient [15]

normalization of streamwise distance, the following general form of equation was derived in the present study.

$$x_m/d = \frac{(C/d + 0.5)^2}{0.45} + 1.7 \tag{2}$$

Figure 10 shows the present experimental data of x_m with some others measured by Marumo et al. [7] and by Fujita et al. [12] in comparison with Eq. (2). Equation (2) agrees very well with the plotted experimental data. This equation clearly indicates that x_m is a function of C/d. Therefore, it is confirmed again that the ratio C/d is an important factor affecting the heat transfer characteristics.

Turbulent Characteristics

Statistical examination was again given to the two signals of streamwise and normal velocity components. In the present study, the measurements were made for the cases of different sizes of cylinders at the positions of the same value of (x/d).

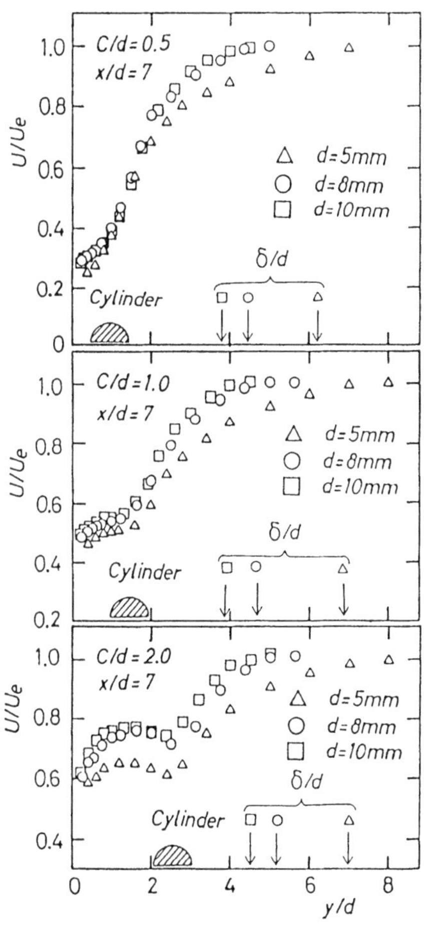

Fig. 11. Distribution of mean velocity [15]

Figures 11, 12 and 13 show normal distributions of averaged velocity, turbulence intensities of two velocity components and turbulence shear stress, respectively. They were obtained for the cases of three different cylinder diameters but for the same clearance ratio. Although the data were obtained at five streamwise stations of different (x/d) ratio, only the data obtained at $x/d = 7$ were plotted in these figures just as an example. At this position, noticeable enhancement of heat transfer and reduction of skin friction occur so that noticeable dissimilarity exists between the momentum transfer and heat transfer. The chosen streamwise position is rather close to the cylinder so that the outer edge of the cylinder wake has not yet reached completely to the boundary layer edge, especially in the case of the smallest clearance ratio. Thus, some difference is noticeable in the outer region among the

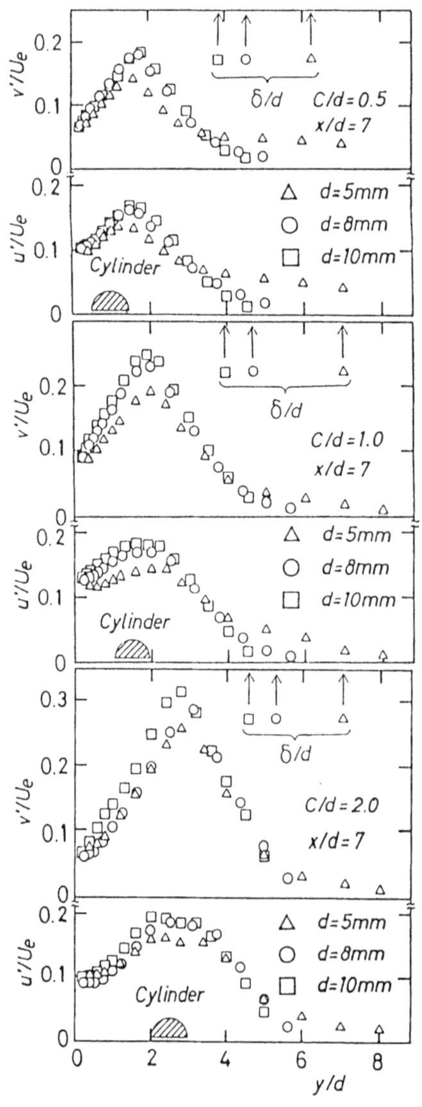

Fig. 12. Distribution of turbulent intensity [15]

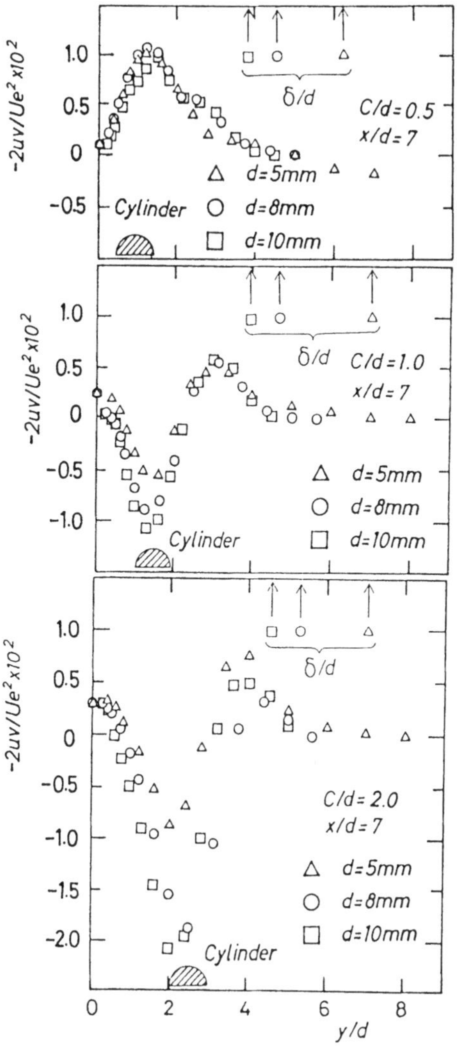

Fig. 13. Distribution of turbulent shear stress

plotted results for different cylinder diameters. However, all the profiles obtained for three different cylinder diameters show close resemblance in shape with each other in the inner half of the boundary layer. This would be the basis of systematic change of the distributions of heat transfer coefficient and skin friction coefficient plotted against (x/d) with a variation of C/d. Therefore, the ratio C/d is regarded to be an important parameter also from the view points of averaged velocity and of turbulence fields.

Quadrant Analysis

In the present study, $u-v$ plane quadrant analysis [17] was again applied to the instantaneously measured u and v signals. In order to classify the instantaneous fluid motions in the $u - v$ plane, an indicator function I_i was defined as

$$I_1 = \begin{cases} 1 \text{ (when the signs of } u \text{ and } v \text{ correspond to the } i\text{th quadrant} \\ \quad \text{of the } u - v \text{ plane} \\ 0 \text{ (otherwise)} \end{cases}$$

Instantaneous fluid motions are classified into one of the following four quadrants as follows:

1st quadrant ($u > 0$, $v > 0$); outward interaction
2nd quadrant ($u < 0$, $v > 0$); ejection-like motion
3rd quadrant ($u < 0$, $v < 0$); wallward interaction
4th quadrant ($u > 0$, $v < 0$); sweep-like motion

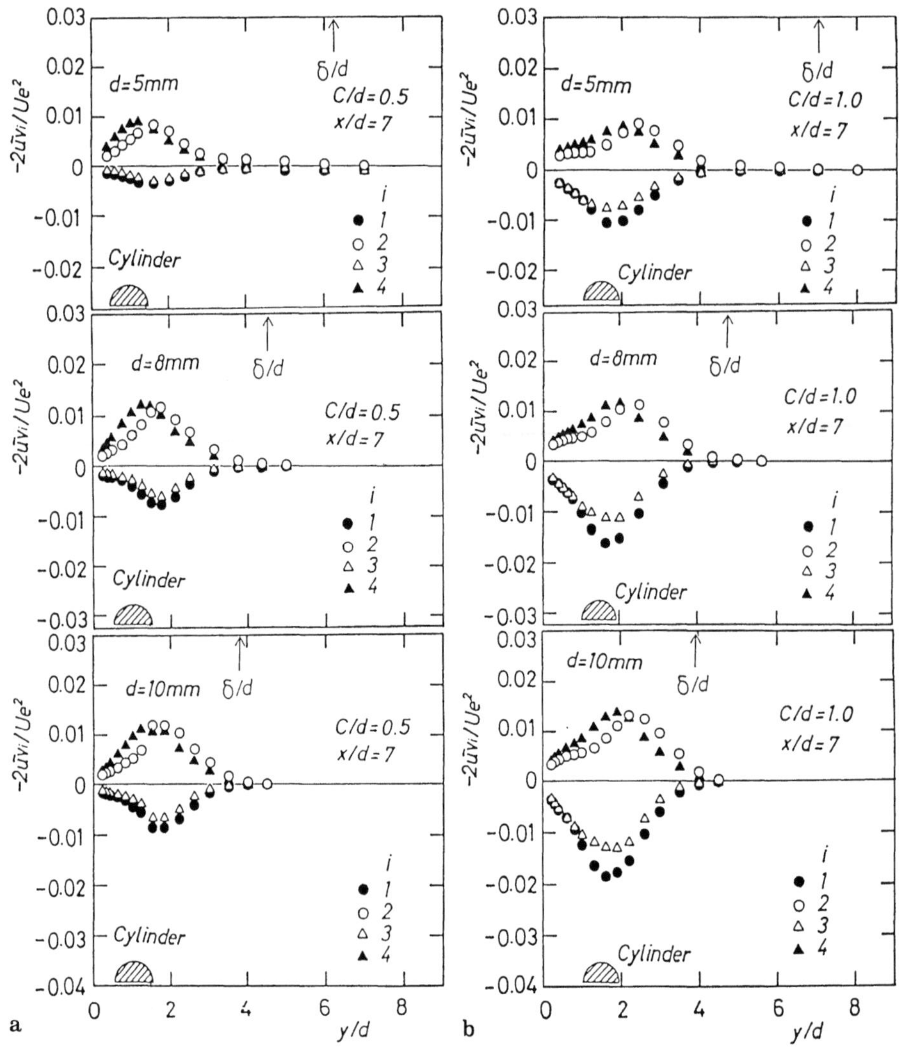

Fig. 14 a–c. Fractional contribution from the ith quadrant

Fluctuations classified into the 2nd and 4th quadrants contribute positively to $-\overline{uv}$ and those in the 1st and 3rd quadrants negatively to $-\overline{uv}$. In the previous study [10], the intensification of the two-types of interactive fluid motions (outward interactions and wallward interactions) caused by the introduction of cylinder was found to cause the enhancement of heat transfer while they lead to the reduction in skin friction. Therefore, interactive fluid motions stored in the 1st and 3rd quadrants were concluded to incur the dissimilarity between the turbulent heat and momentum transfer if they are intensified by the presence of cylinder.

Figure 14 shows an example of how the fractional contribution from each type of fluid motion to the Reynolds shear stress, $-\overline{uv}$, distributes throughout the boundary layer. These data were obtained again at the position of $x/d = 7$ for each

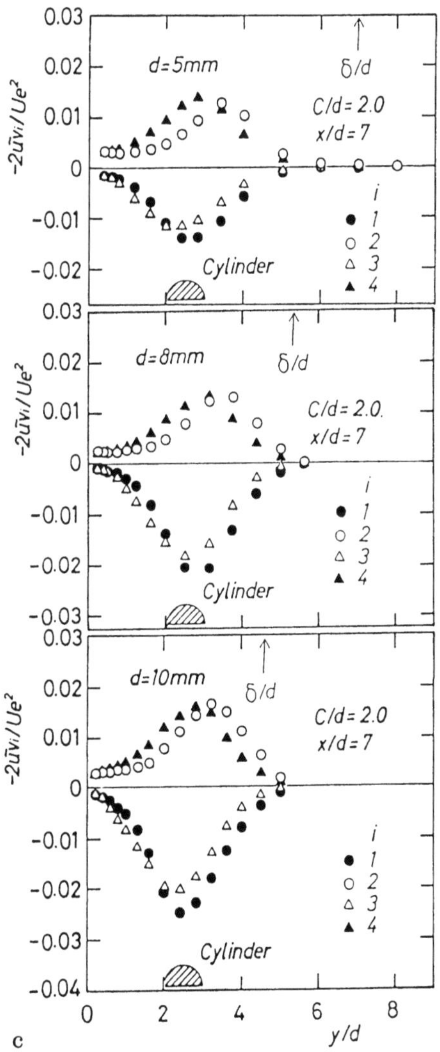

c

Fig. 14. *(Continued)*

clearance ratio. The fractional contribution from each type of fluid motion to the product uv is defined as

$$\widetilde{uv}_i = \overline{I_i uv}$$

where the overbar indicates an average over all the sampled data.

Comparison among each figure for different cylinder diameter but for the same clearance ratio C/d indicates that the shape of the distribution of the plotted fractional contribution from each type of elementary coherent fluid motion to the turbulent shear stress varies systematically with a change in C/d. The same was observed in the case of other values of C/d ratio and at other positions of different (x/d) ratio. Thus, it is clear that C/d is the most important factor affecting the shape of the distribution of the fractional contribution from elementary coherent fluid motion, therefore, affecting the manner of the dissimilarity between the heat transfer and momentum transfer in this disturbed boundary layer.

Concluding Remarks

First, general features of the relaxation process in the present type of disturbed boundary layer was reviewed based on the previously reported data. The present study was carried out to clarify what is most important among the factors affecting the shapes of the streamwise distributions of skin friction and heat transfer coefficients for the turbulent boundary layer disturbed by a cylinder. The measurement of skin friction and heat transfer coefficient was performed for different cylinder diameters and for various values of the clearance between the cylinder and the flat plate. In many cases, the existence of dissimilarity between the momentum transfer and heat transfer was confirmed. These experimental data also showed that the cylinder diameter is a good scale to normalize the streamwise distance in the comparison of the distributions of C_f and h_x data, and that the most important factor affecting the shapes of the distributions is the clearance-to-diameter ratio. Difference in the boundary layer thickness between the flow and thermal fields does not cause severe dissimilarity between the heat and momentum transport compared to the dissimilarity caused by the introduction of cylinder in this disturbed boundary layer. Simultaneous measurement of streamwise and normal components of velocity fluctuations was also carried out. The normal distribution of mean velocities, turbulent characteristics and some features of coherent structures were also found to significantly depend on the clearance-to-diameter ratio.

References

1. Marumo, E., Suzuki, K., Sato, T. (1978): J. Fluid Mech. **87**, 121–141
2. Marumo, E., Suzuki, K., Sasaki, T., Sato, T. (1980): Trans. JSME, **46-407**, 1211–1219
3. Marumo, E., Suzuki, K., Sasaki, T., Kinuta, H., Sato, T. (1980): Trans. JSME, **46-407**, 1220–1228
4. Tsiolakis, E. P., Krause, E., Muller, U. R. (1983): Proc. 4th Symp. Turbulent Shear Flows, pp. 204–215

5. Suzuki, K., Tanaka, J., Marumo, E., Sato, T. (1980): Trans. JSME, **46-408**, 1220–1228
6. Savill, A. M., Mumford, J. C. (1988): J. Fluid Mech. **191**, 389–418
7. Marumo, E., Suzuki, K., Sato, T. (1985): Int. J. Heat and Fluid Flow **6**, 241–248
8. Fujita, H., Takahama, H., Yamashita, R. (1976): Trans. JSME, **42-361**, 2828–2836
9. Kawaguchi, Y., Suzuki, K., Sato, T. (1985): Int. J. Heat and Fluid Flow, **6**, 249–255
10. Kawaguchi, Y., Matumori, Y., Suzuki, K. (1984): Proc. 9th Bennial Symp. on Turbulence, pp. 28.1–28.10
11. Suzuki, H., Suzuki, K., Sato, T. (1988): Int. J. Heat Mass Transfer **31**, 259–265
12. Fujita, H., Takahama, H., Eki, H. (1981): Trans. JSME, **47-414**, 317–325
13. Johnson, H. A., Rubesin, M. W. (1949): Trans. ASME, **71**, 447–456
14. Patel, V. C. (1965): J. Fluid Mech. **23**, 185–280
15. Suzuki, H., Kikkawa, Y., Kigawa, H., Suzuki, K. (1989): Trans. JSME, **55-519**, 3333–3340
16. Bearman, P. W., Zdravkovich, M. M. (1978): J. Fluid Mech. **89**, 33–47
17. Lu, S. S., Willmarth, W. W. (1973): J. Fluid Mech. **60**, 481–511

Some Characteristics of Bypass Transition in a Heated Boundary Layer

K. H. Sohn[1], J. E. O'Brien[2] and E. Reshotko[1]

[1] Department of Mechanical and Aerospace Engineering, Case Western Reserve University
[2] NASA Lewis Research Center

Abstract

Experimental measurements of both mean and conditionally sampled characteristics of laminar, transitional and low Reynolds number turbulent boundary layers on a heated flat plate are presented. Measurements were obtained in air over a range of freestream turbulence intensities from 0.3% to 6% with a freestream velocity of 30.5 m/s and zero pressure gradient. Conditional sampling performed in the transitional boundary layers indicate the existence of a near-wall drop in intermittency, especially pronounced at low intermittencies. Non-turbulent intervals were observed to possess large levels of low-frequency unsteadiness, and turbulent intervals had peak intensities as much as 50% higher than were measured at fully turbulent stations. Heat transfer results were consistent with results of previous researchers and Reynolds analogy factors were found to be well predicted by laminar and turbulent correlations which accounted for unheated starting length. A small dependence of the turbulent Reynolds analogy factors on freestream turbulence level was observed. Laminar boundary layer spectra indicated selective amplification of unstable frequencies. These instabilities appear to play a dominant role in the transition process only for the lowest freestream turbulence level studied, however.

Introduction

Boundary layer transition in a low-disturbance environment is characterized by a well-documented progression of initial amplification of two-dimensional Tollmien-Schlichting waves followed by the formation of three-dimensional structures and subsequent development of turbulent spots. Linear stability theory has been used to predict the critical Reynolds number above which amplification of small (linear) disturbances may occur, leading to the progression described above. Morkovin (1978) introduced the term bypass transition to describe the transition process which occurs in the presence of larger (nonlinear) disturbances. In the bypass transition process, the formation of turbulent spots may occur without the initial amplification of linear waves. Bypass transition occurs in high-disturbance flows such as those present in gas turbines.

Previous studies of the influence of freestream turbulence on boundary layer transition were primarily aimed at measuring mean characteristics of transitioning boundary layers (e.g., Blair 1982; Wang et al. 1985). The present study, along with several other very recent studies (Blair 1988; Kuan and Wang 1988; Kim et al. 1989), focuses additionally on determination of the separate statistics of the respective turbulent and non-turbulent parts of these intermittent boundary layers using

Turbulent Shear Flows 7

conditional sampling techniques. Limited results of conditional sampling in a transitional boundary layer were first reported by Arnal et al. (1978). Detailed studies of individual turbulent spot structure have also been performed using artificially generated turbulent spots in otherwise laminar boundary layers (Antonia et al. 1981; Cantwell et al. 1978; Wygnanski et al. 1976). Results of these studies, while elegant, are only applicable in the very early stages of transition, before coalescence of multiple spots occurs. The turbulent spots of the present study arose and merged "naturally" as a result of freestream turbulence.

The purpose of this study was to measure some of the characteristics of the boundary layer transition process on a heated flat plate for conditions which span a range of freestream turbulence levels. The highest freestream turbulence levels studied were unambiguous bypass cases since transition occurred at Reynolds numbers lower than the minimum critical Reynolds number predicted from linear stability theory. Detailed probing of the pre-transitional boundary layers was not possible in these cases, however, since transition occurred very close to the leading edge of the test surface. Intermediate turbulence levels produced post-critical transitions. These cases may still be termed bypass, however, in the sense that they do not follow the classical progression described above for low-disturbance environments. Detailed measurements presented in the present study will focus on these cases. The lowest freestream turbulence case studied produced a transition process which was dominated by the formation and subsequent breakdown of TS-type waves (see Suder et al. 1988).

The Experiments

The experimental program was carried out in a closed-loop wind tunnel. This tunnel is similar in design to the tunnel of Blair et al. (1981). The tunnel test section is rectangular in cross section and measures 15.2 cm high, 68.6 cm wide and 152.4 cm long. The smooth floor of the test section, which was uniformly heated using an inconel foil heater (0.025 mm foil thickness), serves as the measurement test surface. Surface temperatures were monitored by means of thermocouples spot-welded to the back side of the foil, which was bonded to a 5.08 cm thickness of polyurethane foam. At the entrance to the test section, a series of two upstream-facing scoops were used to bleed off the boundary layer which develops in the contraction nozzle. This arrangement results in a short unheated starting length (3.5 cm). The boundary layer which grows on the test surface is not intentionally excited in any way, other than by the freestream turbulence, and transition occurs naturally, with random arrival of turbulent spots at a given measurement station. Upstream of the test section and nozzle, a flow-conditioning section serves to remove any large-scale nonuniformities introduced by the blower and to reduce the freestream turbulence level. At the downstream end of the flow-conditioning section and upstream of the nozzle contraction, turbulence grids (rectangular-bar, 62% open area) could be positioned in order to set the freestream turbulence level in the test section. Six freestream turbulence conditions were available, ranging from 0.3% to 6%. The roof of the test section was adjustable in order to allow establishment of a desired streamwise pressure gradient. The experiments described herein were carried out with a freestream velocity of 30.5 m/s and zero pressure gradient. Further details on the facility may be obtained from Suder et al. (1988).

Boundary layer measurements of mean and fluctuating velocity and temperature were obtained along the tunnel centerline. Additional profile and intermittency measurements obtained at selected off-centerline locations verified tunnel spanwise uniformity. Surface measurements of mean heat flux and temperature were also acquired. Both overall and conditionally sampled measurements were made. Velocity and temperature profiles were acquired using a commercially available boundary-layer hot wire probe (TSI model 1218). For the present study, boundary layer velocity and temperature profiles were obtained at streamwise locations in the range 12.7 cm $< x <$ 50.8 cm. Typical boundary layer thicknesses were of the order of 6 mm and the minimum y value for each profile was about 0.15 mm.

Steady-state tunnel operating conditions were monitored by means of a multichannel data acquisition system which provided a control-room display of steady-state quantities such as temperatures, pressures, freestream velocity, etc. in engineering units with updates approximately every two seconds. Time-averaged velocity data (mean and rms) were acquired using a separate precision averaging digital multimeter set for 200 averages for each data point. High-speed data acquisition was accomplished using a multichannel 12-bit digital waveform recorder. The sampling rate was set at 50 kHz and 32 kbytes of data were stored for each waveform. Spectral data were acquired using a 1024-point 12-bit digital FFT analyzer with a sampling rate of 5 kHz. Each spectral curve represents at least 100 averages.

Results

Detailed heat transfer and boundary layer measurements have been obtained for six different levels of freestream turbulence. Nominal values of freestream turbulence intensity and integral length scale, Λ, for the various grid configurations are listed in Table 1. Detailed profiles presented in the present paper will focus on results obtained using grid 1. This case was chosen as the focus because the transition zone

Table 1. Nominal grid turbulence characteristics

Grid #	u'/u_e(%)	Λ (cm)
.5	0.6	0.8
1	1.0	1.3
2	2.0	1.5
3	4.2	2.1
4	5.0	2.3

for this turbulence level overlaps the region in which detailed measurements were obtained. Further details of the freestream turbulence measurements are available in Suder et al. (1988).

Momentum Boundary Layer

Typical development of the momentum boundary layer is illustrated in Fig. 1a, b. In Fig. 1a, normalized mean velocity profiles are plotted in Blasius coordinates for eight streamwise measurement stations along the tunnel centerline. The effect of boundary layer transition is evident from the increasing deviation of the profiles

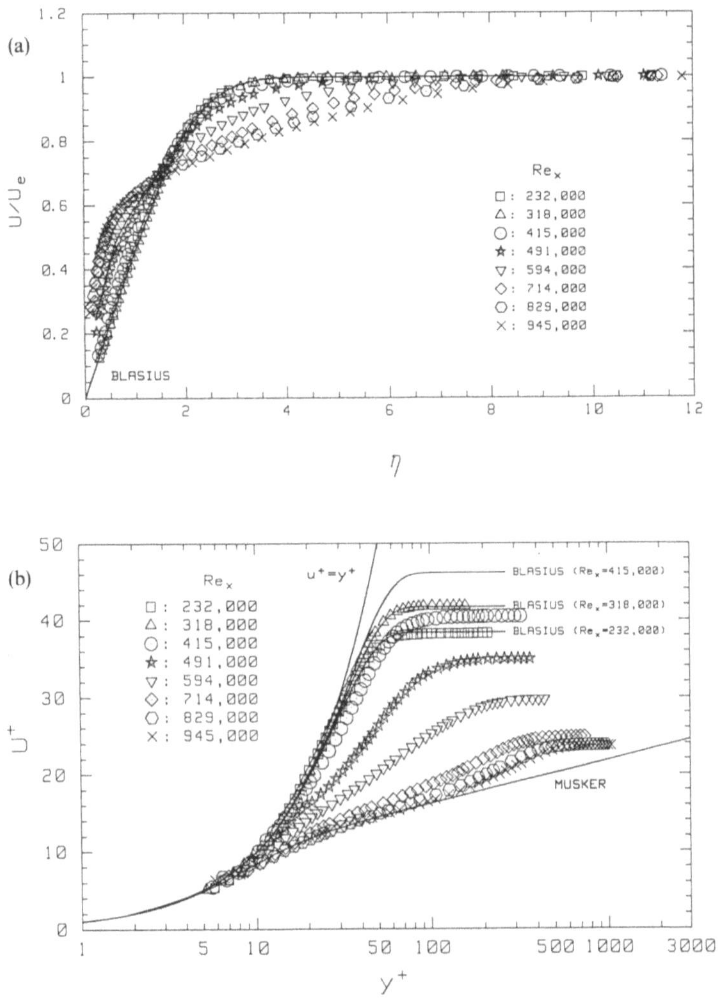

Fig. 1 a, b. Mean velocity profiles, grid 1; **a** Blasius coordinates; **b** wall units

from the theoretical laminar Blasius curve (solid line) at the downstream measurement stations. Excellent agreement of the laminar data with the Blasius curve is notable. The same profiles plotted in wall units are shown in Fig. 1b. In this figure, solid curves representing $u^+ = y^+$, the Blasius solution corresponding to three of the upstream measurement stations, and the Musker (1979) continuous law of the wall for a fully turbulent boundary layer are shown along with the experimental data. These profiles span nearly the entire range from laminar to fully turbulent flow. Note the excellent agreement of the two farthest upstream profiles with the Blasius curves. Values for u_τ required to construct this plot were inferred from the near-wall data points ($y^+ < 10$) where, in all cases, the data are tightly bounded by $u^+ = y^+$ and the Musker (1979) continuous law of the wall. In addition, the laminar values were checked against laminar theory and the fully turbulent cases were analyzed using a Clauser fit technique. A plot of the experimentally determined skin friction coefficients, C_f, versus momentum thickness Reynolds number for the entire set of freestream turbulence values is shown in Fig. 2. Also shown in the figure are solid lines corresponding to laminar theory and a fully turbulent correlation (Schlichting 1979). Transition is observed to occur at increasingly lower values of Reynolds number as the freestream turbulence level is increased.

Conditionally sampled boundary layer profiles were also acquired. The procedure used for accomplishing the necessary laminar/turbulent decisions is illustrated in Fig. 3. The top trace in the figure is a small portion of a digitally recorded near-wall velocity record (grid 1, $y \simeq 0.15$ mm). The velocity trace reveals some of the interesting characteristics of an intermittent boundary layer. For example, upon passage of a turbulent spot, the velocity achieves a level characteristic of a turbulent boundary layer very quickly, whereas following the return to laminar

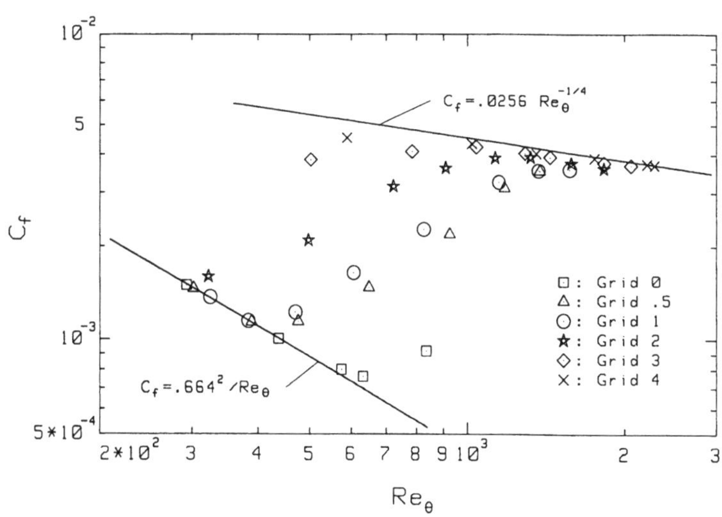

Fig. 2. Skin friction coefficients

transport, there is a relatively long time period required for relaxation to the "steady" laminar level. This behavior is indicative of the much higher transport efficiency associated with turbulence. It also indicates a potential stumbling block for simple intermittency-weighted transition models which assume Blasius-type behavior during laminar periods and fully-developed turbulent behavior during turbulent periods. The second trace in Fig. 3 is the square of the second derivative of the velocity trace. Nine-point smoothing is then applied, resulting in the third trace after which an appropriate threshold value is selected which results in the bottom trace, the indicator function. The square of the first derivative is also treated similarly in order to eliminate false laminar indications caused by second derivative zero-crossings. Attempts were made to adopt a completely numerical method to quantify the threshold selection process using the cumulative distribution function approach described by Hedley and Keffer (1974). This method proved only partially successful and fine tuning of the threshold values was required at each individual measurement location (x and y) in order to obtain an accurate indicator function as determined by careful direct comparison of each indicator function with the actual intermittent waveforms.

Determination of the indicator function allows for calculation of an intermittency value, Γ, (and its variation through the boundary layer) and for discrimination between turbulent and non-turbulent portions of the velocity record. Boundary layer intermittency profiles obtained with grids 1 and 2 are shown in Fig. 4a, b along with the error-function distribution of Klebanoff (1955) for a fully developed turbulent boundary layer. Consistent with recent results obtained by Kuan and Wang (1988), these profiles do not decrease monotonically with y. Instead, a peak is observed at $y/\delta^* \simeq 1$, with lower intermittencies very near the wall. For the farther downstream, higher intermittency cases, the peak is followed by a plateau of nearly constant intermittency. This near-wall drop-off in intermittency is consistent with turbulent spot elevation cross-sectional shapes which exhibit leading and trailing-edge overhangs as observed by Cantwell et al. (1978) and others. The data of Blair (1988), acquired in a moderately accelerating boundary layer, do not exhibit a well-defined near-wall intermittency drop-off, however.

Conditionally sampled mean velocity profiles obtained in the intermittent boundary layer for grid 1 are shown in Fig. 5a–c. In each of the figures, three profiles are shown. The non-turbulent profiles represent the mean of the velocities obtained during times when the indicator function was zero. It should be noted that these times included the post-burst relaxation period during which the transport is laminar, but the magnitude of the instantaneous velocities may be more characteristic of the turbulent part of the flow (see Fig. 3). For comparison, the corresponding Blasius curves are also shown. The middle profile shown in each case represents the overall velocity profile as determined from a direct time average of the digitally recorded data. The turbulent profiles were obtained from an average of the velocity data acquired during times when the indicator function was equal to one.

Low-intermittency non-turbulent profiles agree well with corresponding Blasius profiles. As the intermittency increases, however, the non-turbulent profiles of Fig. 5 increasingly deviate from the Blasius curves, falling well below the Blasius values for the highest intermittency case (Fig. 5c). The turbulent profiles have the appearance

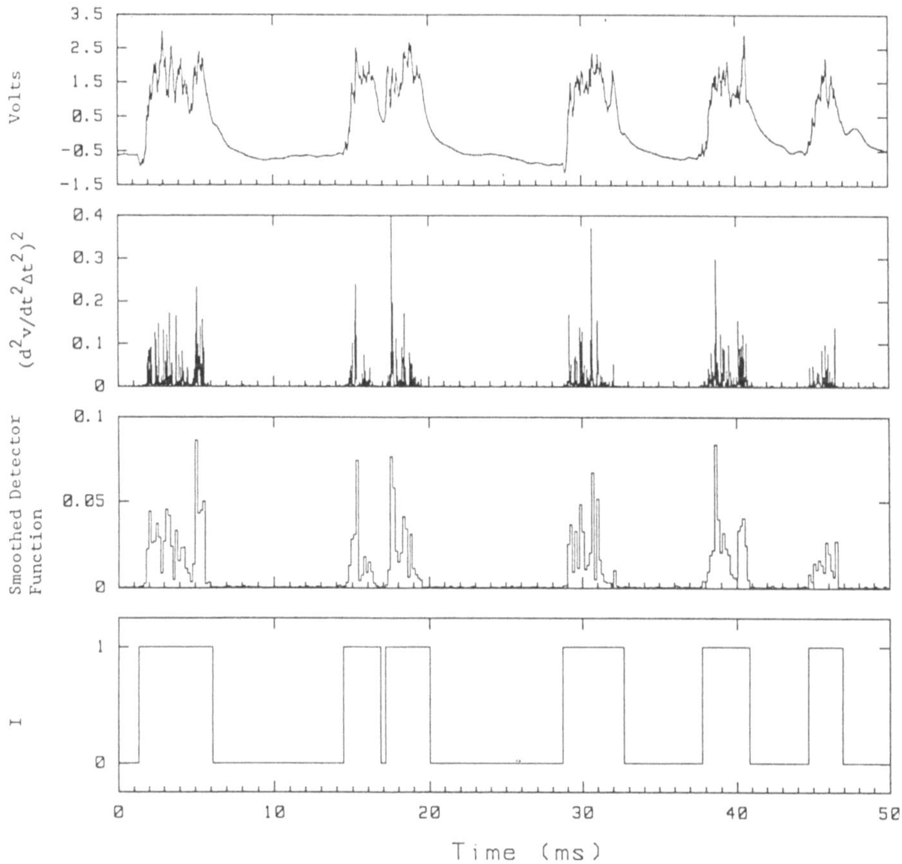

Fig. 3. Technique for determination of indicator function

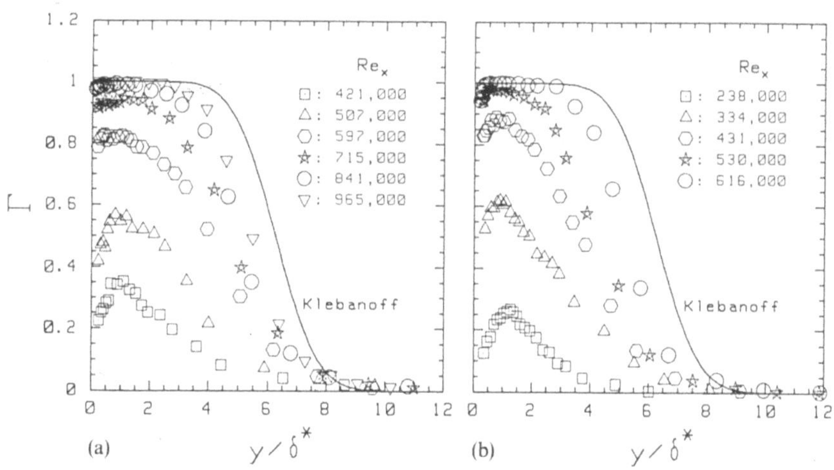

Fig. 4 a, b. Boundary layer intermittency profiles; **a** grid 1; **b** grid 2

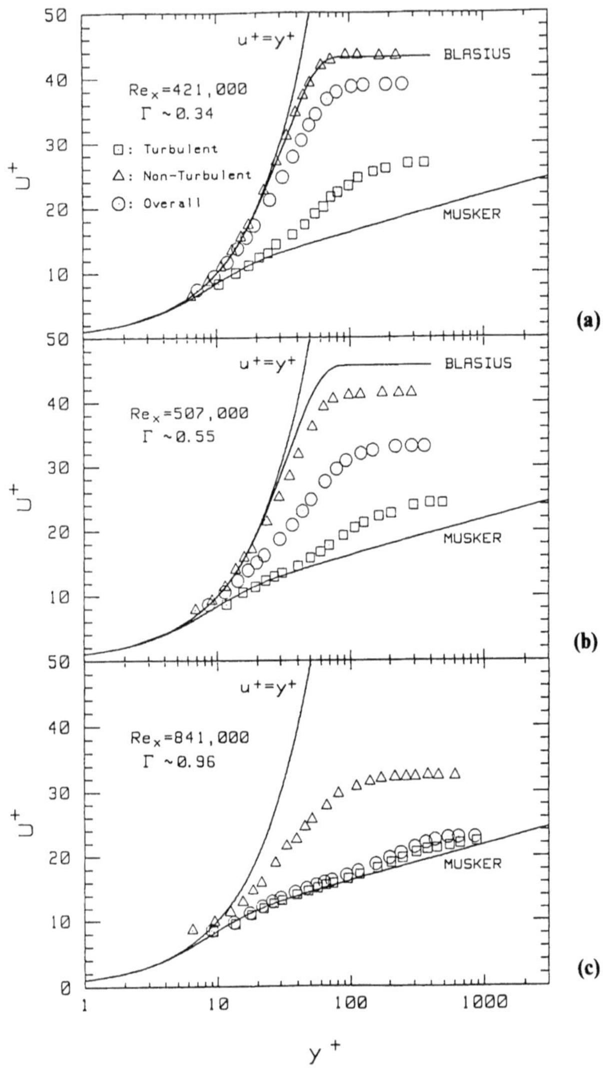

Fig. 5a–c. Conditionally sampled mean velocity profiles, grid 1

of low Reynolds number turbulent boundary layer with a large wake region, especially at low intermittencies.

Boundary layer profiles of the overall apparent streamwise rms velocity for grid 1 are presented in Fig. 6a, b. Recall that the mean profiles for this case exhibited essentially the entire range of behavior from laminar to fully turbulent. The farthest upstream profiles are typical for laminar boundary layers in the presence of freestream turbulence with a peak unsteadiness value occurring at $y/\delta^* \simeq 1.3$ or $y^+ \simeq 30$. A double peak in the rms profiles occurs at measurement stations located in the transition region. It is especially pronounced in the later stages of transition ($\Gamma \simeq 0.8$) and lingers even into the immature stages of the turbulent boundary layer. This double peak will be discussed further in conjunction with presentation of the

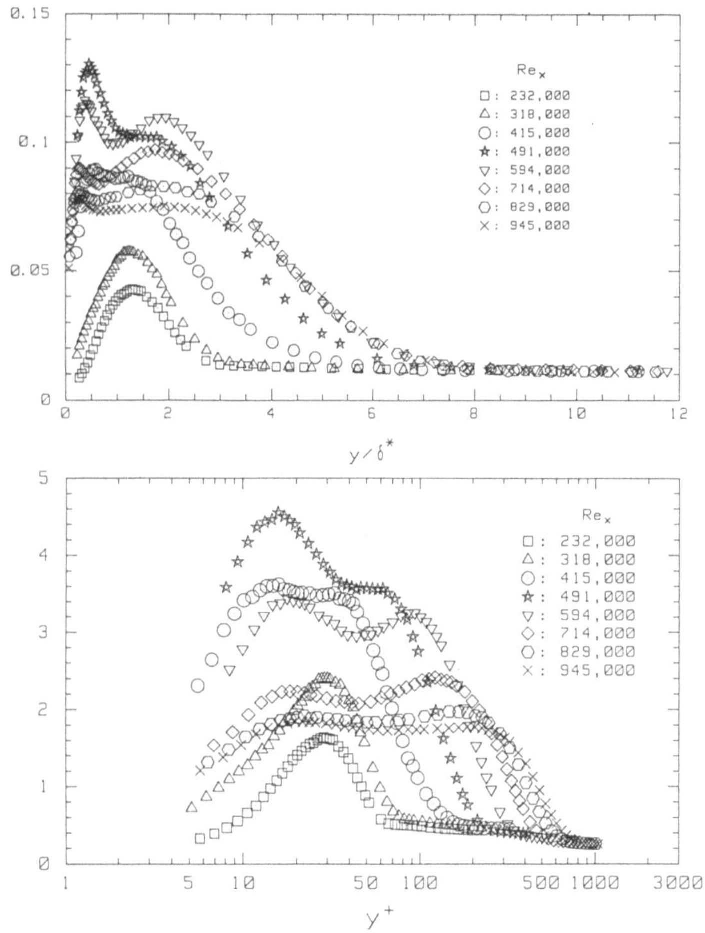

a

b

Fig. 6 a, b. Overall streamwise rms velocity profiles, grid 1; **a** outer variables; **b** wall units

conditionally sampled turbulence intensity profiles. In addition, the magnitude of the near-wall peak is much larger in the transition region than farther downstream. This behavior has been shown to be typical for transitioning boundary layers (Arnal et al. 1978; Suder et al. 1988; Kuan and Wang 1988; Blair 1988; Kim et al. 1989).

Note that the magnitude of the near-wall ($y^+ \simeq 17$) peak for the furthest downstream, turbulent boundary layer case shown in Fig. 6b ($Re_x = 945{,}000$) is too low. The peak value of u'/u_τ expected for a fully turbulent boundary layer is in the range of 2.5 to 3.0. This value was not measured due to eddy averaging over the length of the standard hot wire used in the present study. A wire length of $l^+ < 20$ is required to properly resolve the small-scale near-wall turbulence (Ligrani and Bradshaw 1987). For the conditions of the present study, a custom submicron wire would have been required. Fortunately, the turbulent length scales present in the transitional turbulent spots are much larger than the viscous length scales of a fully turbulent boundary layer. Consequently, the near-wall turbulence measurements acquired in the transition region should be more accurate than those acquired in the fully turbulent region.

Conditionally sampled streamwise turbulence intensity (rms) profiles are shown in Fig. 7a–c for the same conditions as for Fig. 5a–c. These rms values were calculated from the digitally stored velocity records relative to the respective mean values for the turbulent and non-turbulent parts of the flow, thereby eliminating the mean-step rms contribution associated with switching between the turbulent and non-turbulent levels. Reminiscent of the laminar overall rms profiles presented in Fig. 6a for the farthest upstream stations, the non-turbulent conditionally sampled rms profiles of Fig. 7a–c have a rounded shape and exhibit a peak at $y/\delta^* \simeq 1.3$. The

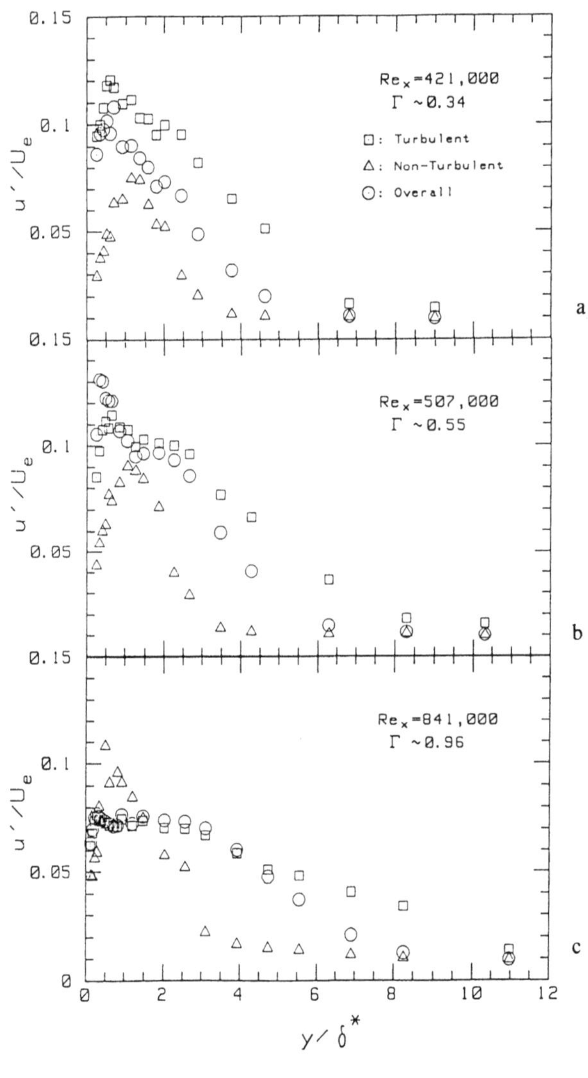

Fig. 7 a–c. Conditionally sampled streamwise rms velocity profiles grid 1

peak magnitudes are greater, however, than were observed for any of the strictly laminar profiles and increase significantly with downstream distance, exceeding both the turbulent and the overall profiles at $y/\delta^* \cong 1$ and $Re_x \geqq 597{,}000$. This observation was supported by direct inspection of the intermittent waveforms which revealed high levels of low-frequency unsteadiness in the non-turbulent intervals between turbulent spots, especially pronounced near $y/\delta^* \cong 1$. The turbulent part of the rms profiles also have a shape which is similar to the shape expected for a fully turbulent boundary layer, with a peak very near the wall ($y/\delta^* \simeq 0.5$) followed by a relatively flat region and a final drop off to the freestream value outside of the boundary layer. The magnitude of the near-wall turbulence intensity, however, is highest for the lowest intermittency case, and decreases significantly with downstream distance. This result indicates that transport processes occurring in the turbulent spots will not be well modeled by standard turbulence models used in equilibrium turbulent boundary layers.

Values of overall apparent turbulence intensity in excess of a direct intermittency-weighted average of the respective turbulent and non-turbulent parts are observed both very near the wall ($y/\delta^* \simeq 0.6$) and also farther out in the boundary layer ($2 < y/\delta^* < 4$), indicating the presence of a mean-step contribution to the overall rms value. This mean-step contribution is associated with the significantly different mean values of the turbulent and non-turbulent velocities at these y locations. In fact, the overall rms value actually exceeds the turbulent rms value for several near-wall stations (Fig. 7b). At the near-wall stations, the velocity records are characterized by large positive excursions during the passage of a turbulent spot and correspondingly large apparent overall rms values are observed. Similarly, at y measurement stations located farther out in the boundary layer, large negative velocity excursions occur during the passage of a turbulent spot, resulting in a large mean-step contribution to the second peak in the overall rms profiles observed at these y locations. Hedley and Keffer (1974) provided the following formula for quantitative assessment of the mean-step contribution to the overall apparent streamwise rms velocity:

$$\overline{u_0'^2} = \Gamma \overline{u_t'^2} + (1 - \Gamma)\overline{u_l'^2} + \Gamma(1 - \Gamma)(\overline{U}_t - \overline{U}_l)^2$$

where Γ is the time-average intermittency function. The mean-step contribution is given by the last term in the expression. The magnitude of this term is largest for intermediate values of intermittency and for large values of $\overline{U}_t - \overline{U}_l$. Thus the contribution of this term is zero at the "crossover" elevation ($y/\delta^* \simeq 1.2$) in the boundary layer where the mean velocities of the turbulent and non-turbulent parts of the flow are equal.

Thermal Boundary Layer

A representative series of boundary layer temperature profiles corresponding to the velocity profiles of Fig. 1 is shown in Fig. 8. A solid line representing the theoretical laminar temperature profile corresponding to the farthest upstream station,

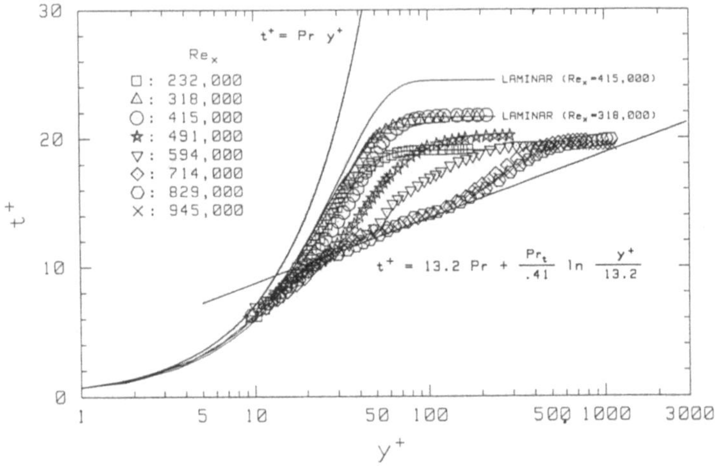

Fig. 8. Mean temperature profiles in wall units, grid 1

including a correction for unheated starting length is also shown. These data exhibit behavior which is qualitatively similar to the velocity profiles. A turbulent Prandtl number of 0.9 and a conduction layer thickness of 13.2 were assumed in obtaining the fully turbulent line shown in the figure. A slightly higher value of Pr_t may be required for a best fit to the early post-transition turbulent profiles (see Wang et al. 1985). In order to check for energy closure, enthalpy thickness values were calculated from the boundary layer temperature profiles and compared to values obtained from direct integration of the wall heat flux. This comparison is shown in Fig. 9. The agreement is excellent, indicating consistency between the wall

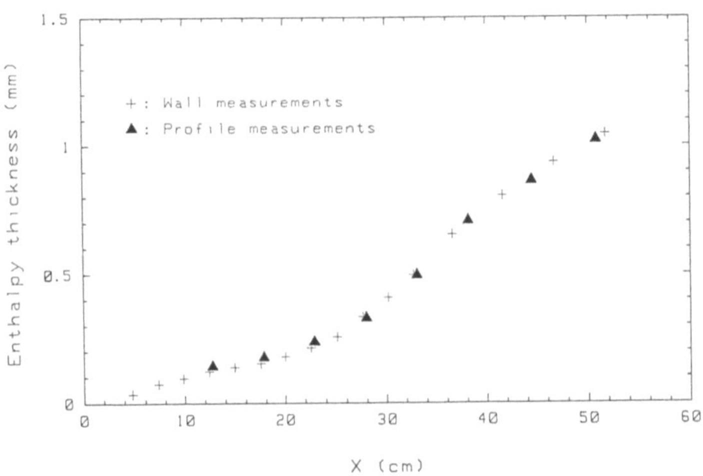

Fig. 9. Enthalpy thickness and energy closure verification, grid 1

measurements and the profile measurements. Thermal boundary layer development is also depicted.

Stanton number distributions, which were obtained directly from wall heat flux and temperature measurements, are presented in Fig. 10. A correction for radiation heat loss was also included. These data agree very well with the data of Blair (1983), both in terms of magnitude and transition location. The lower solid line is the theoretical laminar Stanton number including a correction for unheated starting length. The upper solid line is the turbulent correlation given by Kays and Crawford (1980).

A quantity which combines information from both the momentum and thermal boundary layers is the Reynolds analogy factor, $2St/C_f$. A plot of experimentally determined Reynolds analogy factors is presented in Fig. 11. The solid curves shown on the figure represent expected Reynolds analogy factors for the laminar and turbulent regimes, accounting for the thermal boundary conditions and unheated starting length. Appropriate laminar theoretical results and turbulent correlations from Kays and Crawford (1980) were combined in order to obtain the curves shown. In the laminar region, the Stanton number is very sensitive to thermal boundary condition and the uniform wall heat flux of the present experiment results in a Stanton number augmentation of 36% (Kays and Crawford 1980) over the uniform wall temperature case. Furthermore, the unheated starting length of the test surface produces an additional augmentation, especially important for the upstream stations. Consequently, Reynolds analogy factors as high as 2.0 were observed in the laminar region. Note the excellent agreement of the laminar data with the laminar prediction. For turbulent boundary layers, heat transfer results are much less sensitive to both the thermal boundary condition and the unheated starting length, and values much closer to the "standard" $Pr^{-2/3}$ value (1.27 for $Pr = 0.7$) are observed. The data for the high-turbulence cases are generally better predicted by $Pr^{-2/3}$, while the turbulent heat transfer data for the lower freestream turbulence

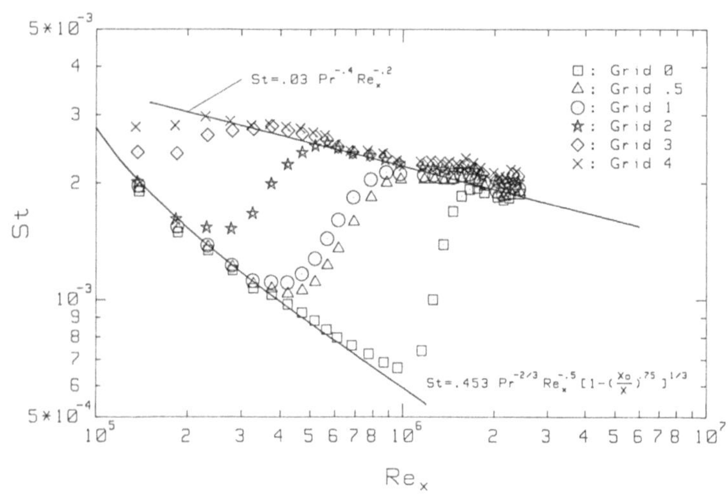

Fig. 10. Stanton number distributions—effect of freestream turbulence

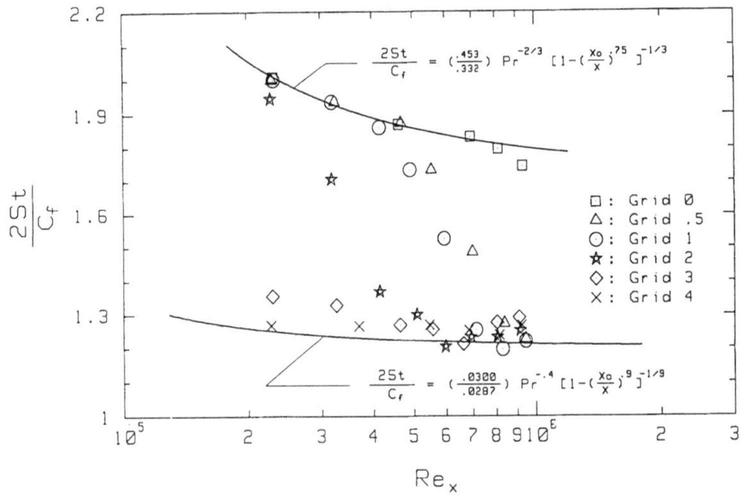

Fig. 11. Reynolds analogy factors

cases are closer to the lower Kays and Crawford curve. This slight increase of turbulent Reynolds analogy factor with freestream turbulence is consistent with the results of Blair (1983).

Spectra

One-dimensional boundary layer spectra were also acquired. A series of power spectral density curves for grid 1 is presented in Fig. 12. Recall that the farthest upstream measurement stations for grid 1 exhibit Blasius-type mean profiles and no evidence of turbulence. Note also that even the farthest upstream measurement location ($Re_x = 232{,}000$; $Re_{\delta^*} \simeq 840$) is post-critical in terms of linear stability theory. The approximate range of unstable frequencies for this Reynolds number is indicated on the figure. All of the spectra presented in the figure were obtained with the hot wire at its closest proximity to the test surface ($y \simeq 0.13$ mm) and no conditional sampling was applied.

Within the unstable frequency range, a broadband hump is observed in the power spectrum. In addition, a narrow peak is observed at 930 Hz. This peak was not related to any intentional excitation and was also present in grid 0 and grid .5 spectra and is therefore not grid related. It also did not appear in spectra obtained at the maximum rms elevation. For the grid 0 case, selective amplification appeared to be centered near this 930 Hz frequency, with the peak value shifted to slightly lower frequencies at farther downstream stations. For the grid 1 case, however, preferred bandwidth amplification is more broadband and is centered in the 1200–1300 Hz range, again shifting to slightly lower frequencies with increasing downstream distance as expected from linear stability theory. These spectra indicate that receptivity and selective amplification of disturbance frequencies in the range

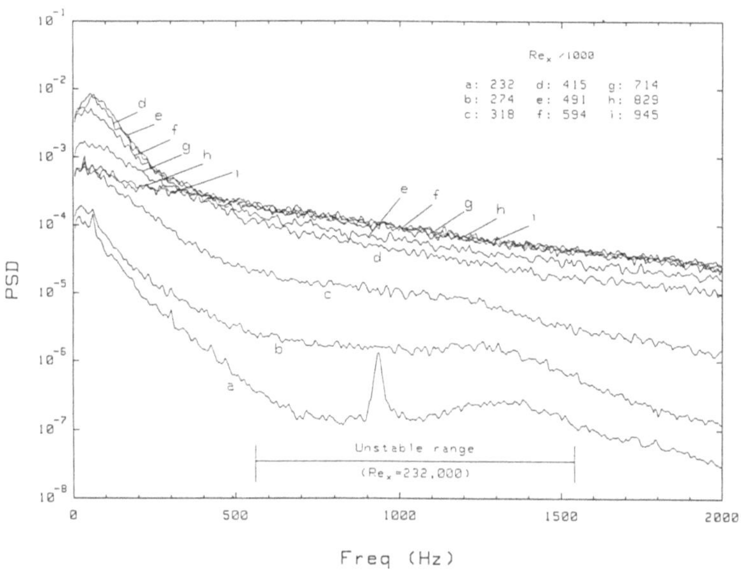

Fig. 12. Wall spectra through transition, grid 1

predicted by linear stability theory can occur in laminar boundary layers perturbed by moderate levels (i.e. 1%) of freestream turbulence. The waves generated in this process are localized very near the wall, however, and do not seem to play a dominant role in the ultimate transition process at this level of freestream turbulence. With low disturbance levels (e.g., grid 0), however, the selective amplification of TS-type waves was clearly shown to be the dominant mechanism in the process leading to the formation of turbulent spots (Suder et al. 1988).

Spectra obtained farther out in the laminar boundary layers at elevations corresponding to the maximum observed rms values do not show any clear evidence of selective amplification. This result is consistent with the observations of Blair (1988). Apparently the large-amplitude, low-frequency unsteadliness present at the maximum rms elevation in the perturbed laminar boundary layers is not related to the unstable frequencies predicted by linear stability theory.

At streamwise measurement locations where intermittency is first observed (e.g., curve *d* in Fig. 12) the power spectrum is characterized by broadband high-frequency content and a distinct low-frequency hump (< 200 Hz) associated with passage of turbulent spots. Further downstream, as the intermittency approaches 1.0, this low-frequency hump disappears.

Summary and Conclusions

As part of an ongoing research program focusing on the boundary layer transition process in the presence of freestream disturbances, detailed measurements of the momentum and thermal boundary layers and surface heat transfer on a heated flat plate have been obtained. Conditional sampling was applied as a means of segregating the turbulent and non-turbulent statistics of the intermittent boundary layers.

Excellent agreement of the measured boundary layer mean velocity profiles with laminar theory and turbulent correlations was observed, along with intermediate transitional profiles. Intermittency profiles obtained in the transitional boundary layers exhibited a peak at $y/\delta^* \simeq 1$, with lower intermittencies near the wall. These observations are consistent with turbulent spot shapes observed in previous studies which indicated a maximum spot length away from the wall. Non-turbulent conditionally sampled mean velocity profiles had Blasius-type shapes, but were observed to deviate increasingly from corresponding Blasius profiles with increasing intermittency. Turbulent conditionally sampled mean profiles had the appearance of low-Reynolds number turbulent boundary layer with a large wake region. Conditionally sampled rms velocity profiles revealed high levels of low-frequency unsteadiness in the non-turbulent part of the intermittent boundary layers. The turbulent part of these transitional boundary layers possessed peak turbulence intensities as much as 50% higher than corresponding fully turbulent values. Thermal boundary layer temperature and surface heat transfer measurements obtained in this study were in excellent agreement with previous studies. Measured Reynolds analogy factors were found to be well predicted by combining appropriate correlations for the respective laminar and turbulent regimes, including corrections for unheated starting length. Boundary layer spectra indicated that some selective amplification of disturbances in the linear instability frequency range predicted was occurring before the onset of turbulent spot formation. For the 1% freestream turbulence case, however, the waves involved were localized very near the wall and do not seem to play a role in the ultimate breakdown to turbulence.

References

Antonia, R. A., Chambers, A. J., Sokolov, M., Van Atta, C. W. (1981): Simultaneous temperature and velocity measurements in the plane of symmetry of a transitional turbulent spot. J. Fluid Mech. **108**, 317

Arnal, D., Juillen, J. C., Michel, R. (1978): Experimental analysis and computation of the onset and development of the boundary layer transition. NASA TM-75325

Blair, M. F., Bailey, D. A., Schlinker, R. H. (1981): Development of a large-scale wind tunnel for the simulation of turbomachinery airfoil boundary layers. ASME J. Engr. for Power **103**, 678

Blair, M. F. (1982): Influence of free-stream turbulence on boundary layer transition in favorable pressure gradients. ASME J. Engr. for Power **104**, 743

Blair, M. F. (1983): Influence of free-stream turbulence on turbulent boundary layer heat transfer and mean profile development, Parts I & II. ASME J. Heat Transfer **105**, 33

Blair, M. F. (1988): Bypass-mode boundary layer transition in accelerating flows. J. Fluid Mech. (To appear)

Cantwell, B. J., Coles, D., Dimotakis, P. E. (1978): Structure and entrainment in the plane of symmetry of a turbulent spot. J. Fluid Mech. **87**, 641

Hedley, T. B., Keffer, J. F. (1974): Turbulent/non-turbulent decisions in an intermittent flow. J. Fluid Mech. **64**, Part 4, 625

Kays, W. M. Crawford, M. E. (1980): *Convective Heat and Mass Transfer*, 2nd ed., McGraw-Hill, New York

Kim, J., Simon, T. W., Kestoras, M. (1989): Fluid mechanics and heat transfer measurements in transitional boundary layers conditionally sampled on intermittency. To be presented at the 1989 ASME National Heat Transfer Conference, Philadelphia

Klebanoff, P. S. (1955): Characteristics of turbulence in a boundary layer with zero pressure gradient. NACA Report No. 1247

Kuan, C. L., Wang, T. W. (1988): Some intermittent behavior of transitional boundary layers. Exp Ther Fluid Sci (To Appear)

Ligrani, P. M., Bradshaw, P. (1987): Spatial resolution and measurement of turbulence in the viscous sublayer using subminiature hot-wire sensors. Exp Fluids **5**, 407

Morkovin, M. V. (1978): Instability, transition to turbulence and predictability. NATO AGARDograph No. 236

Musker, A. J. (1979): Explicit expression for the smooth wall velocity distribution in a turbulent boundary layer. AIAA Journal **17**, 655

Schlichting, H. (1979): *Boundary Layer Theory*, 7th ed. McGraw Hill, New York

Suder, K. L., O'Brien, J. E. Reshotko, E. (1988): Experimental study of bypass transition in a boundary layer. NASA TM 100913

Wang, T., Simon, T. W., Buddhavarapu, J. (1985): Heat transfer and fluid mechanics measurements in transitional boundary layer flows. ASME Journal of Engr. for Gas Turbines and Power **107**, 1007

Wygnanski, I., Sokolov, M., Friedman, D. (1976): On a turbulent spot in a laminar boundary layer. J. Fluid Mech. **78**, Part 4, 785

Turbulent Characteristics Inside a Turbulent Spot in a Plane Poiseuille Flow

Dan S. Henningson[1] and John Kim[2]

[1] Department of Mathematics, MIT, Cambridge MA 02139, USA
[2] NASA Ames Research Center, Moffett Field, CA 94035, USA

Abstract

Turbulence characteristics inside a turbulent spot in plane Poiseuille flow are investigated by analyzing a database obtained from a direct simulation. The spot area is divided into two distinct regions—a turbulent area and a wave area. It is found that the flow structures inside the turbulent area have strong resemblance to those found in fully-developed turbulent channel flow. A suitably defined mean and r.m.s. fluctuations as well as the internal shear-layer structures are found to be similar to the turbulent counterpart. The r.m.s. fluctuations and Reynolds stress are found to be higher in the wave area, and the shear-layer structures are similar to those observed in the secondary instability of two-dimensional Tollmien–Schlichting waves.

1. Introduction

Transition to turbulence in plane channel flow has previously been studied both as a simple model problem for more complicated shear shows and as an interesting problem in its own right. Traditional vibrating-ribbon-induced transition was studied by Nishioka et al. (1976) where they found that similar three-dimensional breakdown occurs as in the boundary layer (Klebanoff et al. 1962; Kovasznay et al. 1962). Two-dimensional Tollmien–Schlichting (T–S) waves grow to finite amplitude and become unstable to small oblique waves. In the process intense shear layers and associated lambda vortex structures are created which quickly break down to turbulence. Secondary instability analysis, in which the mean plus the two-dimensional wave is taken as the basic flow, has been able to theoretically predict the onset of three-dimensionality (Orszag and Patera 1983; Herbert 1984). The secondary instability gives a critical Reynolds number of about 1000, based on the channel half-height, h, and certerline velocity, U_{CL}, for plane channel flows. This is a much lower critical Reynolds number than that for linear disturbances, 5722 (Orszag 1971).

When a localized disturbance is introduced into the flow, waves are created in the form of a wave packet (Gaster and Grant 1975). In addition to the waves it is found that, due to the inherent three-dimensionality of the localized disturbance, the normal velocity induces normal vorticity through a *lift-up effect* (Landahl 1975; Gustavsson 1978; Henningson 1988). That is, a major part of the horizontal momentum is retained as a fluid particle is displaced vertically, thus providing a

Turbulent Shear Flows 7

contribution to the horizontal velocity components if there is a mean shear present. Henningson (1988) pointed out that the lift-up part of the horizontal velocity is directly proportional to the normal vorticity component and showed that it was mainly associated with the creation of shear-layers in the flow. Henningson et al. (1990) studied the development of localized disturbances further using numerical simulations of the Navier–Stokes equations. They found that the non-linear development was dominated by classical wavenumber summations, with a majority of the non-linearly induced energy in Fourier modes associated with small linear amplification.

When a localized disturbance breaks down the result is a propagating region of turbulence in an otherwise laminar flow, i.e. a turbulent spot. The critical Reynolds number for spot formation is dependent on the amplitude of the disturbance. For large amplitudes a turbulent spot can be created well below the linear critical Reynolds number, both in the boundary layer (Elder 1962) and in plane Poiseuille flow (Carlson et al. 1982). Turbulent spots in boundary layers have a self-similar arrow-head shape pointing in the downstream direction, with wave packets trailing the upstream edges (Riley and Gad-el-Hak 1985). Inside the turbulent region characteristics such as r.m.s. values and shear-layer structure are found to be similar to fully developed turbulent boundary layers (Johansson et al. 1987). Turbulent spots in plane Poiseuille flow have also been investigated (Carlson et al. 1982; Alavyoon et al. 1986). The early spot has a downstream pointing arrow-head that quickly reverses itself as the turbulence generation in the front part terminates. Around the spot waves are found, most clearly seen at the sides or wingtips. Henningson and Alfredsson (1987) used hot-film anemometry to study the velocity field associated with the waves. Using ensemble-averaged mean velocity profiles obtained from their measurements they found that modified mean flow outside the wingtips was destabilized, indicating that the spreading of the spot was consistent with the idea of *growth by destabilization* proposed by Gad-el-Hak et al. (1981).

In order to obtain more complete data on the velocity field outside the spot as well as the turbulence inside, a numerical simulation of the plane Poiseuille spot was carried out at NASA Ames Research Center (Henningson et al. 1987). The numerical results showed that features such as propagation velocities, spreading angle and wave characteristics were in good agreement with the previous experimental findings. Further analysis by Henningson (1989) of the numerically obtained wingtip wave field showed that the instability causing the waves growth was of the cross-flow type (see Gregory et al. 1955). In this paper results from the simulation will be analyzed in order to compare the turbulence characteristics inside the Poiseuille spot with those of fully developed turbulent channel flow.

2. Mean Velocity Fields

The present investigation will concentrate on the fully-developed spot obtained from Henningson et al. (1987) at a non-dimensional time of 258, unless otherwise noted, (all quantities are made non-dimensional using appropriate combinations of

h and U_{CL}, and x,y,z refer to the streamwise, normal and spanwise directions, respectively). This particular time was chosen because the spot has already acquired all of the characteristics of the experimentally observed spot, but it had not become large enough for substantial interference by the boundary of the computational domain. Because the development of the Poiseuille spot, both experimentally and computationally, has been found to be approximated self-similar, conical coordinates are introduced to take advantage of this fact. The coordinates $\zeta = x/t$ and $\zeta = z/t$ will be used when comparisons are made between temporal velocity variations obtained from experiments and spatial variations from the simulation. Note also that the Reynolds number in the present study, as in most of the previously mentioned Poiseuille spot investigations, is 1500.

Because of the large cost of computation, no attempt was made to run several cases to obtain ensemble-averaged statistics. In order to compare the simulation results from this single realization to the ensemble-averaged results obtained from experiments, a mean velocity has to be defined. Analysis of the fluctuating components also requires a properly defined mean. Several different spatial averages were tried. The best results were obtained using Gaussian filters. A fairly wide Gaussian filter was applied in the horizontal directions which resulted in a smooth velocity field that was taken as the mean. The filter width was determined such that the filtering process removed small- and medium-scale structures in the turbulent region without altering the largest scales of the spot.

In Figs. 1a and 1b, an ensemble-averaged field from Klingmann et al. (1989), and that obtained from a Gaussian-averaged velocity field are shown. Agreement between the two results are remarkable; note, for instance, the sharp transition region with the dip in the velocity, the arrow head front part of the disturbance, and the speed-up of the laminar fluid around the spot. The agreement is not only qualitatively good but corresponds for the most part also quantitatively. The shape of the interface in the wingtip area is somewhat different, however, possibly owing to slight non-conical similarity of that area (Klingmann et al. 1989, assumed conical similarity in order to transform their temporal signals into spatial structures). Although not shown here, it should be noted that averages from the experiments and the simulations at early times do not agree. Note that due to the spanwise symmetry, only half of the spot is shown when horizontal planes are plotted.

The Gaussian-averaged velocity fields chosen as the representation of the mean velocity are shown in Figs. 1b through 1e. Figures 1b and 1c show the streamwise and the spanwise mean velocities at the centerline, $y = 0$. Note that the increase in the streamwise velocity around the spot is associated with a spanwise velocity towards the spot downstream of the turbulent region and a velocity outwards at the wingtips of the spot. Thus the laminar fluid is found to flow around the spot, which acts as a partial flow blockage. Inside the spot the streamwise mean is lower than in the laminar region and the spanwise velocity is everywhere directed outwards. Figures 1d and 1e show the streamwise and spanwise velocities closer to the wall. Both velocities have similar features to those at the centerline. However, the streamwise velocity inside the spot is now higher than in the laminar region as well as shifted upstream. The region where the spanwise velocity is directed outward have also moved considerably further upstream compared to the correspondingly feature at

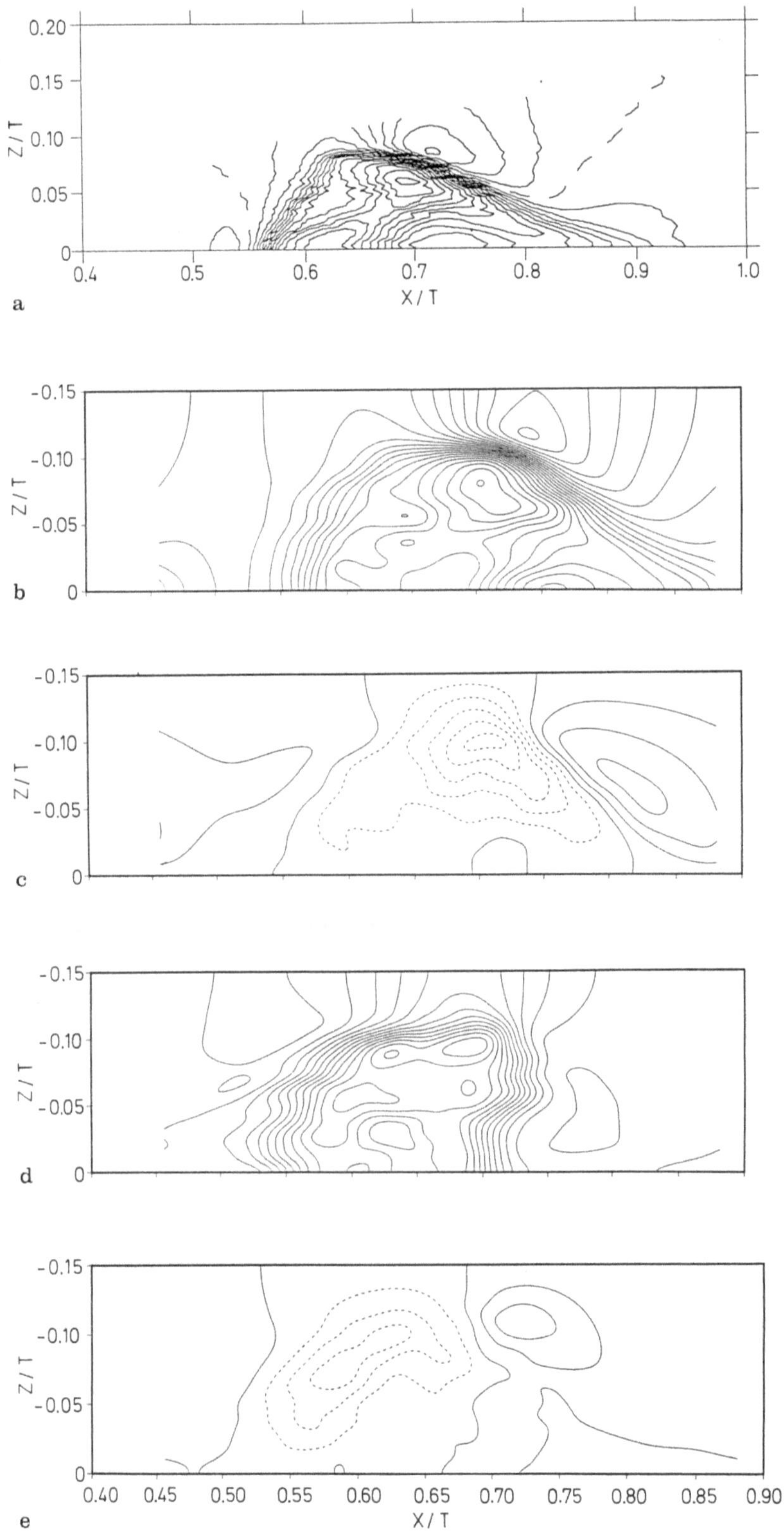

a
Z / T
0.20
0.15
0.10
0.05
0
X / T
0.4
0.5
0.6
0.7
0.8
0.9
1.0
b
c
d
e
-0.15
-0.10
-0.05
0.40
0.45
0.50
0.55
0.60
0.65
0.70
0.75
0.80
0.85
0.90

the centerline. Thus the typical characteristics of the spot close to the walls remains the same as in the central part of the channel, but with an upstream shift.

3. Characteristics of the Turbulent Part

The fluctuating velocity can be defined as the instantaneous velocity minus the spatially-averaged mean velocity. The streamwise fluctuations close to the wall can be seen in Fig. 2a. In the interior of the spot we see the low and high speed streaks in the interior of the spot typical of near wall turbulence, and at the spanwise wingtips we find the characteristic oblique wavepacket. From the fluctuations the r.m.s. values can be computed using the same Gaussian filter used for the mean velocity. The r.m.s. fluctuation of the normal component for $y=0$ is shown in Fig. 2b. It is non-zero in the turbulent area and rises to twice its average value in

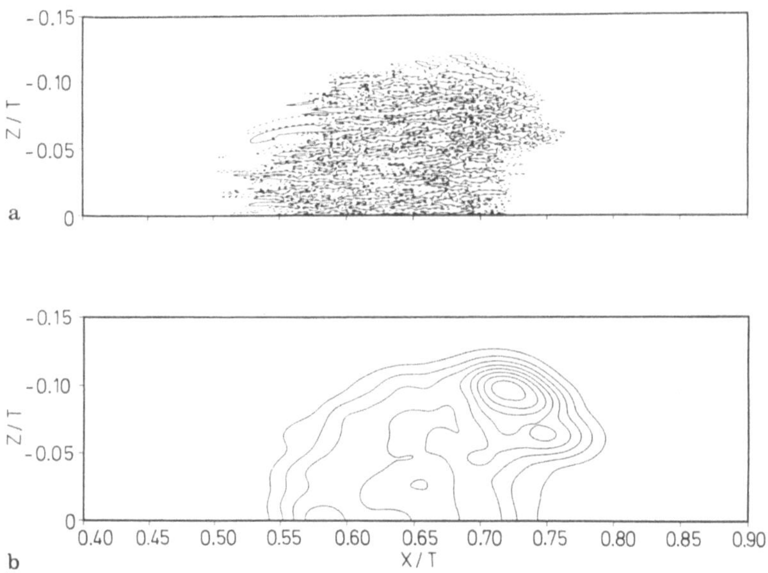

Fig. 2 a. ± 0.04 contours of the streamwise disturbance velocity at $y=-0.83$. **b** Contours of constant normal r.m.s. velocity at the centerline. Contour spacing is 0.01, with the outermost contour representing the 0.01 level

◀

Fig. 1 a–e. Contours of constant mean velocities. **a** Experimentally ensemble-averaged streamwise velocity at the centerline using 100 different spots at each measurement position, from the experiments of Klingmann et al. (1989). **b** Streamwise mean velocity at the centerline. **c** Spanwise mean velocity at the centerline. **d** Streamwise mean velocity at $y=0.83$. **e** Spanwise mean velocity at $y=0.83$. Contour spacing is 0.02 in all cases with the dashed lines indicating negative contours

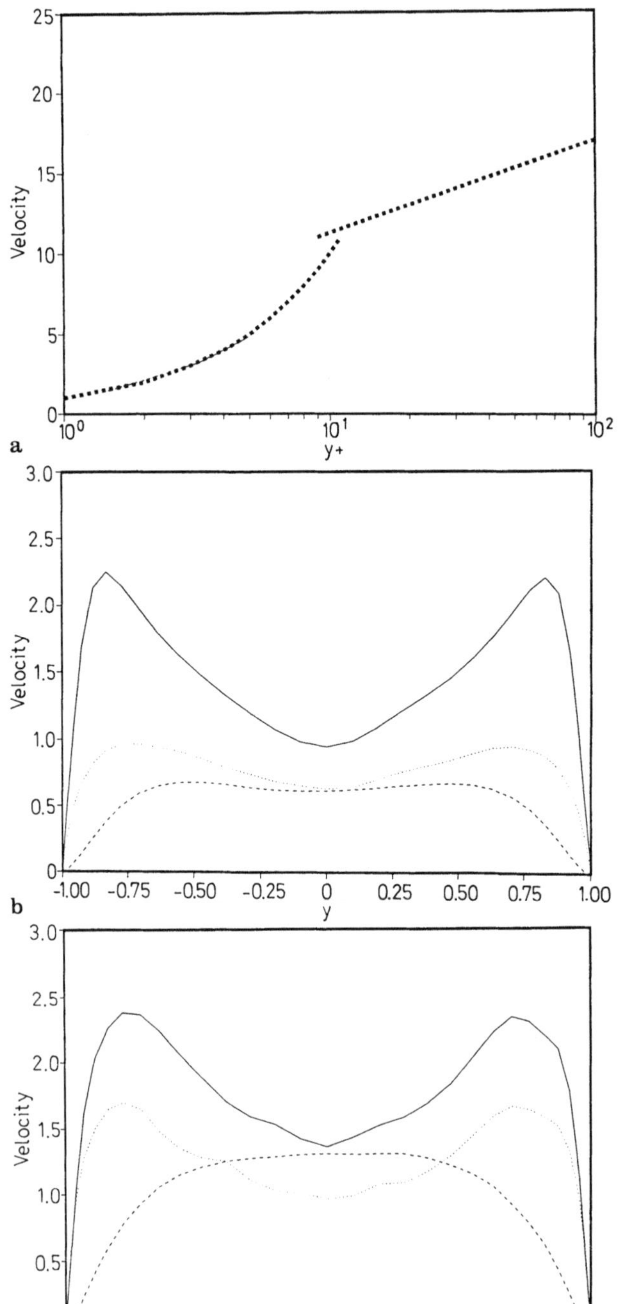

Fig. 3 a–d. Normal dependence of turbulence quantities plotted in viscous wall units. **a** Mean velocity in the turbulent region obtained from the lower half of the channel; dashed curves is $u^+ = y^+$ and $u^+ = 2.5 \ln y^+ + 5.5$. **b** R.m.s. velocities averaged over the turbulent part. **c** R.m.s. velocities averaged over the wave area (full line u_{rms}, dashed line v_{rms} and dotted line w_{rms}). **d** Reynolds shear stress in the turbulent (full line) and wave area (dashed line, dotted in $uv = -y^+$)

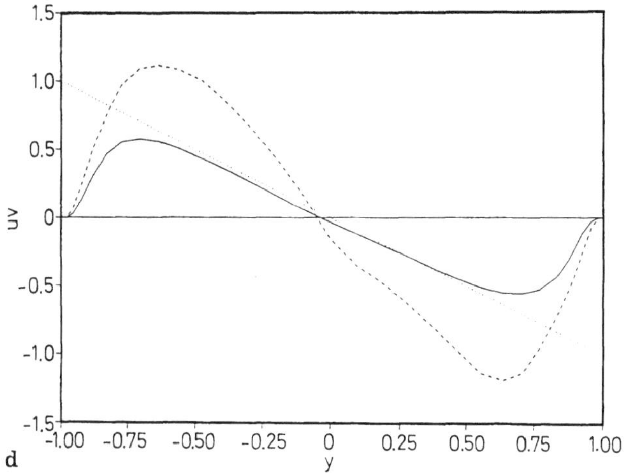

Fig. 3 (*Continued*)

the wingtip region as a consequence of the high amplitude wavepacket. It is apparent from this figure that the turbulence statistics reflect the difference in the behavior of the spot in the region where waves are found compared to the area where there are no waves. Instead of averaging these regions together when the overall statistics were computed, the averaging was done separately in the two regions. Thus, an area inside the spot defined by $0.58 < \xi < 0.70$ and $-0.08 < \zeta < 0.0$, will be referred to as the turbulent area, while an area including $0.70 < \xi < 0.74$ and $-0.12 < \zeta - 0.08$ will be referred to as the wave area.

3.1 Turbulence Statistics

In order to compare turbulence characteristics inside the spot with those in fully developed channel flow turbulence, one would like to scale the flow variables with the inner-wall scaling, i.e., with the friction velocity $u_\tau = \sqrt{\nu dU/dy|_{\text{wall}}}$ and the kinematic viscosity ν. A mean value of u_τ inside the spot was obtained by averaging it in the turbulent area. It was found to be 0.05. The y-dependence of the mean streamwise velocity in the turbulent area is shown in Fig. 3a. The dotted curve is $u^+ = y^+$ and the line $u^+ = 2.5 \ln (y^+) + 5.5$, where the superscript $^+$ denotes a quantity normalized by the inner-wall scaling. The log-region is very small for the present case because of the low Reynolds number ($Re_\tau = u_\tau h/\nu \approx 70$; note that the undisturbed laminar value is about 50). Considering the low Reynolds number the shape is in good agreement with that of fully-developed turbulent channel flow (Kim et al. 1987). The y-dependence of the r.m.s. values of the three velocity components averaged over the turbulent area are shown in Fig. 3b. They are also in good agreement with those corresponding to the fully-developed turbulent channel flow, except that the peak in the streamwise velocity close to the wall appears to be a bit low. To contrast the findings in the turbulent area with those in the wave area, the y-dependence of the r.m.s. values in the wave area is shown in Fig. 3c. All of the fluctuating components are larger in the wave area. The normal r.m.s. velocity, in particular, has a different shape somewhat similar to the mode shape of the least stable Orr–Sommerfeld mode. Figure 3d shows the Reynolds shear stress both in the turbulent and the wave area. In the turbulent area it shows the behavior expected from a fully-developed channel flow; in the wave area it is much higher, indicating higher turbulence production in that region.

3.2 Turbulence Structure

Another characteristic of wall-bounded turbulent shear flows is the sharp shear-layer structures found in the near-wall region. In order to examine the turbulent structures inside the spot, the VISA (Variable-Internal-Space-Average) technique was applied. This technique has been used by Kim (1985) and Johansson, Alfredsson and Kim (1987) to detect shear-layers in fully-developed turbulent channel flows. An event is considered to have occurred when the local streamwise variance exceeds a threshold, i.e.

$$\frac{1}{L}\int_{-L/2}^{L/2} u^2(x-s)ds - \left(\frac{1}{L}\int_{-L/2}^{L/2} u(x-s)ds\right)^2 > ku_{\rm rms}^2$$

where L is the averaging length, which for our case was about 200 viscous length units, and k was chosen to be 0.8. Islands of high variance of streamwise velocity were identified at $y^+ = 12$ at both walls. An ensemble average velocity field of a typical event was obtained by aligning the maximum local variance in the islands for every individual event detected and then summing the events. The results obtained using this procedure on all events detected in the turbulent area is given in Fig. 4. Figure 4a shows the typical inclined shear-layer structure associated with the lift-up of low speed fluid at the downstream half, and the downward sweep at the upstream half of the shear-layer. The spanwise structure of the shear layer is shown in Fig. 4b. The spanwise distance between the outer loops is slightly above 100 viscous units, indicating that the average spacing between the low- and high-speed streaks appears to be the same as that in a fully-developed channel flow (Kim et al. 1987). These shear-layer structures are in good agreement with those found in fully-developed turbulent channel flow. The ensemble average in the wave area shows a similar shear-layer structure in the (x, y)-plane, however, the horizontal structure (Fig. 4c) clearly reflects the wave-like characteristics. In addition to the oblique character of the primary waves a secondary structure of peaks and valleys can be found along curves of constant phase of the primary waves. Thus the breakdown of the primary waves in the transition area is associated with shear layers much in the same way as the secondary breakdown seen in vibrating ribbon generated waves (Klebanoff et al. 1962; Nishioka et al. 1976). This suggests that a similar process is responsible for the breakdown of the large amplitude waves in the spot case as in the two-dimensional vibrating ribbon case.

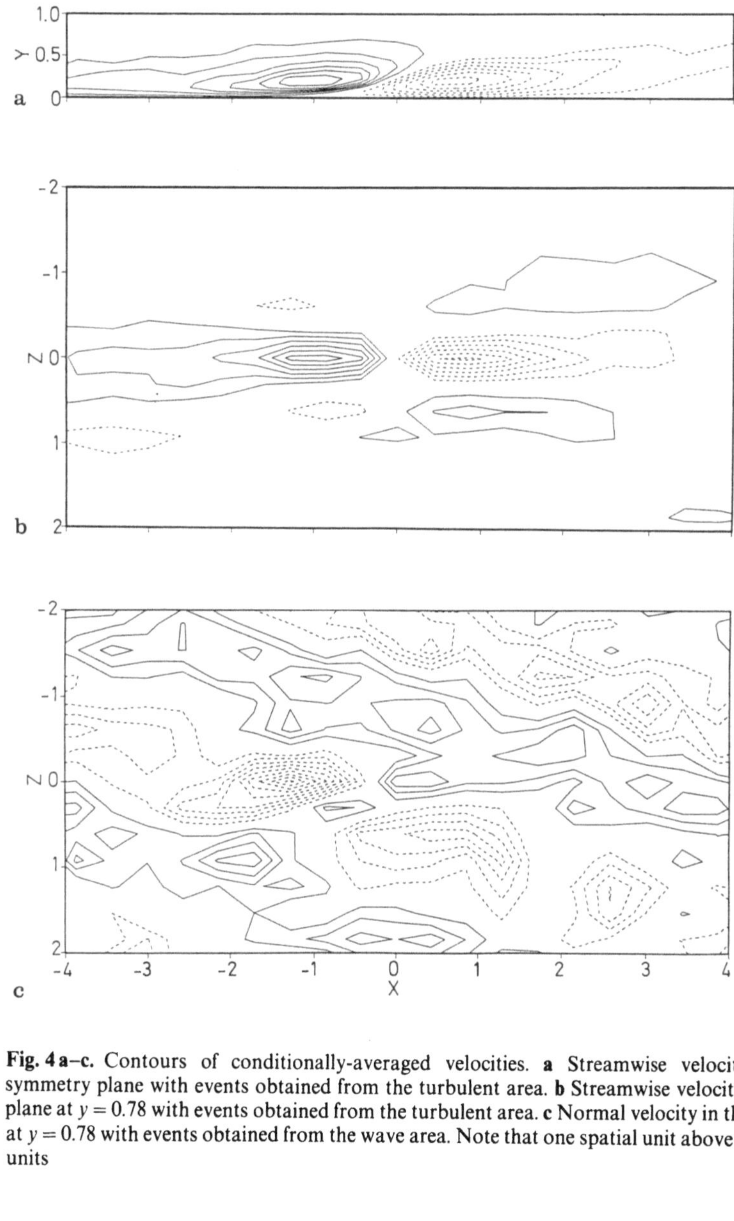

Fig. 4 a–c. Contours of conditionally-averaged velocities. **a** Streamwise velocity in the normal symmetry plane with events obtained from the turbulent area. **b** Streamwise velocity in the horizontal plane at $y = 0.78$ with events obtained from the turbulent area. **c** Normal velocity in the horizontal plane at $y = 0.78$ with events obtained from the wave area. Note that one spatial unit above is about 70 viscous units

4. Discussion and Conclusions

One observation that can be made from the comparison between the simulation and experiments is that the fully-developed spot seems to be independent of the initial conditions. A comparison between the early stages of development in the two cases does not show the agreement seen in the later stages. These findings agree with other investigators: for instance, Elder (1962) found that a fully developed turbulent spot in the Blasius boundary layer in independent of the initial disturbance as long as the disturbance is large enough to produce a spot.

A flow field of the fully-developed spot was chosen and investigated in detail. Non-turbulent fluid outside the spot was found the flow around the turbulent part, indicating that the turbulent part of the spot acts as a partial flow blockage. Inside the spot, the mean streamwise velocity had a turbulent character with the lower maximum than in the laminar region, while the spanwise velocity was always directed outward. Closer to the wall the mean fluid motion is similar to that in the central part of the channel except for a upstream shift of the flow pattern.

The turbulence characteristics of the flow inside the spot were also examined. The general conclusion is the same for the Poiseuille spot as for the turbulent spot in the Blasius boundary layer (Johansson et al. 1987), i.e. the interior of the spot behaves as if it were part of a fully-developed turbulent flow. The mean profile has the turbulent characteristics, although the logarithmic region is very short because of the very low Reynolds number. The profiles of the r.m.s. values and the Reynolds shear stress also agree with those found in fully-developed channel flow. One should note, however, that the mean motion is not two-dimensional inside the spot, thus exact comparisons are not possible. The structures usually found in turbulent flows were also present in the turbulent part of the spot. Close to the wall we have the typical low- and high-speed streaks with approximately the same spacing (about 100 viscous units) found in fully-developed turbulent channel flow. The ensemble-averaged turbulent structures obtained from the VISA technique also show good agreement with the fully developed case. The shear layer in the (x,y)-midplane displays the associated lift-up of low-speed fluid in front of the structure and the sweep of higher speed fluid toward the wall at the rear of the structure.

The r.m.s. variations were also computed for the wingtip area of the spot. It was found that all fluctuating components had higher amplitudes than those in the fully developed region and that the Reynolds stress was twice as high in that area. The breakdown process of the waves was also illuminated in the present investigation. It was found to be associated with periodic shear-layers in the cross wave direction, similar to the secondary instability in two-dimensional T–S waves (Kovasznay et al. 1962). These findings add to the picture presented by Henningson (1989). He found that the waves originated in the unstable mean flow outside the wingtip and that the energy for the wave growth was drawn from that mean. The large Reynolds stress in the area of the waves found in the present investigation show that the breakdown of the waves add to the turbulent fluctuations of the spot, and thus enhances its spreading.

Acknowledgments: We wish to thank Henrik Alfredsson for providing the VISA analysis programs and for valuable assistance in their implementation. Part of the present work was done during the 1988 Center for Turbulent Research Summer Program at NASA Ames Research Center. The first author also acknowledges the support of the Aeronautical Research Institute of Sweden (FFA) during the course of this work.

5. References

Alavyoon, F., Henningson, D. S., Alfredsson, P. H. (1986): Turbulent spots in plane Poiseuille flow—flow visualization. Phys. Fluids **29**, 1328–1331

Carlson, D. R., Widnall, S. E., Peeters, M. F. (1982): A flow visualization study of transition in plane Poiseuille flow. J. Fluid Mech. **121**, 487–505

Elder, J. W. (1962): An experimental investigation of turbulent spots and breakdown to turbulence. J Fluid Mech. **9**, 235–246

Gad-el-Hak, M., Blackwelder, R. F., Riley, J. J. (1981): On the growth of turbulent regions in laminar boundary layers. J. Fluid Mech. **110**, 73–95

Gaster, M., Grant, I. (1975): An experimental investigation of the formation and development of a wave packet in a laminar boundary layer. Proc. Roy. Soc. Lond. A **347**, 253–269

Gregory, N., Stuart, J. T., Walker, W. S. (1955): On the stability of three-dimensional boundary layers with application to the flow due to a rotating disc. Proc. R. Soc. Lond. A **248**, 155–199

Gustavsson, L. H. (1978): On the evolution of disturbances in boundary layer flows. Ph.D. Thesis, Dept. of Mech., R. Inst. Tech., Stockholm, Sweden

Henningson, D. S. (1989): Wave growth and spreading of a turbulent spot in plane Poiseuille flow. Phys. Fluids A **1**, 1876–1882

Henningson, D. S. (1988): The inviscid initial value problem for a piecewise linear mean flow. Stud. Appl. Math. **78**, 31–56

Henningson, D. S., Alfredsson, P. H. (1987): The wave structure of turbulent spots in plane Poiseuille flow. J. Fluid Mech. **178**, 405–421

Henningson, D. S., Spalart, P., Kim, J. (1987): Numerical simulations of turbulent spots in plane Poiseuille and boundary layer flows. Phys. Fluids **30**, 2914–2917

Henningson, D. S., Johansson, A. V., Lundbladh, A. (1990): On the evolution of localized disturbances in laminar shear flows. *Laminar-Turbulent Transition* (D. Arnal & R. Michel Eds.), 279–284, Springer

Herbert, T. (1984): Modes of secondary instability in plane Poiseuille flow. In *Turbulence and chaotic phenomena in fluids.* (T. Tatsumi, ed.) 53–58, Elsevier

Johansson, A. V., Alfredsson, P. H., Kim, J. (1987): Shear-layer structure in near-wall turbulence. *Proc. of CTR summer program, NASA Ames, Stanford*, 237–251

Johansson, A. V., Her, J. Y., Haritonidis, J. H. (1987): On the generation of high-amplitude wall-pressure peaks in turbulent boundary layers and spots. J. Fluid Mech. **175**, 119–142

Kim, J. (1985): Turbulence structure associated with the bursting event. Phys. Fluids **28**, 52

Kim, J., Moin, P., Moser, R. (1987): Turbulence statistics in fully developed channel flow at low Reynolds number. J. Fluid Mech. **177**, 133–166

Klebanoff, P. S., Tidstrom, K. D., Sargent, L. M. (1962): The three dimensional nature of boundary layer transition. J. Fluid Mech. **12**, 1–34

Klingmann, B., Alfredsson, P. H., Henningson, D. S. (1989): An experimental study of the velocity field of turbulent spots in plane Poiseuille flow. *Advances in Turbulence 2* (H.-H. Fernholz, H. E. Fiedler, eds.), 9–14, Springer

Kovasznay, L. S. G., Komoda, H., Vasudeva, B. R. (1962): Detailed flow field in transition. In *Proceeding of the 1962 heat transfer and fluid mechanics institute*, Univ. of Wash., 1–26

Landahl, M. T. (1975): Wave breakdown and turbulence. SIAM J. Appl. Math. **28**, 735–756

Nishioka, N., Iida, S., Ichikawa, Y. (1976): An experimental investigation of the stability of plane Poiseuille flow. J. Fluid Mech. **72**, 731–751

Orszag, S. A. (1971): Accurate solution of the Orr-Sommerfeld stability equation. J. Fluid Mech. **50**, 689–703

Orszag, S. A., Patera, A. T. (1983): Secondary instability of wall bounded shear flows. J. Fluid Mech. **128**, 347–385

Riley, J. J., Gad-el-Hak, M. (1985): The dynamics of turbulent spots. In *Frontiers of Fluid Mechanics* (S. M. Davis, J. L. Lumley, eds.), 123–155, Springer

Part III

Free Shear Flows

Introduction

L. J. S. Bradbury

Department of Mechanical Engineering, Plymouth Polytechnic, Plymouth, Devon, UK

The six papers in this section on free shear flows cover a wide range of phenomena. The first four papers study essentially organised structures that occur in a number of free flows and it is only the last two papers that are directed towards studies of the structure of fully turbulent flows.

The paper by C. H. K. Williamson stands alone from the other papers in this session in being concerned with the classical problem of vortex shedding from circular cylinders at comparatively low Reynolds numbers where the Strouhal number is still a strong function of Reynolds number. The work is largely experimental and discusses discontinuities that occur in the relationship between Strouhal number and Reynolds number. It is argued that the first discontinuity that has been observed at a Reynolds number of about 60 is due to inclination of the vortex shedding and that this inclination is determined by end conditions. By manipulating the end conditions in a variety of ways, it is possible to ensure parallel shedding and, under these circumstances, the discontinuity is removed. The author also puts forward a simple empirical expression for collapsing Strouhal number data in which the vortex shedding is inclined. From the point of view of turbulence, the discontinuities at higher Reynolds number are rather more interesting and it seems these are due to the development of three-dimensional cells in the shedding pattern.

The next paper in this collection is due to Meiburg et al. and is concerned primarily with numerical simulations of the formation of vortex rings in an axisymmetric jet subjected to both axial and circumferential periodic variations in the strength of the vortex sheet shed from the nozzle lip. These numerical simulations are visually entertaining and they are discussed in rather the same manner as experimental flow visualisation data might be discussed. The onset of vortex pairing is apparently demonstrated as is the formation of streamwise vorticity from circumferential variations in vortex sheet strength. The results here presented provide an interesting glimpse of the structural character of the flow. Since, however, the complete time- and space-resolved velocity field is available, one looks forward to the authors providing some quantitative analysis of these data in a future publication.

The paper of Miyata et al. is an experimental study of the influence of a thin plate oscillating in the nozzle of a two-dimensional plane jet. The Strouhal number of the forced oscillations was in the region of 0.01, which corresponds to a very long wavelength of oscillation compared to the jet width. The study thus amounts essentially to an exploration of the effects of flapping the jet up and down. The measurements are mainly concerned with the rate of spreading of the jet up to a distance of 25 nozzle widths downstream. The results show that the rate of spread of the jet is apparently strongly affected by the oscillations and, in the fully turbulent flow beyond 10 nozzle widths downstream, it is shown that the rate is up to five times greater than that of an undisturbed jet. However, these results are based on mean velocity measurements and, if one studies the results of the phased averaged velocity profiles contained within the paper, these do not seem to show the same dramatic effect on the rate of increase of the jet width. It is possible that part of the effect is simple due to oscillating the jet profile up and down in a quasi-steady manner. To illustrate this, we may assume a simple Gaussian expression for the velocity profile and imagine this profile to be oscillated transversely through a distance a. The time averaged velocity profile would then be given by

$$\overline{U(t)} = \frac{U_0}{2\pi} \int_{2\pi}^{2\pi} e^{-0.6932\left(\frac{y + a\sin t}{\delta}\right)^2} dt$$

where δ is the half-width of the undisturbed jet and U_0 is the centre-line velocity of the undisturbed jet. Figure 1 shows this time averaged profile compared with the steady jet profile for the case when $a = \delta$. The figure also shows the apparent turbulent intensities produced by the oscillations. It is not suggested that this is the sole cause of the apparent increase in the rate of spread of the jet but it is clearly making a contribution which it would be interesting to quantify more precisely than is possible here.

The paper by Fric and Roshko is a very interesting flow visualisation study of the flow around an axisymmetric jet issuing transversely into a cross stream. Interest in flows of this sort has generally been motivated by their importance to the aerodynamics of VTOL aircraft and it has to be said that the current attention on the periodic structures that have been observed in small scale tests is not of any real significance to the VTOL problem. Nevertheless, it is very gratifying to find that this

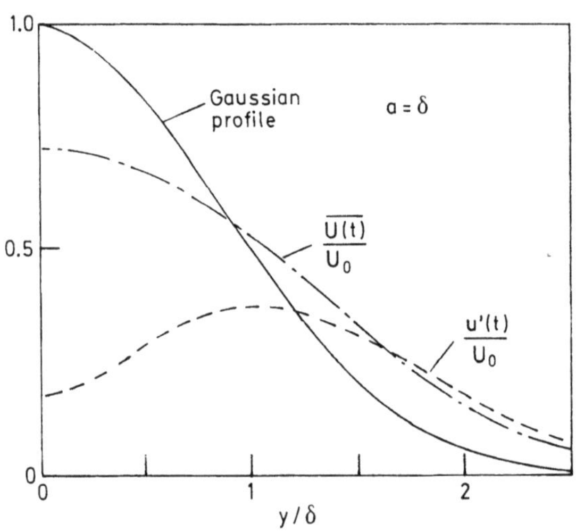

Fig. 1. Quasi-steady oscillations of a plane jet velocity profile

paper lays to rest some of the common misconceptions that surround the nature of flows past transverse jets. In particular, the discussion about the sources of vorticity contained in the flow downstream of the jet (the use of the word "wake" should be avoided) is very welcome. The writer attempted some similar flow visualisation studies some years ago and an attempt was made to suppress the boundary layer separations from the wall around the jet by the use of suction. However, the work was part of a final year student project and the results were inconclusive, as is often the case with student projects! Considering the entirely different characteristics of the flow downstream of a jet and the wake of a solid cylinder, I still find it surprising to find periodic structures present that have a Strouhal number not too different from that associated with the vortex shedding frequency behind a solid cylinder. I hope work on this topic will be continued because it raises some taxing questions in fundamental fluid mechanics.

The two final papers in this group examine the effects of curvature on the structure of free shear layers. Plesniak and Johnston were able to set up both stabilising and destabilising flow curvatures whereas Johnson and Hancock studied mainly stabilising curvature effects but with the additional effects of flow divergence. These papers must be seen as part of a continuing study of the effects of increasingly complex strain rates on turbulence structures and there is no point in reiterating the discussion of the results which are reasonably comprehensive in both cases. Their data will provide interesting material for turbulence modellers to mull over.

Three-Dimensional Aspects and Transition of the Wake of a Circular Cylinder

C. H. K. Williamson*

Mechanical and Aerospace Engineering, Upson Hall, Cornell University, Ithaca, NY 14853, USA

Abstract

Three-dimensionality in the wake of a cylinder is an intrinsic feature, not only in the transition regime when small-scales appear, but also when the vortex shedding is laminar. Certain features associated with this laminar regime also have relevance to the flow at higher (turbulent) Reynolds numbers.

In the laminar shedding regime (for $Re < 180$) two characteristics of the wake, which have received a great deal of debate in the past, are found to be related to each other; in the present paper the existence of a discontinuity in the Strouhal–Reynolds number relationship is explained by a transition between different modes of oblique vortex shedding. Oblique shedding is shown to be caused by the end boundary conditions in the case of cylinders of even hundreds of diameters in length. By manipulating the end conditions, parallel shedding can be induced, which then results in a completely-continuous Strouhal curve. It is also shown that this curve is "universal" in that all the oblique-shedding and parallel-shedding frequency data can be collapsed onto one continuous curve using a simple relationship. The case of parallel shedding represents *two-dimensional* vortex shedding, and therefore the "universal" Strouhal curve is compared with data from two-dimensional numerical computation.

The transition to three-dimensionality in the wake, at higher Reynolds numbers, is shown to involve two successive stages, each of which corresponds with a discontinuity in both the character of wake formation and also the S–Re relationship. The first discontinuity (near $Re = 180$) is associated with the inception of vortex loops, and it is hysteretic. The second discontinuity (between $Re = 230$–260) corresponds with a change to a finer-scale streamwise vortex structure, and in this case there is no hysteresis. There are fundamental differences between the development of three-dimensionality in this separated wake as compared with that found in other "unseparated" shear flows.

Finally, it is shown (in the laminar shedding regime) that a larger-scale wake evolves far downstream of the cylinder (and at roughly the same downstream location whether the shedding be parallel or oblique). In the case of the oblique shedding, the large-scale structure involves oblique instability "waves" at an angle of typically around twice the (upstream) oblique-shedding angle.

Nomenclature

L/D	Length/Diameter of cylinder
Re	Reynolds number
S	Strouhal number
θ	Oblique vortex shedding angle
U_c	Convection speed of the vortices downstream
U_∞	Free stream speed

* *PREVIOUSLY AT*: Graduate Aeronautical Laboratories, California Institute of Technology, Pasadena CA 91125, USA

Turbulent Shear Flows 7

1. Introduction

A number of investigations have recently been concerned with the general problem of the development of three-dimensional structure in turbulent shear flows, and with the corresponding implications for mixing in such flows. The form and development of such structures can often best be studied in the transition region, where smaller scales have not yet developed.

There are, however, features of the flow around a bluff body even in the laminar vortex shedding regime (in this case for Re from 49 to 178) that are three-dimensional, and have a bearing on the flow at higher (turbulent) Reynolds numbers. One form of three-dimensionality is the phenomenon of oblique shedding, which is shown to be induced by the end conditions, and which is also found to have a direct influence on the frequency of vortex shedding. Such an effect could, at least in part, explain some of the scatter in the measurements of Strouhal numbers in the literature (of the order of 20%), since the first extensive measurements by Roshko (1954). By a slight change in the end conditions, a Strouhal–Reynolds number curve for parallel shedding, or *two-dimensional shedding*, can be defined. The definition of such a curve is of particular importance in a comparison with two-dimensional numerical computations. With the advent of greater computational power to study turbulent shear flows, it would seem to be of some interest to define these experimental results carefully, particularly in the laminar flow regime, so that careful comparisons can be made.

When the wake undergoes transition to smaller-scale three-dimensionality at higher Reynolds numbers, vortex loops and streamwise vortices appear. In this case it is clearly of interest to determine the origin of such structures. One might question whether the development of streamwise vorticity in the separated wake is similar to such development in other free shear flows, or whether the transition to turbulence in the separating shear layers is necessarily involved in the formation of streamwise structures. These and other aspects of the transition regime are considered in the present work.

This paper comprises basically three parts: in Sect. 2 we shall consider the main three-dimensional aspects associated with the laminar vortex shedding regime; in Sect. 3 we will study the transition within the near wake to small-scale three-dimensionality, at higher Reynolds numbers. In Sect. 4 follows a brief discussion of other secondary aspects of three-dimensionality in the laminar vortex shedding regime.

2. Three-Dimensional Aspects of the Laminar Vortex Shedding Regime

2.1 Oblique and Parallel Modes of Vortex Shedding

The longstanding problem of the origin of discontinuities in the Strouhal–Reynolds number relationship has received a number of explanations in the past, since the original observation of such a discontinuity by Tritton (1959). For example, different modes of vortex shedding have been attributed to different forms of wake instability (Tritton 1959), to shear in the oncoming free stream (Gaster 1971), to differences in free stream turbulence (Berger and Wille 1972), and to changes in the importance of vorticity diffusion in the near wake (Gerrard 1978). More recently, Sreenivasan (1985) interpreted "Strouhal discontinuities" as involved in the route to chaos in the cylinder wake. A convincing case from some careful experiments was given by Van Atta and Gharib (1987) that some discontinuities could be the result of cylinder vibration. A particularly significant aspect of their work was the suggestion that a continuous S–Re relationship may exist without the presence of cylinder vibration. Their work was well supported by the 2-D numerical computations of Karniadakis and Triantafyllou (1989), who showed that in the absence of external forcing their Strouhal curve was indeed continuous. In fact, there exists strong evidence in the literature to support the suggestions that shear in the free stream and also cylinder vibration can cause S–Re discontinuities. However, the question still remains as to whether a discontinuity would appear in the absence of such effects.

In the present work it is found that the S–Re relationship that is measured here has a single discontinuity which is not due to any of the above suggested mechanisms, but rather is due to a pronounced three-dimensional effect; namely the phenomenon of oblique vortex shedding (a result which was originally shown in Williamson (1988a)). It was found that three different cylinders all produced a discontinuity at the same Reynolds number ($Re = 64$) as shown in Fig. 1a, providing strong evidence that, in this case, cylinder vibration is not involved. For these measurements, the cylinders were placed in a circular 6 in. test section of a wind tunnel, and two hot wires in the wake were used to measure oblique shedding angles. A great deal of care was taken to dampen the end constraints of the cylinders. (Van Atta and Gharib (1987) had demonstrated this important aspect earlier). It is immediately clear from Fig. 1b that oblique shedding is involved in an explanation for the discontinuity, because the measured angles of oblique shedding θ also show a discontinuity at $Re = 64$.

The shedding mode at $Re > 64$ involves a single frequency (f_L) right across the span (with the exception of small low-frequency cells adjacent to the ends, as found by Gerich and Eckelmann 1982), thus giving periodic velocity spectra as shown typically later in Fig. 8a. The mode at $Re < 64$ involves the appearance of a cell of slightly higher frequency (f_u) over the central main portions of the span, which is then sandwiched between cells of frequency f_L (shown schematically in Fig. 1 a). In this case, the spanwise cells cause quasi-periodic spectra, beat frequencies, and also the appearance of "vortex dislocations" between cell boundaries when the vortices in each cell are out of phase (and these features are discussed in detail in Williamson 1989). The main new result is that the Strouhal discontinuity is shown to be the direct result of a change in the modes of oblique shedding.

It is found (in the absence of other influences) that oblique shedding is caused by the end effects inducing a boundary condition of oblique shedding, whose influence spreads across the span from each end after the start of motion. This leads to a pattern of shedding in the form of a "chevron" as shown in Fig. 2a for $Re > 64$, with a symmetry about the centre of the span. (This visualisation was carried out by towing a cylinder through water in our XY towing tank, using fluorescent dyes washed off

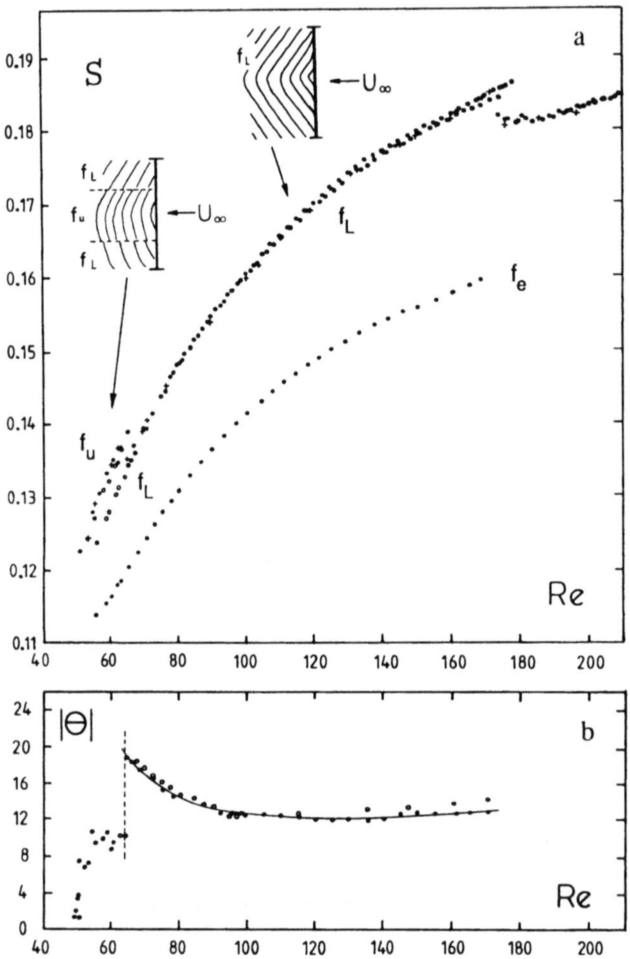

Fig. 1 a,b. Strouhal numbers and oblique shedding angles as a function of Reynolds number. **a** S versus Re: ○, $D = 0.104$ cm. ($L/D = 140$); ●, $D = 0.061$ cm ($L/D = 200$); +, $D = 0.051$ cm. ($L/D = 240$). **b** $|\theta|$ versus Re. ($|\theta|$ is the absolute value of oblique shedding angle.) ○ is for $-\theta$, ● is for $+\theta$. (From the 0.104 cm cylinder.) $L/D = 123$

the cylinder surface.) The above observations lead to a fundamental question: *Given that oblique shedding is caused by the end conditions, can we then manipulate the ends to induce parallel shedding?* It was found that indeed this could be done, in this case by angling inwards the leading edges of the endplates (an idea that was triggered by work with yawed cylinders of Ramberg 1983). It is important to note that there are other ways to achieve the same end; Eisenlohr and Eckelmann (1989) also induce parallel shedding in their case in a particularly neat fashion, by ending their cylinder span with larger coaxial cylinders. Parallel shedding is induced in the present case by reducing the base pressure at the ends (when the endplates are angled), and this has the effect of changing the vortex formation near the ends so that the vortices are no longer angled back. A boundary condition of parallel shedding leads to parallel shedding over the whole span, and is demonstrated for example in Fig. 2b for

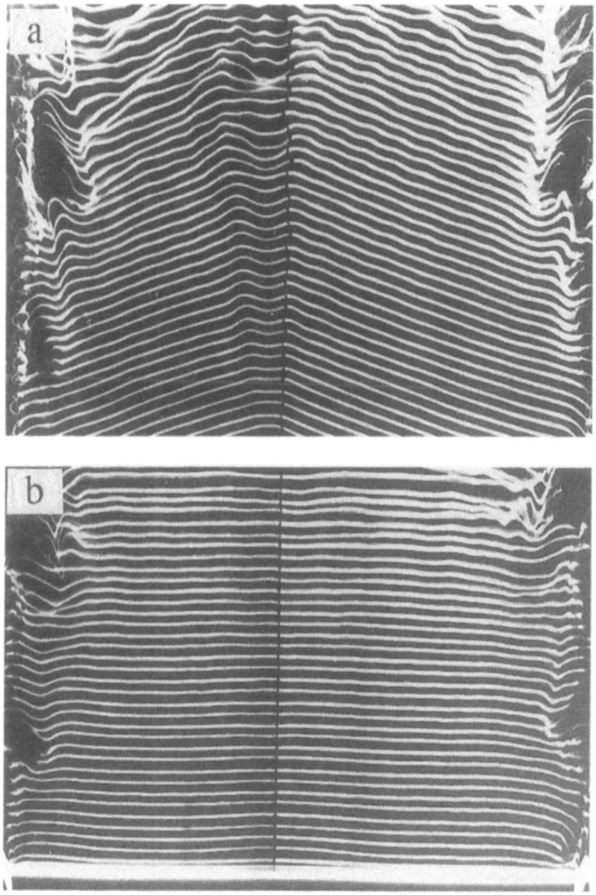

Fig. 2 a, b. Visualization of the different modes of oblique vortex shedding. Flow is upwards. Distance travelled by cylinder = $600D$. **a** Oblique shedding mode (high Re mode, $Re > 64$) with the "chevron" pattern of oblique vortices, at $Re = 85$. $L/D = 140$. **b** Parallel shedding mode, $Re = 110$, $L/D = 130$

$Re = 110$ and $L/D = 130$. This is in marked contrast with the oblique shedding mode in Fig. 2a.

We can thus arrange for either oblique shedding or parallel shedding to occur by manipulating the end boundary conditions, and the visualization of these modes of Fig. 2 are shown in schematic form in Fig. 3. Included also in this figure are other possibilities of oblique shedding which are the result of having *different end conditions*. In case (b), an oblique-shedding boundary condition at the left-hand end, and a parallel-shedding boundary condition at the right-hand end leads to one angle of shedding ($+\theta°$) across the whole span (rather than the "chevron-pattern" shown in (a)). Similarly, if we arrange the end conditions of (b) the other way round, this leads to the opposite angle of shedding across the whole span, i.e. we now induce the angle $-\theta°$ as shown in the schematic of (c). The "Parallel shedding condition" marked in Fig. 3 signifies an end boundary condition which, if applied at both ends, will yield parallel shedding across the whole span. (Similarly for the "oblique

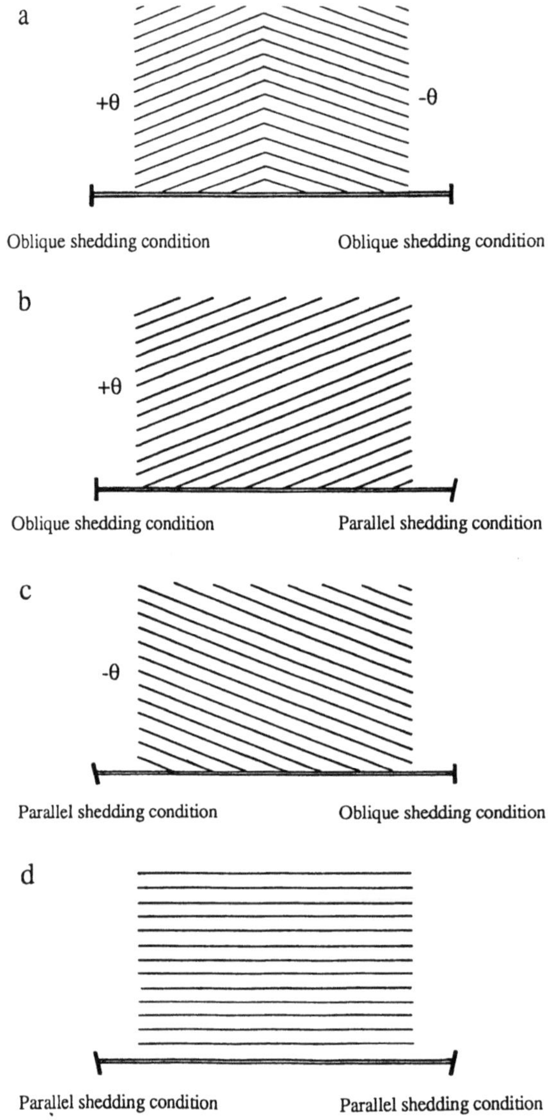

Fig. 3a–d. Schematic of different oblique vortex patterns dependant on boundary conditions. The "Parallel shedding condition" marked in Fig. 3 signifies an end boundary condition which, if applied at both ends, will yield parallel shedding across the whole span. (Similarly for the "oblique shedding condition")

shedding condition".) Some interesting conclusions about conditions necessary for oblique and parallel shedding can now be made:

1) For oblique shedding to occur it is *not necessary* to have different end boundary conditions.
2) For parallel shedding, it is *necessary but not sufficient* to have both end conditions the same. (We must also arrange for the base pressure in the end regions to be sufficiently low.)
3) If oblique shedding occurs and the end conditions are the same, then a *chevron pattern of oblique shedding* will occur, rather than simply one angle of shedding across the span.

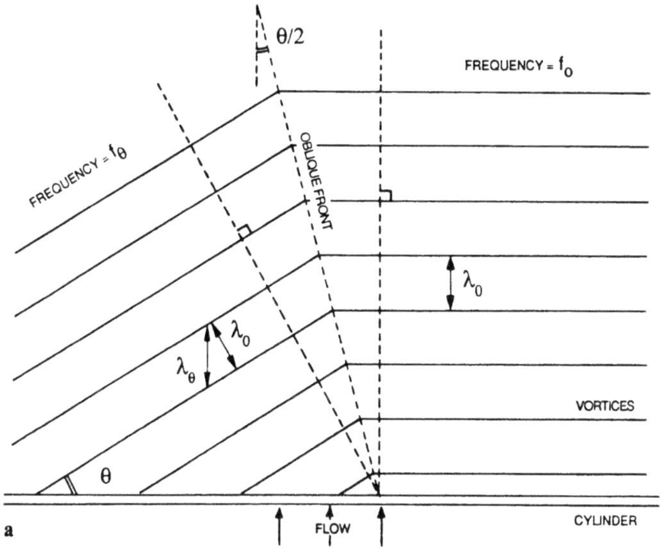

a

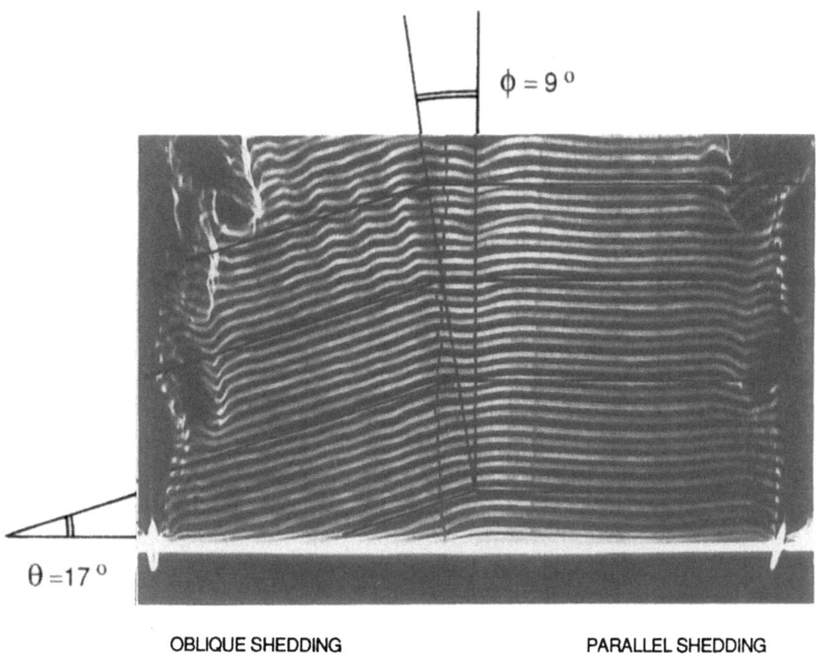

b

Fig. 4 a, b. Geometry of an oblique vortex front. **a** Sketch of the oblique front dividing a spanwise region of oblique vortices from a region of parallel vortices. **b** Visualisation of the oblique front. Flow is upwards. Distance travelled by cylinder $= 500\,D$. $Re = 90$. $L/D = 130$

4) If one end has a boundary condition to induce oblique shedding while the other end has a condition to induce parallel shedding, then the *oblique shedding will take over the whole span* (at one single oblique angle, say $+\theta$).

It should also be noted that none of the oblique shedding patterns in Fig. 3a,b or c develop immediately from the start of motion in the *XY* Towing Tank. The vortices *initially shed parallel to the cylinder* regardless of the end conditions. Gradually, as the cylinder is towed down the tank, there is a region of oblique shedding due to the end conditions, which spreads into the region of initially-parallel shedding behind an "oblique front". It can be shown that this front will have an angle $\phi = \theta/2$ as shown in Fig. 4a, i.e. the angle of the oblique front will be half the oblique shedding angle. This result comes from measurements in Williamson (1989) which show that, at a given Reynolds number, the wavelength normal to the vortex axes are constant, irrespective of the oblique vortex angle. Furthermore, using simple geometric considerations with the configuration in Fig. 4a, we can predict the time it takes for oblique shedding end conditions to impose oblique shedding across the whole span as in Fig. 2a. This result can be given more conveniently in terms of the distance travelled by the cylinder (x) from the start of motion:

$$\frac{x}{D} = \frac{(L/D)}{\theta(U_c/U_\infty)}$$

where the shedding angle θ is small, and where the term (U_c/U_∞) is the normalised convection speed of the vortices downstream. Typically, for a cylinder with $L/D =$ 100, and using typical values for θ and (U_c/U_∞) this equation suggests that we need to move the cylinder around 450 diameters before oblique shedding in the "chevron pattern" will cover the whole span. (This is roughly equivalent to 90 cycles of vortex shedding, which is surprisingly large considering we might expect otherwise that the wake will have reached its asymptotic form after about 3–4 cycles!) An even more extreme example is given in the case of different end conditions, shown for example in Fig. 3b, where we should expect (for $L/D = 100$) that oblique shedding will cover the whole span only after 900 diameters of travel.

Finally, experiments in the towing tank show that indeed these are reasonable figures for the cylinder travel in order that the wake should assume its final form. Typical flow visualisation is shown in Fig. 4b where, after 500 diameters of travel, the oblique front has reached the centre of the span, and has an angle approximately half the oblique shedding angle. Ultimately, after sufficient cylinder travel in the towing tank, the shedding will be at one oblique angle $+\theta$ ($= 17°$) across the whole span, as shown diagramatically in Fig. 3b.

2.2 A "Universal" and Continuous *S–Re* Relationship for Two-Dimensional Shedding

In the case of parallel shedding, the Strouhal–Reynolds number curve becomes completely continuous (and it should be noted that the velocity spectra throughout the laminar regime are periodic) because there is no change of modes as is found for the oblique shedding. This result is demonstrated by the solid symbols of Fig. 5. Further support for these results comes from data for parallel shedding from a very different facility, namely the XY water towing tank, from which there was found to be close aggreement with the wind tunnel data. It should also be observed that the results are independant of L/D, and further data not plotted here for $L/D = 250$ support this conclusion. Finally, we define this curve of Strouhal data as "universal" (which was a concept introduced in Williamson 1988a) specifically in the sense that the oblique shedding data (S_θ) can be collapsed onto the parallel shedding data (S_0) by the transformation

$$S_0 = S_\theta / \cos\theta$$

where θ is the angle of oblique shedding. (The use of such a formula was suggested to the author by Garry L. Brown, private communication). The collapse of this data is shown clearly in Fig. 5. It should be mentioned here that an experimentally-obtained continuous S–Re relationship *does not imply* that it will lie on the "universal" curve, unless the shedding is parallel to the cylinder.

One can imagine several alternative ways of inducing parallel shedding by manipulating the end conditions so that the flow near the ends (the base pressure and shedding frequency) matches that flow over the rest of the span (for example, consider the technique used by Eisenlohr and Eckelmann (1989) to obtain parallel shedding). It seems most probable that, in whatever fashion the end conditions are

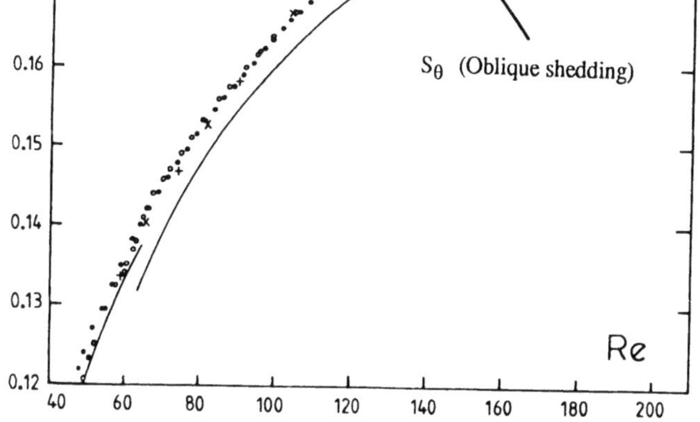

Fig. 5. The "universal" Strouhal–Reynolds number curve. Wind tunnel data: ●, parallel-shedding data ($D = 0.104$ cm, $L/D = 90$); ○, collapsed oblique-shedding data using the formula: $S_0 = S_\theta/\cos\theta$ ($D = 0.104$ cm, $L/D = 123$); ——, curves through the original S_θ data. X–Y Towing Tank data: ×, $D = 0.657$ cm, $L/D = 72$; +, $D = 0.328$ cm, $L/D = 140$

arranged to yield parallel shedding over long cylinder spans, the resulting S–Re relationships will agree closely. There are indications already that this is the case, from a preliminary comparison with the measurements of Eisenlohr and Eckelmann (though at the time of writing their exact data points or curve fit are not available). A least squares curve fit to the present "universal" Strouhal curve is given by:

$$S = A/Re + B + CRe,$$

where $A = -3.3265$, $B = 0.1816$, and $C = 1.600 \times 10^{-4}$, or less accurately by

$$S = A/Re + B$$

where $A = -5.11$ and $B = 0.218$.

Finally, the case of parallel shedding is an *experimental realisation of two-dimensional shedding*, in that the flow at all cross sections are in phase (except close to the cylinder ends) and there is an absence of axial flow in the vortices (Williamson 1989). For this reason, some two-dimensional numerical computations of Strouhal numbers are compared with the present experimental data in Fig. 6, although the agreement amongst the majority of the results is not close. It is perhaps a challenge to determine the causes of this disparity, a question which is at present being

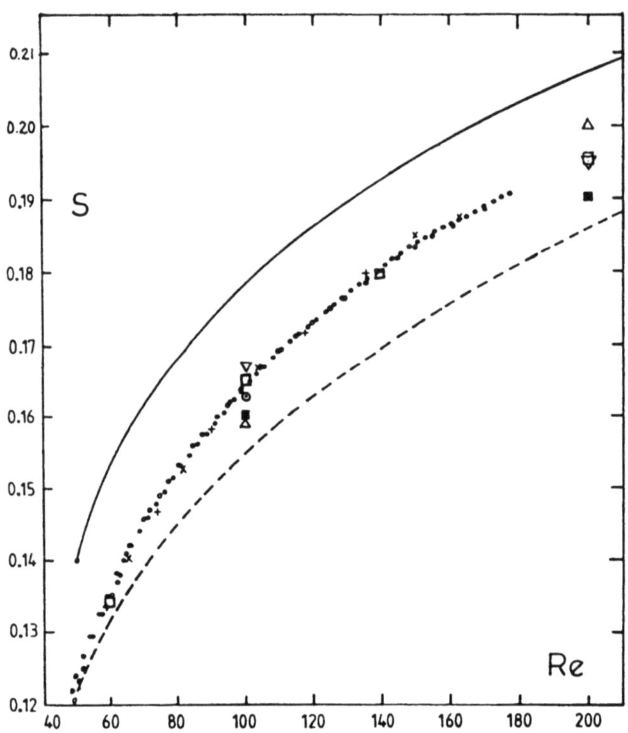

Fig. 6. Comparison of the "universal" Strouhal–Reynolds number curve for two-dimensional vortex shedding, with data from two-dimensional numerical simulations. Experimental results: symbols as in Fig. 3. Numerical results:——Karniadakis and Triantafylou (1989); – – – –Sa and Chang (1988); △ Braza et al. (1986); □ Lecointe and Piquet (1988); ■ Martinez (1987); ▽ Koschel et al. (1989)

investigated at Caltech and elsewhere. It seems that both the upstream and downstream boundary conditions of the computational domain are of great importance in such calculations.

3. The Transition to Small-Scale Three-Dimensionality in the Near Wake

In the case where the shear layers separate from the surface of a bluff body, it is of interest to study how the transition to three-dimensionality and the resulting streamwise vorticity structures differ from those which occur for a free shear layer (e.g. Bernal and Roshko), or for an "unseparated" wake downstream of a splitter plate (e.g. Meiburg and Lasheras).

The inception of three-dimensionality, in the case of the cylinder wake, involves two successive stages, each of which corresponds with a discontinuity in the Strouhal–Reynolds number relationship, as is shown in Fig. 7 (which has also been demonstrated in Williamson 1988b). The first discontinuity (at around *Re* of 180) is associated with the inception of vortex loops (mode A), and it is hysteretic. The second discontinuity (between $Re = 230$ to 260) corresponds with a change to a

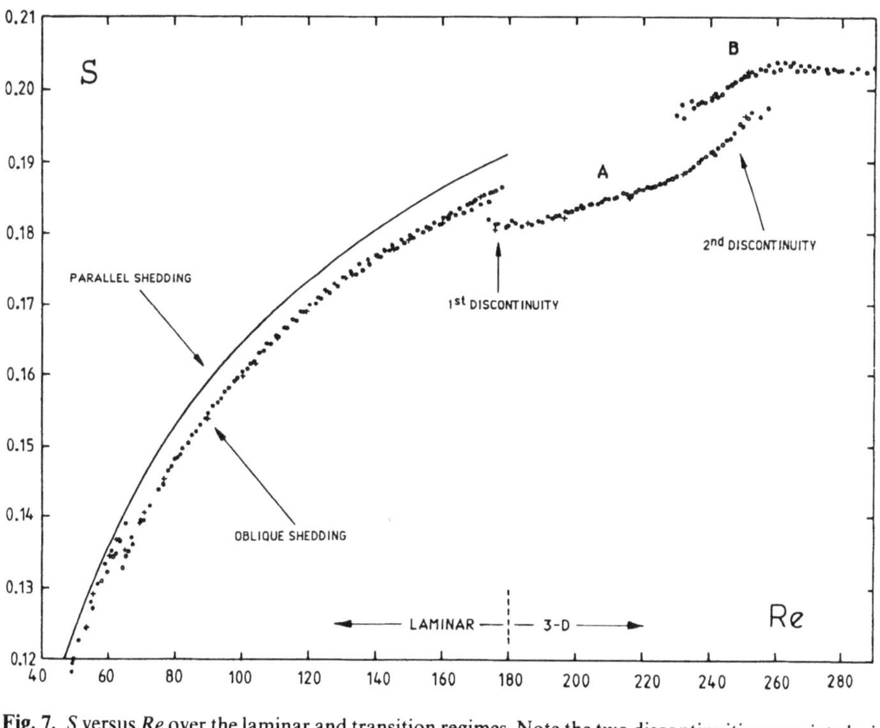

Fig. 7. *S* versus *Re* over the laminar and transition regimes. Note the two discontinuities associated with the inception of different forms of three-dimensional structures. Legend: ○, $D = 0.104$ cm ($L/D = 140$); ●, $D = 0.061$ cm ($L/D = 200$); + $D = 0.051$ cm ($L/D = 240$)

finer-scale streamwise vortex structure (mode B). At this discontinuity there is no hysteresis, and it is suggested that two modes of vortex shedding alternate in time. The character of these discontinuities and the form of the vortex structure are investigated in the present work.

Above $Re = 178$, the behaviour of the spectra corresponding to each of the two Strouhal discontinuities (marked in Fig. 7) is distinctly different. In the case of the first discontinuity, there is a small overlap of the Strouhal curves near $Re = 180$, and also a hysteresis, which is demonstrated by the spectra in Fig. 8a. The sharply peaked spectrum was obtained by increasing Re up to 172.8, and this spectrum reflects the presence of laminar vortex shedding; in this case we are on the (upper) laminar Strouhal curve (see Fig. 7). The broad spectrum in Fig. 8a was the result of decreasing Re down to 172.8, and it is associated with the less-ordered velocity fluctuations caused by the inception of vortex loops; we are now on the lower Strouhal curve marked A (in Fig. 7). Some confidence in the curves of the Strouhal frequency data in Fig. 7 is assured by the fact that there is close agreement between experiments using three different cylinders.

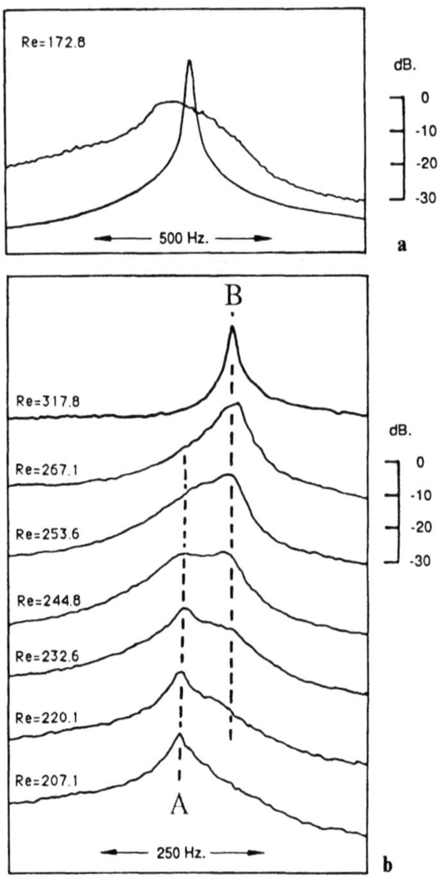

Fig. 8 a,b. Spectra of velocity fluctuations. **a** Spectra at first discontinuity ($Re = 172.8$). **b** Spectra at second discontinuity ($Re = 207$ to 318)

The second discontinuity is distinctly different from the first. It not only extends over a larger range of *Re* (from 230 to 260), but involves the existence of two peaks in the spectra simultaneously, as shown in Fig. 8b. Consequently, the flow behaviour does not exhibit a hysteresis, as found for the first discontinuity. We can see that there is a gradual shift of energy from the lower peak to the upper peak of the spectra (i.e. from mode A to mode B), as *Re* is increased (moving upwards in this figure). This corresponds, in Fig. 7, to a transfer of energy from the Strouhal frequencies

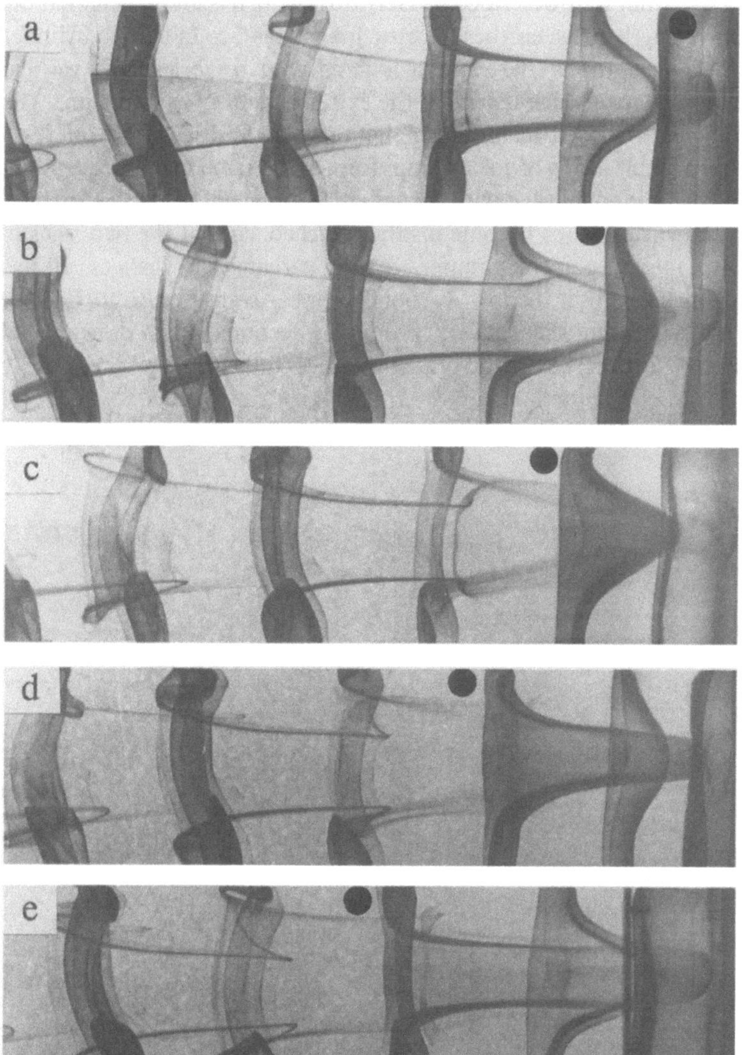

Fig. 9. Generation of vortex loops and pairs of counter-rotating streamwise vortices. This is mode A, after the first discontinuity, $Re \approx 180$. (This corresponds with Strouhal curve A in Fig. 7.) Cylinder is vertical, and at extreme right of each photograph. Flow is to the left. Length of span visualised = 3 Diameters

associated with curve A to those frequencies associated with curve B. It is suggested (and there is further evidence to show) that the lower-frequency mode of shedding alternates in time with the upper-frequency mode; the flow is neither locked into one mode nor the other, and thereby peaks corresponding to both modes appear in the long-time averaged spectra simultaneously.

Some results of flow visualization in the *XY* towing tank are shown in Figs. 9 and 10. The sequence of photos in Fig. 9 shows the typical development of vortex loops when *Re* is roughly between 178 and 250, corresponding to our mode A. These loops are similar to those waves first observed by Hama (1957), and are possibly the same as the "fingers of dye" that were described by Gerrard (1978). It is found here that the characteristic wavelength between these vortex loops was around three cylinder diameters. In the photographs, the camera is fixed with respect to the vertical cylinder, whose downstream edge is seen at the extreme right of each picture. The process of loop generation is self-sustaining in that there is a feedback from one loop to the next, so that a whole string of vortex loops form at the same spanwise position. Referring to the vortex loop marked with a solid circle in the photograph sequence, we can see that the vortex loops become highly stretched, so that the two sides of each loop evolve into a pair of counter-rotating streamwise vortices. It has previously been suggested that the low Reynolds number transition to turbulence must occur in the separating shear layers (producing secondary two-dimensional "Kelvin–Helmholtz" vortices) before the primary vortices fully form and break away

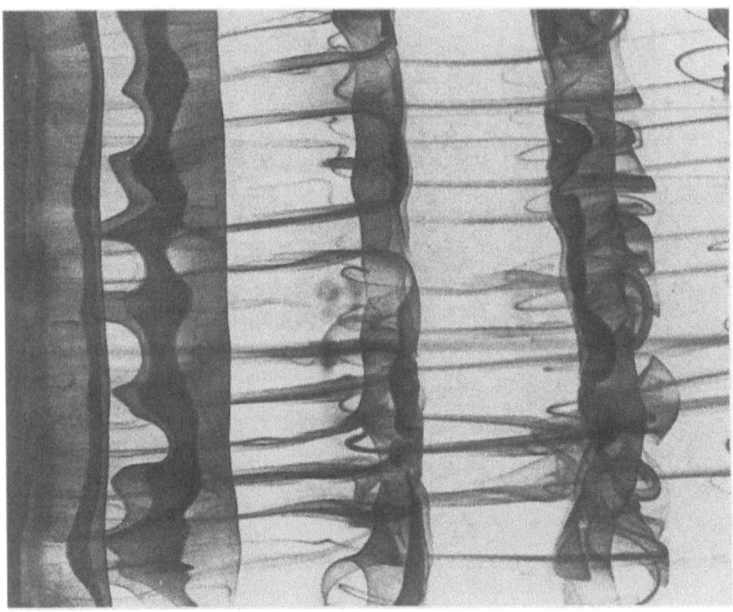

Fig. 10. Generation of fine-scale streamwise vortices. This is mode *B*, after the second discontinuity, $Re = 285$. (This corresponds with Strouhal numbers along curve *B* in Fig. 7.) Cylinder is vertical and at extreme right, and flow is to the left. Length of span visualised is about 6.5 diameters

from the near wake (Roshko 1954). It has also been supposed that streamwise vortices in the cylinder wake originate from the three-dimensional stretching of those shear layer instability vortices (Wei and Smith 1987). However, those vortices in the shear layer begin to appear at a Reynolds number of around 1000, and therefore do not seem to be involved in the generation of the streamwise vortices at the lower Reynolds numbers of the transition regime (i.e. *Re* around 200–300). In the present work, it is found that the primary vortices roll up first, and subsequently deform during the process of shedding to create the vortex loops, and thereby also generate the streamwise vortices.

At higher Reynolds numbers, finer-scale streamwise vortices are formed, as shown typically in Fig. 10. The primary vortices now deform with a markedly smaller spanwise wavelength during shedding, and leave vorticity in a high strain-rate region (between primary vortex structures), where three-dimensional distortions are rapidly amplified. The wavelength between these more vigorous streamwise vortex pairs is approximately one diameter, and remains so up to at least $Re = 800$. In other flows, such as free shear layers (e.g. Bernal and Roshko 1986) and the "unseparated" wake from a splitter plate (e.g. Meiburg and Lasheras 1988), a

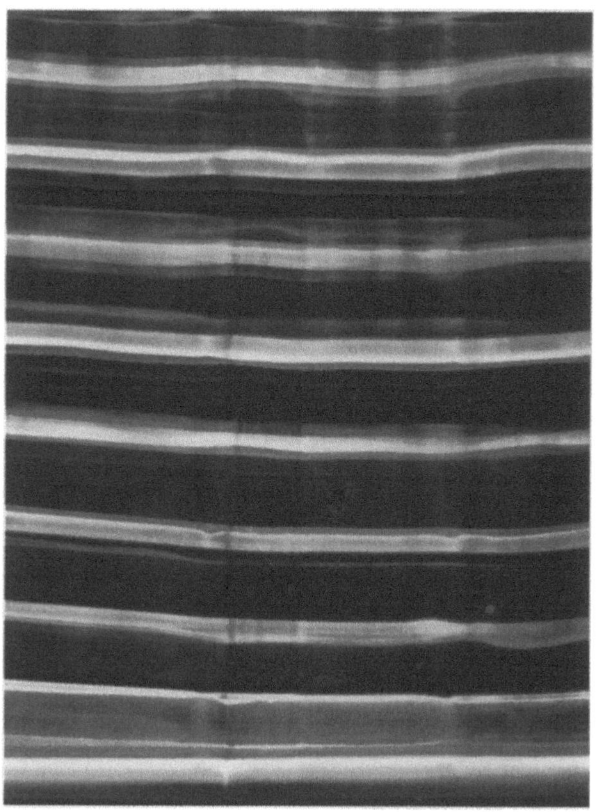

Fig. 11. Colour spanwise visualisation of laminar vortex shedding. $Re = 100$. Flow is upwards. Cylinder is shown as the yellow horizontal line at the bottom of the photograph. $Re = 100$

basic mechanism in the formation of streamwise vortices is the deformation and intense streamwise stretching of the spanwise vorticity that resides in the "braid" region between the major spanwise vortices. The cylinder wake (at low *Re*) seems to be different, in that the streamwise vortices are formed mainly from "core" vorticity; i.e. from vorticity which is originally in the primary vortices. Core vorticity is pulled into the braid region when the vortices deform during the process of shedding. *Thereafter*, the vorticity is highly strained in the regions between the primary vortices, as it is in other shear flows. The process by which the wake forms behind the cylinder is clearly different from the other shear flows (mentioned above) because the cylinder itself interacts with the formation of the wake vortices.

Finally, some flow visualisation in colour shown in Figs. 11, 12 and 13 demonstrates clearly the length scales associated with each mode of vortex shedding. It should be noted that each of these photographs are to exactly the same scale. The laminar mode of sheeding is shown in Fig. 11, and may be compared with the large-scale vortex loops of mode A (3-D) shedding in Fig. 12 and also with the fine-scale streamwise vortex structures of mode B (3-D) shedding in Fig. 13. In particular, the laminar primary (Karman) vortices contrast greatly with the turbulent primary vortices that are shown in Fig. 13, when the fine-scale streamwise vortex pairs are wound around the primary vortex cores, and are stretched into thin filaments.

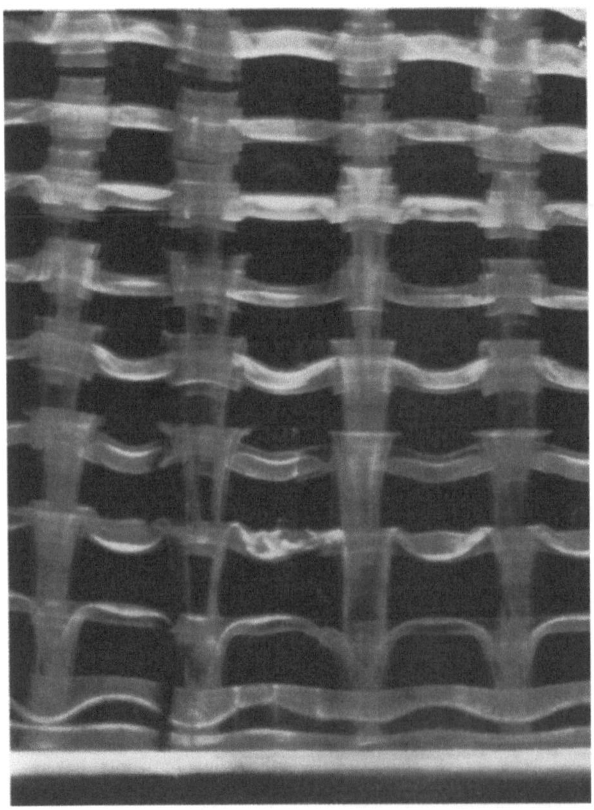

Fig. 12. Colour spanwise visualisation of mode A (3-D) shedding. $Re = 200$. This photograph shows the initial spanwise instability which leads to the formation of large-scale vortex loops. (To the same scale as Fig. 11.) Flow is upwards

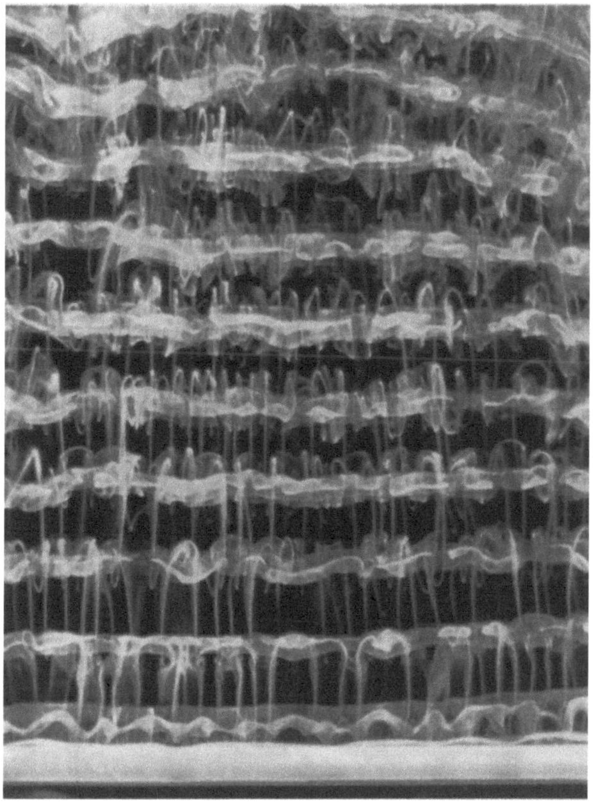

Fig. 13. Colour spanwise visualisation of mode B (3-D) shedding. $Re = 270$. This photograph shows the deformation of a shedding vortex and the formation of the fine-scale streamwise vortices. (To the same scale as Fig. 11.) Flow is upwards

4. Three-Dimensional Secondary Instability of the Far Wake in the Laminar Shedding Regime

As the wake from the cylinder travels downstream, so a larger-scale wake evolves out of the original vorticity distribution by diffusion and amalgamation (and also possibly by subharmonic vortex interactions, though this is not necessarily the case). This large-scale downstream wake structure occurs in both the laminar and 3-D turbulent shedding regimes. Recent studies have been made of the large-scale structure; from experiment by Cimbala et al. (1988), and from numerical computation by Meiburg (1987). Cimbala et al. found that the far-wake structure does not depend directly on the scale or frequency of Karman vortices, but that the growth of this structure is due to the hydrodynamic instability of the developing mean wake profile, and that its scale can reasonably be predicted using linear stability analysis of the profile. Three-dimensionality in the far wake, in the form of a cellular (or "honeycomb") structure seen in a plan view of the wake was also observed strikingly, and it was suggested that this structure is a result of an interaction of oblique disturbances with the two-dimensional (parallel with the cylinder) instability in the far wake. The interaction was suggested to be in the form of a parametric subharmonic resonance similar to that put forward by Pierrehumbert and Widnall (1982) for shear layers, and by Herbert (see Herbert 1988) for boundary-layer flows. Observations in the present study have some similarity with the above results.

In the present work it is found, from observations in planview, that a large-scale wake structure develops out of the original shed vorticity, and at around the same location downstream of the cylinder whether the shedding at the body is oblique or parallel. When there is parallel shedding, the downstream large-scale wake evolves into predominantly a parallel (to the cylinder) structure, and some waviness in the spanwise direction was found although a spanwise-periodic form did not necessarily

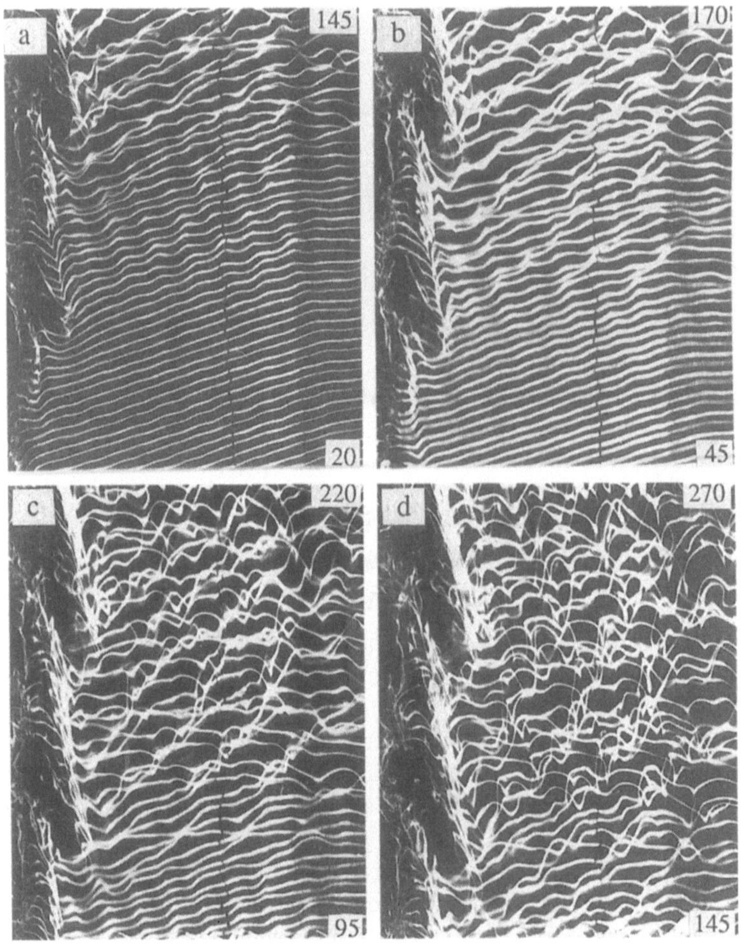

Fig. 14a–d. Evolution of large-scale instability in the far wake. $Re = 100$. Following the oblique shedding at the cylinder, the wake downstream develops "waves" at an angle typically around twice the original oblique angle. The numbers show the relevant values of x/D. "The letters a–d simply show the correct order of the photographs"

occur. However, in the case of oblique vortex shedding, the downstream structure involves the evolution of what seem to be oblique instability waves at an angle typically of around twice the original oblique shedding angle. This can be seen in the sequence of Fig. 14, and in the more close-up visualisation in Fig. 15. (The large structures to the left of each photograph are end effects and have been discussed in Williamson (1989)). In these visualisations we can see that the original oblique Karman vortices (or the dye that now remains after the vorticity has diffused at a faster rate) are distorted in a wavy fashion along lines, or along what we could perhaps term as oblique instability waves, which are slanted one way only (the same way as the original shedding angle). Also, the lower halves of Fig. 14c and 14d show that there is some repeatability in the three-dimensional pattern, in the direction

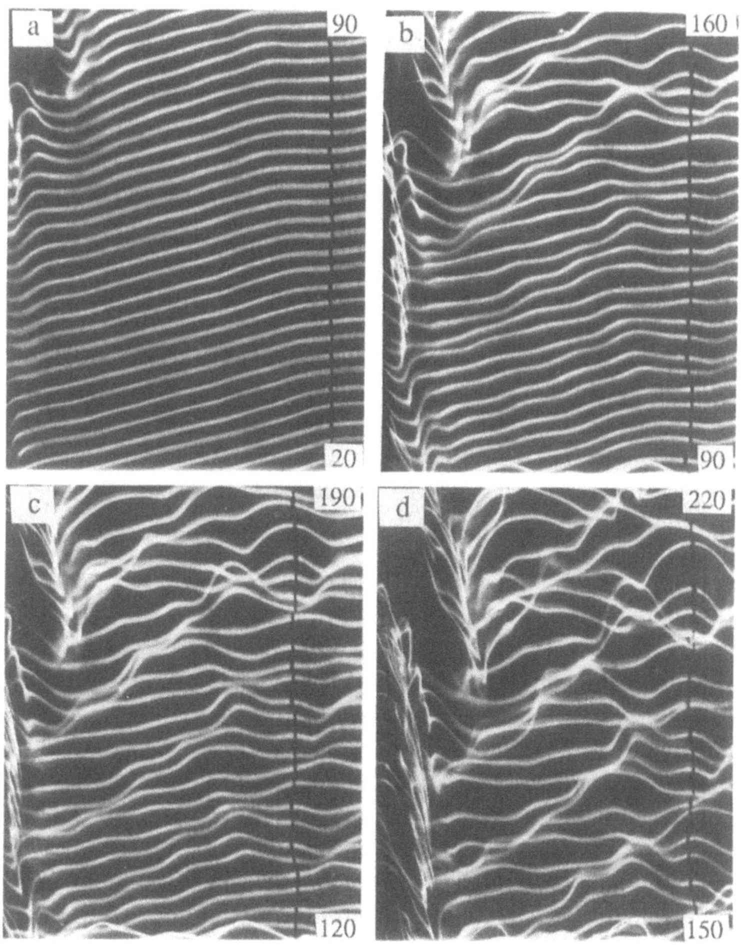

Fig. 15 a–d. Close-up views of evolution of large-scale instability in the far wake. $Re = 100$. This sequence shows a more close-up view than Fig. 14, and is for similar experimental conditions. The numbers show the relevant values of x/D. "The letters a–d simply show the correct order of the photographs"

parallel to the cylinder (in this sense like those observations from Cimbala et al.). It seems quite possible that these oblique instability "waves" are the result of an interaction between the original oblique vortex shedding distribution (after it has diffused, etc.) with a large-scale two-dimensional instability as discussed by Cimbala et al. The suggestion here is that the origin of the oblique disturbances is in fact the oblique shedding from the cylinder which, after interaction with two-dimensional disturbances, can then cause other oblique angles to show up in planview in the far wake (such as the oblique instability waves found here). Finally, it should be noted that the present observations of this structure was carried out in a towing tank under extremely "quiet" conditions, and therefore suggests that such three-dimensional structures in the far wake do not need to be triggered by oblique disturbances related to the test facility, but can develop intrinsically out of the original shed vorticity distribution.

5. Conclusions

It is clear from the present work that there are intrinsic three-dimensional aspects of the cylinder wake in the laminar vortex shedding regime, as well as in the transition regime at higher Reynolds numbers ($Re > 180$).

In the laminar regime, the longstanding question as to the origin of discontinuities in the S–Re relationship is explained by the transition between different modes of oblique vortex shedding. It is found that the end effects influence the flow over spans of hundreds of diameters in length, and are the cause of oblique shedding itself. However, by a slight manipulation of the end conditions we can induce parallel shedding over the whole span. The resulting Strouhal–Reynolds number curve is not only completely-continuous but also *universal*, in the sense that the oblique-shedding data (S_θ) can be collapsed onto the parallel-shedding data (S_0) by the simple transformation

$$S_0 = S_\theta / \cos\theta$$

where θ is the oblique shedding angle. The parallel-shedding condition is an experimental realisation of *two-dimensional* shedding, and for this reason comparison is made with two-dimensional computation, although the disparity between the results is not, as yet, understood. There has been some question in the past as to whether oblique or parallel shedding is the intrinsic experimental mode, but the present work shows that both cases are intrinsic, and are simply solutions to different problems, because the end boundary conditions are different in each case.

The transition to smaller scales at higher Re involves two stages where different modes of three-dimensionality appear, and for each mode their onset is characterised by a discontinuity in the S–Re relationship. The first discontinuity corresponds with the inception of large-scale vortex loops (mode A), while the second discontinuity is associated with finer-scale streamwise vortices (mode B). The streamwise structures at these low Re are not the result of transition in the separating shear layers as previously supposed, because the latter instability appears at much higher *Re*. Also, the formation of streamwise vorticity in the present cylinder wake is different from the case with other shear flows, because here it originates in the first instance *primarily* from "core" vorticity, rather than from vorticity in the "braid" region. The above results for the transition regime form part of a larger and more comprehensive piece of research, which is being completed.

Further observations of secondary aspects for the laminar shedding regime show that a large-scale wake can form far downstream of the cylinder, as has been found before. In the present case, when there is oblique shedding at the cylinder, this large-scale downstream structure involves instability waves which are also oblique, but typically at around twice the original angle of oblique shedding.

Acknowledgements: The author would like to thank Anatol Roshko for valuable discussions, and Garry L. Brown (Director of the Aeronautical Research Laboratories, Melbourne, Australia) for his suggestions and immense enthusiasm, and particularly for his suggestion for a $\cos\theta$ relationship for the oblique shedding frequencies.

This work was supported by an O.N.R. contract No. N00014-84-K-0618.

References

Berger, E., Wille, R. (1972): Periodic flow phenomena. Ann. Rev. Fluid Mech. **4**, 313

Bernal, L. P., Roshko, A. (1986): Streamwise vortex structure in plane mixing layers. J. Fluid Mech. **170**, 499

Braza, M., Chassaing, P., Ha Minh, H. (1986): Numerical study and analysis of the pressure and velocity fields in the near wake of a cylinder. J. Fluid Mech. **165**, 79

Cimbala, J. M., Nagib, H. M., Roshko, A. (1988): Large structure in the far wakes of two-dimensional bluff bodies. J. Fluid Mech. **190**, 265

Eisenlohr, H., Eckelmann, H. (1989): Vortex splitting and its consequences in the vortex street wake of cylinders at low Reynolds number. Phys. Fluids **A1**, 189

Gaster, M. (1971): Vortex shedding from circular cylinders at low Reynolds numbers. J. Fluid Mech. **46**, 749

Gerich, D., Eckelmann, H. (1982): Influence of end plates and free ends on the shedding frequency of cylinders. J. Fluid Mech. **122**, 109

Gerrard, J. H. (1978): The wakes of cylindrical bluff bodies at low Reynolds number. Phil. Trans. Roy. Soc. **A288**, 351

Hama, F. R. (1957): Three-dimensional vortex pattern behind a circular cylinder. J. Aerosp. Sci. **24**, 156

Herbert, T. (1988): Secondary instability of boundary layers. Ann. Rev. Fluid Mech. **20**, 487

Karniadakis, G. E., Triantafyllou, G. S. (1989): Frequency selection and asymptotic states in laminar wakes. J. Fluid Mech. **199**, 441

Koschel, W., Lotzerich, M., Vornberger, A. (1989): Solution on unstructured grids for the Euler and Navier–Stokes equations. AGARD, CP-437, Vol. 1, 26–1

Lecointe, Y., Piquet, J.: Numerical methods for unsteady flows. VKI Lectures 1988–7. Also Lecointe, Y.: private communication

Martinez, G. (1979): Character of the flow around a circular cylinder at moderate Re. PhD. Thesis, Institut. Nat. Polytec. Toulouse

Meiburg, E. (1987): On the role of subharmonic perturbations in the far wake. J. Fluid Mech. **177**, 83

Meiburg, E., Lasheras, J. (1988): Experimental and numerical investigation of the three-dimensional transition in plane wakes. J. Fluid Mech. **190**, 1

Pierrehumbert, R., Widnall, S. (1982): The two- and three-dimensional instabilities of a spatially periodic shear layer. J. Fluid Mech. **114**, 59

Ramberg, S. E. (1983): The effects of yaw and finite length upon the vortex wakes of stationary and vibrating cylinders. J. Fluid Mech. **128**, 81

Roshko, A. (1954): On the development of turbulent wakes from vortex streets. NACA Report 1191
Sa, J.-Y., Chang, K.-S. (1988): Shedding pattern of the near-wake vortices behind a circular cylinder. J. Fluid Mech. (To appear)
Sreenivasan, K. R. (1985): Transition and turbulence in fluid flows and low-dimensional chaos. In *Frontiers in Fluid Mechanics* (S. H. Davis, J. L. Lumley, ed.), pp. 41–66. Springer
Tritton, D. J. (1959): Experiments on the flow past a circular cylinder at low Reynolds numbers. J. Fluid Mech. **6**, 547
Van Atta, C., Gharib, M. (1987): Ordered and chaotic vortex streets behind circular cylinders at low Reynolds numbers. J. Fluid Mech. **174**, 113
Williamson, C. H. K. (1988a): Defining a universal and continuous Strouhal–Reynolds number relationship for the laminar vortex shedding of a circular cylinder. Phys. Fluids **31**, 2742
Williamson, C. H. K. (1988b): The existence of two stages in the transition to three-dimensionality of a cylinder wake. Phys. Fluids **31**, 3165
Williamson, C. H. K. (1989): Oblique and parallel modes of vortex shedding in the wake of a cylinder at low Reynolds numbers. J. Fluid Mech. **206**, 579

Experimental and Numerical Analysis of the Three-Dimensional Evolution of an Axisymmetric Jet

Eckart Meiburg[1], Juan C. Lasheras[2], and James E. Martin[1]

[1] Center for Fluid Mechanics, Division of Applied Mathematics, Brown University, Providence, RI 02912, USA
[2] Department of Mechanical Engineering University of Southern California Los Angeles, CA 90089-1453, USA

Abstract

We study the three-dimensional evolution of a transitional axisymmetric jet subjected to periodic perturbations both in the streamwise and in the circumferential direction. The combined analysis of flow visualization experiments and inviscid vortex dynamics simulations provides a detailed picture of the processes leading to the concentration, reorientation, and stretching of the vorticity. A single perturbation in the streamwise direction leads to the formation of vortex rings, while a free stagnation point forms in the downstream half of the braid region between successive vortices. If we also introduce a subharmonic perturbation in the streamwise direction, neighboring vortices proceed towards a pairing process. In addition, it is shown that when the jet is also subjected to a sinusoidal perturbation in the azimuthal direction, counterrotating pairs of streamwise vortex tubes are formed in the braid regions, and the cores of the vortex rings develop a wavy dislocation. We discuss the importance of global and local induction for the evolution and interaction of these three-dimensional instability modes.

1. Introduction

Axisymmetric jets occur in a variety of technical applications such as in engines, chemical reactors, and in combustion processes. The structures evolving in these flows are intimately linked to features like flame stability, noise generation, and the effectiveness of mixing. At the same time, axisymmetric jets represent a prototype flow for studying fundamental aspects of transitional and turbulent flows. As a result, there is considerable interest both from the fundamental and from the application oriented point of view to achieve a better understanding of, and eventually control over, the fluid mechanical processes involved.

It is well known from stability theoretical, experimental, and numerical work (Crow and Champagne 1971; Yule 1978; Plaschko 1979; Zaman and Hussain 1980; Hussain and Zaman 1980, 1981; Hussain and Clark 1981; Dimotakis et al. 1983; Cohen and Wygnanski 1987a, b; Agui and Hesselink 1988; Tso and Hussain 1989) that the vorticity layer emanating from the nozzle rolls up into vortex rings, which can undergo subsequent pairing, fractional pairing, and tearing processes. Furthermore, helical modes have been observed and analyzed as well. During the evolution of the primary roll-up instability, the braid regions between these vortex rings become largely depleted of vorticity, in a manner similar to plane shear layers, while most of the vorticity becomes concentrated in the large-scale vortex structures.

Turbulent Shear Flows 7

In addition, streamwise structures have been observed in the jets as well, both in the region close to the vortex rings and in the braid region (e.g. Becker and Masaro 1968; Agui and Hesselink 1988). While it appears that these longitudinal structures are counterrotating pairs of streamwise vortices, a complete understanding of the generation and evolution of these vortical structures and their interaction with the vortex rings has not yet been achieved. Some insight into the mechanisms governing the three-dimensional evolution of the jet can be gained from recent work performed on plane mixing layers. Numerical calculations (Corcos and Sherman 1984; Corcos and Lin 1984; Lin and Corcos 1984; Ashurst and Meiburg 1988) as well as experimental results (Lasheras et al. 1986; Lasheras and Choi 1988) indicate that the three-dimensional structures originate in the braid region between the Kelvin–Helmholtz rollers rather than in the rollers themselves, as suggested by the linear theory of Pierrehumbert and Widnall (1982), thus lending support to the model first suggested by Bernal and Roshko (1986). However, it is known that the vortex rings evolving in the axisymmetric jet have distinctly different instability modes from the plane rollers (Widnall and Sullivan 1973; Widnall et al. 1974; Widnall and Tsai 1977), so that it is not clear whether the plane mixing layer results are fully applicable to axisymmetric jets.

The vortex filament technique (Leonard 1985) has already proven to be an efficient tool for analyzing the concentration, reorientation, and stretching of vorticity in three-dimensional free plane shear layers and wakes. Comparisons between inviscid vortex dynamics simulations and flow visualization experiments (Ashurst and Meiburg 1988; Lasheras and Choi 1988; Meiburg and Lasheras 1988; Lasheras and Meiburg 1990) provided some detailed insight into the transitional three-dimensional evolution of these flows. These studies demonstrate that, even at moderate Reynolds numbers, the transitional evolution of free shear flows is dominated by inviscid mechanisms. Therefore, we expect to gain insight into the origin and evolution of the three-dimensional effects in axisymmetric jets by modeling them on the basis of their vorticity dynamics. Consequently, in this paper we have used these techniques in combination with flow visualization experiments to analyze the three-dimensional evolution of jet flows under combined axial and azimuthal forcing.

2. Experimental Technique

Our base flow consists of two laminar co-axial jets of air. The primary jet instability is triggered through the addition of periodic pressure pulses in the axial direction along the central jet. This acoustic forcing of the central jet leads to a sinusoidal variation in the strength of its boundary layer, which in turn causes a strongly periodic axisymmetric roll-up when the two streams merge.

In addition, following a technique similar to the one used in plane shear flows (Lasheras and Choi 1988; Meiburg and Lasheras 1988), the three-dimensional instability is forced by introducing a sinusoidal corrugation of the nozzle lip. The nozzle has an average diameter of 5 cm, while the corrugation amplitude is 2 mm (an order of magnitude smaller than the thickness of the boundary layers). The central jet has velocities in the range of 40–50 cm/s, while the outer jet, which merely serves to stabilize the flow pattern, typically has a velocity one order of magnitude smaller.

In order to visualize the three-dimensional evolution of the flow, we have modified a technique used to visualize flames which was developed by Chen and Roquemore (1986). The dry air stream forming the central jet is bubbled through a $TiCl_4$ solution until it becomes saturated with $TiCl_4$ vapor. Subsequently, this saturated dry air passes through a settling chamber equipped with the standard array of honeycomb and screens, which is followed by an axisymmetric nozzle. The outer co-flowing air is intentionally kept saturated with the H_2O vapor typically present in the laboratory environment. As the two co-flowing jets meet at the nozzle lip, the $TiCl_4$ vapors of the central jet react with the H_2O vapors of the outer stream. Thus, only at the interface between the two streams, the reaction

$$TiCl_4 + 2H_2O \longrightarrow TiO_2 + 4HCl$$

results in the generation of submicron TiO_2 particles. The diffusivity of these particles is small enough to ensure throughout the characteristic residence time in the test section that they only occupy the position of the interface. Therefore, the interface can subsequently be visualized by light scattering of these particles induced by planes of laser light that can be placed either at a given distance from the nozzle lip or traversing with the convective velocity of the vortex rings. The interface separating the central and co-flowing streams will appear in the visualization as a bright region within the dark ambient fluid.

3. Numerical Technique

The non-divergent nature of the velocity field in incompressible flows, along with the definition of vorticity, allows for a complete description of the kinematics of the flow in the form of the Biot–Savart law (Batchelor 1967). Using the theorems of Kelvin and Helmholtz for inviscid dynamics and following the general concepts reviewed by Leonard (1985), vortex filaments are used for the representation of the vorticity field. Each filament is represented by a number of node points along its centerline, through which a cubic spline is fitted to give it a smooth shape. The Biot–Savart law is evaluated assuming an invariant vorticity distribution around the filament centerline, based on the functional form

$$p(|\vec{\mathbf{x}} - \vec{\mathbf{r}}|) = \frac{3\alpha\sigma^2}{4\pi(|\vec{\mathbf{x}} - \vec{\mathbf{r}}|^2 + \alpha\sigma^2)^{5/2}}$$

where we use the notation of Leonard (1985). Here $\vec{r}(s)$ denotes the position of the vortex filament centerline, with s being the arclength. Incorporation of this vorticity distribution into the Biot–Savart law then allows us to obtain the velocity $\vec{\mathbf{u}}$ at any position $\vec{\mathbf{x}}$ by integrating over the arclength of all N filaments in the flowfield

$$\vec{\mathbf{u}}(\vec{\mathbf{x}},t) = -\frac{1}{4\pi}\sum_{i=1}^{N}\Gamma_i\int_{C_i}\frac{[\vec{\mathbf{x}}-\vec{\mathbf{r}}_i(s,t)]\times(\partial\vec{\mathbf{r}}_i(s,t)/\partial s)}{\{[\vec{\mathbf{x}}-\vec{\mathbf{r}}_i(s,t)]^2+\alpha\sigma_i^2\}^{3/2}}\,ds$$

where Γ_i denotes the strength of the *i*th vortex filament. For the numerical simulation, we limit ourselves to the temporally growing problem, i.e., our flow is periodic in the streamwise direction. Consequently, the Biot–Savart integration has to be carried out over the periodic images of the vortex filaments as well, as will be discussed below. Since we furthermore assume periodicity in the circumferential direction, we only have to discretize one section of the axisymmetric shear layer, provided that we extend the integration over the periodic images in the azimuthal direction as well. We take the velocity difference between the centerline and infinity as our characteristic velocity. The thickness of the axisymmetric shear layer, defined as the velocity jump divided by the maximum slope of the velocity, serves as the characteristic length scale, with results in the filament radius $\sigma = 0.5$. The Biot–Savart integration is carried out with second order accuracy both in space and in time by employing the predictor–corrector time-stepping scheme and the trapezoidal rule for spatial integration, respectively. We typically discretize the axisymmetric shear layer into 59 filaments per wavelength of initially 60 nodes each. As the flow develops a three-dimensional structure, the vortex filaments undergo considerable stretching. To maintain an adequate resolution, additional nodes are introduced, based on a criterion involving distance and curvature. Furthermore, the time-step is repeatedly reduced as local acceleration effects increase. The filament core radius decreases as its arclength increases to conserve its total volume. A more detailed discussion of the numerical method can be found in Ashurst and Meiburg (1988).

4. Results

In the numerical simulation, we investigate the evolution of an axisymmetric jet with a ratio of potential core radius R to shear layer thickness of 5. If we define the momentum thickness θ of the jet shear layer as

$$\theta = \int_0^\infty \frac{U}{U_0}\left(1-\frac{U}{U_0}\right)dr$$

the above vorticity distribution results in a value R/θ of approximately 25. Figure 1 shows how the velocity profile converges as we extend the Biot–Savart integration over successively more periodic images in the streamwise direction. We observe that by taking into account three images each in the upstream and the downstream direction, the velocity profile is within three per cent of the fully converged value, which would be obtained for infinitely many images. Consequently, to limit the computational effort, we carry along three images in the calculation. Using

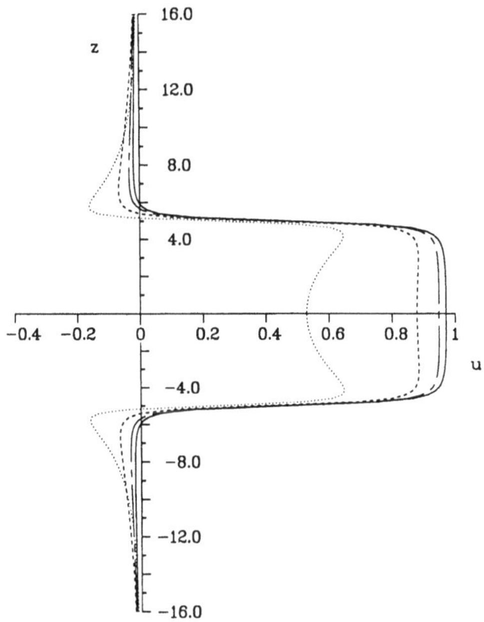

Fig. 1. Convergence of the velocity profile as the number of periodic images in the streamwise direction is increased in the Biot-Savart integration:........closest image only, --- one periodic image, —·— two periodic images, ——— three periodic images

Michalke and Hermann's (1982) data for the inviscid growth rates of spatially evolving jets, we can estimate the wavelength of maximum growth for our temporally growing case as approximately 2π by applying Gaster's (1962) transformation.

4.1 The Axisymmetric Case

In order to duplicate the acoustic forcing applied to the jet in the experiment, we modulate the strength of the vortex filaments periodically in the streamwise direction. The amplitude of the perturbation circulation represents five per cent of that due to the base flow. Figure 2 demonstrates the roll-up of the voriticity layer with time resulting from this purely axisymmetric perturbation. Shown is a perspective view of the vortex filaments as they begin to form a vortex ring. As the vorticity increasingly concentrates in these rings, the braid regions in between them develop a free stagnation point whose neighborhood becomes more and more depleted of vorticity. In the case of a tamporally growing plane shear layer, the free stagnation point would be located at the center between the evolving large-scale vortices. The axisymmetric nature of the present problem, however, shifts the stagnation point slightly in the downstrean direction. The behavior is similar to that observed for a plane wake (Meiburg and Lasheras 1988). As a result, the immediate upstream neighborhood of an emerging vortex ring is subjected to more intense depletion of vorticity than the downstream side. This is clearly visible in the vorticity contour plots (Fig. 3) which show a stronger vorticity braid on the downstream side than on the upstream side of the emerging vortex ring.

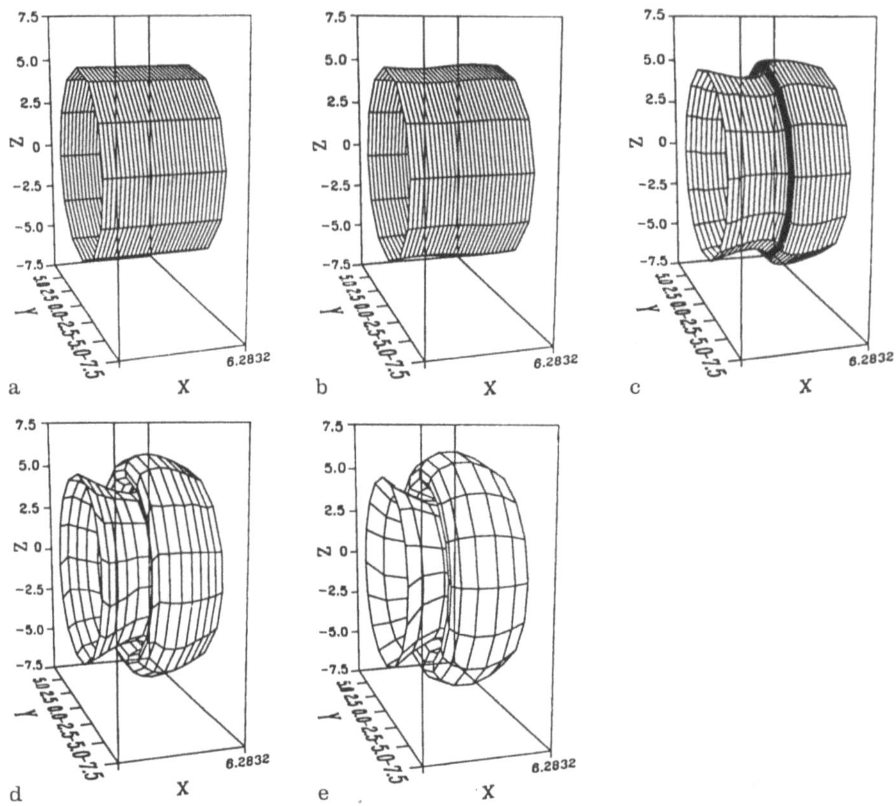

Fig. 2a–e. Axisymmetric evolution of a round jet due to a wavy circulation distribution. Shown are the vortex filaments at times 0.05, 4.0, 8.0, 12.0, 16.0. Only every fourth node along the filaments is plotted. Notice the emergence of a concentrated vortex ring

Several experimental observations indicate the occurrence of a vortex pairing process before significant streamwise vorticity is formed. This pairing mechanism represents a subharmonic instability of the same kind as the one known to be responsible for the growth of a mixing layer in the streamwise direction. In order to duplicate this subharmonic behavior in the calculation, we extend our control volume over two wavelengths of the basic circulation perturbation described above. In addition, we superimpose a subharmonic perturbation which affects the radial position of the vortex filament centerline. Its phase with respect to the basic wave is such that the filaments that eventually form one of the two rings will be displaced towards larger radii, whereas the ones forming the second ring are now located closer to the jet axis. In this way, the smaller one of the two rings is located in a region of higher velocity in the streamwise direction, which causes it to travel faster than the larger vortex ring. Consequently, the smaller vortex ring tends to take over the larger one by moving through its center—a phenomenon known as the leapfrogging mechanism—and thus proceeds towards a pairing process. The time-dependent evolution is shown in the perspective views of Fig. 4 and the vorticity concentration

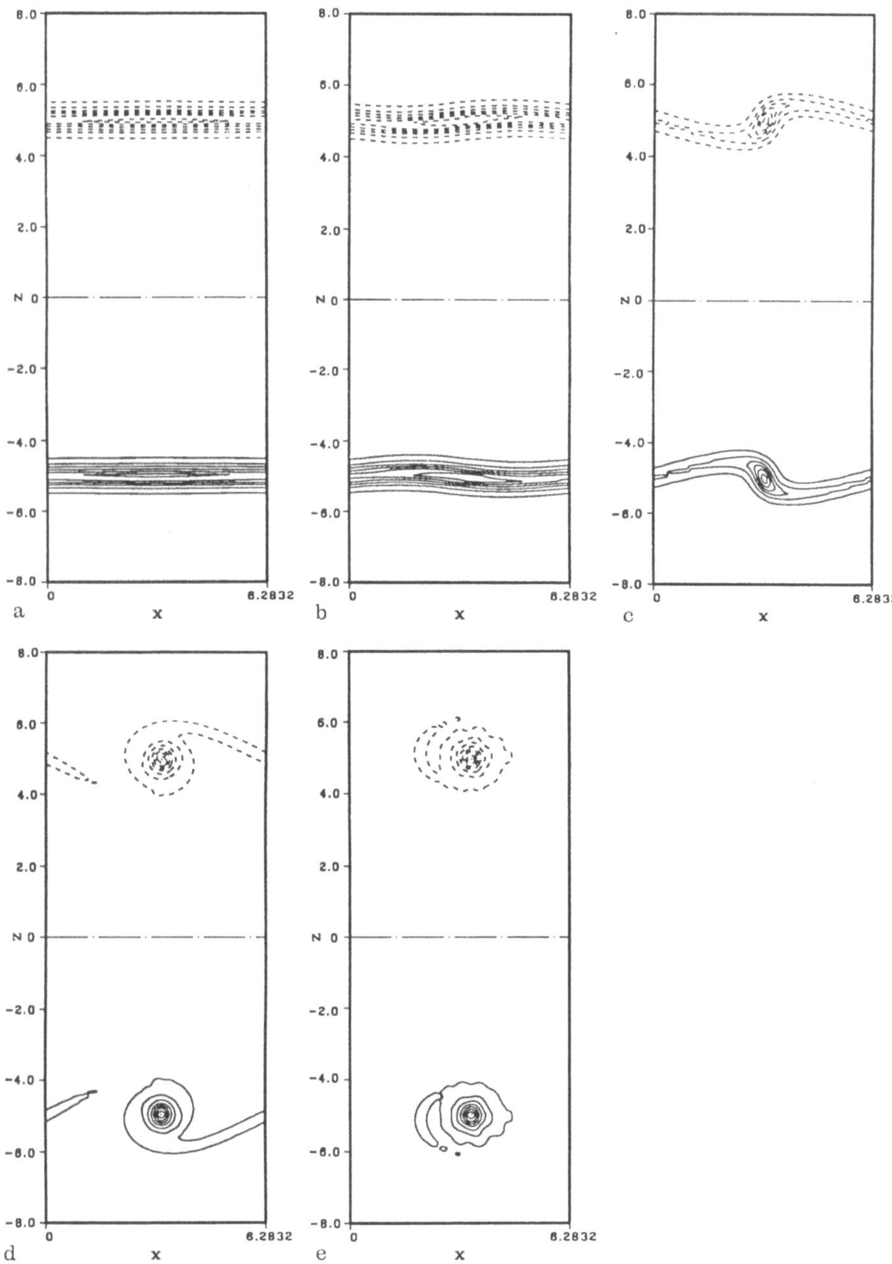

Fig. 3a–e. Contour plots of the azimuthal vorticity for the case and the times shown in Fig. 2. The upstream neighborhood of the vortex ring experiences the strongest depletion of vorticity due to strong extensional strain, which makes it a prime candidate for the emergence of streamwise vorticity

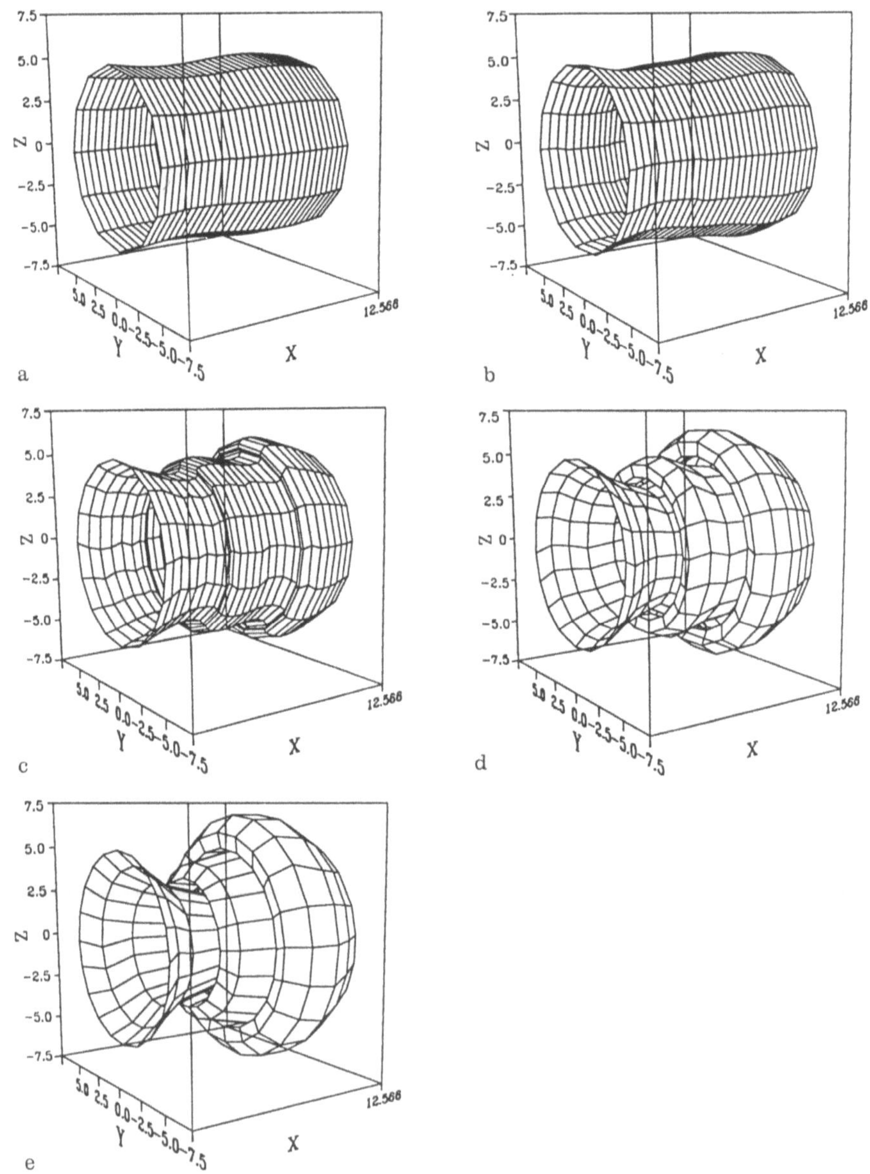

Fig. 4 a–e. Vortex filament configuration for the subharmonic axisymmetric case at times 0.05, 4.0, 8.0, 12.0, 16.0. Observe the two vortex rings proceeding towards a pairing

contours of Fig. 5. From the streamwise spacing of the circular filaments we recognize that the region in between the vortices that eventually will pair is subject to the most intense strain, and thus experiences the strongest depletion of vorticity, which makes it a prime candidate for the generation of streamwise vorticity if we allow for a three-dimensional evolution.

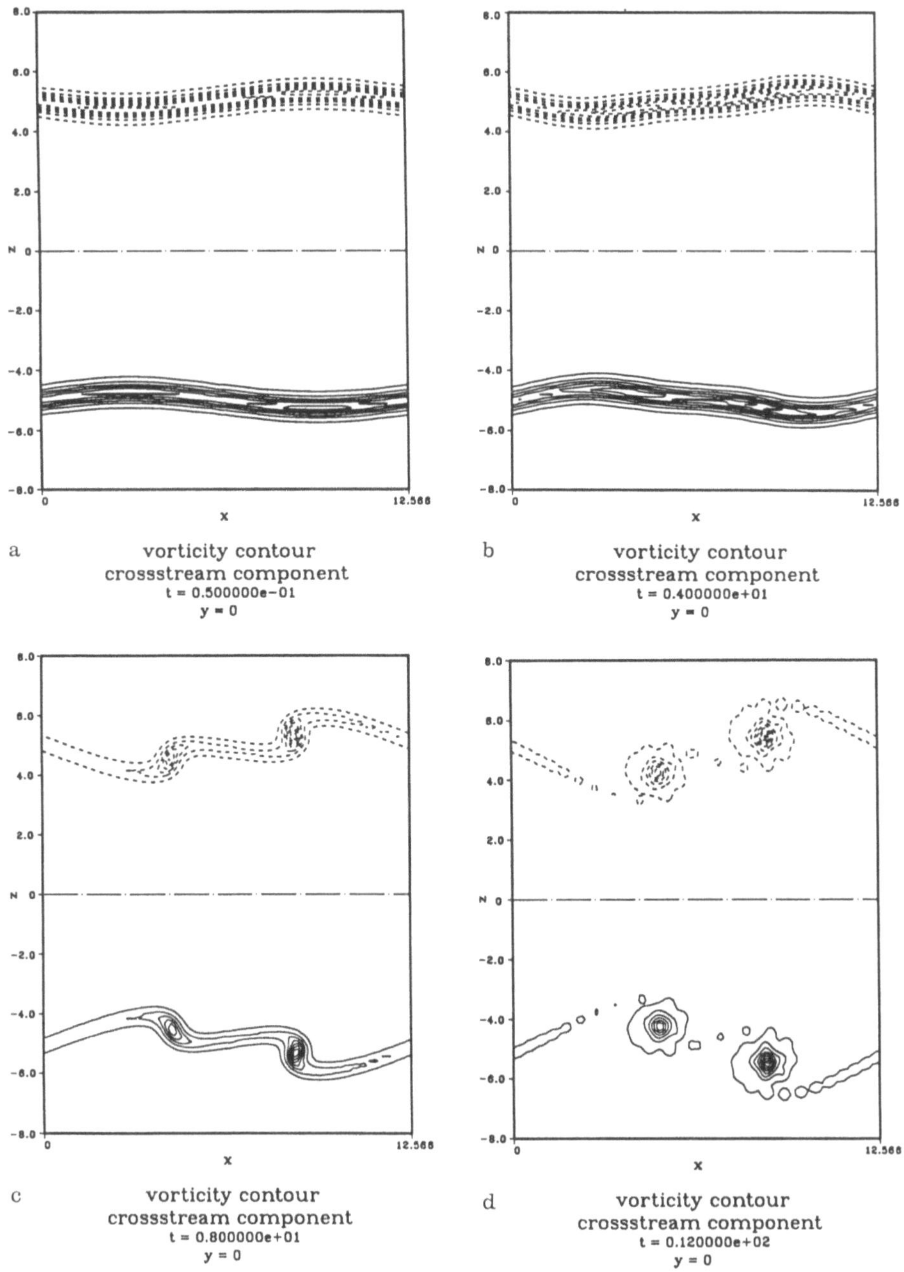

Fig. 5 a–e. Contour plots of the azimuthal vorticity for the case and the times shown in Fig. 4

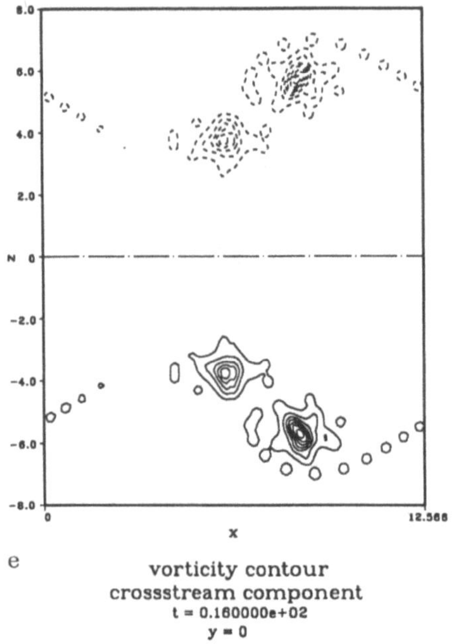

Fig. 5. (*Continued*)

4.2 The Three-Dimensional Case

Figure 6 shows the results of a calculation in which we introduce a perturbation in the circumferential direction in addition to the basic streamwise perturbation. This azimuthal disturbance periodically displaces the filament centerline in the radial direction, thus duplicating the effect of the nozzle corrugation. The calculation shown in Fig. 6 corresponds to an azimuthal wavenumber of 5. Both the side views and the streamwise views demonstrate the roll-up of the axisymmetric shear layer into vortex rings. For clarity, we only show those filament sections located at $y > 0$ in the side view. The simulation reveals two mechanisms for the generation of streamwise vorticity: 1. The reorientation and stretching of the braid vorticity close to the free stagnation point between consecutive vortex rings, and 2. The instability of the vortex rings themselves. As we had predicted on the basis of the axisymmetric calculations, those filaments just upstream of the vortex rings, i.e., in the downstream half of the braid region, acquire the largest streamwise vorticity component as a

Fig. 6. Side view and front view of the vortex filaments for the three-dimensional evolution of the axisymmetric jet at times 0.625, 7.812, 9.688, 12.813, 15.938. Observe the growing wavy deformation of the vortex rings, which causes them to acquire a streamwise component. At the same time, streamwise vorticity is generated in the braid region due to the amplification of small perturbations in the global strain field set up by the vortex rings. For clarity, the side view extends over two consecutive wavelengths, and only the vortex filament section located at $y > 0$ are shown

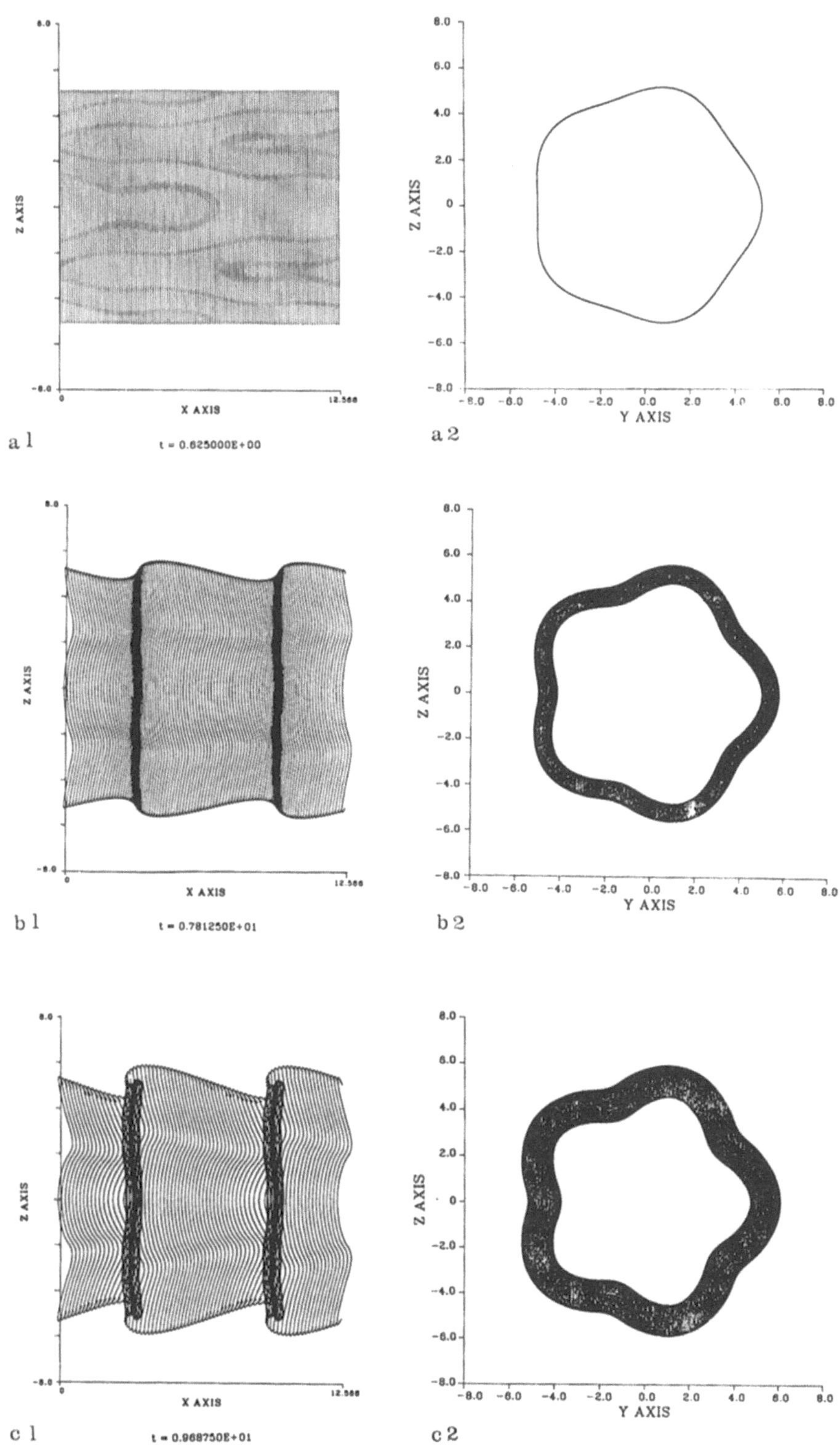

Z AXIS
X AXIS
8.0
-8.0
0
12.566
a 1
t = 0.625000E+00
Y AXIS
8.0
6.0
4.0
2.0
0
-2.0
-4.0
-6.0
-8.0
a 2
b 1
t = 0.781250E+01
b 2
c 1
t = 0.968750E+01
c 2

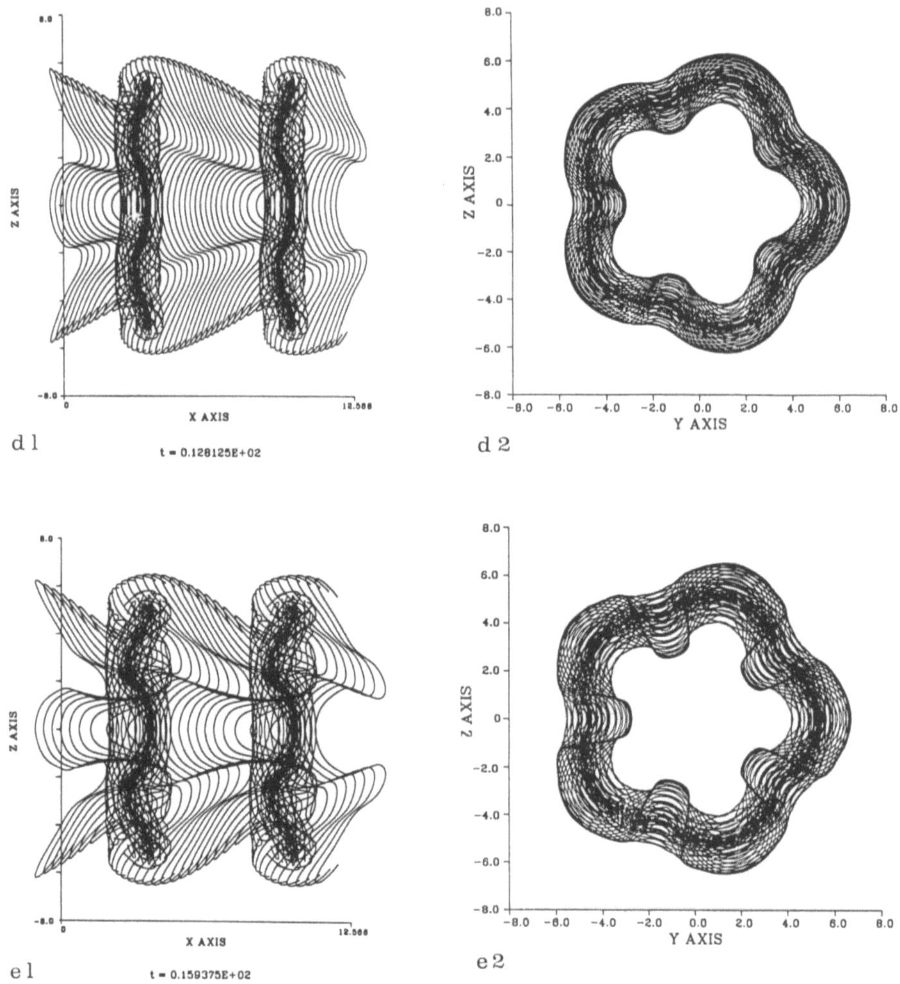

Fig. 6. (*Continued*)

consequence of the strong extensional strain field. Furthermore, the question of the sign of the streamwise vorticity in the different regions of the flow field is of interest. There are two different scenarios: In general, we would expect those segments of the vortex filaments farther away from the jet centerline to travel at lower velocities, as the streamwise velocity decreases with distance from the jet centerline (global induction). However, we notice that the radius of curvature of the vortex filaments is smaller in those regions where the filaments are farther away from the centerline. That means that the accelerating effect of local induction should be expected to be more important in those regions. Consequently, we have a competition between these two mechanisms. Figure 6 shows that, for the present flow parameters, in the braid region the filament sections farther away from the jet axis trail, which indicates that the global induction effect is dominant there. However, in the ring region the

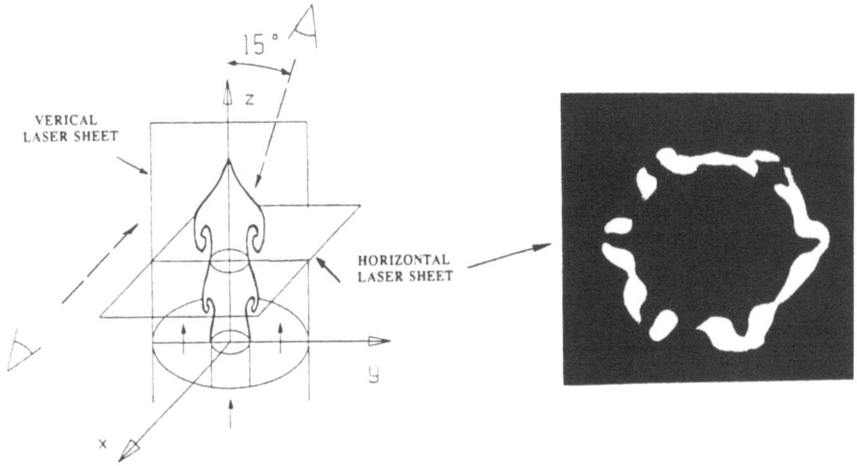

Fig. 7. Horizontal cross-cut of the jet at $x = 10$ cm. The light scattered by the TiO_2 particles marking the interface appears white in the visualization. Note the periodic folding of the interface as a result of the counterrotating pairs of streamwise vortex tubes also seen in Fig. 6

situation is opposite, and the outer ring regions travel at a higher velocity. We conclude that the local induction effect dominates in the ring region. Consequently, the streamwise vorticity at a given circumferential position is of opposite sign in the braid region and in the ring region.

We furthermore recognize that the streamwise braid vorticity is oriented such that it will cause the wavy radial displacement of the vortex rings to grow. Hence there is a strong coupling between the emerging vortex rings and the streamwise braid vorticity: While the vortex rings set up the global strain field that amplifies small streamwise perturbations in the braid region, the streamwise vorticity created in this way amplifies wavy displacements of the vortex rings.

The above mentioned flow features become also manifest in the experimental flow visualization (Fig. 7). A laser sheet that cuts the jet horizontally clearly shows the periodic displacement of the jet interface (white in the figure) away from the jet's axis, resulting in a lobe structure of the white fluid, while in between these lobes ambient fluid (dark) is entrained into the jet. The presence of the counterrotating pairs of streamwise vortex tubes, shown in Fig. 6, becomes manifest in the periodic folding of the interface as one moves along the azimuthal direction. This flow visualization corresponds to six wavelengths in the circumferential direction.

This work is supported by the National Science Foundation under grant #MSM-8809438 (to EM) and by DARPA under URI contract #N00014-86-K0754 to Brown University, as well as by grants from the United Technologies Corporation and the US-Spain Joint Committee for Scientific Cooperation (to JCL). The San Diego Supercomputer Center is providing computing time on its CRAY-X/MP.

References

Agui, J. C., Hesselink, L. (1988): Flow visualization and numerical analysis of a coflowing jet: a three-dimensional approach. J. Fluid Mech. **191**, 19

Ashurst, W. T., Meiburg, E. (1988): Three-dimensional shear layers via vortex dynamics. J. Fluid Mech. **189**, 87

Batchelor, G. K. (1967): An Introduction to Fluid Mechanics. Cambridge University Press

Becker, H. A., Masaro, T. A. (1968): Vortex evolution in a round jet. J. Fluid Mech. **31**, 435

Bernal, L. P., Roshko, A. (1986): Streamwise vortex structures in plane mixing layers. J. Fluid Mech. **170**, 499

Chen, L. D., Roquemore, W. M. (1986): Visualization of jet flames. Comb. and Flame **66**, 81

Cohen, J., Wygnanski, I. (1987a): The evolution of instabilities in the axisymmetric jet. Part 1. The linear growth of disturbances near the nozzle. J. Fluid Mech. **176**, 191

Cohen, J., Wygnanski, I. (1987b): The evolution of instabilities in the axisymmetric jet. Part 2. The flow resulting from the interaction between two waves. J. Fluid Mech. **176**, 221

Corcos, G.M., Lin, S. J. (1984): The mixing layer: deterministic models of a turbulent flow. Part 1: Introduction and the two-dimensional flow. J. Fluid Mech. **139**, 29

Corcos, G. M., Sherman, F. S. (1984): The mixing layer: deterministic models of a turbulent flow. Part 2: The origin of the three-dimensional motion. J. Fluid Mech. **139**, 67

Crow, S. C., Champagne, F. H. (1971): Orderly structure in jet turbulence. J. Fluid Mech. **48**, 547

Dimotakis, P. E., Miake-Lye, R. C., Papantoniou, D. A. (1983): Structure and dynamics of round turbulent jets. Phys. Fluids **26**(11), 3185

Gaster, M. (1962): A note on the relation between temporally-increasing and spatially-increasing disturbances in hydrodynamic stability. J. Fluid Mech. **14**, 222

Hussain, A. K. M. F., Clark, A. R. (1981): On the coherent structure of the axisymmetric mixing layer: a flow visualization study. J. Fluid Mech. **104**, 263

Hussain, A. K. M. F., Zaman, K. B. M. Q. (1980): Vortex pairing in a circular jet under controlled excitation. Part 2. Coherent structure dynamics. J. Fluid Mech. **101**, 493

Hussain, A. K. M. F., Zaman, K. B. M. Q. (1981): The 'preferred' modes of the axisymmetric jet. J. Fluid Mech. **110**, 39

Lasheras, J. C., Cho, J. S., Maxworthy, T. (1986): On the origin and evolution of streamwise vortical structures in a plane free shear-layer. J. Fluid Mech. **172**, 231

Lasheras, J. C., Choi, H. (1988): Three-dimensional instability of a plane, free shear layer: an experimental study of the formation and evolution of streamwise vortices. J. Fluid Mech. **189**, 53

Lasheras, J. C., Meiburg, E. (1990): Three-dimensional vorticity modes in the wake of a flat plate. Phys. Fluids A. **2**, 371

Leonard, A. (1985): Computing three-dimensional incompressible flows with vortex elements. Ann. Rev. Fluid Mech. **17**, 523

Lin, S. J., Corcos, G. M. (1984): The mixing layer: deterministic models of a turbulent flow. Part 3: The effect of plane strain on the dynamics of streamwise vortices. J. Fluid Mech. **141**, 139

Meiburg, E., Lasheras, J. C. (1988): Experimental and numerical investigation of the three-dimensional transition in plane wakes. J. Fluid Mech. **190**, 1

Michalke, A., Hermann, G. (1982): On the inviscid instability of a circular jet with external flow. J. Fluid Mech. **114**, 343

Pierrehumbert, R. T., Widnall, S. E. (1982): The two- and three-dimensional instabilities of a spatially periodic shear layer. J. Fluid Mech. **114**, 59

Plaschko, P. (1979): Helical instabilities of slowly diverging jets. J. Fluid Mech. **92**, 209

Tso, J., Hussain, F. (1989): Organized motions in a fully developed turbulent axisymmetric jet. J. Fluid Mech. **203**, 425

Widnall, S. E., Sullivan (1973): On the stability of vortex rings. Proc. R. Soc. Lond. A. **332**, 335

Widnall, S. E., Bliss, D. B., Tsai, C.-Y. (1974): The instability of short waves on a vortex ring. J. Fluid Mech. **60**, 35

Widnall, S. E., Tsai, C.-Y. (1977): The instability of the thin vortex ring of constant vorticity. Philos. Trans. R. Soc. Lond. Ser. A. **287**, 27

Yule, A. J. (1978) Large-scale structure in the mixing layer of a round jet. J. Fluid Mech. **89**, 413

Zaman, K. B. M. Q., Hussain, A. K. M. F. (1980): Vortex pairing in a circular jet under controlled excitation. Part 1. General jet response. J. Fluid Mech. **101**, 449

Turbulent Plane Jet Excited Mechanically by an Oscillating Thin Plate in the Potential Core

M. Miyata[1], N. Kurita[1] and I. Nakamura[2]

[1] Department of Mechanical Engineering, Yamanashi University, Takeda 4-3-11, Kofu 400, Japan
[2] Department of Mechanical Engineering, Nagoya University, Furo-cho, Chikusa-ku, Nagoya 464, Japan

Abstract

A two-dimensional rectangular jet excited mechanically by an oscillating thin plate in the jet core region was investigated experimentally. The plate is driven at both ends by stepping motors which enable control of torsion-free rotational oscillation about a mean position set at zero angle of attack. The maximum attack angle is nominally 1.8 degrees for the plates at 15 Hz and 27 Hz and zero for the stationary plate. The stationary blade gives no discernible modification in the downstream region of the jet. The effects of excitation are salient and first appear as a periodic deflection of the jet near the nozzle exit, which is amplified down-stream and gives notable effects on the development of the jet. The rate of increase in jet half width reaches more than 5 times the value for an undisturbed jet at the excitation frequency of 27 Hz. Both conventional and phase averaged mean velocity fields were documented as well as the distributions of the fluctuating quantities.

Nomenclature

d	thickness of plate ($=0.20$ mm)
f	oscillating frequency of plate
hw	jet half width
H	nozzle height
l	width of plate
Q	volumetric flow per unit width
Re_u	unit Reynolds number ($=U_0/\nu$)
U	streamwise component of mean velocity
U_{max}	maximum mean streamwise velocity component in cross-section
U_{min}	minimum mean streamwise velocity component in plate wake
U_0	nozzle exit velocity
$U(t)$	instantaneous streamwise velocity component
u'	r.m.s. intensity of streamwise fluctuating velocity
$-\overline{uv}$	turbulent shear stress
x, y	streamwise and transverse coordinates, respectively. $x=0$ mm corresponds to trailing edge of stationary plate (see Fig. 1).
ν	kinematic viscosity

Introduction

The problem of the response of a jet to periodic excitations has long attracted considerable interest of many researchers, mainly because of its fundamental importance in the problem of flow instability but also because of its practical implications in noise and/or turbulence reduction and in enhanced mixing or spreading rate. An important contribution to this problem was made by Crow and Champagne (1971) who found that an axisymmetric disturbance receives maximum amplification in the jet column at the excitation of Strouhal number of 0.30, the so-called "preferred" mode of an axisymmetric jet. In contrast, Vlasov and Ginevskiy (1967) and Petersen et al. (1974) observed turbulence suppression in a circular jet under excitation at Strouhal numbers one order of magnitude higher than that of the preferred mode.

It is now well known that these phenomena are closely related to the behaviour of the initially rolled-up vortices in the free-shear-layer part of the jet and hence may possibly influence the large scale coherent structure of the turbulent jet downstream. Comprehensive studies in this direction have been made by Zaman and Hussain (1980, 1981) and Hussain and Zaman (1981). A two-dimensional turbulent jet under excitation was studied by Thomas and Goldschmidt (1983, 1986). In most of these investigations, the jets are excited with small, acoustical field perturbations.

Fielder and Korschelt (1979), on the other hand, investigated the two-dimensional jet subjected to a relatively large, local transverse excitation at the nozzle exit by means of an acoustical device and showed that the jet is strongly influenced by excitations at even smaller Strouhal numbers at an appropriate amplitude. They proposed three typical flow configurations of the forced jet according to the forcing frequency. Stimulated by this experiment, Simmons et al. (1981) and Lai and Simmons (1985) studied the effect of a mechanical transverse excitation rather than an acoustical excitation on the development and entrainment of the two-dimensional jet, where a symmetric airfoil, relatively thick with respect to the nozzle height, was oscillated in the jet potential core.

The present study follows nearly the same direction as these experiments and aims at establishing the nature and the extent of controllable characteristics of the turbulent plane jet through a mechanical transverse excitation. To minimize the effect of the plate thickness, we used a thin plate, which oscillates about its axis through a small rotational angle.

Experimental Facility and Instrumentation

A thin plate ("feeler" gauge stock) of 0.20 mm thickness and 12.7 mm width was installed with the leading edge 3 mm downstream from the nozzle exit, on the plane of flow symmetry, as shown schematically in Fig. 1. The blade was tensioned strongly enough to prevent flutter and driven at both ends by stepping motors, which enabled control of the frequency and the attack angle of torsionless rotational oscillation about a mean position set at zero angle of attack. The maximum attack

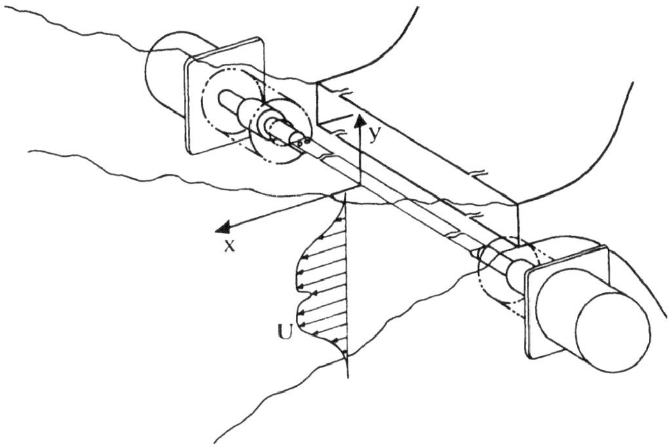

Fig. 1. Schematic view of test rig

angle used was nominally 1.8 degrees for blades at 15 Hz and 27 Hz and zero for the stationary blade.

We used two kinds of contraction nozzle, both 240 mm wide but with different nozzle heights of 20 mm (Nozzle 1) and 10 mm (Nozzle 2). Each nozzle had a parallel portion nearly as long as the nozzle height at the small end. The boundary layer at the nozzle exit is essentially laminar as shown in Fig. 2 and the momentum thickness for Nozzle 1 is 0.113 mm at the experimental condition. Strouhal numbers of excitation based on the nozzle height are thus one order of magnitude smaller than that of the "preferred" or "jet column" mode and larger than those studied by Lai and Simmons (1985). The flow in the test section including the contraction nozzle was isolated from the laboratory drafts by $2 \times 2\,m^2$ side plates, the blade going through these plates. The effects of excitation thus introduced on the downstream

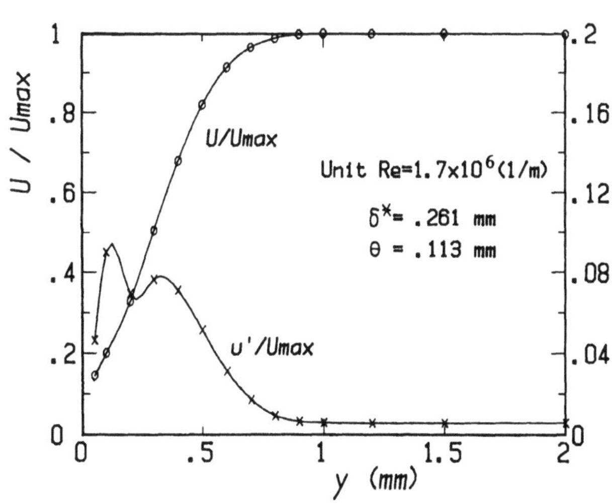

Fig. 2. Boundary layer at nozzle exit: nozzle 1

development of the jet, as well as the manner in which the jet is affected by the disturbance in the downstream region ($x/H < 25$) were investigated experimentally. The downstream restriction of the measuring range was encountered because of the lack of two-dimensionality of the excited jet which shows a large spreading rate.

Most of the measurements were made under a constant unit Reynolds number of $1.7 \times 10^6\,\mathrm{m}^{-1}$ ($25\,\mathrm{m/s} < U_0 < 28\,\mathrm{m/s}$), using a computer-controlled automated data acquisition and processing system. To minimize the measuring time and hence the effect of drift, hot-wire signals were first recorded on a magnetic tape and processed afterward. A typical measuring time is about 20 minutes per one traverse. Most of the data presented here were obtained with hot-wire I-probes which have a 1 mm unplated heated length of 5 μm tungsten wire and which were operated with a constant-temperature anemometer. Limited measurements were made using X-probes having the same wires as the I-probe and a quartz coated split-film probe of 152 μm in diameter. The latter was used to confirm the results obtained by I- and X-probes, especially in the outer intermittent region of the excited jet.

Experimental results presented below were obtained at each streamwise location over the lower part of the jet, after checking the symmetry with respect to the jet axis of the mean velocity and fluctuation intensity distributions of the blade wake and the jet as a whole.

Results and Discussion

Figure 3 shows distributions of normalized mean velocity defect $(U_{max} - U)/(U_{max} - U_{min})$ and fluctuating intensity u'/U_{max} in the wakes of the oscillating and the stationary blades at $x = 3$ mm, both obtained at a lower unit Reynolds number to

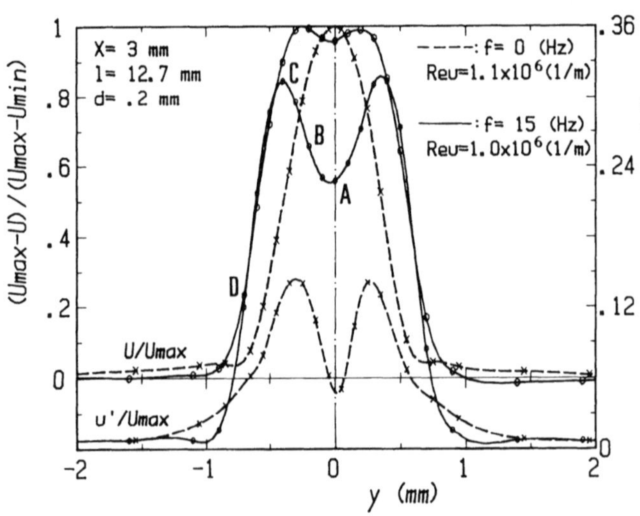

Fig. 3. Mean velocity defect and fluctuating intensity distributions in wakes of oscillating and stationary blades

lessen the effect of natural transition of the wake. Comparing with Gaussian velocity defect profile for the stationary case, the wake of the oscillating blade was thickened on both sides by about 0.2 mm, which is exactly half the oscillating amplitude of the trailing edge of the blade.

Instantaneous velocity signals obtained at typical positions A, B, C and D in the u' distribution of Fig. 3 are shown in the left hand side of Fig. 4, together with calculated wave forms at the corresponding points assuming that a Gaussian velocity defect profile of the stationary blade wake is displaced transversely, without any deformation, in accordance with the periodic motion of the blade.

A temporal variation of the trailing edge displacement of the blade is shown in the uppermost right hand side of Fig. 4. It is seen that the fundamental wave form of the perturbation is well reproduced by the quasisteady wake-oscillation model. At the higher Reynolds number at which the remainder of the measurements were made, a high frequency fluctuation component (~ 11 kHz) arising from natural transition of the blade wake is superimposed on these wave forms.

Mean velocity distributions of the jet just downstream from the blade, e.g. at $x/H = 0.25$, do not differ from each other except for the above mentioned thickening of the blade wake which exists in the central part of the otherwise uniform potential

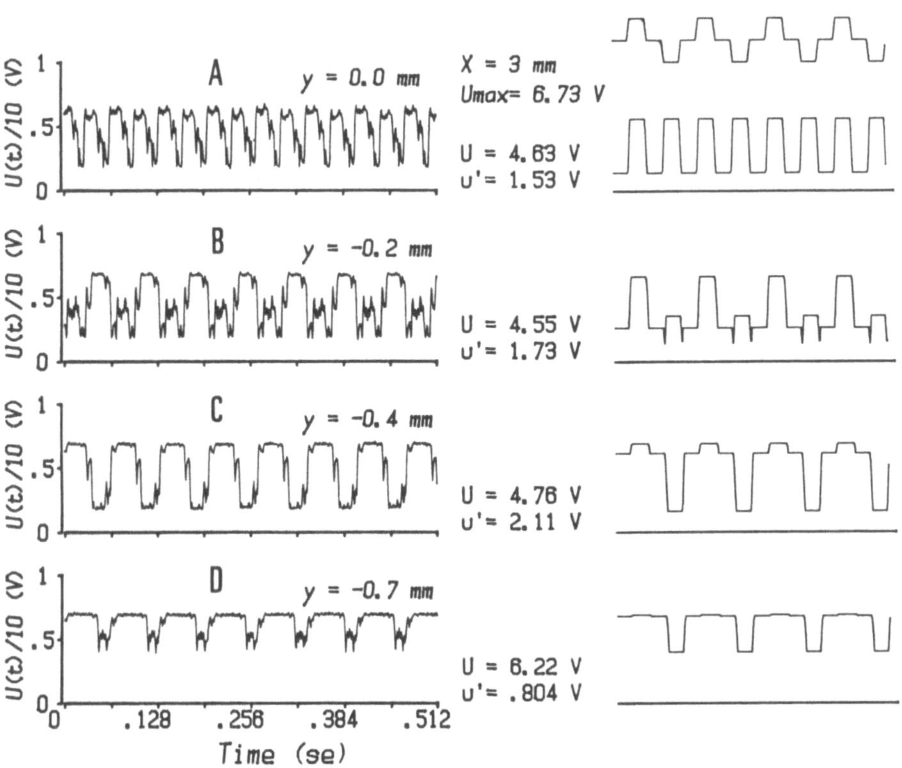

Fig. 4. Instantaneous velocity signals in wake of oscillating blade at $f = 15$ Hz. Left: hot-wire signals at points A, B, C and D in Fig. 3. Right: wave forms simulated by oscillating wake model

core. On the other hand, the distribution of fluctuation intensity u'/U_{max} clearly shows a higher peak in the free-shear-layer part of the excited jet.

The streamwise variation of mean velocity distributions in the near jet field for the excitation frequency of 15 Hz is shown compared with that for the stationary blade in Fig. 5a, b. It is seen that the oscillation of the blade enhances turbulent mixing in the blade wake and has shortened the potential core of the jet, leading to a faster transition to a fully developed flow. The distribution at $x = 80$ mm ($x/h = 4$) for the excited jet is mostly Gaussian, although there remains clearly the effect of the blade wake for the stationary case which persists till $x = 100$ mm ($x/h = 5$).

The effects of excitation appear more definitely in the development of the jet far from the nozzle. Figure 6 shows the streamwise variations of jet half width h_w defined as a point which gives $U/U_{max} = 0.5$. As shown in the figure, two stationary blades

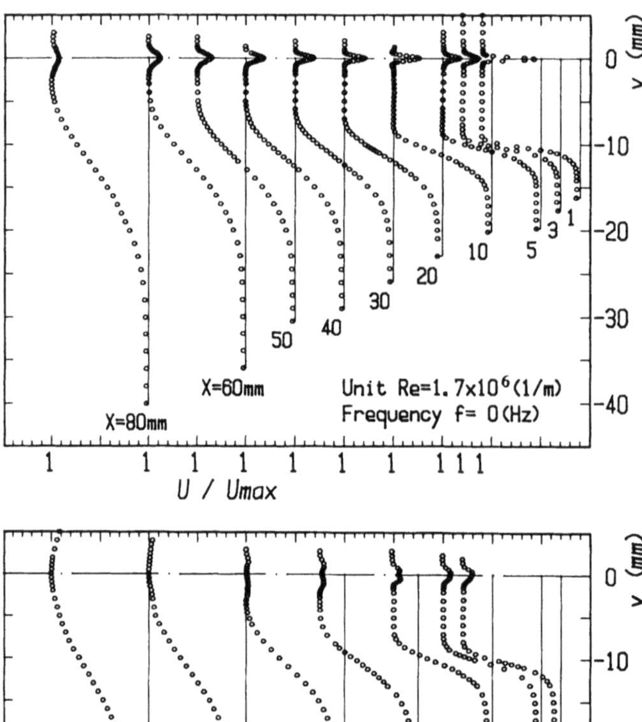

a

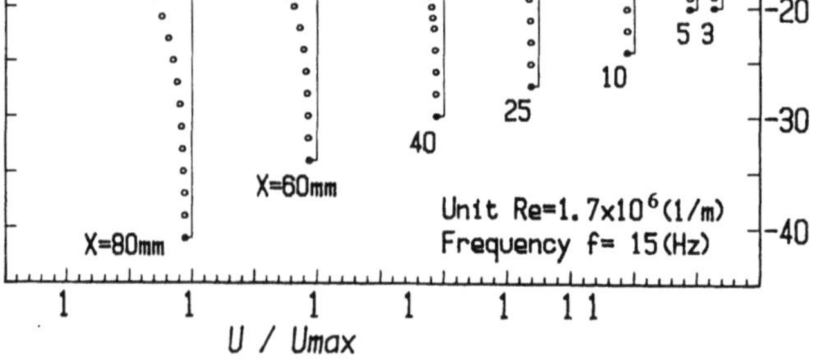

b

Fig. 5 a, b. Streamwise variation of mean velocity profile in near jet field: nozzle 1. **a** $f = 0$ Hz; **b** $f = 15$ Hz

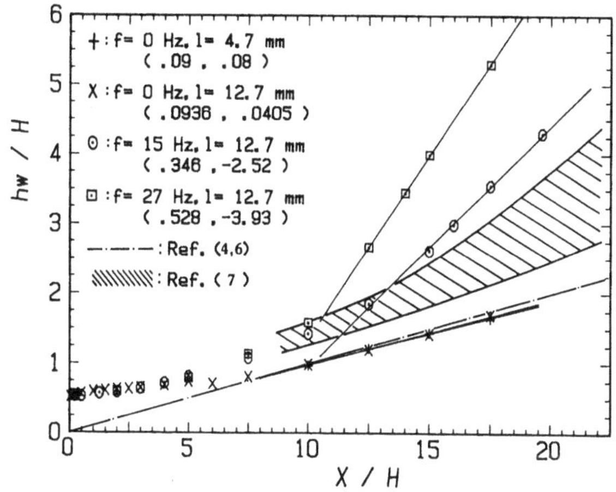

Fig. 6. Streamwise variation of jet half width. Figures in parentheses indicate slope and intersection of straight lines respectively: nozzle 1

with different plate width have a negligible effect on the development of the jet and the spreading rate, i.e. the slope dh_w/dx (which is shown by the first figures in each parentheses) in the linear developing region nearly coincides with that for an undisturbed plane jet (Gutmark and Wygnanski 1976; Ishigaki 1980) shown by a dotted line with a slope 0.10. On the other hand, the excited jet shows a very large spreading rate; the slope dh_w/dx in the developed region for 15 Hz excitation is nearly 4 times the value at $f = 0$ Hz and attains an even larger value of about 5 times at $f = 27$ Hz. The shaded region in the figure indicates the zone of data collapse by Lai and Simmons (1985). Their results for several excitation conditions gave approximately the same spreading rate in the region far from the nozzle exit, which is very close to the value for $f = 27$ Hz in the present experiments. It is interesting to note that for all conditions the jet half width does not change at all in the near jet field ($x/H < 5$); a rapid increase in hw starts at a position where the maximum axial velocity had decreased to nearly half the exit velocity (see Fig. 14), and the excited jets have nearly the same hypothetical origins.

Figure 7 shows the streamwise variation of entrainment by the jet defined as $Q/Q_0 - 1$, where $Q = \int_{-b}^{b} U\,dy$ and Q_0 is the volume flux per unit width at the nozzle

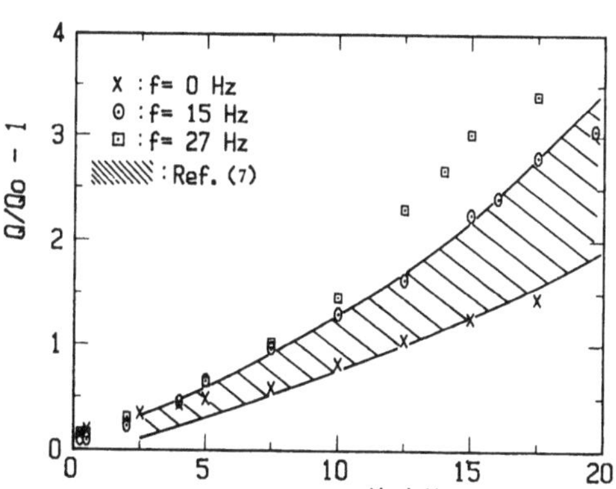

Fig. 7. Streamwise variation of entrainment: nozzle 1

exit. The integral limit b is conveniently taken as $2.5\,h_w$. Enhancement of the entrainment is very effective and it reaches 3 times the value of that for a non-disturbed jet at $x/H = 15$ for an excitation of 27 Hz. A faster increase of the entrainment than that obtained by Lai and Simmons (1985) corresponds to a faster rapid increase of the jet half width shown in Fig. 6.

Non-dimensional mean velocity and fluctuating intensity distributions for the stationary blade are shown in Fig. 8. The mean velocity profile follows the same Gaussian distribution as a typical steady plane jet (Everitt and Robins 1978) and the profile of fluctuating intensity also approaches that of an undisturbed self-preserving plane jet within the range of measurements. On the other hand, the excited jet shows a characteristic departure from these standard distributions as shown in Fig. 9a–c which show the normalized distributions of mean velocity, fluctuating intensity and turbulent shear stress, respectively for an excitation of 27 Hz. The mean velocity profile is flattened near the jet center line, which is also observed by Lai et al. (1981), and has a much wider lingering outer region. The fluctuating intensity profile has a higher peak which occurs relatively nearer to the jet axis with increasing downstream distance. The peak value has not yet saturated at the last measuring position and tends to decrease slowly further downstream. A similar feature can be observed for the turbulent shear stress distributions as shown in Fig. 9c. The distribution at $x = 50$ mm ($x/h = 5$) is nearly the same as that for the stationary blade, which was found to agree well with that for the jet without the blade but shows substantially smaller values than that of Everitt and Robins (1978). The peak value attains 4 times the size of that for the non-disturbed case at $x = 200$ mm ($x/h = 20$).

Because the excited jet shows a large entrainment, there must exist a relatively large incoming flow in the outer region of the jet where the streamwise velocity is

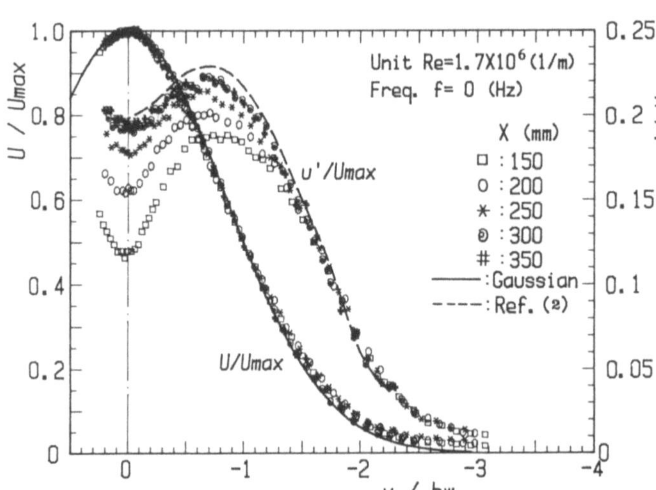

Fig. 8. Non-dimensional velocity and fluctuating intensity distributions for stationary blade: nozzle 1

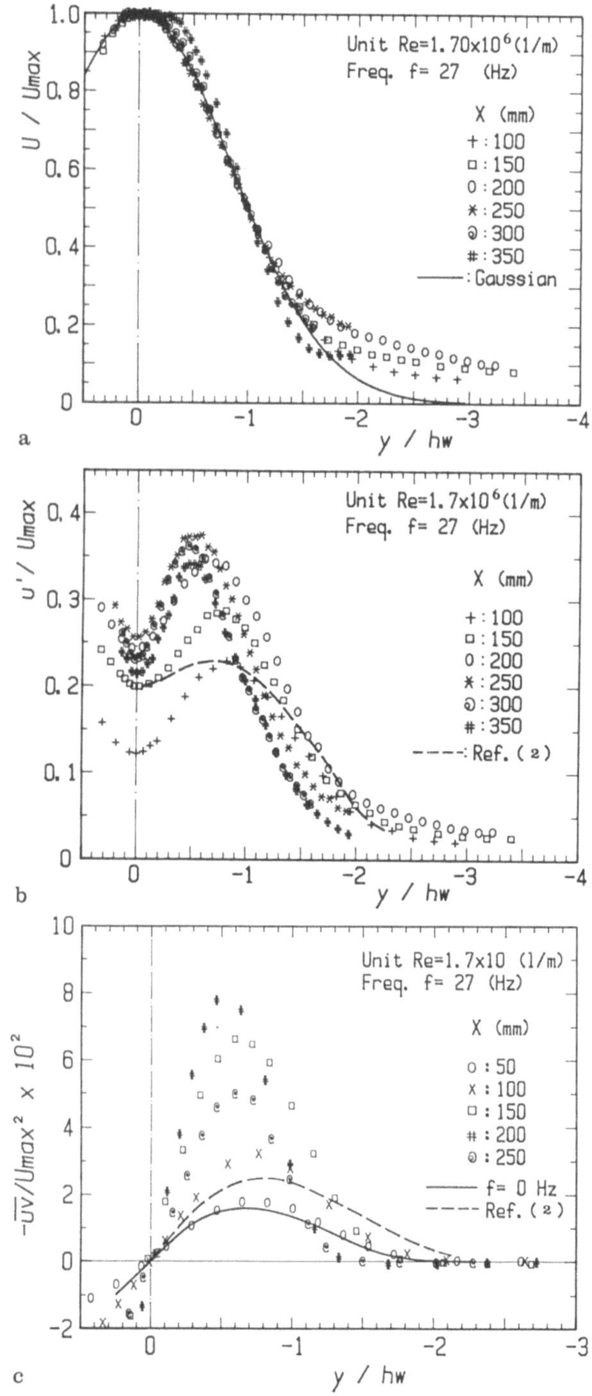

Fig. 9 a–c. Non-dimensional distributions of mean velocity **a**; fluctuating intensity **b**; nozzle 1, and turbulent shear stress **c**; nozzle 2 for excitation at $f = 27$ Hz small, which may render the hot-wire data obtained there ambiguous. To examine this point, we made measurements using a split-film probe at several downstream sections. The split-film data showed generally good agreement with those obtained by the hot-wire probes.

Instantaneous velocity signals at u' peak points of the jet at $x = 0$ mm ($x/h = 0$), 150 mm ($x/h = 7.5$) and 300 mm ($x/h = 15$) cross-sections for the excitation frequency of 27 Hz are shown in Fig. 10. The upper reference signal in each diagram was obtained at the u' peak point of the blade wake corresponding to the position C in Fig. 3. The lower one shows a 1 kHz low-pass filtered signal recorded at the measuring position simultaneously with the reference signal. As can be seen clearly, the lower signal has a distinct fundamental frequency equal to that of the reference signal, and the phase difference between each two signals is approximately 180 degrees for $x = 0$ mm and $x = 300$ mm and 0 for $x = 150$ mm.

Recalling that the signs of mean velocity gradient dU/dy at the reference and the measuring u' peak points are opposite, i.e. positive for the former and negative for the latter, and that the fundamental mode of the reference signal is caused by the oscillating blade wake, these results indicate that the jet deflects periodically according to the oscillating motion of the blade—a flapping motion which was

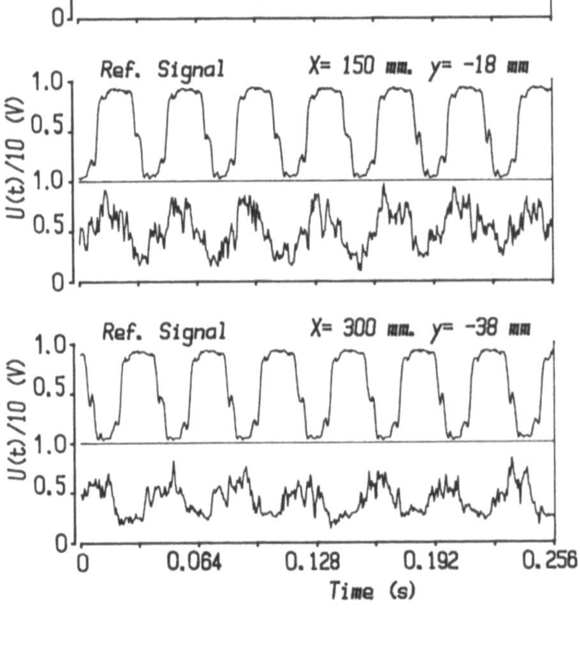

Fig. 10. Instantaneous velocity signals at reference and measuring u' peak positions at $f = 27$ Hz: nozzle 1

visually demonstrated by Fiedler and Korschelt (1979) for an acoustically excited plane jet.

Figure 11 shows typical non-filtered hot-wire signals in the outer part of the jet excited at 27 Hz, in the cross-section $x = 300$ mm. They show in common very distinct turbulent and non-turbulent regimes which appear alternately with roughly

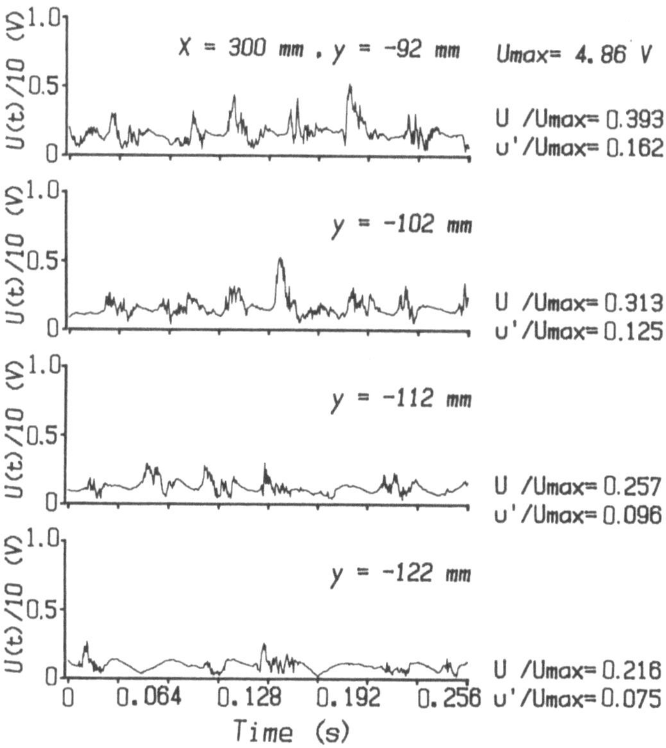

Fig. 11. Instantaneous velocity signals in the outer part of the jet ($h_w = 80$ mm) at $f = 27$ Hz: nozzle 1

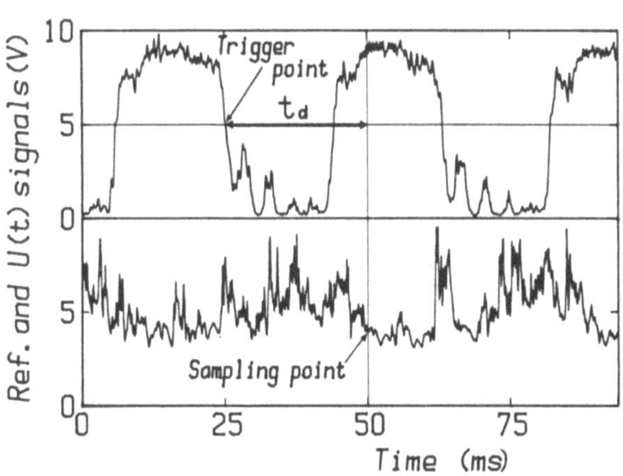

Fig. 12. Sampling timing of phase averaging

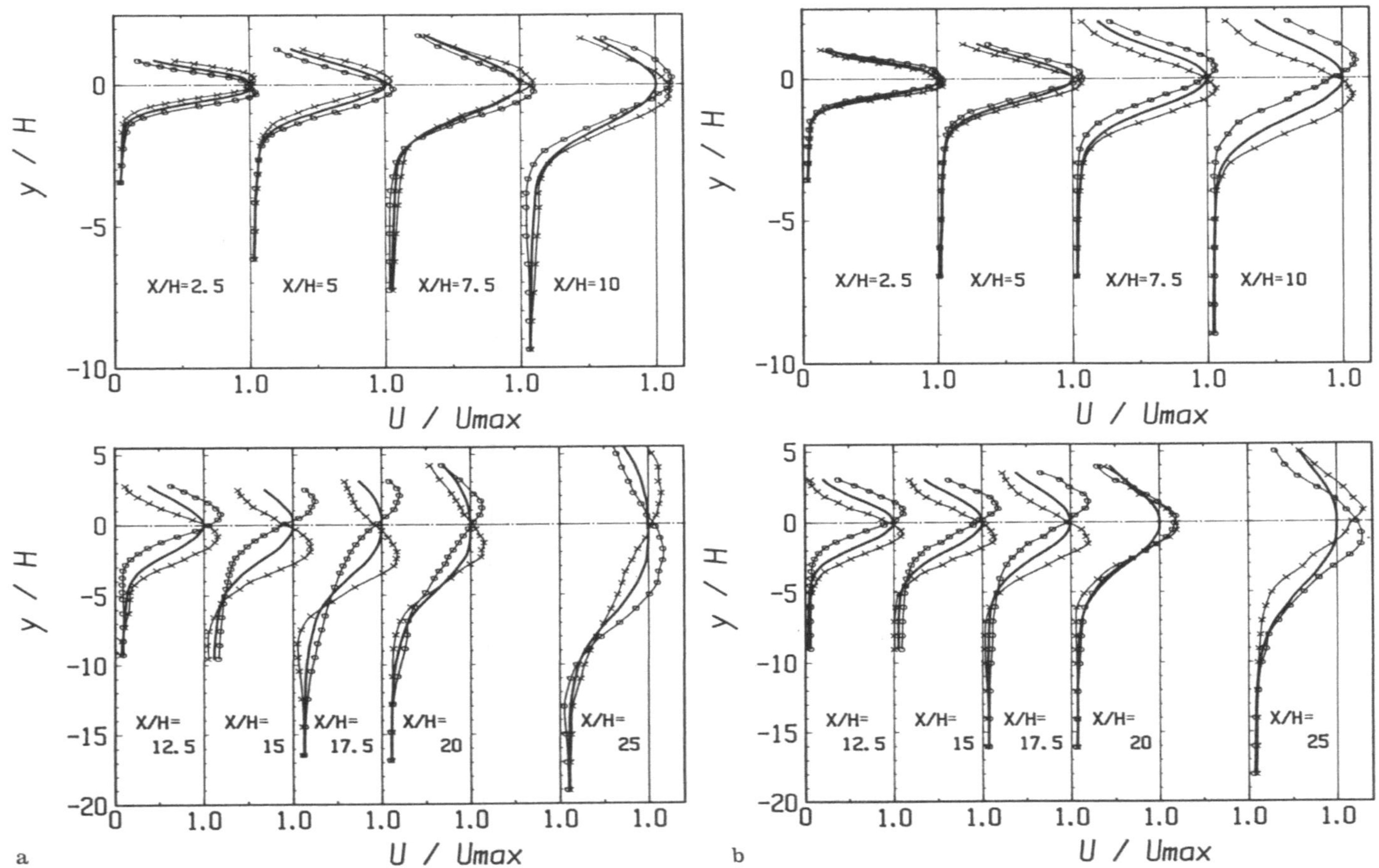

Fig. 13 a, b. Streamwise variations of phase-averaged mean velocity profiles: nozzle 2. **a** $f = 15$ Hz; **b** $f = 27$ Hz. ○: phase 0 degrees; ×: phase 180 degrees. Thick lines show conventional mean velocity profiles the same period as that of the excitation (0.037 sec in this case). Furthermore, the intermittency factor seems to be much smaller compared with that for an undisturbed jet at the point which gives the same U/U_{max}, showing that the intermittent region penetrates deeper into the jet core region. These characteristics are consistent with the above mentioned flapping motion of the excited jet.

To confirm the implied flapping motion of the jet and to see how the jet is affected by the periodic excitation, phase averaging measurements were made by sampling the hot-wire signal of a measuring probe at several fixed phases of the reference signal. The phase is selected so that the angle of attack of the oscillating blade is (nearly) zero, a maximum or a minimum. For experimental convenience, phase zero is defined as the instant when the output of the reference signal crosses half the maximum value with negative slope (Fig. 12). Two kinds of phase averaging schemes were executed; averaging the data at a fixed delay time *td* after each triggering; and averaging the wave form data during one cycle of the reference signal. The two methods gave practically the same results and the data given below were obtained by the former scheme.

Figure 13 shows the streamwise variation of phase averaged mean velocity profiles thus obtained for $f = 15$ Hz and 27 Hz excitations at the phases of 0 and 180 degrees, along with the conventional mean velocity profiles which were obtained from the same trains of recorded signals at each streamwise location. As can be seen clearly, the phase averaged mean velocity profile at each streamwise position shows a skewed Gaussian-like distribution and the jet undergoes a cyclic transverse displacement with an increasing amplitude downstream. The phase averaged maximum velocity in each cross-section is larger than the conventional mean maximum velocity. It is interesting to note that the decay of the phase averaged maximum velocity nearly coincides with that for a non-disturbed jet in the range $x/H < 20$ as shown in Fig. 14, which shows the streamwise decay of the conventional and phase averaged maximum velocity of the oscillating and the stationary jets.

Phase averaged mean streamlines at the phases of 0, 90, 180 and 270 degrees calculated from the results of Fig. 13 and the corresponding velocity profiles at the

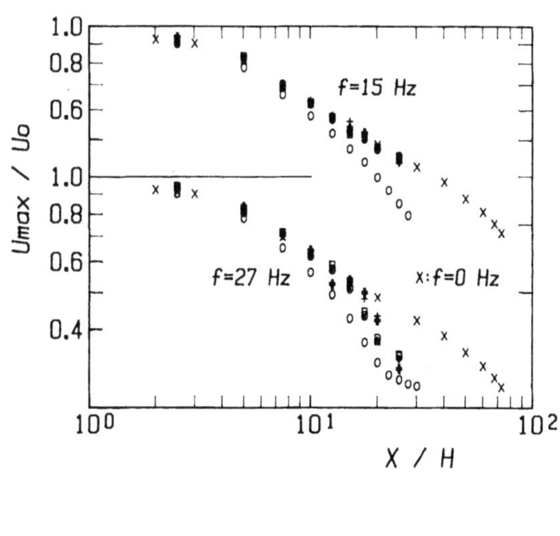

Fig. 14. Streamwise decay of maximum velocity: nozzle 2. ○: Conventional mean maximum velocity; others: phase averaged maximum velocity

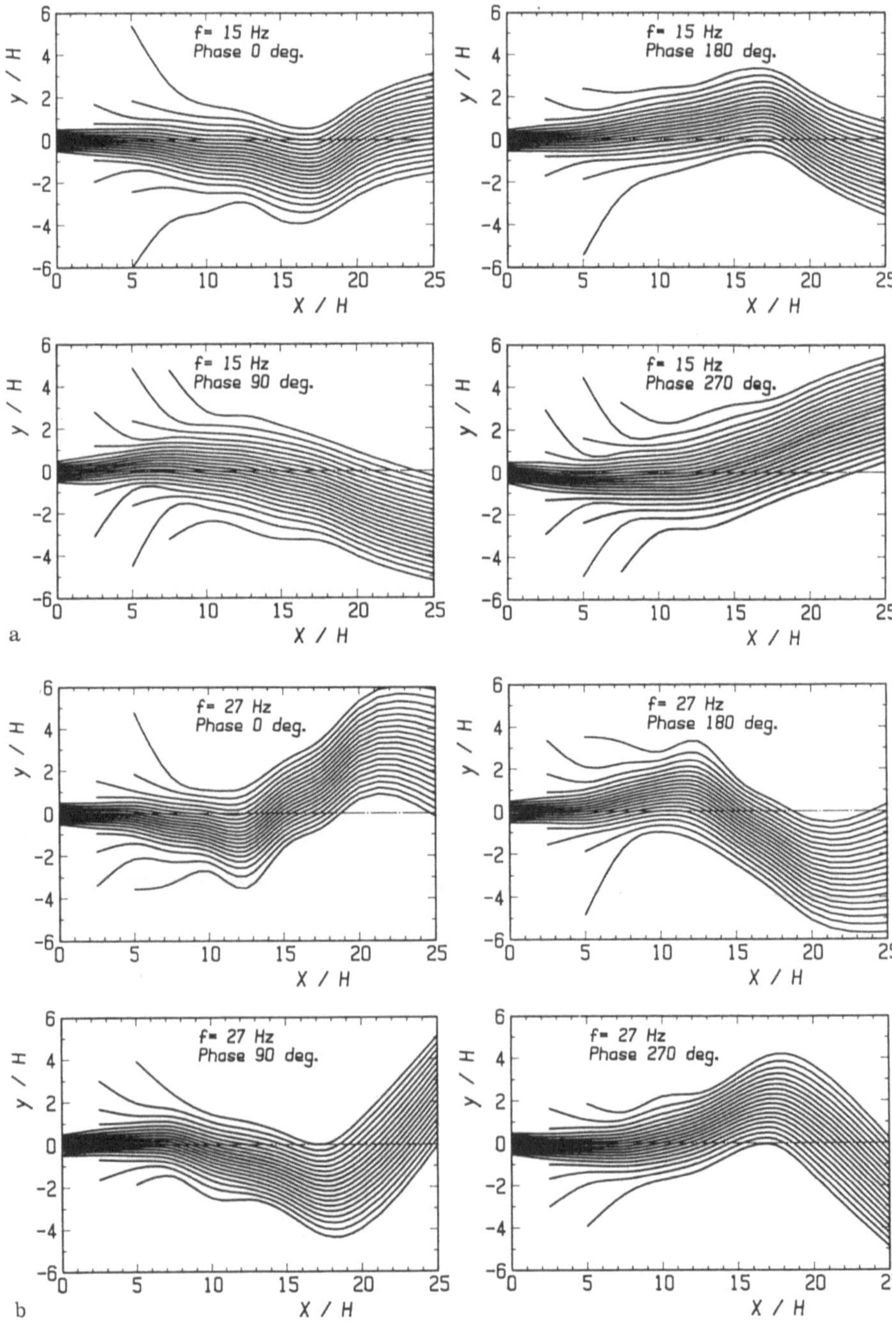

Fig. 15 a, b. Phase-averaged mean streamlines: nozzle 2. **a** $f = 15$ Hz; **b** $f = 27$ Hz phases of 90 and 270 degrees are shown in Fig. 15a, b, indicating that the jet evidently flows downstream in a zigzag manner. The results show also that the wave length or the phase velocity of cyclic deflection decreases downstream with an increasing amplitude. A larger spreading rate and higher peak in the u' distribution of the excited jet mentioned above can thus be attributed mainly to the flapping motion of the jet.

The pattern of the phase-averaged streamlines at the same phase of both excitation frequencies is very similar to each other, although the phase velocity and amplification of deflecting amplitude at 27 Hz are larger than that at 15 Hz. It should be noted that the position of the rapid increase in hw nearly coincides with the point where the flapping motion leads by half a wave length.

Concluding Remarks

The effects of the transverse excitation by a small amount of rotational oscillation of a thin plate in the potential core are salient and appear primarily as a periodic deflection of the jet, which is amplified down-stream and gives notable effects on the development and the distributions of mean velocity and turbulent quantities of the two-dimensional jet. The spreading rate of the excited jet reaches more than 5 times the value for an unexcited jet at the excitation frequency of 27 Hz. The phase averaged mean streamlines clearly show a flapping motion of the jet and the decay of the phase averaged maximum velocity agrees well with that for an undisturbed plane jet in the jet near field of $x/H < 20$.

References

Crow, S. C., Champagne, F. H. (1971): Orderly structure in jet turbulence. J. Fluid Mech. **48**, 547

Everitt, K. W., Robins, A. G. (1978): The development and structure of turbulent plane jets. J. Fluid Mech. **88**, 563

Fiedler, H., Korshelt, D. (1979): The two-dimensional jet with periodic initial condition. Proc. 2nd Turbulent Shear Flows, 8.18

Gutmark, E., Wygnaski, I. (1976): The planar turbulent jet. J. Fluid Mech. **73**, 465

Hussain, A. K. M. F., Zaman, K. B. M. Q. (1981): The preferred mode of axisymmetric jet. J. Fluid Mech. **110**, 39

Ishigaki, H. (1980): A numerical calculation of turbulent jet. Japan Soc. Aero. Spa. Sci. **28**, 56

Lai, J. C. S., Simmons, J. M. (1985): Instantaneous velocity measurements in a vane-excited plane jet. AIAA J. **23**, 1157

Petersen, R. A., Kaplan, R. E. Laufer, J. (1974): N. A. S. A. Contractor Rep. no. 134733

Simmons, J. M., Lai, J. C. S., Platzer, M. F. (1981): Jet excitation by an oscillating vane. AIAA J. **19**, 673

Thomas, F. O., Goldschmidt, V. W. (1983): Interaction of an acoustic disturbance and a two-dimensional jet: Experimental data. J. Fluids Engrg. **105**, 134

Thomas, F. O., Goldschmidt, V. W. (1986): Acoustically induced enhancement of widening and fluctuation intensity in a two-dimensional turbulent jet. J. Fluids Engrg. **108**, 331

Vlasov, Y. E. V., Ginevskiy, A. S. (1974): N.A.S.A. TTF-15, 721

Zaman, K. B. M. Q., Hussain, A. K. M. F. (1980): Vortex pairing in a circular jet under controlled excitation. Part 1. General jet response. J. Fluid Mech. **101**, 449

Zaman, K. B. M. Q., Hussain, A. K. M. F. (1981): Turbulence suppression in free shear flows by controlled excitation. J. Fluid Mech. **103**, 133

Structure in the Near Field of the Transverse Jet

T. F. Fric† and A. Roshko

Graduate Aeronautical Laboratories, California Institute of Technology, Pasadena, CA 91125, USA

Abstract

Photographs of a jet issuing from a wall into a crossflow display the four types of vortical structures which exist in the near field: namely, the jet shear layer vortices, the nascent far field vortex pair, the near-wall horseshoe vortices, and a system of vortices in the wake of the jet. It is shown that the wake vorticity is not "shed" from the jet but is formed from vorticity which originated in the wall boundary layer. The sources of vorticity for the other types of structures are also briefly discussed.

Nomenclature

D_c	circular cylinder diameter
D_j	jet diameter
f_{sep}	crossflow boundary layer separation power spectral peak frequency
f_w	wake power spectral peak frequency
Re_{cf}	nominal* crossflow Reynolds number $= U_{cf} D_j/\nu$ or $U_{cf} D_c/\nu$
Re_j	nominal* jet Reynolds number $= VR \cdot Re_{cf}$
St_{sep}	crossflow boundary layer separation Strouhal number $= f_{sep} D_j/U_{cf}$
St_w	wake Strouhal number $= f_w D_j/U_{cf}$
U_{cf}	nominal* crossflow velocity
VR	nominal* jet to crossflow velocity ratio
X, Y, Z	crossflow, lateral, and jet directions, respectively
δ_{cf}	nominal* crossflow boundary layer displacement thickness at $X/D_j = 0$
δ_j	nominal* jet axisymmetric shear layer displacement thickness at $Z/D_j \approx 0$

*nominal refers to values measured with either no crossflow or no jet, as appropriate.

1. Introduction

Jet injection is used widely in many technical applications, often to mix the injected fluid with that of the crossflow into which it is injected. Our interest in the experiments reported here is in the physics and structure of turbulent entrainment

†Present address: GE Corporate Research and Development, P. O. Box 8, Schenectady, NY 12301

Turbulent Shear Flows 7

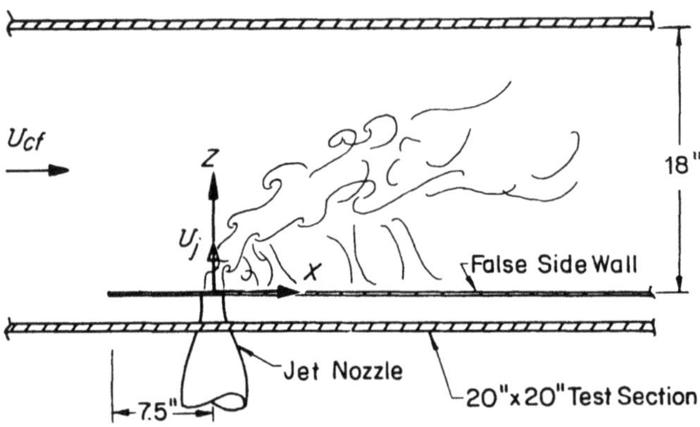

Fig. 1. Experimental set-up

and mixing in the near field of the incompressible transverse jet. The near field is defined somewhat arbitrarily to include the first few jet diameters of the flow, where most of the jet's bending occurs and where the crossflow/jet interaction dynamics is the most complex.

There have been numerous prior investigations of both the near and far fields of the transverse jet. Experimental, analytical, and computational work in the far field has concentrated on either the measuring of (e.g. Fearn and Weston 1974) or the modelling of (e.g. Broadwell and Breidenthal 1984; Karagozian and Greber 1984) the counterrotating pair of vortices, which comprises the transverse jet's far field. The near field has also received considerable attention (e.g. McMahon et al. 1971; Moussa et al. 1977; Foss 1980). Although many observations have been made, and models have been constructed, there is still no complete understanding of the structure in the near field of the transverse jet.

The experiments were performed in GALCIT's 20in by 20in open-return low-speed wind tunnel. Figure 1 shows the experimental set-up. To afford better control of the boundary layer on the wall from which the jet issues, a false side wall was used. In this paper, the false side wall is referred to as the crossflow wall, and the boundary layer on this wall as the crossflow boundary layer. The rms turbulence level of the crossflow was about 0.2%. The jet, supplied by a 1.5in diameter nozzle had a nearly top hat velocity profile in the absence of crossflow. The nozzle was mounted flush with the crossflow wall. The jet was supplied by a centrifugal blower; jet turbulence intensity levels of 0.5% to 1% were achieved with the appropriate use of flow management upstream of the nozzle. Flow conditions of $2 \leqq VR \leqq 10$ and $3800 \leqq Re_{cf} \leqq 11400$ (correspondingly, $7600 \leqq Re_j \leqq 114{,}000$) were investigated.

2. The Dominant Near Field Vortical Structures

The dominant vortical structures observed in the transverse jet's near field are the following:

a) Distorted shear layer ring vortices at the circumference of the bending jet.

b) The inception of the counterrotating pair of vortices which eventually dominates the far field jet structure.
c) A system of horseshoe or collar vortices at the crossflow wall.
d) A system of wake vortices nearly aligned with the initial jet direction.

The smoke-wire flow visualization technique (Corke et al. 1977) with strobe lighting was used (except where noted). Examples of each of these structures are shown in the figures which follow.

Figure 2 shows the bending jet's distorted ring vortices. The approaching streaklines, visualized with a smoke-wire upstream in the XZ plane, are entrained into the leading edge shear layer of the jet. As a result of axial flow along the cores of the distorted vortex rings, the smoke is seen coming out of that plane. In this case, $\delta_j/D_j \approx 0.03$. Such a shear layer roll-up is analogous to what is observed in a free jet. Features similar to these ring vortices are seen over the range of VR and Re_{cf} studied here.

Results indicate that the counterrotating pair of vortices, which is universally considered to dominate the mean far field structure of the jet, begins forming quite early. For example, Fig. 3 shows a nearly planar slice of the flow at $X/D_j = 1$. Smoke begins in the crossflow boundary layer, is entrained away from the wall at the lee side of the jet, and then visualizes a structure which suggests a counterrotating vortex pair. Such a structure is observed for the complete range of VR and Re_{cf}.

Due to the adverse pressure gradient just ahead of the jet on the crossflow wall, the approaching laminar boundary layer separates and forms a system of horseshoe vortices. In Fig. 4, smoke begins in the crossflow boundary layer. Here $\delta_{cf}/D_j \approx 0.06$.

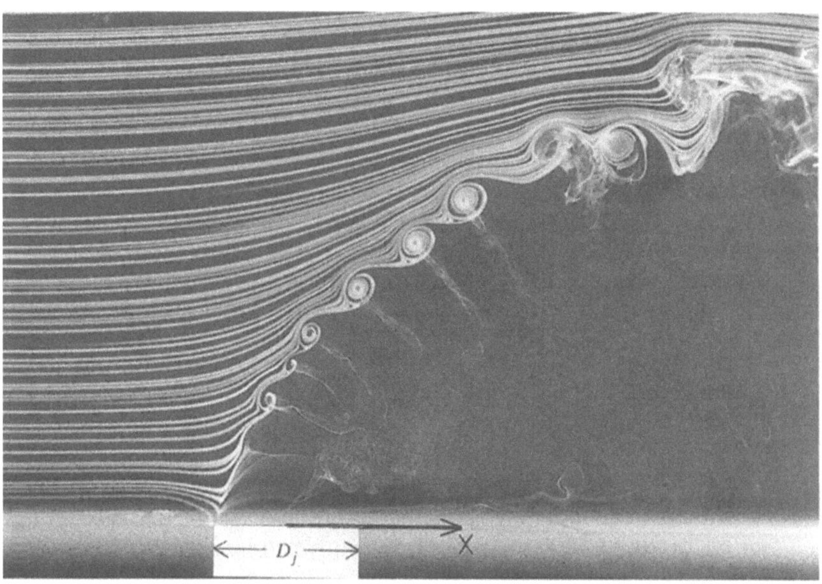

Fig. 2. Jet's distorted ring vortices. In this case, the vortices tilt clockwise. The jet issues from the bottom, as indicated, and the crossflow is from left to right. $VR = 2$, $Re_{cf} = 3800$

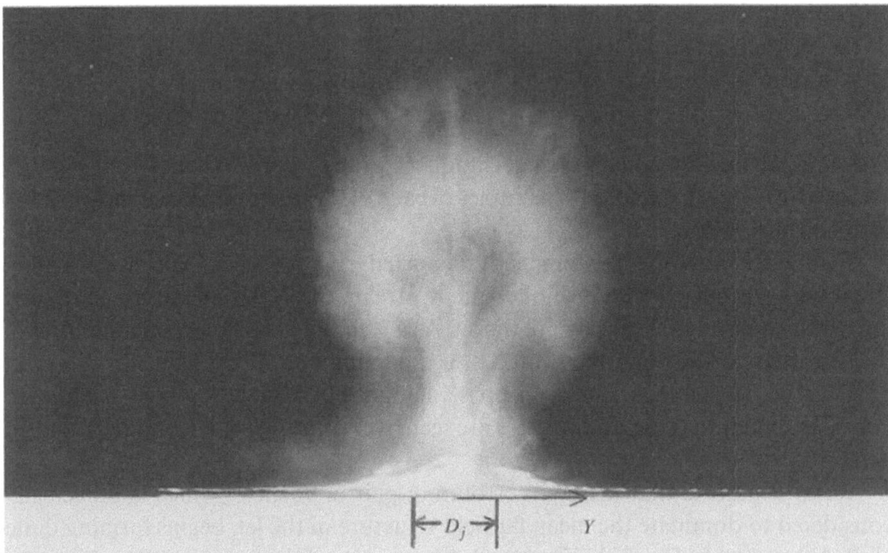

Fig. 3. Early indication of the counterrotating vortex pair. Looking upstream, a vertical "sheet" of light is at $X/D_j = 1$. $VR = 4$, $Re_{cf} = 3800$

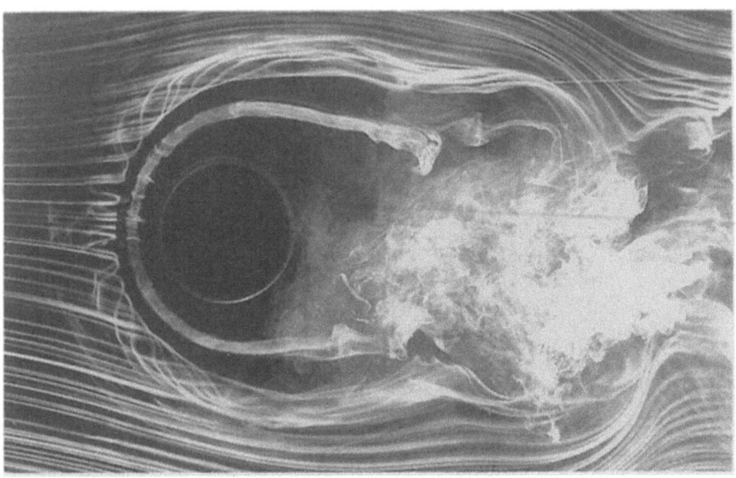

Fig. 4. Horseshoe vortex system. The jet, near the center of the photograph, issues directly at the viewer. $VR = 2$, $Re_{cf} = 3800$

The near wall flow around a wall-mounted circular cylinder also forms such structures. The upstream portions of the horseshoe vortices are visually similar for all VR and Re_{cf}, but such is not the case for the downstream portions (legs) of these structures. For $X/D_j \gtrsim 0$, the near wake, whose structure strongly depends on VR, affects the dynamics of the horseshoes' legs. In Fig. 2, cross sections of two horseshoe vortices are faintly visible just upstream of the jet.

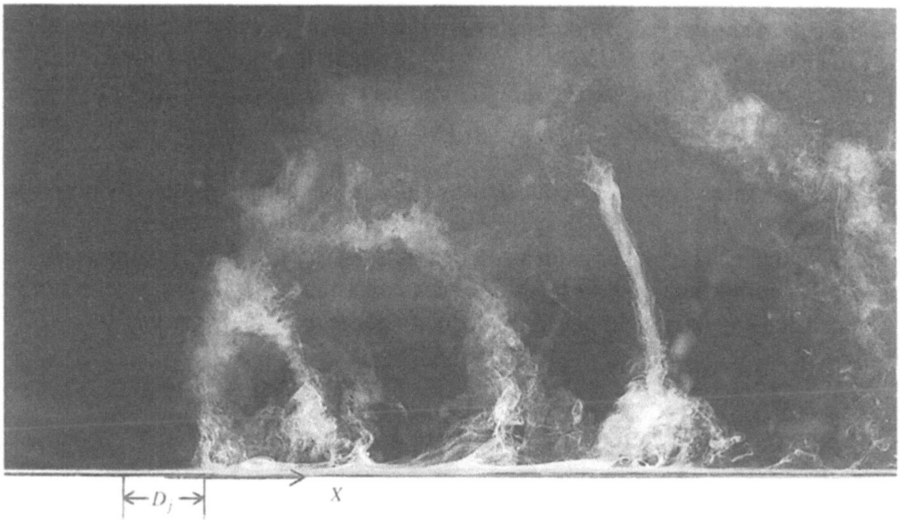

Fig. 5. Wake vortices in side view. The smoke is initially in the crossflow boundary layer. $VR = 4$, $Re_{cf} = 3800$

Fig. 6. Wake vortices in cross section. The jet, near the left side of the photograph, issues directly at the viewer. The smoke-wire is at $Z/D_j = 0.5$. $VR = 4$, $Re_{cf} = 11400$

Possibly the most intriguing and least understood of the structures in the near field are the wake vortices. Figure 5 shows a side view of the wake structures, while in Fig. 6 one sees a nearly cross sectional view. Kuzo and Roshko (1984) have observed that such wake structures persist several hundred jet diameters downstream. A wake exists for all VR and Re_{cf} studied, but its regularity and qualitative appearance varies with VR and Re_{cf}.

The structures displayed in our photographs have been either directly observed or indirectly inferred previously in the literature (cf. references mentioned previously or elsewhere in this paper and also their reference lists). The dynamics of the structures' formation, however, has not been consistently explained, at least not in terms of the constraints imposed by vorticity generation mechanisms.

3. Sources of Vorticity and Structure Formation

Since the near field is dominated by several distinct types of vortical structures, it is of interest to try to understand their origin and formation. In particular, the source and subsequent dynamics of the vorticity leading to the structures presented in Sect. 2 are described here.

The vorticity equation for a Newtonian, barotropic, and incompressible flow with no body forces is given by

$$\frac{D\boldsymbol{\omega}}{Dt} = \boldsymbol{\omega}\cdot\nabla\mathbf{u} + \nu\nabla^2\boldsymbol{\omega}, \tag{1}$$

where $\boldsymbol{\omega} = \nabla \times \mathbf{u}$ is the vorticity. As is emphasized by Morton (1984), Eq. (1) does not explicitly contain any vorticity generation terms; it shows only that vorticity is convected, stretched, turned, and diffused. New vorticity can only enter a flow through imposed initial conditions and/or wall boundary conditions; there are no sources of new vorticity within the flow. This is a fundamental point when considering the formation of the wake structures in particular, as will be seen presently.

Defining the vorticity flux out of a wall as $\hat{\mathbf{n}}\cdot\mathbf{J}_0$, where $\mathbf{J}_0 = -\nu(\nabla\boldsymbol{\omega})_0$ is the vorticity flux tensor at the wall and $\hat{\mathbf{n}}$ is the wall-normal unit vector, it can be shown (Wu et al. 1987) that for nonaccelerating and nonrotating surfaces

$$\rho\hat{\mathbf{n}}\cdot\mathbf{J}_0 = -\hat{\mathbf{n}} \times (\nabla p)_0 - \hat{\mathbf{n}}[\hat{\mathbf{n}}\cdot(\nabla \times \boldsymbol{\tau}_0)] + (\hat{\mathbf{n}} \times \boldsymbol{\tau}_0)\cdot\nabla\hat{\mathbf{n}}. \tag{2}$$

The first term on the right-hand side of Eq. (2) represents the vorticity source term due to a wall pressure gradient. The second term accounts for the gradient of wall-normal vorticity due to a wall shear stress $\boldsymbol{\tau}_0$ with a nonzero $\nabla \times \boldsymbol{\tau}_0$ wall-normal component. Finally, the third term accounts for the effects of wall curvature transverse to the wall shear stress direction. With the above discussion in mind, the sources of vorticity for the four kinds of structures mentioned in Sect. 2 can be identified.

The source of the distorted ring vorticity of the jet is the boundary layer vorticity in the nozzle supplying the jet flow, again directly analogous to a free jet. The initial nearly axisymmetric shear layer leaving the jet becomes unstable and forms the ring-like vortices.

We believe that the counterrotating vortex pair evolves from the shear layer vorticity of the jet, i.e. its ultimate source is in the vorticity issuing from the nozzle. In fact, the numerical results of Coehlo and Hunt (1989) suggest that this process has already begun within the boundary layer of the nozzle; the effects of the crossflow are felt within the nozzle. From numerical results, Sykes et al. (1986) have previously suggested that the jet's ring vorticity may evolve into the counterrotating pair.

The source of vorticity for the horseshoe structures is in the approaching crossflow boundary layer and the crossflow wall region near the jet. The separated crossflow boundary layer ahead of the jet supplies the horseshoes with vorticity, and in addition the horseshoe vortices induce pressure gradients on the wall, thereby generating more vorticity.

That no new vorticity can be generated within the flow is essential to the proper description of the wake formation. To our knowledge, in every paper in which the transverse jet's wake formation has been discussed previously, the wake vorticity has been described as vorticity "shed" from the jet and/or as vorticity due to a jet/crossflow interaction analogous to a solid cylinder/crossflow interaction. Since the jet does not offer the crossflow a solid surface on which to generate new vorticity, as a solid cylinder does, such analogies are physically wrong and lead to an incorrect explanation of the wake formation. As will be addressed in Sect. 4, the actual events leading to the appearance of vortices in the wake of the jet appear to be quite different and more complex.

4. Wake Characteristics and Wake Formation Model

This section is devoted to the vortical structures in the jet's wake. The wake is especially interesting because, even though it is qualitatively similar in appearance to a circular cylinder wake, its formation must be very different. The vorticity which is shed from a solid cylinder is vorticity which was generated solely at the surface of the cylinder; the pressure gradient production term in Eq. (2) is responsible. There is no such mechanism in the present case of a flow around a jet, since there is *no* no-slip condition at the jet/crossflow interface. The vorticity in the jet's wake must be vorticity which originated at a solid boundary, either within the nozzle or at the crossflow wall.

To study the wake further, smoke from different sources was used to track fluid. Smoke is considered to be a suitable marker of vorticity for our purposes. Since the effective Schmidt number for smoke is large, vorticity diffuses faster than smoke does; we assume that once the vorticity is tagged, the smoke marks the cores of vortices. In Fig. 7, the jet is tagged by seeding the jet supply with cigarette smoke. A well-defined jet body is seen with no presence of smoke nor, apparently, jet fluid in the wake. Conversely, when the crossflow boundary layer is tagged with smoke, as was seen in Fig. 5, most of the smoke ends up in the wake vortices. These results suggest that the jet vorticity does not contribute to the wake vorticity. Not only

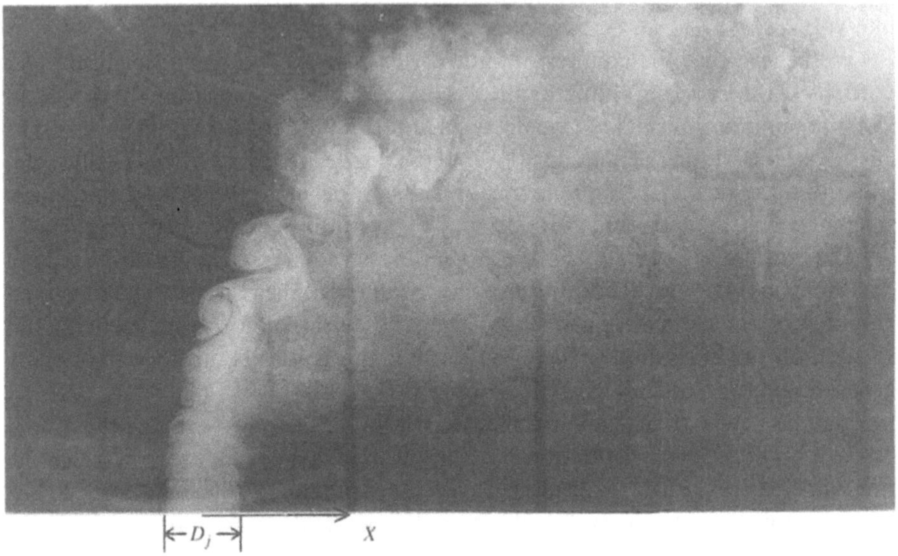

Fig. 7. The jet, issuing from the bottom left-hand portion of the photograph, is tagged with smoke. This view is the same as in Fig. 5. $VR = Re_{cf} = 3800$

does the jet not act like a solid cylinder, not generating any new vorticity, it also does not shed any of its vorticity to the wake. Rather, the source of the wake vorticity is the crossflow boundary layer.

It is clearly of interest to determine whether the wake vortices convect with a characteristic (Strouhal) frequency. Accordingly, frequency measurements were made using a hot-wire anemometer placed in the wake of the jet. The values, shown in Fig. 8, are independent of the hot-wire's location in the wake. These values of St_w, are based on the rms average of 32 individual power spectra, obtained with an HP 3582A spectrum analyzer. In general, the spectral peaks f_w (dominant frequencies) are not as sharply defined here as they would be in the wake of a solid cylinder, and the values of St_w are somewhat lower than they are for a circular cylinder wake at the same Re_{cf}. The sharpness of the spectral peaks varies considerably with VR, but for each Re_{cf} investigated, the sharpest peaks were recorded for $3.5 \lesssim VR \lesssim 4.5$. This is also the range of VR for which the values of St_w appear to be independent of Re_{cf}. Our values of St_w are in the same range as those measured by Moussa et al. (1977).

For the purpose of verifying the source of the wake vorticity, spectra were also also taken near the region denoted by the arrows in Fig. 9, where smoke is tagging the crossflow boundary layer fluid. Upstream of the jet, a portion of the crossflow boundary layer separates and forms a horseshoe vortex. The boundary layer fluid outside the horseshoe structures on the jet's $+Y$ side wraps around the jet and then separates on the its lee side, near the arrows in Fig. 9. The previous separation event, on the $-Y$ side of the jet, has convected further downstream. Figure 10 shows the values for St_{sep} measured near such separation events; the agreement with Strouhal numbers measured in the wake is very good. Also, the sharpest spectral peaks are

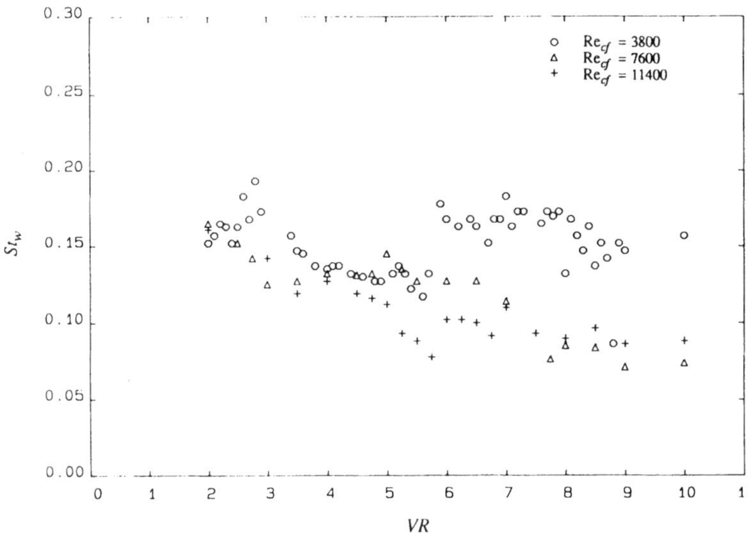

Fig. 8. Wake Strouhal number dependence on velocity ratio at $Re_{cf} = 3800$ (○), 7600 (△) and 11400 (+). Measured at $X/D_j = 3.5$, $Y/D_j = 1.5$, $Z/D_j = 0.5$

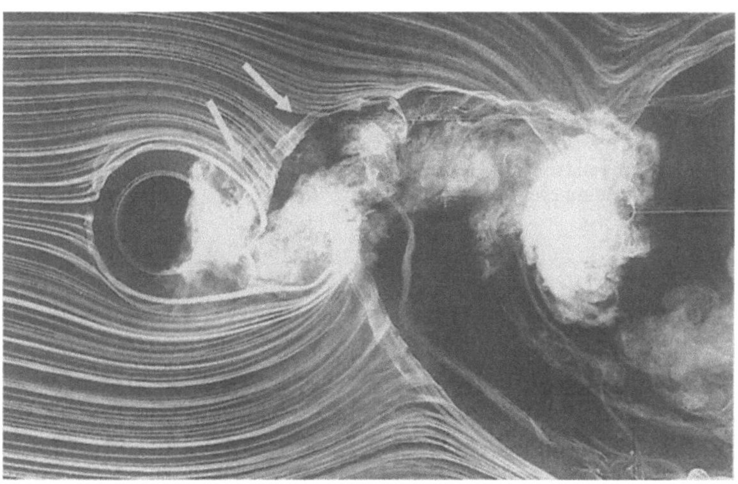

Fig. 9. Separation event. A separation of the crossflow boundary layer is indicated by the arrows. $VR = 4$, $Re_{cf} = 3800$

again seen for $3.5 \lesssim VR \lesssim 4.5$. These results support our conclusion that the source of the wake vorticity is in the crossflow boundary layer; that the characteristic frequencies of the separation events match those of the wake structures implies a connection between the two. The crossflow boundary layer separating at the downstream side of the jet to subsequently feed the wake is consistent with Fig. 5, where smoke seeding the crossflow boundary layer leaves the wall aft of the jet. We

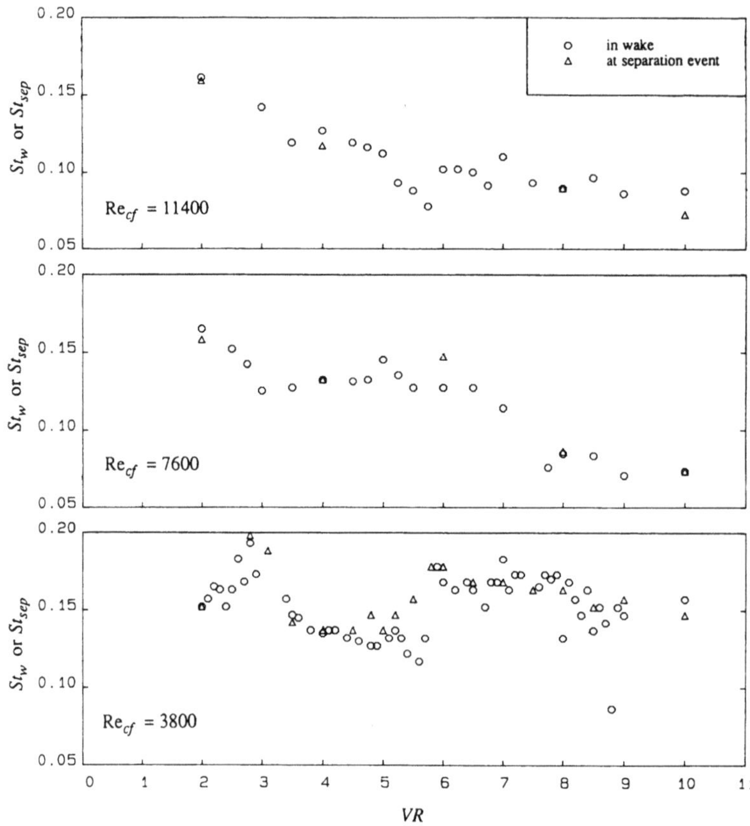

Fig. 10. Comparison of wake (○) and separation events' (△) Strouhal numbers. St_{sep} measured near the cross-flow boundary layer separation events. St_w measured at $X/D_j = 3.5$, $Y/D_j = 1.5$, $Z/D_j = 0.5$

conclude that the crossflow boundary layer separation events represent the inception of the transverse jet's wake formation.

A striking difference between the wake of a cylinder and jet is evident when comparing figure 11 with 6. In both cases, the smoke-wire is located at $Z/D_j = 0.5$ (or, equivalently, $Z/D_c = 0.5$) and $R_{cf} = 11400$. The flow around the jet in Fig. 6 looks nearly potential; the near wake streaklines are closed. Conversely, the flow around the cylinder separates, thereby opening the near wake. The potential-like streakline pattern around the jet has important implications for the crossflow boundary layer flow; the adverse pressure gradient on the lee side of the jet is imposed on the crossflow wall. This adverse pressure gradient is conducive to separating the crossflow boundary layer, near where the separation events have been identified. On the other hand, for the case of the cylinder, which has an open near wake, the crossflow boundary layer does not encounter such an adverse pressure gradient on the lee side; thus there is no analogous separation.

The model presented here for wake formation is based on the crossflow boundary layer separation events as described above. Once separation occurs, the

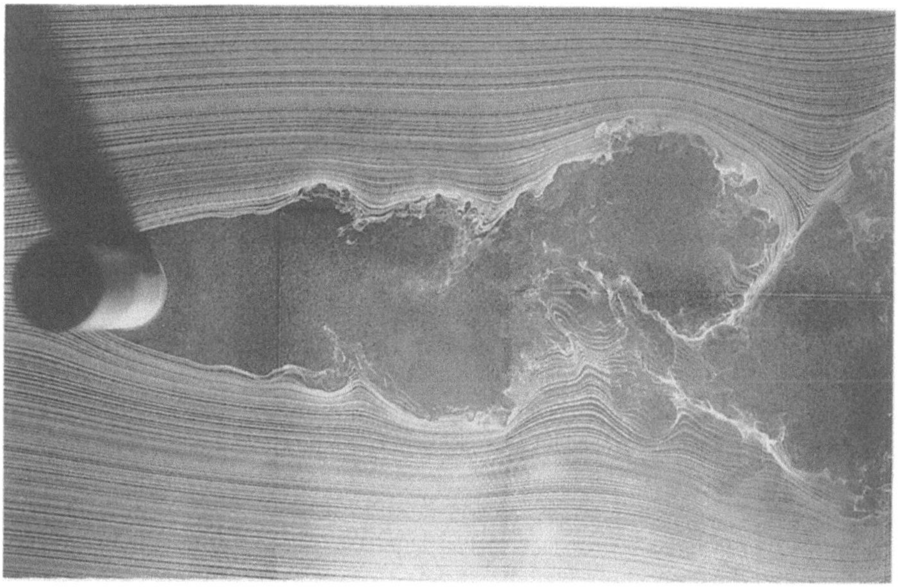

Fig. 11. Circular cylinder wake in cross section. The cylinder is at the left end of the photograph. The aspect ratio of the wall-mounted cylinder is 6. $Re_{cf} = 11400$

vorticity from the crossflow boundary layer is free to convect, stretch, turn, and diffuse, as Eq. (1) shows. That portion closest to the jet is entrained and convected by the jet, thus establishing the connection of the wake vortices with the jet. As these newly formed structures convect downstream, they are free to stretch and turn to form a wake such as those seen in Figs. 5 and 6. In general, the separation events alternate from one side of the jet to the other with $+Y$ crossflow boundary layer fluid feeding wake vorticity of $-Z$ sense, and vice versa. The separation events, in all likelihood, represent only the inception of the wake structures. Vorticity near the wall can be continuously fed into the wake structures as they convect downstream. In addition to the vorticity already present in the crossflow boundary layer, new vorticity is undoubtedly generated due to the newly imposed fluctuating pressure gradients and rotational wall shear stress present at the crossflow wall (refer to Eq. (2)).

5. Summary and Implications

The principal message of this paper is the finding, which we believe is new, that the system of vortices in the wake of a transverse jet is distinctly different in origin and formation from the vortices which are shed from a solid cylinder. In the case of the transverse jet, where the jet/crossflow interface cannot generate new vorticity, the wake vorticity comes from the boundary layer on the wall from which the jet issues.

The crossflow boundary layer separates on the downstream side of the jet because it cannot negotiate the adverse pressure gradient which is imposed on it by the flow around the jet, and its vorticity is incorporated into the wake vortices, which extend from the wall to the bending jet.

The mechanism which determines the frequency at which the wake vortices form is not clear but appears to be different from the shedding mechanism in a solid cylinder wake. The characteristic Strouhal number in the wake of a jet issuing from a wall was found to be consistently lower than that of a cylinder wake. It is of interest to compare the wake structure here to the corresponding case of a jet issuing from a pipe protruding into the crossflow. In the latter case, Moussa et al. (1977) have observed that the jet's wake frequency more closely matches the frequency of a cylinder wake. This implies that the vortex shedding from the pipe controls the structure in the jet's wake. In fact, it is consistent with our case to speculate that the vortices in the wake of a protruding jet are extensions of the vortices shed from the pipe, just as the vortices in the wake of a flush-mounted jet are extensions of vorticity from the crossflow boundary layer.

Other problems of interest include the question of how the wake vortices interact with the other vortex structures: namely, the jet shear layer vortices, the horseshoe vortices, and the nascent far field vortex pair. Consequently, it is not clear whether the wake vortex system (as well as the horseshoe vortex system) has a strong effect on the development of the main jet flow.

That the wake vorticity comes from the crossflow boundary layer fluid and not from the jet fluid is of importance when considering the mixing in this flow. The results suggest that the wake does not add substantially to the mixing of ambient (crossflow) fluid with jet fluid; the wake contains essentially no jet fluid. Broadwell and Breidenthal (1984) note that the transverse jet appears to be a better mixer of jet and ambient fluid than a free jet. The present results imply that the wake may not contribute to this enhanced mixing. The results also suggest that it may be possible to efficiently mix fluid near or in the crossflow boundary layer with the mean crossflow by taking advantage of the wake formation dynamics.

Acknowledgements: We would like to thank Professor H. G. Hornung for many helpful discussions on the issue of vorticity generation. This work was partially supported by ONR contract #N00014-85-K-0646.

References

Broadwell, J. E. Breidenthal, R. E. (1984): Structure and mixing of a transverse jet in incompressible flow. J. Fluid Mech. **148**, 405–412

Coehlo, S. L. V., Hunt, J. C. R. (1989): The dynamics of the near field of strong jets in crossflows. J. Fluid Mech. **200**, 95–120

Corke, T., Koga, D., Drubka, R., Nagib, H. (1977): A new technique for introducing controlled sheets of streaklines in wind tunnels. IEEE Publication 77-CH1251-8 AES

Fearn, R., Weston, R. P. 1974: Vorticity associated with a jet in a cross flow. AIAA J. **12**, 1666–1671

Foss, J. F. 1980: Interaction region phenomena for the jet in a cross-flow problem. Rep. SFB 80/E/161, University of Karlsruhe

Karagozian, A. R., Greber, I. 1984. An analytical model for the vorticity associated with the transverse jet. AIAA 17th Fluid Dyn., Plas. Dyn. & Las. Conf.

Kuzo, D. M., Roshko, A. (1984): Observations on the wake region of the transverse jet. Bull. Am. Phy. Soc. **29**, 1536

McMahon, H. M., Hester, D. D., Palfrey, J. G. (1971): Vortex shedding from a turbulent jet in a cross-wind. J. Fluid Mech. **48**, 73–80

Morton, B. R. (1984): The generation and decay of vorticity. Geophys. Astrophys. Fluid Dyn. **28**, 277–308

Moussa, Z. M., Trischka, J. W., Eskinazi, S. (1971): The near field in the mixing of a round jet with a cross-stream. J. Fluid Mech. **80**, 49–80.

Sykes, R. I., Lewellen, W. S., Parker, S. F. (1986): On the vorticity dynamics of a turbulent jet in a crossflow. J. Fluid Mech. **168**, 393–413

Wu, J. Z., Wu, J. M., Wu, C. J. (1987): A general 3-dimensional viscous compressible theory on the interaction of solid body and vorticity-dilation field. UTSI Report 87/03

Reynolds Stress Evolution in Curved Two-Stream Turbulent Mixing Layers

Michael W. Plesniak and James P. Johnston

Department of Mechanical Engineering, Thermosciences Division, Stanford University, Stanford, CA 94305, USA

Abstract

The effects of moderate streamwise (longitudinal) curvature on the growth, development, mixing and turbulence transport in parallel, co-flowing, two-stream mixing (free-shear) layers have been experimentally investigated. Stabilizing and destabilizing configurations (in the Taylor–Görtler sense) with identical boundary and initial conditions have been studied and compared to a plane mixing layer. Flow visualization and three-component laser Doppler velocimetry (LDV) techniques have been applied to investigate these flows. It was found that destabilizing curvature enhances the streamwise growth rate of the shear layer. Turbulence production, three-dimensionality, and fine-scale mixing are enhanced by destabilizing curvature, as well.

Nomenclature

r	Moment of momentum ratio, $(UR)_{\min}/(UR)_{\max}$ outside shear layer
R	Local radius of streamline curvature, $R_c + Y$
R_c	Mean radius of curvature at mid-profile, $R_c = R_{cl} + Y_0$
R_{cl}	Radius of arc that is centered on splitter plate trailing edge
Re_δ	Reynolds number $= \Delta U \delta / \nu$
U, V, W	Mean velocities along x, y, and z directions
U^*	Normalized mean velocity, $[(UR) - (UR)_{\min}]/[R_c \times \Delta U]$
u', v', w'	Velocity fluctuations
x, y, z	Streamwise, normal and spanwise coordinates
Y_0	Y shift, evaluated at mid-profile, $\Delta U/2$
ΔU	Velocity scale, $(U_0 R_0 - U_i R_i)/R_c$
δ	Slope thickness of the shear layer, based on straight line through the central portion of the layer
η	Normalized cross-stream coordinate for profiles, $(y - Y_0)/\delta$
λ	$(1 + r)/(1 - r)$
ν	Kinematic viscosity
θ	Momentum thickness
σ	Spreading parameter
σ_0	Spreading parameter for single-stream mixing layer
Ω_x	Streamwise component of vorticity

Turbulent Shear Flows 7

$(\ )_i, (\ )_0$	Inner or outer stream's irrotational region
$(UR)_{min}$	UR in irrotational region of lower speed stream
$(UR)_{max}$	UR in irrotational region of higher speed stream
$\langle\ \rangle$	Time averaged quantity

Background

Straight or plane mixing layers have been studied extensively beginning with the early work of Liepmann and Laufer (1947) and Townsend (1956). However, experiments concerning the effects of curvature on the mixing layer have remained scarce. The pioneering work of Margolis and Lumley (1965), Wyngaard (1967) and Wyngaard et al. (1968) showed that curvature affects turbulence shearing stress production and turbulence transport. More recent studies by Castro and Bradshaw (1976), Gibson and Rodi (1981), Gibson and Younis (1983) on single-stream mixing layers subjected to curvature have reported measurements of Reynolds stress, triple products, energy dissipation rate and other statistical quantities. Single-stream layers are known to behave differently than two-stream layers. Single-stream layers, with entrainment from the stagnant zero-velocity side, have characteristic turbulent structures that are different from the turbulence structures in two-streams at low velocity ratio (see Mehta and Westphal 1986).

Only two studies of a mildly curved two-stream mixing layers, Wang (1984) and Minardi et al. (1988), have been reported in the literature in addition to our earlier work, Plesniak and Johnston (1988). Wang's work is a systematic investigation of the interaction of various instability mechanisms due to velocity ratios, curvature effects, and density differences. However, Wang (1984) and Minardi et al. (1988) employed primarily flow visualization techniques and reported no quantitative measurements of the turbulence quantities. Thus except for our study, no data describing the velocity field in a curved two-stream mixing layer currently exists. A complete report of our study containing flow visualization results and velocity field statistics including: triple correlations, skewness and flatness profiles, and velocity spectra is contained in Plesniak and Johnston (1989).

Objectives

The main goal of this program is to gain a fundamental understanding of the physics of curved, turbulent, unforced, free mixing layers at moderate Reynolds numbers and, incidentally, to provide a data base for turbulence modelers. A three-component LDV system allows simultaneous measurement of all three velocity components and allows the entire Reynolds stress tensor to be computed. The streamwise growth rates computed from mean measurements are compared. The interaction of large-scale structures arising from the Kelvin–Helmholtz and Taylor–Görtler instability mechanisms is investigated through several flow visualization techniques. Finally, the interaction of the secondary streamwise vortices (previously observed in plane two-steam mixing layers) with the Taylor–Görtler instability due to destabilizing curvature is investigated with LDV techniques and flow visualization.

Experimental Apparatus and Techniques

All the experiments were performed in a low-speed water channel. A mildly curved ($\delta/R < 0.06$) test section schematically represented in Fig. 1 can be configured for stabilizing or destabilizing curvature with nearly identical upstream conditions. Typically, a velocity ratio of 2:1 is employed with the faster stream flowing at approximately 18 cm/s. The boundary layers on the splitter plate remain laminar up to separation at the sharp trailing edge which is tapered with an included angle of < 4 degrees to a thickness of 0.25 mm, less than the boundary layer thickness at separation. The freestream turbulence intensity of both streams is approximately 1%, including unsteadiness in the channel. Variation of U measured over the central 1/3 of the span ($z = \pm 8$ cm) is less than $\pm 0.5\%$. Shear layer Reynolds number, (Re_δ), based on ΔU and vorticity thickness ranged from 1000 to 7000 (200 to 1400 based on momentum thickness) over the region in which measurements were made.

Velocity measurements are made with a TSI three-beam, two-component LDV system with frequency shifting on both channels. The system is configured to operate in the backscatter mode. A submersible TSI fiber optics probe is positioned so that its measuring volume coincides with that of the 2-component system to obtain the third (spanwise) velocity component. All probe movement is done by a computer-controlled, stepper-motor-driven traversing system which is part of the automated data acquisition and control system.

Results

Flow visualization results are described in detail in our other work, Plesniak and Johnston (1988, 1989). To summarize: *Stabilizing curvature* does not promote mixing; the large-scale spanwise vortical structures remain primarily two-

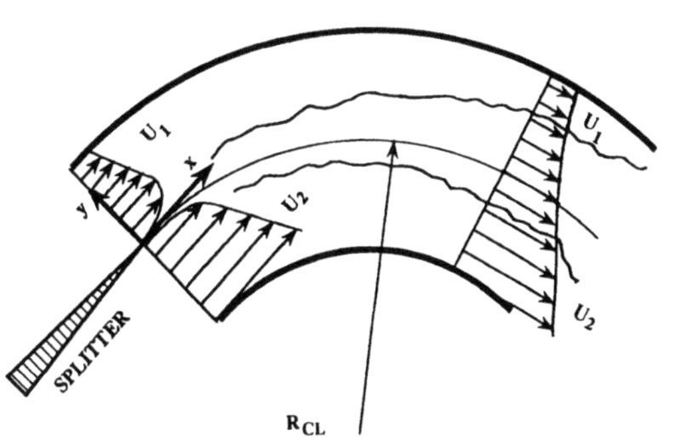

Fig. 1. Schematic of curved shear layer, unstable configuration is pictured

dimensional and less than normal fine-scale three-dimensionality is observed. In contrast, *destabilizing curvature* promotes mixing at all scales as demonstrated by the rapid spread of dye across the layer. Large-scale transport of dye across the layer is observed in videotapes of the destabilizing layer. Hydrogen bubble flow visualization showed small-scale streamwise vortices in the curved layers. These vortices were stronger and more vigorous in the unstable layer.

Counter-rotating streamwise vortex pairs, similar to those observed here, have been observed in a number of experiments in single- and two-stream plane mixing layers with laminar boundary layers on the splitter plate; for example, see Konrad (1976), Bernal (1981), Bernal and Roshko (1986), Lasheras et al. (1986, 1988), Hussain (1986), Ho and Huerre (1984), Bell and Mehta (1989) and others. It has been shown by Huang and Ho (1989) and Bell and Mehta (1989) that spanwise variations in the mean streamwise velocity, U, can be related to the presence of streamwise vortices. In addition, Bell and Mehta documented a direct correspondence of the position and magnitude of peaks in the secondary shear stress, $-\langle u'w' \rangle$, to peaks in the streamwise vorticity, Ω_x. Measurements of the secondary shear stresses have been made and are presented in Fig. 5.

In an attempt to characterize the position and magnitude of the secondary streamwise vortices, mean U-velocity profiles were taken along the span of the mixing layer on the approximate centerline ($y \approx 0$), at various streamwise positions. These results are presented in Fig. 2. In the *unstable case*, variations in U of more than $\pm 7\%$ are observed initially, at $x = 15$ cm. Far downstream, at the last measurement station, $x = 102$ cm, variations of $\pm 4\%$ are observed. Based on the results of Bell and Mehta (1989), variations in U of this magnitude can cause variations of approximately 20% in the shear stresses. Thus, some scatter is expected in shear stresses measured along a single spanwise plane in the unstable layer. The poor repeatability of the U data at $z = 0$ mm suggests that the streamwise vortical structures may slowly meander from side-to-side at a low frequency, or they may shift in z-position occasionally if the upstream disturbance varies in time. Typically, 50 minutes elapsed between the first and last (repeat) measurement in these spanwise profiles.

In contrast, the *plane* and *stable layers* show variations in U across the span of less than $\pm 1\%$ at the downstream measuring stations. The spanwise variations of U in our plane mixing layer are not as great as those reported by Huang and Ho (1989) or by Bell and Mehta (1989). Huang and Ho report maximum variations in U of $\pm 10\%$. Our levels of secondary shear stress, $-\langle u'w' \rangle$ are correspondingly lower than reported in these other experiments. Since the strength and position of the streamwise vortices is thought to be dependent on upstream, randomly located disturbances, it is difficult to make quantitative comparisons between experiments performed in different facilities.

Another feature which may be observed in Fig. 2 is that the wavelength (or distance between counter-rotating vortex pairs) increases with streamwise distance. This increase is consistent with results reported in recent literature from computational and experimental work in plane mixing layers suggesting that the wavelength associated with the streamwise vortices is approximately 2/3 that of the wavelength associated with the primary spanwise vortices. Since the primary, two-

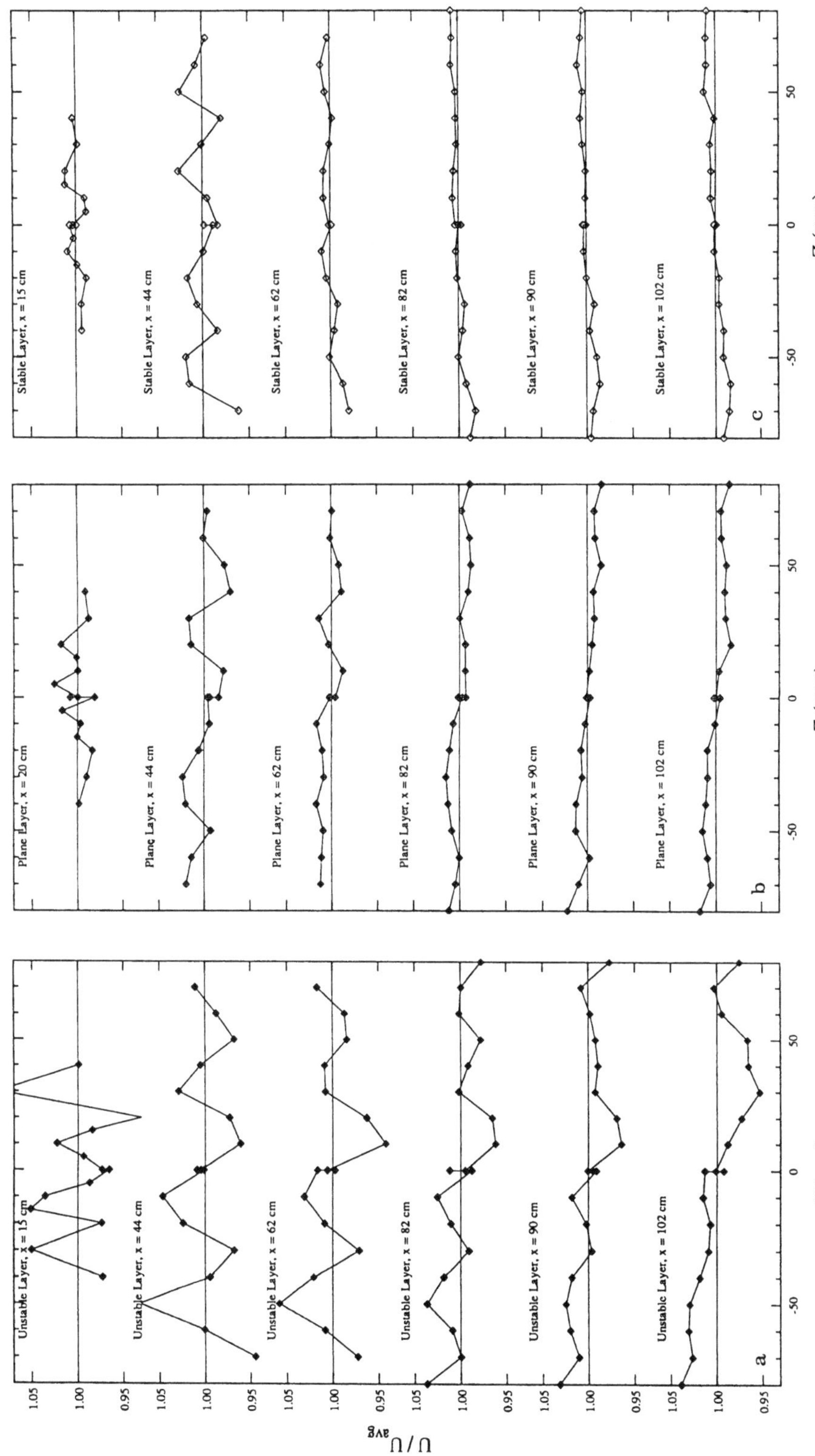

Fig. 2 a–c. Spanwise variation of mean streamwise velocity, U, at various streamwise locations for **(a)** unstable layer, **(b)** plane layer, and **(c)** stable layer dimensional spanwise structures undergo pairings and triple interactions which increase their wavelength as the layer grows downstream, it is not unexpected to observe that the spanwise wavelength has also increased downstream.

Three-component velocity profiles were acquired at seven streamwise locations in the mixing layer. A typical station at $x = 90\,\mathrm{cm}$ is representative of the profiles acquired in the self-similar region of the mixing layer. Several criteria are used to indicate self-similarity. The least stringent is to require linear growth of the shear layer thickness. Next is the criterion that the mean streamwise velocity profiles be self-similar. An even more demanding criterion is that the normal turbulence stresses reach an asymptotic value. The most stringent requirement suggested by the work of Mehta and Westphal (1986) and more recently by Bell and Mehta (1989) is that the primary shear stress, $-\langle u'v' \rangle$, reach an asymptotic limit. However, in light of recent results showing that mixing layers exhibit secondary streamwise vortices, it may be necessary to further impose the requirement that all six (independent) components of the Reynolds stress tensor reach self-similar profile shapes. For the

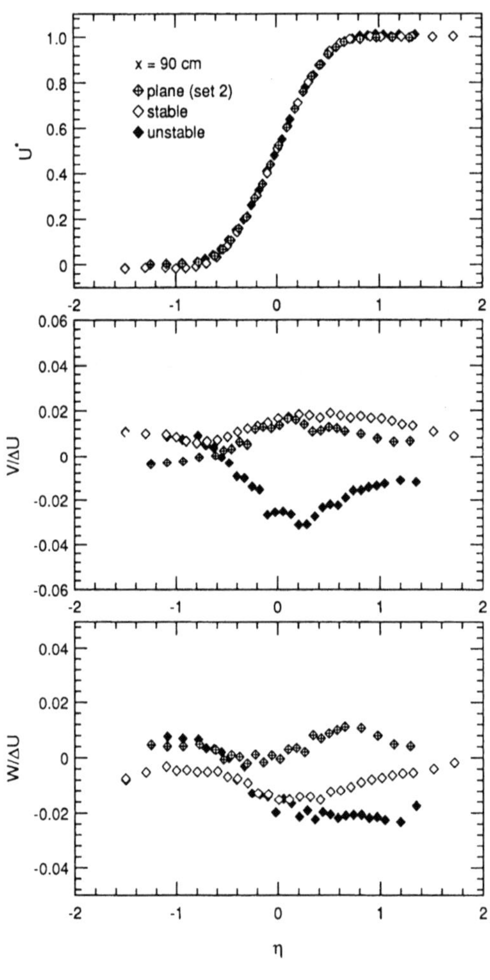

Fig. 3. Normalized mean velocity profiles for unstable, plane, and stable mixing layers at $x = 90\,\mathrm{cm}$

purposes of defining the self-similar regime in the present discussion, we use the following criteria: (i) that the mean and fluctuating (normal stress) velocity profiles attain a self-similar shape, and (ii) that the shear layer exhibit a linear growth.

Figure 3 presents the normalized mean velocity profiles for the unstable, plane, and stable mixing layers at a moment of momentum ratio, $r \cong 0.51$. Due to streamline curvature, the potential flow region in the curved mixing layer does not exhibit a constant velocity. However, the fact that moment of momentum is constant in these regions facilitates its use in the definition of the similarity variable U^*. Note that for a straight layer this parameter reduces to the usual definition of: $U^* = (U - U_{min})/\Delta U$. The abscissa is normalized using the displacement of the shear layer to the low speed side, Y_0, and the maximum slope thickness, δ. Normalized in this manner, the mean streamwise profiles for all three cases achieve a self-similar shape.

The mean normal, V, velocity profiles show a dip in the unstable case (filled symbols), which based on spanwise mean velocity measurements and flow visualization studies, is believed to be due to the velocity field induced by the stationary streamwise vortices. Destabilizing curvature acts to enhance these streamwise structures, and as a result the V-velocity profiles in the unstable case deviate from the flat shape exhibited by the plane and stable cases. The W-component velocity profiles are also typically flatter for the stable and plane cases than the unstable case.

All of the evidence currently available suggests that strict two-dimensionality is not to be expected in mixing layers. However, in the stable and plane layers, where there is a suppression of the streamwise vorticity, the mean flow far downstream appears to be two-dimensional. The slight deviation from a zero-mean W-velocity component expected in such two-dimensional shear layers was due to probe misalignment. Angle errors in the x–z plane as small as one-half degree have a significant effect on W. Slight misalignment of this amount was tolerated to ensure that the probe measuring volumes were all properly mated.

Figure 4 indicates that each component of the normal turbulent stresses is greater by a factor of 1.5 to 2.2 for the unstable layer with respect to the stable layer. The plane layer values fall in between the stable and unstable values, but are closer to the stable case at this particular streamwise location. These stresses reach an asymptotic value downstream of the $x = 62$ cm station. The transverse (v-component) Reynolds stress is most affected by destabilizing curvature. Table 1 lists the maximum values (normalized by ΔU^2) of the normal Reynolds stresses (twice the) turbulent kinetic energy, and primary Reynolds shear stress for the three cases

Table 1. Reynolds stress maximum values

Quantity	Unstable	Plane	Stable	U/P	S/P	U/S
$\langle u'u' \rangle/\Delta U^2$	0.032	0.024	0.021	1.3	0.87	1.5
$\langle v'v' \rangle/\Delta U^2$	0.033	0.020	0.015	1.6	0.75	2.2
$\langle w'w' \rangle/\Delta U^2$	0.024	0.018	0.013	1.3	0.72	1.8
$\langle q'^2 \rangle/\Delta U^2$	0.089	0.062	0.049	1.4	0.79	1.8
$-\langle u'v' \rangle/\Delta U^2$	0.0135	0.009	0.0065	1.5	0.72	2.1

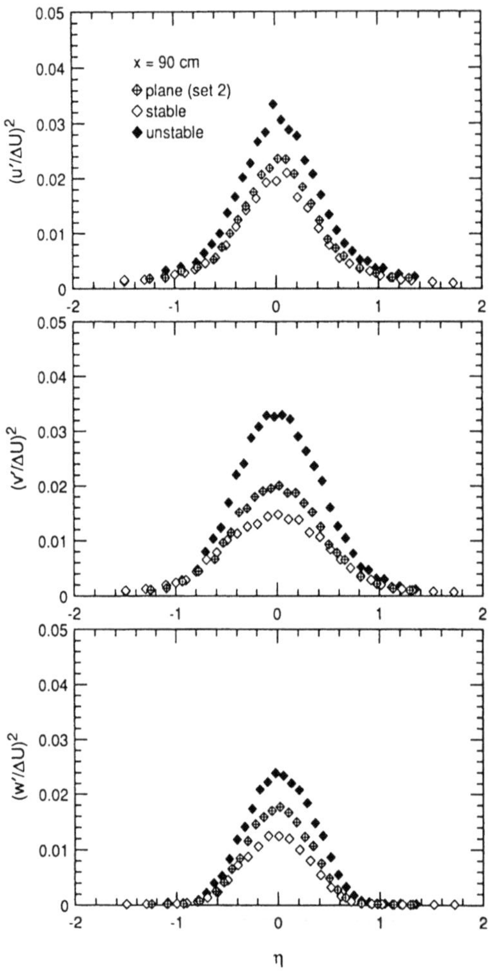

Fig. 4. Normalized Reynolds normal stress profiles for unstable, plane, and stable mixing layers at $x = 90\,\mathrm{cm}$

investigated. The ratios of the unstable to plane (U/P), stable to plane (S/P), and unstable to stable (U/S) maximum values of these quantities are also tabulated.

A comparison of the Reynolds shear stress distribution across the stable and unstable layers at $x = 90\,\mathrm{cm}$ is shown in Fig. 5. The primary shear stress intensity in the unstable layer peaks at a value approximately twice that in the stable layer (see Table 1). This is indicative of increased turbulence production and mixing in the unstable layer. The secondary turbulent shear stresses are an order of magnitude lower than the primary stresses. At this location the $-\langle u'w' \rangle$ distribution for the unstable layer exhibits a negative peak near the centerline, while the stable layer has a flat distribution. As illustrated by Bell and Mehta (1989), the larger peak in $-\langle u'w' \rangle$ in the unstable case compared to the plane and stable cases indicates the presence of stronger streamwise vorticity in the unstable mixing layer.

It should be noted, however, that the shear stress distributions, particularly $-\langle u'w' \rangle$ and $-\langle v'w' \rangle$, are not strictly self-similar and their shapes and magnitudes vary somewhat from station to station. This is due to the spanwise nonuniformity

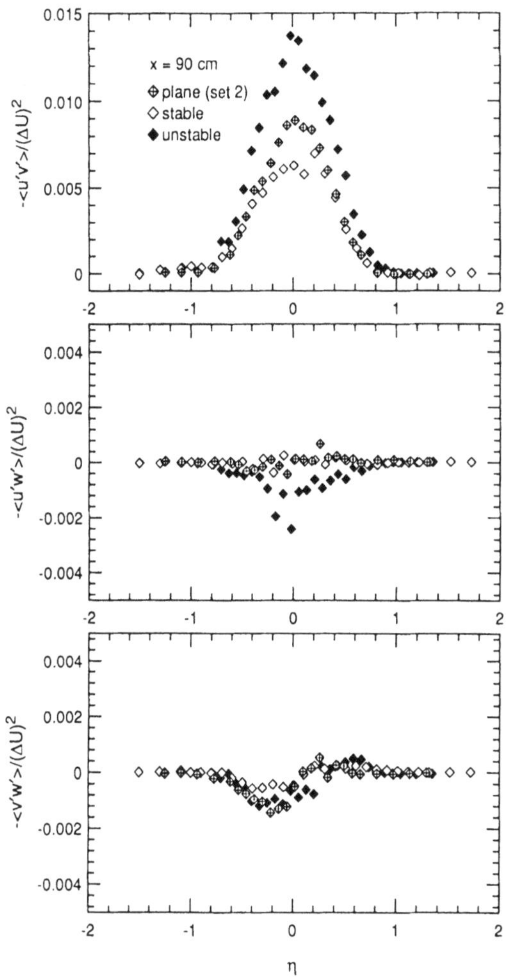

Fig. 5. Normalized Reynolds shear stress profiles for unstable, plane, and stable mixing layers at $x = 90\,\text{cm}$

caused by the stationary streamwise vortices and differences in the relative position of the physical probe measuring volumes and the streamwise vortices. Whether one is measuring between a pair of these counter-rotating structures or through the center of such a structure causes a different profile shape to be realized. Spanwise averages are desirable to obtain a more accurate estimate of the turbulent stress values in the unstable layer. Due to limitations in the traversing system, no three-component spanwise profiles were acquired.

Figure 6 shows the maximum slope (vorticity) thickness versus streamwise distance for the three cases. In the curved flows, this thickness is obtained by fitting a straight line to the central portion of the shear layer plotted as UR vs. Y. The intersection of this line with that representing the constant values of UR in the freestream potential flow defines the slope thickness. The growth rates are computed using a linear least squares curve fit for points downstream of the $x = 62\,\text{cm}$ station where similarity begins (employing the criteria described above). The unstable layer grows at a rate $d\delta/dx = 0.057$, the plane layer grows at a rate $d\delta/dx = 0.043$ while the

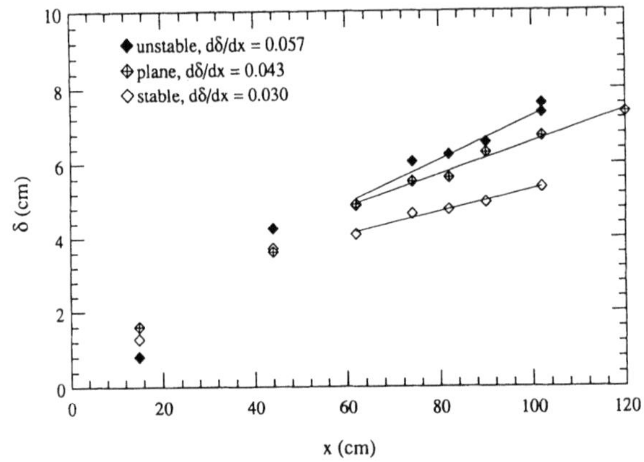

Fig. 6. Streamwise growth of the vorticity thickness for the unstable, plane, and stable layers

stable layer grows at a rate of $d\delta/dx = 0.030$. In each case the moment of momentum ratio, $(UR)_{min}/(UR)_{max} = 0.51$. The growth rate of the unstable layer is twice the growth rate of the stable layer, while the plane layer growth rate is between the two. A plane layer with a velocity ratio of $r = 0.51$ would grow at a rate $d\delta/dx = 0.052$ based on the relation given by Sabin (1965) using $\sigma_0 = 11$, i.e.

$$\sigma/\sigma_0 = \lambda = (1 + r)/(1 - r)$$

and

$$d\delta/dx = \frac{\sqrt{\pi}}{\sigma}$$

Our plane layer growth rate is 17% lower than predicted by Sabin's relation. Similarly, asymptotic limits of the primary Reynolds shear stress, are 18% lower than predictions based on a two-dimensional momentum balance (see Townsend 1956). For a fully developed plane mixing layer, the asymptotic limit $-\langle u'v' \rangle_{max}/\Delta U^2$ is independent of velocity ratio and predicted to be 0.011. In our case, the two-dimensional momentum balance is not strictly applicable due to the spanwise non-uniformity described previously. Bell and Mehta (1989) have reported variations of $-\langle u'v' \rangle_{max}$ of 20 to 30% across the span of a plane mixing layer. A more representative value of $-\langle u'v' \rangle_{max}$ in our flow would require spanwise averaging.

This low Reynolds number plane mixing layer appears to be fully developed and self-similar in that it satisfies the criteria of: invariant mean velocity profiles, asymptotic values of normal Reynolds stresses, and linear growth of vorticity thickness. However, these criteria may not be stringent enough to guarantee full self similarity. In particular, most experiments performed in air attain Reynolds numbers (based on x) at least an order of magnitude higher than achieved here. There are still conflicting views in the literature on the importance of a minimum Reynolds number and the development distance needed to achieve fully-developed mixing layer that is independent of initial conditions.

Figures 7 and 8 represent the streamwise evolution of the maxima of the normal stresses and the shear stresses. The normal stresses in all cases approach constant values beginning with the $x = 62$ cm station. As observed in the typical profile at $x = 90$ cm, the maxima of the normal turbulent stresses in the unstable layer are approximately twice those in the stable layer. There appears to be slightly more

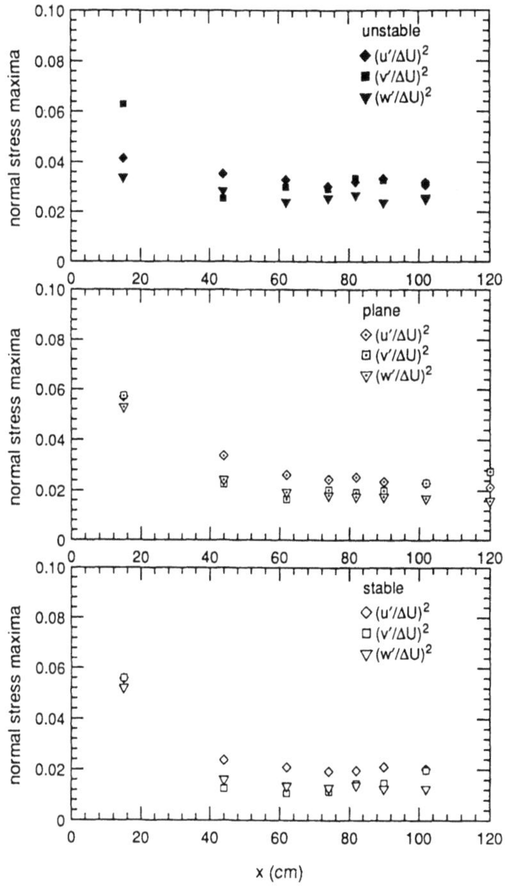

Fig. 7. Streamwise evolution of Reynolds normal stress maxima for the unstable, plane, and stable mixing layers

scatter in the data representing the unstable case. Again this may be due to meandering of the structures (although a long time average is taken with 15,000 samples per point acquired at 25 Hz to converge the second order correlations). More likely, the scatter is due to the change in relative positions of the streamwise vortices and the physical probe position as described previously. Note in Fig. 7 that, the $\langle v'^2 \rangle$-Reynolds stress component in the plane and stable layers exhibits an increasing trend at the last streamwise measuring stations after appearing to have attained an asymptotic limit. A similar rise in the $\langle v'^2 \rangle$-component (not in $\langle u'^2 \rangle$ or $\langle w'^2 \rangle$) has been observed by Wood and Bradshaw (1982) and is attributed by them to side-wall confinement effects. Confinement effects would seem likely to manifest themselves in our unstable case since that mixing layer is thickest at a given streamwise location. However, the unstable layer is also the most fully developed in terms of turbulence dynamics, whereas the stable layer is least developed. Therefore, more complex intercomponent redistribution mechanisms (see Wood and Bradshaw) are present in the unstable layer than in the stable layer. This difference may make the stable layer more sensitive to the wall effects than the unstable layer. Also, strong three-dimensional mixing is present in the unstable layer so that it will probably be less sensitive to confinement effects.

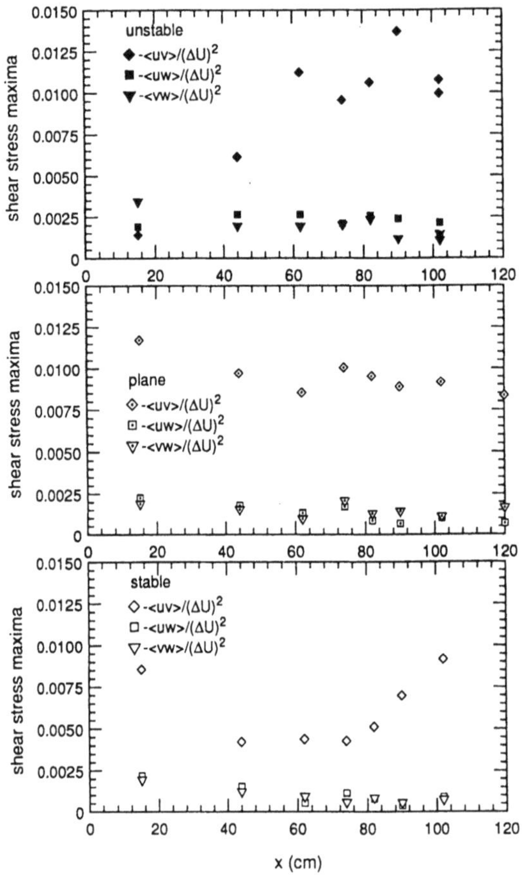

Fig. 8. Streamwise evolution of Reynolds shear stress maxima for the unstable, plane, and stable mixing layers

Figure 8 illustrates the streamwise evolution of the turbulent shear stresses. Several features are presented. First, the levels of the primary shear stress, $-\langle u'v' \rangle$, in the unstable layer are approximately twice those in the stable layer, except at the furthest downstream station, where their magnitude becomes nearly the same. It is presently unknown why the levels are comparable at $x = 102$ cm. The primary shear stress in the unstable and plane layer approaches an asymptotic value, although as discussed previously, there is some scatter due to the spanwise non-uniformity. The stable layer behaves quite differently; it exhibits a monotonic increase of the primary shear stress downstream of $x = 74$ cm. This increase is believed to be associated, in part, with the increase in the $\langle v'^2 \rangle$-Reynolds stress component observed in Fig. 7. However, it is due primarily to the fact that the stable layer is not yet fully developed. The delayed mixing transition causes the stable layer to develop more slowly with streamwise distance. In this respect, the unstable shear layer may be considered more mature or more fully developed than the stable layer at the same streamwise location. However, due to the low Reynolds number of the flows being studied, neither layer can be considered fully self-similar when applying the more stringent requirements that the primary shear stresses should reach an asymptotic value.

Conclusions

Comparison of results obtained for curved mixing layers subjected to mild stabilizing and destabilizing curvature with nearly identical initial and boundary conditions demonstrate the differences in their growth and evolution. Flow visualization results show that small-scale three-dimensionality necessary for efficient mixing is enhanced by destabilizing curvature and inhibited by stabilizing curvature. The velocity data indicate that all the layers achieve self-similarity in the mean streamwise component. Peak values of the fluctuating velocity components reach asymptotic limits downstream of the $x = 62$ cm station. The growth rate of the unstable layer is twice that of the stable layer at the same moment of momentum ratio, r. Similarly, the maxima of normal turbulent stresses and primary and secondary shear stresses in the unstable layer are found to be approximately twice those in the stable layer.

Dips in the mean normal V-velocity profiles and the shapes of the secondary shear stress profiles suggest the presence of strong stationary streamwise vortical structures in the unstable layer. Spanwise surveys of the mean streamwise velocity (U) exhibit variations of up to $\pm 7\%$ in the unstable layer and only $\pm 2\%$ in the stable layer. These results combined with the flow visualization demonstrate that the Taylor–Görtler instability mechanism present in the unstable layer enhances or reinforces the strength of the streamwise vortices which appear (but are weaker) in the plane mixing layer. Some evidence suggests (see Plesniak and Johnston 1989) that the streamwise vortices arise due to a secondary instability in the flow, although rigorous proof does not yet exist. If the secondary instability is T–G-like, then the interaction of the global Taylor–Görtler instability and the secondary instability is responsible for the observed increase in streamwise vorticity.

A mapping of all the components of the Reynolds stress tensor quantitatively shows the differences in turbulence production and transport in the two layers. Such information is necessary for turbulence modelers to accurately account for the effects of extra rates of strain due to curvature in their numerical simulations. In summary, the effects of mild stabilizing and destabilizing longitudinal curvature on the development of turbulent, two-stream mixing layers were experimentally investigated. It was found that curvature significantly enhances or diminishes the transport of turbulence and the development of fine-scale three-dimensionality, depending on its sense.

Follow-up experiments using the same high-speed air flow facility used by Bell and Mehta (1989) are being performed to extend the domain of validity of these results to higher Reynolds numbers. Questions concerning the effects of low Reynolds number, development distance and possible confinement effects encountered in the present study should be answered by these new experiments.

Acknowledgements: This work was supported at various times by the U.S. Air Force Wright Aeronautical Laboratories, the NASA/Stanford Center for Turbulence Research, and currently by the National Science Foundation. Dr. Plesniak wishes to acknowledge the support provided by an Office of Naval Research graduate fellowship, through 1987.

References

Bell, J. H., Mehta, R. D. (1989): Three-dimensional Structure of Plane Mixing Layers. JIAA TR-90, Stanford University, Dept. Aero. and Astro.

Bernal, L. P. (1981): The coherent structure of turbulent mixing layers. Ph.D. Thesis, California Inst. Tech.

Bernal, L. P., Roshko, A. (1986): Streamwise vortex structure in plane mixing layers. J. Fluid Mech. **170**, 499–525

Castro, I. P., Bradshaw, P. (1976): The turbulence structure of a highly curved mixing layer. J. Fluid Mech. **73**, 265–304

Gibson, M. M., Rodi, W. (1981): A Reynolds-stress closure model of turbulence applied to the calculation of a highly curved mixing layer. J. Fluid Mech. **103**, 161–182

Gibson, M. M., Younis, B. A. (1983): Turbulence measurements in a developing mixing layer with mild destabilizing curvature. Exp. Fluids **1**, 23–30

Ho, C-M., Huerre, P. (1984): Perturbed free shear layers. Ann. Rev. Fluid Mech. **16**, 365–424

Hussain, A. K. M. F. (1986): Coherent Structures and Turbulence. J. Fluid Mech. **173**, 303–356

Konrad, J. H. (1976): An experimental investigation of mixing in two-dimensional turbulent shear flows with applications to diffusion-limited chemical reactions. Ph.D. thesis, California Inst. Tech.

Lasheras, J. C., Cho, J. S., Maxworthy, T. (1986): On the origin and evolution of streamwise vortical structures in a plane, free shear layer. J. Fluid Mech. **172**, 231–258

Lasheras, J. C., Choi, H. (1988): 3-D instabilities of a plane free shear layer: An experimental study of the formation and evolution of streamwise vortices. J. Fluid Mech. **189**, 53–86

Liepmann, H. W., Laufer, J. (1947): Investigations of free turbulent mixing. NACA TN No. 1257

Margolis, D. P., Lumley, J. L. (1965): Curved turbulent mixing layer. Phys. Fluids **8**, 1775–1784

Mehta, R. D., Westphal, R. V. (1986): Near-field turbulence properties of single- and two-stream plane mixing layers. Exp. Fluids **4**, 257–266

Minardi, J. E., Skomrock, M. P., von Ohain, M. O., Boehman, L. I. (1988): Final report on the interface instability between two gas streams of different density in a curved flow. UDR-TR-88-94, University of Dayton

Plesniak, M. W., Johnston, J. P. (1988): The effects of stabilizing and destabilizing curvature on a turbulent mixing layer. *Transport Phenomena in Turbulent Flows: Theory, Experiment, and Numerical Simulation* Hirata, M. Kasagi, N. eds, Hemisphere Publishing Corp. 377–390

Plesniak, M. W., Johnston, J. P. (1989): The effects of longitudinal curvature on turbulent two-stream mixing layers. Report MD-54, Thermosciences Div., Dept. of Mechanical Engineering, Stanford University

Sabin, C. M. (1965): An analytical and experimental study of a plane, incompressible, turbulent shear layer with arbitrary velocity ratio and pressure gradient. Trans. ASME, J. Bas. Eng. **87**, 421–428

Townsend, A. A. (1956): The structure of turbulent shear flow. Cambridge University Press

Wang, C. (1984): The effects of curvature on turbulent mixing layers. Ph.D. Thesis, GALCIT, California Inst. Tech.

Wood, D. H., Bradshaw, P. (1982): A turbulent mixing layer constrained by a solid surface. Part 1. Measurements before reaching the surface. J. Fluid Mech. **122**, 57–89

Wyngaard, J. C. (1967): An experimental investigation of the small-scale structure of turbulence in a curved mixing layer. Ph.D. Thesis, Penn. State University

Wyngaard, J. C., Tennekes, H., Lumley, J. L., Margolis, D. P. (1968): Structure of turbulence in a curved mixing layer. Phys. Fluids **11**, 1251–1253

The Effect of Extra Strain Rates of Streamline Curvature and Divergence on Mixing Layers

A. E. Johnson[1] and P. E. Hancock

Department of Mechanical Engineering, University of Surrey, Guildford, Surrey, GU2 5XH, UK
[1] Present address: BP Research, Sunbury Research Centre, Sunbury-on-Thames, Middlesex, TW16 7LN, UK

Abstract

Measurements have been made in two axisymmetric single-stream mixing layers, each subjected simultaneously to curvature and divergence. The boundary layer was turbulent at separation. In the 'very-strongly-strained' layer, the curvature was entirely of the stabilizing sense. The large-scale structures were suppressed but not destroyed. Compared with what would be expected for curvature alone, the effect of divergence was to resist the suppression of the large-scale structures, and to increase the rate of recovery. In the 'moderately-strained' layer, destabilizing curvature followed the stabilizing curvature. The Reynolds stresses and triple-velocity products varied qualitatively as the lagged strains, though, quantitatively, destabilizing effects are unlikely to have been absent in the lagged region of stabilizing curvature. The approach of an undistorted layer to asymptotic stress levels is distinctly non-monotonic.

Nomenclature

U_r	Reference velocity
U_{max}	Maximum U at constant s
R	Radius of curvature, negative for concave (i.e. stabilizing) curvature
s,n	Shear layer curvilinear orthogonal axes, with n positive towards the low-velocity side
x,y,z	Cartesian axes
X,r,ϕ	Cylindrical axes—X aligned with symmetry axis
Γ	Angle between X and tangent to s
U,V,W	Mean velocities in s-n-ϕ directions
u, v, w	Velocity fluctuations in s-n-ϕ directions
q^2	$= u^2 + v^2 + w^2$
Λ	Gradient thickness, $\Lambda = U_{max}/(\partial U/\partial n)_{n=0}$
θ	Momentum thickness
Φ,k	Spectral density, and wavenumber: $u^2 = \int \Phi_{11}(k)dk,\ v^2 = \int \Phi_{22}(k)dk$.

Turbulent Shear Flows 7

Introduction

Shear layers in engineering practice are rarely two-dimensional with nearly parallel streamlines. Generally, the mean streamlines curve and diverge or converge, and give rise to extra rates of strain. Whilst there have been many investigations of 'simple' single-stream mixing layers—i.e. mixing layers subjected to a single predominant strain rate, $\partial U/\partial y$—relatively little attention has been given to the effects of extra strain rates. Extra strain rates are well known to have a much larger effect on Reynolds stresses—by roughly an order of magnitude (Bradshaw 1973, 1975)—than is implied by the explicit extra generation terms in the Reynolds-stress transport equations because of their large effect on the turbulence structure.

The work reported here forms part of a systematic study of the effect of more than one extra strain rate on a single-stream mixing layer, specifically curvature and divergence. Strictly speaking, curvature alone involves two extra strain rates; these are $\partial V/\partial x$ and $\partial U/\partial x$ ($= -\partial V/\partial y$), but these are coupled. In general, two or more extra rates of strain may occur 'in series' or 'in parallel', or both, and it is highly probable that the effect of one *on* the turbulence structure will affect the response *of* the structure to a second. On recent example in this category appears to be the finding by Castro (1988) that the effect of streamline curvature on a mixing layer bounding a closed region of separated flow is much less than it for an isolated mixing layer, presumably because of the effect of the fluctuating strain imposed by the turbulence in the recirculating region.

The effects of streamline curvature on two-dimensional single-stream mixing layers have been investigated experimentally by Castro and Bradshaw (1976) for strong stabilizing curvature and by Gibson and Younis (1983) for moderate destabilizing curvature. Gibson and Rodi (1981) have compared a Reynolds-stress closure model with the results of Castro and Bradshaw, with reasonable agreement. Keffer (1965) found that the spanwise structures of a two-dimensional wake subjected to strong divergence became intensified, consistent with the anticipated effect of stretching the spanwise vorticity. Features of extra strain rate effects, primarily stress ratios, have been calculated by Townsend (1980) and Savill (1987) by means of a rapid-distortion approach.

Measurements have been made in two axisymmetric mixing layers (Figs. 2 and 8)—hereafter referred to as Case 1 and Case 2:

Case 1. The combined effects of 'very strong' divergence and stabilizing curvature.
Case 2. The combined effect of 'moderate' divergence and curvature, the latter changing from a stabilizing to a destabilizing sense.

An axisymmetric flow was selected because the flow is well defined and quantities such as divergence are comparatively easily determined, and because such a flow is much easier to handle computationally than a general three-dimensional flow. There is, however, an implicit drawback in that the amount of the irrotational flow is more restricted than it might be in a non-axisymmetric flow (and more restricted than in a two-dimensional flow of equal shear-layer curvature). Nevertheless Case 1 in particular is of interest in its own right as an impinging jet. A set of 'reference' measurements was also made in an undistorted mixing layer.

In addition, measurements were made in a plane mixing layer subjected to divergence alone by means of a specially-shaped channel. The level of divergence was comparable to that of Case 2. However, the measurements exhibited unexpected features, even in the absence of divergence, and our povisional conclusion is that this may have been due to a feedback effect from the downstream flow. Results of this case will be presented elsewhere, after a more detailed investigation.

It was decided that in the first series of experiments, extra rates of strain should be applied only after the self-preserving conditions had been at least approximately obtained. This has the advantage of 'well established' initial conditions upstream of the curvature and divergence, but has the drawback that the high-velocity side at least is affected by the wall sooner than is desirable. This is exacerbated in the present case by a mixing layer formed from an initially turbulent boundary layer because of the slower approach to an asymptotic state.

Measurement Techniques

In all cases the mixing layer was developed from a boundary layer developed on a purpose-made duct (of 326 mm internal diameter) and the contraction boundary layer bled off, rather than from a boundary layer originating in the wind-tunnel contraction. This was done because of variations of $\pm 30\%$ in wall shear stress around the contraction exit, attributed to Taylor–Görtler vortices developed in the concave region of the contraction. In the newly-formed boundary layer, the variation in wall shear stess just upstream of the separation was $\pm 2\%$. At separation $U_r\theta/\nu \approx 1000$, and the velocity profile was well behaved, though the wake strength was slightly larger than expected, perhaps as a residual effect of the transition trip (Dymo-label strip impressed with "V's"). Wall shear stress was measured by means of Preston tubes. U_r was about 20 m/s throughout, and the streamwise free-stream intensity was about 0.2%, with a characteristically turbulent spectrum. Mixing-layer symmetry was confirmed by making check measurements in several ϕ planes.

Turbulence measurements were made by means of hot-wire anemometry, using standard Dantec P51 and P61 X-wire probes, driven by Prosser anemometers. The outputs were digitized using 10-bit A/D converters, after adding voltage offsets so as to maximise the digital resolution, and linearized. Spectral measurements were made mostly by an (8-bit) fast-Fourier transform, typically on 128 blocks each containing 128 samples. Two sampling rates of 400 Hz and 7 kHz were used at each point. No frequency smoothing was employed, except in so far as not each point is plotted.

The centre line, s, was deduced iteratively from preliminary traverses. Some final traverse lines in Case 1 are not precisely perpendicular to s, the maximum departure being 7°, but the turbulence quantities quoted have been rotated to axes parallel and perpendicular to the tangent to s. The slight non-constancy of s for a traverse is significant only for streamwise gradients.

Results and Discussion

Undistorted Layer

Axisymmetric mixing layers cannot strictly be self preserving because the ratio of shear layer thickness to diameter cannot be constant. However, to a good approximation the profiles of U behaved in a closely self-preserving manner (see also Birch 1980). The transverse mean velocity, V, was not found to be self preserving.

The gradient thickness, Λ (not shown), varies as $0.19\,(s - s_0)$, where s_0 is the virtual origin. Despite the apparent simplicity of planar two-dimensional mixing layers there is still considerable disagreement in such simple parameters as spreading rate. However, the present growth rate is in good agreement with those of Wood and Bradshaw (1982), Castro and Bradshaw (1976), and Hussain and Clark (1981).

The stresses $\overline{u^2}, \overline{v^2}, \overline{w^2}$ and $\overline{uv}$, normalised by the respective maximum, are closely similar, and so are discussed here only in terms of maximum values, shown in Figure 1, normalized by nominal downstream levels. All the stresses are clearly non-monotonic. Spectral measurements (not shown here) indicate structural changes at wave-numbers of the order of $1/\Lambda$. The decrease in stress levels would probably also occur (at least qualitatively) in a two-dimensional *plane* layer because the decrease is slowest where the effects of axisymmetry and merging are likely to become large—i.e. just before the irrotational core ends. It is possible that the 'final' levels may be lower still in a plane layer unaffected by interaction with another layer. A previous distinction between a mixing layer formed from a turbulent boundary and one formed from a laminar boundary layer has been the monotonic growth of the Reynolds stresses for the former case and the distinctly non-monotonic growth for the latter (e.g. Foss 1977; Husain and Hussain 1979). A noticeable feature of Fig. 1 is that the peak in $\overline{u^2}$—the most commonly measured stress—is small, and arguably within measurement error. The peaks are all at roughly the same distance, s, corresponding to about 1000 times the momentum thickness at separation (c.f. Bradshaw 1966). A non-monotonic behaviour would also be consistent with the conclusion—though by no means a proof—that initial conditions do *not* persist for a long distance downstream. One reason for disbelieving the existence of more

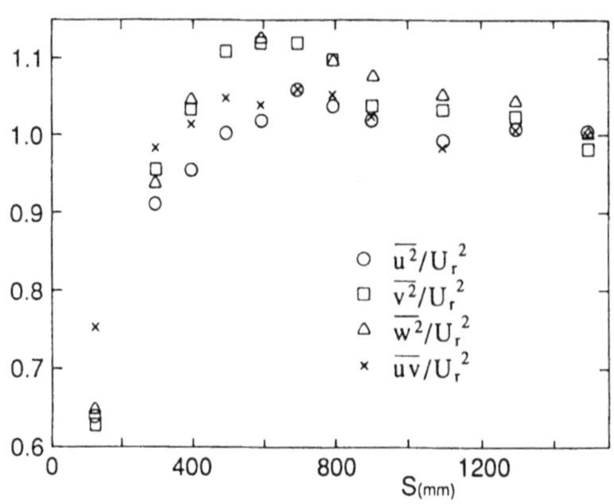

Fig. 1. Variation of maximum Reynolds stresses with s for the undistorted mixing layer, normalised by 'downstream' levels (○ 2.91×10^{-2}; □ 1.52×10^{-2}; △ 1.78×10^{-2}; × 0.94×10^{-2})

than one asymptotic state is that (unlike a wake, say) there is continual energy input to a mixing layer by virtue of the constant velocity difference across it (Kline, private communication). Third- and fourth-order moments (not included here) also varied non-monotonically in a broadly similar manner to the Reynolds stresses. The present results are broadly consistent with the non-monotonic behaviour of the mean-flow spreading rate deduced by Birch (1983) from a survey of previous data.

Case 1

Figure 2 shows the strongly perturbed layer, deflected by the 'wall' perpendicular to X at $X = 996$ mm, as defined by lines of $0.95U_{max}$ and $0.25U_{max}$, and by centre-lines based on $0.67U_{max}$ and $\overline{q^2_{max}}/U_r^2$. At most, the tangents between these two differ by 10°, and the latter runs almost parallel to a centre-line based on stream function, omitted here for clarity. Mean-flow and turbulence quantities can be described unambiguously in terms of a curvilinear, orthogonal, $(s-n)$ coordinate system, with s along the clearly identifiable predominant flow direction, and n perpendicular to s. For convenience, we have taken the centre-line, s, as the velocity contour defined by $U = 0.67U_{max}$, throughout. Figure 2 also indicates the stations at which measurements were made. As discussed later, the effect of the wall becomes significant after about $s = 900$ mm.

The mean velocity profiles, normalized by U_{max} and the gradient thickness, closely follow a self-similar form, and so are not shown. Beyond the position at which $U = U_{max}$ the decrease in U is a result of the flow curvature, and cannot possibly conform to a self-similar form, of course. Figure 3a, b shows the variations of the maximum velocity, U_{max}, and the gradient thickness, Λ, and the extra strain rates, as defined by centre-line values, are shown in Fig. 3c in Cartesian notation for brevity (with x along the centre line, s, and y towards the low-velocity side), and normalized by $-\partial U/\partial n$ on the centre line. A minus sign is included so

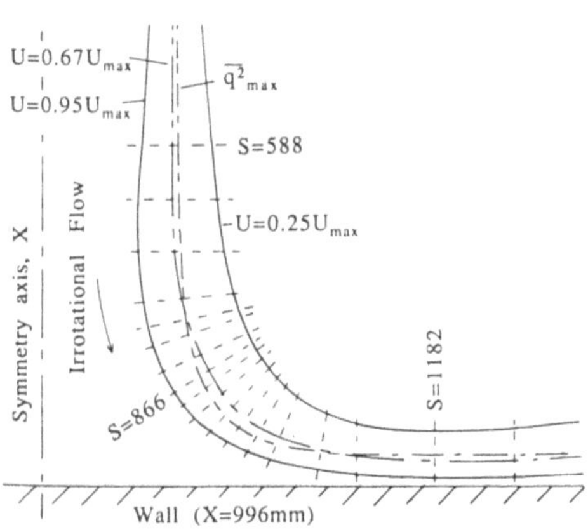

Fig. 2. Illustration of the very-strongly-strained layer

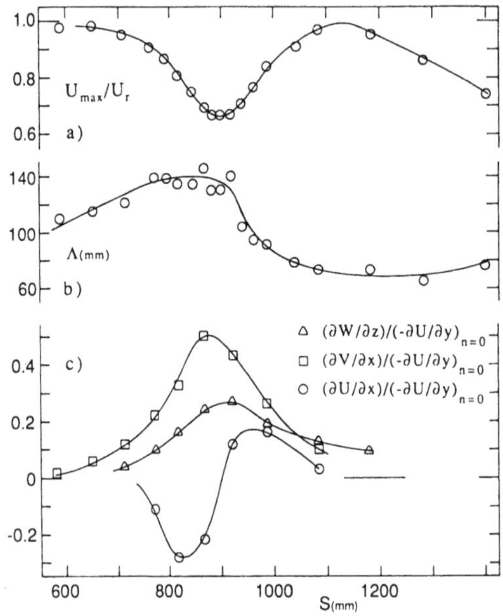

Fig. 3. U_{max}, Λ, and $\partial W/\partial z$, $\partial V/\partial x$ on the centre-line, normalised by $\partial U/\partial y$. $\partial U/\partial x$ is also shown

that the non-dimensional ratios retain the sign of the numerators. In s–n coordinates for an axisymmetric system

$$\frac{\partial U}{\partial x}=\frac{1}{h}\left\{\frac{\partial U}{\partial s}+\frac{V}{R}\right\},\quad \frac{\partial V}{\partial x}=\frac{1}{h}\left\{\frac{\partial V}{\partial s}-\frac{U}{R}\right\},$$

$$\frac{\partial U}{\partial y}=\frac{\partial U}{\partial n},\quad \frac{\partial V}{\partial y}=\frac{\partial V}{\partial n},\quad \frac{\partial W}{\partial z}=\frac{U\sin\Gamma+V\cos\Gamma}{r+n\cos\Gamma},$$

where $h=(1+n/R)$. Here, r is the radius between the axis of symmetry, X, and the centre-line, s. $\Gamma=0$ in an unstrained layer. ($\partial W/\partial z$ is not strictly zero in an axisymmetric layer.) $\partial U/\partial x$ is zero where U_{max} is minimum. The curvature parameter Λ/R is given by $1.5(\partial V/\partial x)/(\partial U/\partial y)_{n=0}$, neglecting $\partial V/\partial s$, which is small. The peak in $(\partial V/\partial x)/(\partial U/\partial y)$ is about 2.5 times larger than the peak in the flow of Castro and Bradshaw. Downstream of about $s=1150$ mm the wall boundary layer begins to reduce U_{max}.

Λ decreases substantially and then very rapidly just after $(\partial V/\partial x)/(\partial U/\partial y)$ has passed its peak and $(\partial U/\partial x)/(\partial U/\partial y)$ has become positive (Fig. 3c), though the decrease is not nearly so marked if it is compared with the convergence of the mean streamlines, indicated by $\partial V/\partial y<0$; from continuity $\partial V/\partial y=-\partial U/\partial x-\partial W/\partial z$. $(\partial W/\partial z)/(\partial U/\partial y)$ first increases because of the increase in $(\partial W/\partial z)$, and decreases because of the decrease in shear layer thickness.

As found by Castro and Bradshaw and Gibson and Younis the transverse distribution of Reynolds stresses (not shown) are of broadly similar shape but the levels change substantially. Figure 4 shows the peak values of $\overline{u^2}, \overline{v^2}, \overline{w^2}$ and $\overline{uv}$, together with those for the unstrained mixing layer, indicated by broken lines. Stress levels at the centre line are close to these maxima. The effect on the shear stress is particularly dramatic, changing sign just downstream of maximum curvature, and

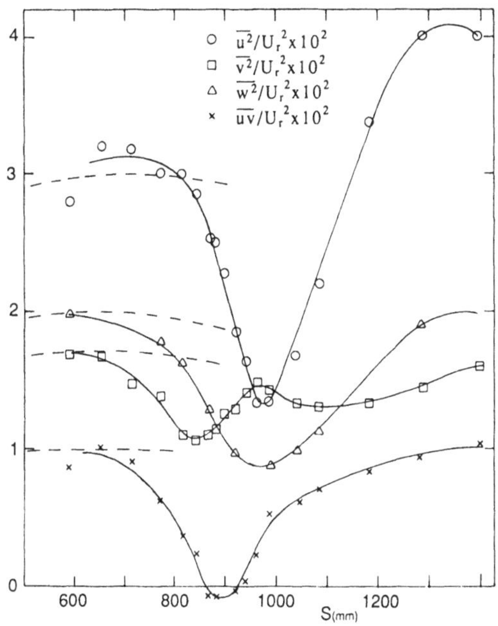

Fig. 4. Variation of maximum Reynolds stresses with s for Case 1. Broken lines are for undistorted layer

remaining negative for about half a shear layer width. The shear strain rate remains positive, though. The minimum coincides almost exactly with the minimum in U_{max}. The normal stresses $\overline{u^2}$ and $\overline{w^2}$ are also reduced dramatically, the minima being distinctly downstream of the minimum in shear stress by about one shear-layer thickness. The double minimum in $\overline{v^2}$ and its rise above the minimum in $\overline{u^2}$, rather than (say) remaining below $\overline{w^2}$, as observed by Castro and Bradshaw, is particularly interesting. ($\overline{v^2}$ and $\overline{uv}$ are not highly dependent on centre-line definition.) The rise after the first minimum in $\overline{v^2}$ is probably due in part to the cessation of negative production of $\overline{v^2}$ following the collapse in $\overline{uv}$, to positive production from the negative $\partial V/n$, and to divergence. Changes in $\overline{u^2}$, $\overline{v^2}$, $\overline{w^2}$ and $\overline{uv}$ are of course also determined by (the so-called 'rapid' and 'return-to-isotropy') pressure-strain redistribution, and by turbulent diffusion. The latter part of the (first) rise in $\overline{v^2}$ is identifiably caused by the wall. The generation terms in the transport equations for $\overline{u^2}$, $\overline{v^2}$, $\overline{w^2}$, and $\overline{uv}$ are, respectively,

$$\overline{u^2}: \quad -\overline{u^2}\frac{1}{h}\left\{\frac{\partial U}{\partial s}+\frac{V}{R}\right\}-\overline{uv}\left\{\frac{\partial U}{\partial n}+\frac{1}{h}\frac{U}{R}\right\},$$

$$\overline{v^2}: \quad -\overline{uv}\frac{1}{h}\left\{\frac{\partial V}{\partial s}-2\frac{U}{R}\right\}-\overline{v^2}\frac{\partial V}{\partial n},$$

$$\overline{w^2}: \quad -\overline{w^2}\frac{U\sin\Gamma+V\cos\Gamma}{r+n\cos\Gamma},$$

$$\overline{uv}: \quad -\overline{u^2}\frac{1}{h}\left\{\frac{\partial V}{\partial s}-2\frac{U}{R}\right\}-\overline{v^2}\left\{\frac{\partial U}{\partial n}+\frac{1}{h}\frac{U}{R}\right\}-\overline{uv}\frac{1}{h}\left\{\frac{\partial U}{\partial s}+\frac{V}{R}+h\frac{\partial V}{\partial n}\right\}.$$

A simple argument for the decrease in $\overline{uv}$, applicable to stable or unstable curvature, is that a fluid particle with a positive u-fluctuation will be less inclined to turn than a particle with a negative u-fluctuation but subject to the same radial force. Thus, we would expect positive u to be associated with negative v, while in the undistorted layer $\overline{uv}$ is positive. In destabilizing flows this 'centrifugal effect' would appear to be outweighed by destabilizing effects. The radial-acceleration term in the instantaneous momentum equation for $V + v$, namely $(U + u)^2/R$, leads to the generation term $2\overline{u^2}\,U/Rh$ in the transport equation for $\overline{uv}$; as a proportion of $\overline{v^2}\partial U/\partial y$ it is $2(\overline{u^2}/\overline{v^2}) \times (\partial V/\partial x)/(\partial U/y)$ which increases with the strain ratio. Terms in the shear-stress transport equation are also discussed by Gillis and Johnston (1983).

The curvature has become small by $s = 1150$, though there is still some readjustment of the stresses, most noticeably $\overline{u^2}$ which 'overshoots' its undistorted level by roughly 25%. Figure 5 compares the present measurements with those of Castro and Bradshaw, where the streamwise coordinate has been measured with respect to the position at which $\overline{u^2}$ is minimum. A notable feature is that the minima in the direct stress levels are broadly not very different from those of Castro and Bradshaw. The streamwise extent over which the strains are applied are about the same in both cases, the higher strain rates of the present case being primarily because of the larger shear-layer thickness. However, the decreases and rises in $\overline{u^2}$, $\overline{w^2}$ and $\overline{uv}$ are considerably quicker than in the case of Castro and Bradshaw, presumably because of the divergence, though it is surprising that $\overline{v^2}$ rises at about the same rate, rather than more quickly still, as might be expected from vortex-stretching

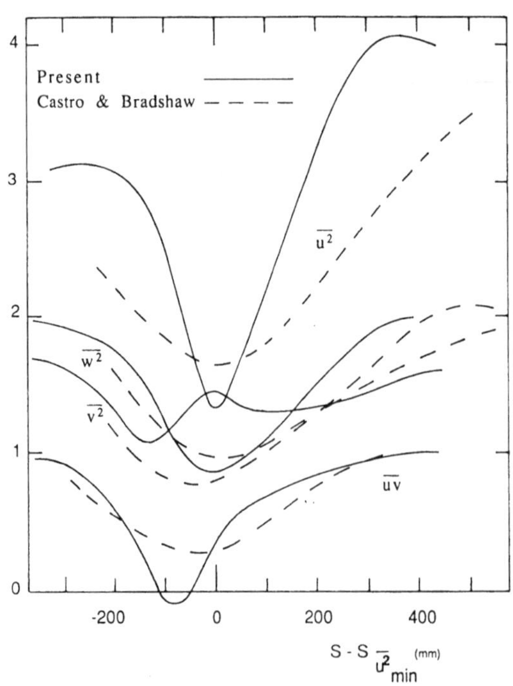

Fig. 5. Comparison of the present maximum stress levels with those of Castro and Bradshaw (1976)

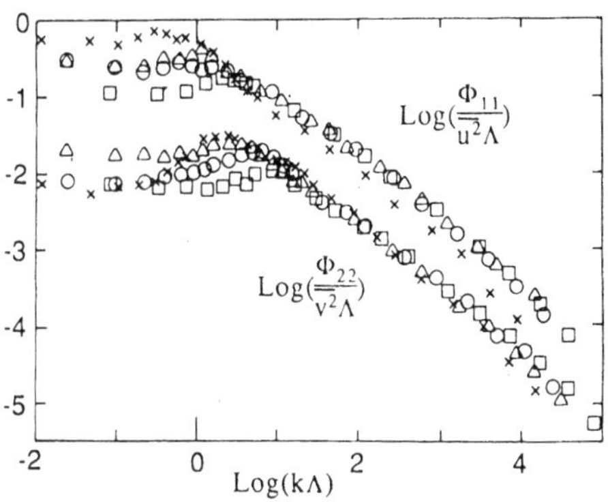

Fig. 6. Φ_{11} and Φ_{22} at $s = 588$, ○; 918, □; 987, △; and 1083, ×. Φ_{22} is displaced one decade down

arguments (e.g. Keffer 1965). Overall, though, the maxima in $\overline{u^2}$, etc. occur much sooner after the position at which curvature has become zero. However, the levels of the maxima are not very different, suggesting the main effect of divergence is to resist suppression and increase the rate of recovery.

Spectral measurements of u- and v-fluctuations were made at eight stations. Samples are shown in Fig. 6 for the centre-line position. $k = \omega \Lambda / U$, where ω is the angular frequency, and we have taken the local velocity, U, as the convection velocity. Compared with the spectra at $s = 588$ mm, where the strain effects are small, strain effects are most evident in the low wave-number range, defined for the moment as those below which $k\Lambda = 3$, and hence on the large-scale structures. The spectrum of v is distinctly suppressed at $s = (816, 866, \text{and})\ 918$, and is amplified at $s = 987$, and 1083 mm. The spectrum of u is also suppressed and amplified, but not so markedly. The suppression of the low wave-numbers of v- and u-fluctuations coincides, respectively, with the collapse of the shear stress and the minimum in $\overline{u^2}$, and the amplification in each coincides, respectively, with the peak in $\overline{v^2}$ and the recovery of $\overline{uv}$. However, the rise at low wavenumbers of u and v *above* the unstrained levels is consistent with the results of Wood and Bradshaw (1982, 1984) in an unstrained mixing layer approaching a (plane) wall. They observed, unexpectedly, an increase in $\overline{v^2}$ prior to the mixing layer meeting the wall, whereas the constraint $v = 0$ suggests v would decrease as the wall is approached. They point out that the increase can be explained plausibly in terms of the effect of the wall on the pressure-strain redistribution of normal stresses. Their results suggest that, here, in the absence of a wall effect, the peak in $\overline{v^2}$ would have been just below the minimum in $\overline{u^2}$.

The triple products, representing the transport of the turbulence by the turbulence itself, are all strongly affected. $\overline{v^3}$, $\overline{u^2v}$, $\overline{v^2u}$ and $\overline{u^3}$ are represented here in terms of their peak levels on the high and low-velocity sides, in Fig. 7. $\overline{v^3}$ shows a sustained negligible level over the whole of the layer where the shear stress reaches zero, and $\overline{u^2v}$ becomes very small. Both quantities approach zero sooner on the high velocity side, unlike the peak levels of $\overline{v^2u}$ which become small together, and later. $\overline{u^3}$ on the low-velocity side is the slowest to react. The marked changes on the high-velocity side at about $s = 1000$ mm is caused by the wall. The rapid recovery in each

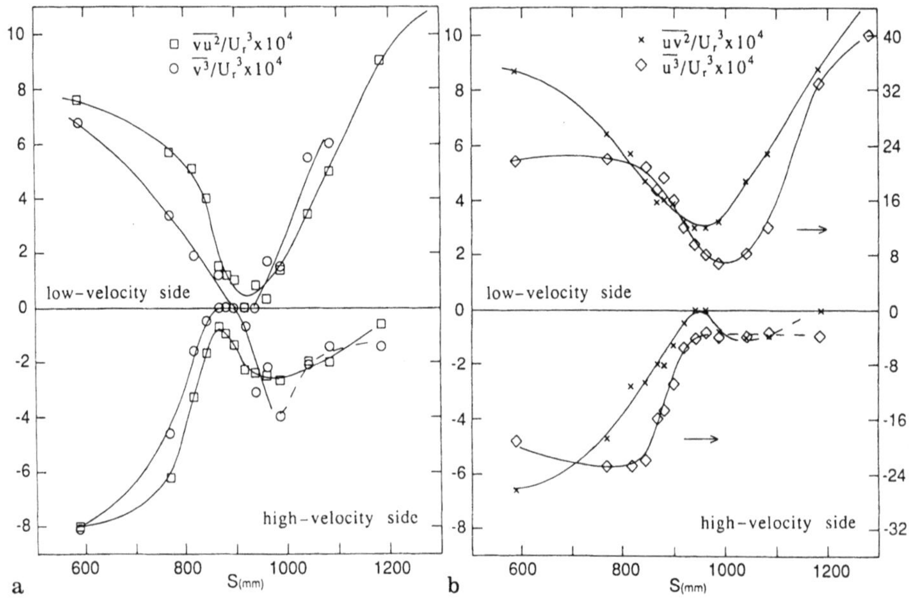

Fig. 7a. Variation of maximum and minimum $\overline{vu^2}$ and $\overline{v^3}$. **b** Variation of maximum and minimum $\overline{uv^2}$ and $\overline{u^3}$

of the triple products as the curvature is relaxed—and before the wall effect becomes significant—is consistent with large-scale structures being heavily suppressed rather than destroyed.

Waves arising from the stabilizing effect of curvature, analogous to those that arise in stably stratified flows, would be expected to be in the approximate range $0 < k\Lambda < 2$, and are likely to be most evident in the intermittent region on the high-velocity side of the layer. However, changes in the spectra at fixed n/Λ are accountable in terms of the lateral movement of the stress profiles with respect to the present centre line, and to the suppression and amplification discussed above. Thus if waves are present it appears their amplitudes are small, perhaps because the wavenumber at which they are expected is not constant over the layer, or because the divergence inhibits waves. Smits et al. (1979) found that Taylor–Görtler vortices were not present in a boundary layer subjected to divergence as well as destabilizing curvature.

Case 2

The mixing layer, defined as for Case 1, is shown in Fig. 8, where distortion is applied by means of a centre-body. In Case 2 the magnitudes of the extra strains (Fig. 9) are about 0.1 times those of Case 1. Over the region of concave curvature, which ends at $s = 580$ mm, the layer is qualitatively as if the centre-body shape had been conical—

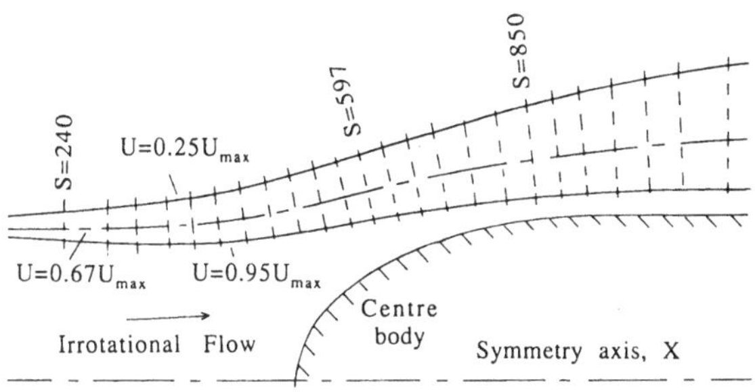

Fig. 8. Illustration of the moderately-strained layer

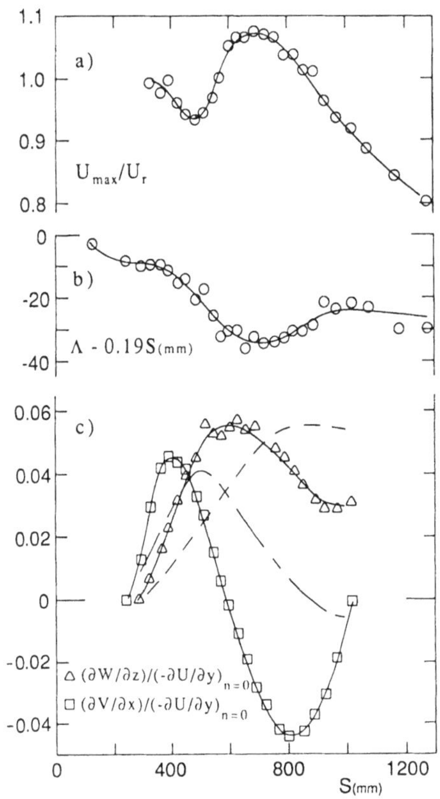

Fig. 9. U_{max}, $0.19s$-Λ, and $\partial W/\partial z$ and $\partial V/\partial x$ on the centre-line, normalized by $-\partial U/\partial y$. Broken lines show the lagged strains: $\partial W/\partial z$, — — — ; and $\partial V/\partial x$, ——-——

i.e. a milder version of Case 1—though the strains are not in precisely the same relative relationship, partly because the curvature is forced to zero sooner than it would be for a conical centre body. Downstream of approximately $s = 600$ mm the effect of the wall starts to become significant, as discussed for Case 1, and U_{max} is reduced by the centre-body boundary layer for $s > 850$ mm. For compactness, Λ, is shown in Fig. 9b as the departure from $0.19\,s$.

As for Case 1, the mean velocity profiles and the Reynolds stress profiles are closely self similar when scaled on Λ and the respective maximum. The stresses are discussed here in terms of maximum values, shown in Fig. 10. The extra strains are

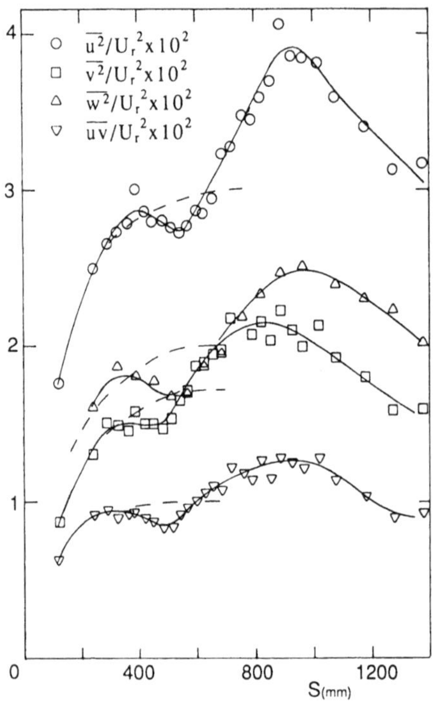

Fig. 10. Variation of maximum Reynolds stresses with s for Case 2. Broken lines are for undistorted layer

imposed earlier in this flow, well before 'self preservation' is reached, the stresses still rising as the strains become significant; the effect of the strains most likely begins before the (first) peaks in the stresses are reached. The decrease in $\overline{v^2}$ is distinctly less than the other stresses, and the subsequent rise is earlier, by roughly a shear-layer thickness. Both effects are presumably because of $\partial W/\partial z$ tending to increase v through stretching of z-component vorticity. Redistribution via pressure strain then leads to the rise in the other stresses. In Case 1, the first minimum in $\overline{v^2}$ was also roughly one shear-layer thickness earlier than the minima in $\overline{u^2}$ and $\overline{w^2}$. The rise in $\overline{v^2}$ (Case 2) starts upstream of where the wall is expected to have an effect. $(\overline{uv})_{\max}/\overline{q^2_{\max}}$, shown in Fig. 11, varies only slightly along the whole layer, though clearly, the stress ratios are not precisely constant. In the absence of divergence the decreases in the stresses would be much larger than observed here, so in the present case the effects of curvature and divergence are roughly compensating (fortuitously, because of the particular combination of strain rates), at least as far as is evident from

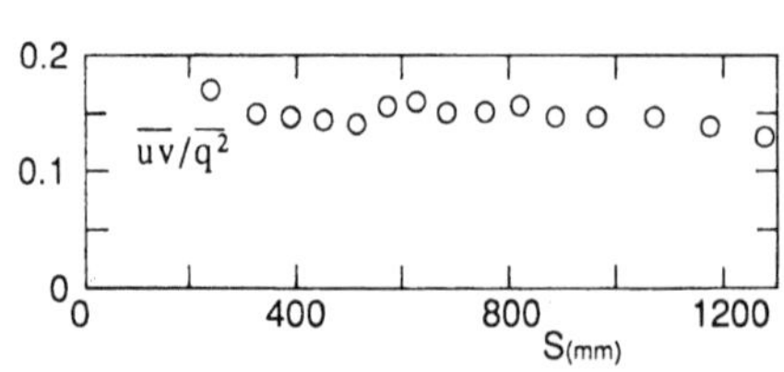

Fig. 11. Variation with s of $\overline{uv}/\overline{q^2}$ on centre line

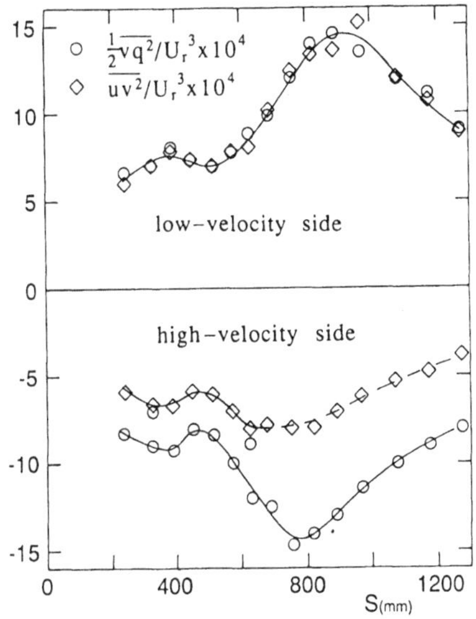

Fig. 12. Variation with s of maximum and minimum $0.5\overline{vq^2}$ and $\overline{uv^2}$

the Reynolds stresses and triple products (Fig. 12). Figure 9 also shows the effective strain rates according to the 'lag' equation (Bradshaw 1975)

$$X\frac{d}{dx}e_{\text{eff}} = e - e_{\text{eff}}, \tag{1}$$

where $e = (\partial V/\partial x)$ and $(\partial W/\partial z)$, and the 'life time', X, has been taken as $\sim 2\Lambda$ for each strain. The minima in the shear stress and kinetic energy are closely coincident with the peak in the lagged $(\partial V/\partial x)$.

Though the layer is affected by the wall once the curvature has changed sign, the substantial rises in stresses to peaks at $s \approx 900$ mm are much larger than observed by Wood and Bradshaw in an unstrained layer (equivalently, as far as $s = 1100$, at least). Thus the rises are largely a result of the effect of divergence, assuming the lagged effect of destabilizing curvature is small. $\overline{v^2}$ is the first to reach its peak, again roughly a shear-layer thickness ahead of $\overline{u^2}$ and $\overline{w^2}$.

Qualitatively, the turbulence quantities depart from the undistorted levels roughly as

$$1 + \alpha_1(\partial V/\partial x)/(\partial U/\partial y) + \alpha_2(\partial W/\partial z)/(\partial U/\partial y). \tag{2}$$

α_1 and α_2 are 'constants', where each is generally *not* independent of other extra strains. The effect of divergence in the presence of stabilizing curvature appears to be substantially stronger than it is in the presence of destabilizing curvature. However, the conclusion of Smits and Joubert (1982) from boundary-layer measurements of divergence enhancing the effect stabilizing curvature clearly does not apply here. (A better form for the divergence part is $\alpha_2(\overline{w^2}/\overline{uv})(\partial W/\partial z)/(\partial U/\partial y)$ as it represents the ratio of production of $\overline{w^2}$ to the production of $\overline{u^2}$. The quantity $\overline{w^2}/\overline{uv}$, however, is approximately constant at about 2, and about 1.6 in Gibson and Younis's flow, though their high $\overline{v^2}$ is a cause for concern.) Downstream of $s \approx 900$ mm the decreases of the stresses are probably partly through a reduction in divergence strain rate and partly through the effect of the wall.

The triple products are summarised here in terms of peak values of $\overline{uv^2}$ and $0.5\overline{vq^2}$, as shown in Fig. 12. These respond broadly as the Reynolds stresses, except that the effect of the wall is clearly evident in $\overline{uv^2}$ on the high-velocity side for $s > 600$ mm (and indicated by a broken line). Interestingly, near $s = 400$ mm, the triple products dip towards zero sooner on the high velocity side, as in Case 1.

Conclusion

In the undistorted layer the Reynolds stresses and higher-order moments approached nominally asymptotic 'self preserving' levels in a distinctly *non-monotonic* manner. The peaks occurred at a distance of about 1000 times the momentum thickness at separation, and were evident primarily because of the large diameter of the mixing layer.

In Case 1 the large-scale structures were heavily suppressed, leading to particularly large decreases in the shear stress (which is more dependent on the large scales than are the contributions to turbulent kinetic energy) and the triple products. Compared with the results of Castro and Bradshaw, the effect of divergence appears to be at first to resist the suppression of the larger scales, and later to increase their rate of recovery, though the peaks in $\overline{q^2}$ and $\overline{uv}$ were comparable with those of Castro and Bradshaw as the curvature was relaxed. Divergence also led to an early rise in $\overline{v^2}$. The effect of the wall was consistent with Wood and Bradshaw. No evidence for the analogue of stably-stratified buoyancy waves was found.

In Case 2 the response of Reynolds stresses and triple products was qualitatively as $1 + \alpha_1(\partial V/\partial x)/(\partial U/\partial y) + \alpha_2(\partial W/\partial z)/(\partial U/\partial y)$, where $(\partial V/\partial x)$ and $(\partial W/\partial z)$ are, here, effective strains (Eq. 1), though the coefficients are not constant. The effect of divergence in the presence of destabilizing curvature appears to be rather less than it is in the presence of stabilizing curvature. On the basis of effective strains, the destabilizing curvature was largely 'cancelled' by the lagged stabilizing curvature, though the physical mechanisms of destabilizing curvature are unlikely to have been cancelled.

Acknowledgement: The authors are grateful to the SERC and to Dr I. P. Castro, who was a co-holder of the contract with PEH. PEH is also grateful to the Robert Blackburn Trust of the Royal Aeronautical Society for substantial financial support to attend the Symposium.

References

Birch, S. (1980): Planar mixing layer data evaluation, Stanford Conference on Complex Turbulent Flows
Birch, S. (1983): The effects of initial conditions on high Reynolds number jets. AIAA paper 83-1681
Bradshaw, P. (1966): The effect of initial conditions on the development of a free shear layer J. Fluid Mech. **26**, 225

Bradshaw, P. (1973) Effects of streamline curvature on turbulent flow AGARDograph 169
Bradshaw, P. (1975): Review—Complex Flows. J. Fluids Eng. **97**, 146
Castro, I. P. (1988): Free-stream turbulence effects on separated shear layers, Zaric Memorial Seminar, Dubrovnic
Castro, I. P., Bradshaw, P. (1976): The turbulence structure of a highly curved mixing layer. J. Fluid Mech. **73**, 265
Foss, J. F. (1977): The effects of laminar/turbulent boundary layer state on the development of a plane mixing layer, 1st Turb. Shear Flows Symp. Penn. State Univ.
Gibson, M. M., Rodi, W. (1981): A Reynolds-stress closure model of turbulence applied to the calculation of a highly curved mixing layer. J. Fluid Mech. **103**, 161
Gibson, M. M., Younis, B. A. (1983): A Reynolds-stress closure model of turbulence applied to the calculation of a highly curved mixing layer. Expts Fluids **1**, 23
Gillis, J. C., Johnston, J. P. (1983): Turbulent boundary-layer flow and structure on a convex wall and its development on a flat wall. J. Fluid Mech. **135**, 123
Husain, Z. D., Hussain, A. K. M. F. (1979): Axisymmetric mixing layer: Influence of initial and boundary conditions. A.I.A.A. **17**, 48
Hussain, A. K. M. F., Clark, A. R. (1981) On the coherent structure of the axisymmetric mixing layer: a flow-visualization study. J. Fluid Mech. **104**, 263
Keffer, J. F. (1965): The uniform distortion of a turbulent wake. J. Fluid Mech. **22**, 135
Savill, A. M. (1987): Recent developments in rapid distortion theory. Ann. Rev. Fluid Mech. **19**, 531
Smits, A. J., Eaton, J. A., Bradshaw, P. (1979): The response of a turbulent boundary layer to lateral divergence J. Fluid Mech. **94**, 243
Smits, A. J., Joubert P. N. (1982): Turbulent boundary layers on bodies of revolution J. Ship Res. **26**, 135
Townsend, A. A. (1980): The response of sheared turbulence to additional distortion. J. Fluid Mech. **98**, 171
Wood, D. H., Bradshaw, P. (1982): A turbulent mixing layer constrained by a solid surface. Part 1. Measurements before reaching the surface. J. Fluid Mech. **122**, 57
Wood, D. H., Bradshaw, P. (1984): A turbulent mixing layer constrained by a solid surface. Part 2. Measurements in the wall-bounded flow. J. Fluid Mech. **139**, 347

Part IV

Reacting Flows

Introductory Remarks

W. P. Jones

Introduction

Chemical reaction in turbulent flows occurs in a variety of different circumstances; examples are the burning of hydrocarbon fuels in furnaces and combustion chambers, the formation of pollutants such as photochemical smog in the atmosphere and the reactions associated with the production of chemical products in reactors. The interaction between turbulent motions and chemical reaction can be usefully classified by two limiting criteria. The first criterion relates to the nature of the reaction process itself, namely whether or not the time scales associated with reaction are short or long compared with those of convective and diffusive transport processes. [It is important to note here that chemical reaction introduces additional time scales only to the problem.] If the reaction time scales are short compared with those of diffusion and convective processes, then this is often referred to as 'fast chemistry. In combustion it is usually the case that reaction time scales are either short or at least comparable with transport time scales so that the 'slow chemistry' case is not, in these circumstances, particularly relevant.

The second criterion involves distinguishing between premixed and non-premixed reaction where in the former case reactants are (perfectly) premixed prior to reaction whereas in the latter case, the reactants enter the reaction zone (or combustion chamber) in separate streams. The Otto cycle reciprocating engine is an important application involving premixed combustion, whereas large furnaces and gas turbine combustion chambers operate essentially in the non-premixed regime.

The interaction between turbulence and reaction is of paramount importance in combustion and the complexity is probably at its greatest for the strongly exothermic reactions involved in the burning of hydrocarbon fuels (and hydrogen) in air. In this situation the heat release and associated rise in temperature results in large changes in the fluid density which in turn strongly couples the mixture composition, temperature and flow fields. In contrast the formation and emission of combustion generated pollutants such as oxides of nitrogen, involves somewhat less complexity in that they are usually present only in small quantities and are inactive in the sense that they do not influence either the mean flow and turbulence fields or heat releasing reactions. The complexities of turbulence-chemical reaction interactions arise essentially because the time scales implicit in reaction rate

Turbulent Shear Flows 7

expressions are very short—typically of the order 10^{-4} seconds in hydrocarbon-air flames—with the consequence that reaction occurs within thin layers; reaction zone thicknesses are very small compared with the typical overall dimensions of a turbulent flame. This is true for both premixed and non-premixed cases and is a source of considerable difficulty both from the experimental and the theoretical viewpoint. From the experimental view the time-space resolution required to measure the properties of reaction zones in turbulent flames represents a major limitation though some progress is being made with laser diagnostic techniques. For modelling, the consequences of 'fast' reaction represent both a major difficulty and a means of simplification. The difficulty arises because, in general, it is necessary to describe the coupling between reaction, molecular diffusion and the convecting turbulence velocity field. The highly non-linear nature of this coupling and the consequential range of turbulence time and length scales involved is poorly understood. On the other hand, considerable simplifications arise if it can be presumed that reaction layers are infinitesimally thin compared with the smallest turbulence length scales with the implication that overall reaction rates are controlled by 'mixing' processes. For non-premixed (diffusion) flames the assumption can be invoked in the context of the conserved scalar formalism, Bilger [1], whereby the gas properties can be related to mixture fraction (a strictly conserved scalar quantity) via either chemical equilibrium results or, alternatively, to laminar diffusion flame computations or measurements. In the case of premixed flames the assumption can be introduced such that reaction is represented by an eddy break-up type approximation with overall reaction rate being controlled by the rate of mixing of unburnt mixture with hot products, Bray, Moss, Libby [2]. For hydrocarbon-air diffusion flames the equilibrium approach gives excessively high CO levels on the rich side of stoichiometric with typical maximum values being around 20% compared with measured maxima of about 6%. While this does not appear to have a large influence on either predicted mean temperatures or the flow field, it is clearly unrealistic. The laminar flamelet model, on the other hand, does give much more realistic results. In this approach the (instantaneous) dependence of gas composition, temperature and density on mixture fraction and some measure of the local rate of strain or flame stretch is presumed to be the same as that prevailing in a laminar flame. It is usual to characterise the rate of strain by the 'scalar' (mixture fraction variance) dissipation rate and to obtain the laminar flame properties from computations of counter-flow diffusion flames. In this way the chemistry is essentially uncoupled from the turbulent flow field computation though a turbulence model is still required to determine the joint probability density function of mixture fraction and the scalar dissipation rate. The laminar flamelet approach has also been widely used for premixed combustion, though in this case the mixture fraction must be replaced by a reaction progress variable and an alternative measure of rate of flame stretch must be provided.

The class of turbulent flames for which the flamelet approach is applicable is currently a matter of some controversy. If the local flame thickness is smaller than the smallest turbulence scales, i.e. the Kolmogorov length scale, then the flamelet approach is most certainly applicable. However, while there is necessarily some uncertainty, this seems unlikely to be the case for non-premixed combustion at the high turbulence Reynolds numbers prevailing in practical systems. Nevertheless, even in this latter case where the use of flamelets may not be strictly justifiable the method does yield plausible CO levels and is probably to be preferred to the chemical equilibrium approach at least for hydrocarbon fuels. In the case of premixed flames, however, combustion more often appears to fall within the laminar-flamelet regime and the approach is consequently more justified though there remains considerable difficulty in relating the basic laminar flamelets to the turbulence field. For both premixed and non-premixed combustion, the fast chemistry/laminar flamelet approach is clearly not applicable, however, in situations where chemical kinetics play a rate controlling role; examples include ignition and extinction phenomena and the emissions of combustion generated pollutants.

Of the papers presented in this chapter four fall into the category of non-premixed combustion while three are concerned with premixed combustion. The papers by Chen and Kollmann and by Gonzalez and Borghi describe alternative approaches to modelling diffusion flames and those of Lee, Metcalfe and Hussain and by Elgobashi and Nomura present the results of direct numerical simulations of isothermal non-premixed combustion. The papers by Rogg and by Liu, Lenze and Leuckel concern, respectively, modelling and experimental studies of premixed combustion and the final paper of Veynante, Lacas, Maistret and Candel presents a model for non-uniformly premixed combustion.

Chen and Kollmann use the joint probability density function (pdf) evolution equation approach to the modelling of non-premixed combustion. The merit of the pdf method is that no modelling is required for chemical reaction terms; the central difficulty is the provision of a 'mixing' model necessary to describe the effects of molecular diffusion. Current popular models may be viewed as generalisations of Curl's coalescence/dispersion model and the corresponding discrete nature of the implied mixing process renders them unsuitable for and inconsistent with the fast chemistry or laminar flamelet regimes of turbulent combustion. To overcome this Chen and Kollmann propose a modified coalescence/dispersion model conditioned by the reaction zone. Mixing and reaction are presumed to occur in a number of substeps with the result that fluid 'particles' lying initially either side of the reaction zone pass through the reaction zone and react in the mixing process. Computations using a Monte-Carlo solution technique demonstrate clearly that modelling the combined effects of mixing and reaction is essential for fast chemical reaction and that, in this case, reaction zone conditioning produces physically plausible results. Future work will presumably involve a combination of the proposed mixing model with a realistic reaction mechanism and subsequent comparison of predicted and measured results.

The paper by Gonzalez and Borghi describes a Lagrangian approach to turbulent combustion modelling. The starting point is the previously proposed Lagrangian I.E.M. model [3] in which diffusional processes are represented by an 'exchange with the mean'. Single step chemistry is used and the method is developed for non-premixed combustion though the authors note that, in principle, it can be applied to be premixed case also. The I.E.M. model is extended through the introduction of a time scale distribution with the aim of describing what the authors term 'sudden combustion'. This relates to reaction, the initial period of which involves an ignition delay period with negligible heat release followed by a rapid consumption of reactants (fast chemistry) to form combustion products. The essential modelling assumptions invoked are the form of the pdf's for the conserved scalar and for the time scale and the method of determining the ignition time. The application of the model to turbulent jet methane-air flames near extinction produces results in reasonable agreement with measurements though when extinction is predicted to occur the model is incapable of reproducing the observed re-ignition further downstream. The authors attribute this to the time scale distribution for which further refinement is needed.

The papers by Elgobashi and Nomura and by Lee, Metcalfe and Hussain both involve direct numerical simulations (DNS) of non-premixed, constant density, isothermal, turbulent reacting flows. Elgobashi and Nomura simulate chemical reaction in two homogeneous shear flows with differing initial strain numbers and decaying isotropic turbulence using a $96 \times 96 \times 96$ grid. The corresponding Reynolds number based on the Taylor scale was 25 and single step infinitely fast chemistry is assumed. Even at this relatively low Reynolds number the grid is, as the authors note, insufficiently fine to resolve the finest scale Kolmogorov motions.

As expected, the results show that the conserved scalar gradient has maximum values in the region of the stoichiometric isoscalar and that its direction is normal to that surface. The evolution of the conserved scalar pdf in each of the three flows is consistent with the corresponding different eddy turnover times. Lee, Metcalfe and Hussain describe the results of a DNS of the temporally evolving reacting mixing layer using a $64 \times 64 \times 64$ grid. A binary single step reaction is studied and reaction rates range from the slow chemistry, small Damkohler regime to the infinite rate case. A number of interesting results on the vortex structure of the mixing layer are described and the chemical reaction zone is marked by very steep gradients of vorticity and reactant concentration. Work currently underway is aimed at providing better resolution of small-scale vortices and information on the statistics of turbulent reacting flow.

The paper by Rogg is concerned with modelling of premixed turbulent combustion in a boundary layer with injection of cold reactants through a porous plate. A flamelet approach is adopted and the mean reaction rate is expressed in terms of the burning velocity of the unstrained laminar flame, the mean progress variable and the mean flame surface to volume ratio. For this latter quantity a modelled transport equation is utilised; the form of this appears plausible but it is a pity that more details of its derivation and basis are not given, particularly as it appears novel. The explicit effects of strain on reaction rate are neglected and turbulent transport is represented by a more or less standard form of the $k-\varepsilon$ model. Application of the model to a nitrogen diluted, slightly lean propane/air flame leads to a computed mean temperature profile in excellent agreement with that measured. Further applications of the model in other configurations are clearly desirable to establish the validity of the flamelet surface density equation.

Liu, Lenze and Leuckel present the results of an experimental study on the influence of turbulence levels on premixed combustion. The configuration is that of a stagnation flame and the CH_4-H_2-air mixtures used provided a fourfold variation in laminar burning velocity. A range of turbulence intensities were generated by means of perforated plates with differing blockage ratios. Two component L.d.a. measurements of the velocity characteristics indicate that the turbulence in the approach flow is unaffected by the turbulent flame-front. The velocity pdf's within the flame display a bimodal form which is consistent with the mixture comprising, alternatively hot products and cold reactants and thus with thin flame/laminar flamelet combustion. The turbulence levels increase within the flame with two maxima arising at positions corresponding roughly to those of the maxima in the mean rate of strain. These increased turbulence levels presumably result from the increased energy production rates associated with the increased mean rate of strain. The turbulent burning velocities of more than fifty stagnation flames have been measured and are shown to be well correlated by a linear function of the rms turbulence velocity fluctuation.

The final paper in this chapter by Veynante, Lacas, Maistret and Candel addresses the important problem of modelling non-uniformly premixed combustion. The method adopted involves a laminar flamelet approach and is based on the Coherent Flame Model. A single step irreversible reaction is considered and a central feature of the approach is the use of a mean flame surface density which is used to determine the global consumption rate of each reactant. The basic model is extended to combine premixed and non-premixed elements by writing two balance equations for each reactant; one represents the mass fraction of reactant present in premixed form and the other the mass fraction of reactant in the burnt gases and burning as a diffusion flame. Equations for the flame surface densities for each of the premixed and diffusion flame elements are also provided. The complete model is then applied to the turbulent flame arising in the mixing layer developing between parallel streams of the two reactants. Regions of both premixed and diffusion flame burning are evident in the calculated results. In the future the authors hope to validate their model by comparison with measurements in turbulent flames.

The papers in this chapter show that progress is being made, albeit slowly, on the problems of turbulent reacting flows. It is clear, however, that the subject remains an important and outstanding scientific challenge and that many problems remain. For example, if for turbulent flames the fast reaction assumption is relaxed— and this is necessary in a number of cases—then there appears to be no demonstrated satisfactory model currently available. The pdf evolution equation method is one viable approach but a satisfactory 'mixing' model is essential. Other approaches are also possible but what is clearly needed is a more detailed comparison of predictions with experimental measurements on aspects where chemical kinetics plays a more important rate controlling influence, than has hitherto been possible. With current developments in optical diagnostic techniques much more detailed information on turbulent flame structure is likely to become available in the near future and such comparisons will then be possible. A direct numerical simulation of a reacting flow with realistic chemistry and heat release could also provide invaluable information. It is necessary, however, that the simulated Reynolds number be sufficiently large for there to be an appreciable separation between the scales of the energy containing and dissipative (Kolmogorov) turbulent motions. Unfortunately, however, the resulting computer storage requirement appears to be beyond that available with the largest current supercomputers. With the present rapid developments in computer technology, this situation is likely to change in the future.

References

1. Bilgev, R. W. (1980): Turbulent flows with non-premixed reactants. In: Libby, P. A., Williams, F. A. (eds), Turbulent Reacting Flows, Springer-Verlag, Topics in Applied Physics, **44**, 65
2. Bray, K. N. C., Libby, P. A., Masuya, G. and Moss, J. B. (1981): Turbulence energy production in premixed turbulent flames. Comb. Sci. Tech. **25**, 127.
3. Borghi, R. (1988): Turbulent combustion modelling. Prog. Energy Combust. Sci. **14**, 245

Mixing Models for Turbulent Flows with Exothermic Reactions*

J.-Y. Chen and W. Kollmann[1]

Combustion Research Facility, Sandia National Laboratories, Livermore, CA 94551, USA

Abstract

The potential influence of age-biased sampling for mixing as suggested by Pope (1982) was investigated for turbulent reacting and nonreacting jet flows with a modified coalescence/dispersion (C/D) model. Age-biased sampling is found to have little influence on the predicted scalar statistics up to fourth moments. To account for the combined effects of mixing and chemical reaction on the joint probability density function (pdf), the modified C/D model is further improved by conditioning the mixing process on the existence of the reaction zone. Monte Carlo simulation of turbulent reacting jet flames with an ideal flame-sheet chemical model demonstrates that the proposed model is capable of predicting physically correct results in the flame-sheet combustion regime.

1. Introduction

Pdf (probability density function) methods for predicting turbulent flows with strongly exothermic reactions have been applied successfully for turbulent jet flows (Pope and Correa 1986; Correa et al. 1988; Chen and Kollmann 1988, 1989; Chen et al. 1989; Haworth et al. 1988). The main emphasis so far has been the development and incorporation of simplified chemical models into the Monte Carlo pdf simulation so that the nature of non-equilibrium chemistry can be captured. The results obtained in these previous investigations strongly suggest that the current Monte Carlo simulation of the combined effects of chemical reaction and the mixing process on the pdf has serious deficiencies, which could lead to unphysical results. Although the chemical reaction terms appear in closed form in the pdf equation, the interactions between the mixing process and chemical reaction have not been investigated in the past; consequently, the dynamics of turbulent flames may not be properly simulated. Modeling of these interactions, as well as the molecular mixing term, is a crucial issue for the pdf method. In this paper, we consider simple reaction schemes to explore the important interactions between chemical reaction and the mixing process. Possible generalizations and modifications of existing Monte Carlo procedures to account for interactions between chemical reaction and the mixing

* Research supported by the United States Department of Energy, Office of Basic Energy Sciences, Division of Chemical Sciences.

[1] Department of Mechanical Engineering, U.C. Davis

Turbulent Shear Flows 7

process are constructed and analyzed. Performance of these Monte Carlo procedures is evaluated by studies of flows with fast and slow chemical reactions.

2. Pdf Equation and Mixing Models

The statistics of the scalar field can be described by a single point pdf $\tilde{P}(\psi_1,\ldots,\psi_n;\mathbf{x},t)$ of the thermo-chemical variables $\phi_1,\ldots,\phi_n$. The single point pdf $\tilde{P}(\psi_1,\ldots,\psi_n;\mathbf{x},t)$ satisfies the following transport equation

$$\langle\rho\rangle\partial_t\tilde{P}+\langle\rho\rangle\tilde{v}_\alpha\partial_\alpha\tilde{P}+\partial_\alpha(\langle\rho\rangle\langle v''_\alpha|\phi_i=\psi_i\rangle\tilde{P})$$
$$=-\langle\rho\rangle\sum_{i=1}^{N}\partial_{\psi_i}\{S_i(\psi_1,\ldots,\psi_N)\tilde{P}\}-\langle\rho\rangle\sum_{i=1}^{N}\sum_{j=1}^{N}\partial^2_{\psi_i\psi_j}(\langle\varepsilon_{ij}|\phi_k=\psi_k\rangle\tilde{P}), \tag{1}$$

where $v''_\alpha\equiv v_\alpha-\tilde{v}_\alpha$ is the velocity fluctuations and $\varepsilon_{ij}\equiv\Gamma\nabla\phi_i\cdot\nabla\phi_j$ is the scalar dissipation. Traditional finite difference schemes are not feasible for solving this pdf equation as the dimension of scalar becomes large. For multi-species problems, a computationally efficient stochastic technique, i.e., the Monte Carlo simulation, can be used for obtaining statistical solutions for the pdf equation. In the Monte Carlo simulation, the pdf is represented by an ensemble of statistical representations (particles) per grid cell, and the evolution of these particles is obtained by simulating the effects of various terms in Eq. (1) on the pdf via their corresponding stochastic processes. Two distinct modes of stochastic processes are identified: (1) the influence of turbulent mean and fluctuation convection (terms on LHS) can be simulated by exchanging particles among neighbor cells resulting in changes of pdf in the *physical* space; (2) terms on RHS stand for chemical reaction and scalar dissipation, and their effects on pdf can be simulated by moving particles in the *composition* space according to chemical kinetics and a closure scheme for the scalar dissipation. Modeling the scalar dissipation (i.e., the mixing process) is a crucial issue in the pdf approach and several mixing models will be considered.

2.1 Coalescence/Dispersion (C/D) Model

The fundamental concept of the coalescence/dispersion (C/D) model is based on the idea of pairwise interaction of fluid parcels in a quasi-homogeneous neighborhood, such as a grid cell (Curl 1963). Constraints, based on physical arguments, are often imposed on the model to prevent unphysical results. For conservation of mass, the mean values of species are required to remain unchanged by the mixing process. Also, based on the fact that mixing increases the degree of homogeneity, the variances of scalars must decrease after mixing. Other physical constraints, such as a Gaussian distribution for homogeneous decaying turbulence at large time, can be important in providing guidance for model improvements.

A general form of the C/D model can be expressed as

$$-\sum_{i=1}^{N}\sum_{j=1}^{N}\partial^2_{\psi_i\psi_j}(\langle\varepsilon_{ij}|\phi_k=\psi_k\rangle\tilde{P})$$
$$\cong\frac{C_D}{\tau}\left\{\int_{D_N}..\int d\psi'\int_{D_N}..\int d\psi''\tilde{P}(\psi')\tilde{P}(\psi'')T(\psi',\psi''|\psi)-\tilde{P}\right\},\qquad(2)$$

where ψ without subscript denotes the set of variables $\psi_1,\ldots,\psi_n$ and $d\psi$ the product of all $d\psi_i$. In Curl's model (1963), the integrand is expressed as

$$T(\psi',\psi''|\psi)=\delta\left(\psi-\frac{1}{2}(\psi'+\psi'')\right).\qquad(3)$$

It is known that this model does not permit a continuous pdf distribution to be developed if the initial pdf has discrete delta spikes (Pope 1982). Janicka et al. (1979) modified Curl's model to remedy this deficiency as

$$T(\psi_i',\psi_i'',|\psi_i)=\begin{cases}\dfrac{1}{|\psi_i''-\psi_i'|} & \text{for } \psi_i'\leqq\psi_i\leqq\psi_i'' \text{ or } \psi_i''\leqq\psi_i\leqq\psi_i';\\ 0 & \text{otherwise,}\end{cases}\qquad(4)$$

with $C_D=6.0$. The scalar domain D_N is the set of all realizable values of $\{\phi_1,\ldots\phi_n\}$ and must be convex. The required convexity ensures that all possible states after mixing are within the scalar domain. The time scale τ for this process (i.e., the reciprocal of mixing frequency) is taken as the integral turbulence time scale, $\tau=\tilde{k}/\tilde{\varepsilon}$.

It is known that the original and modified C/D models are incorrect in the short time limit and lead to unbounded flatness and superskewness (normalized fourth and sixth moments which are equal to 3 and 15 for a Gaussian pdf) values in decaying turbulence as time goes to infinity (Pope 1982; Kosaly 1986; Kosaly and Givi 1987). Besides these deficiencies, additional modeling of the interactions between chemical reaction and molecular mixing is required as these two processes are intimately related to each other. In what follows, we will use the modified C/D model by Janicka et al. (1979) as the basis to explore possible modifications for removing some of the deficiencies noted above.

2.2 Coalescence/Dispersion Model with Age Conditioning

In the Monte Carlo simulation, the equivalent stochastic process for Eq. (2) is carried out by randomly selecting a number of pairs of particles in a grid cell and then mixing them to a random degree. Hence, the probability of a particle being selected for mixing is equal among all the particles within a grid cell. Pope (1982) suggested biasing the sampling probability of particles according to the time (age) elapsed since their last mixing event. The age is reset to zero after each mixing event. By properly biasing the sampling probability, reasonable values of flatness (3.7 vs. 3) and superskewness (29.5 vs. 15) have been obtained with the C/D model in a homogeneous decaying turbulence. Whether or not these improvements for high-order statistics will have a significant impact on the lower statistics (first two moments) in turbulent free shear flows with/without chemical reaction is a subject of interest.

To implement the biasing sampling, an additional scalar carrying the age information is included in the modified C/D model. The normalized age, s, is defined to be the time between mixing events divided by the turbulence integral time scale. Let us denote $\tilde{P}^*(\psi, s; \mathbf{x}, t)$ the joint pdf of $\psi_1, \ldots, \psi_N$, as well as s. The original pdf $\tilde{P}(\psi; \mathbf{x}, t)$ can be recovered as

$$\tilde{P}(\psi; \mathbf{x}, t) = \int ds \tilde{P}^*(\psi, s; \mathbf{x}, t) \tag{5}$$

and the pdf for the normalized age, $P_s(s)$, is given by

$$P_s(s; \mathbf{x}, t) = \int d\psi \tilde{P}^*(\psi, s; \mathbf{x}, t). \tag{6}$$

Sampling bias can be introduced through a biasing function $z(s)$ which is defined as the relative probability of sampling a particle with age s, and it is normalized by

$$\int ds\, z(s) P_s(s) = 1. \tag{7}$$

Now the pdf of finding two particles with ψ' and ψ'' with age-biased sampling can be formulated as

$$\tilde{P}^{**}(\psi'; \mathbf{x}, t) \tilde{P}^{**}(\psi''; \mathbf{x}, t) d\psi' d\psi'', \tag{8}$$

where

$$\tilde{P}^{**}(\psi; \mathbf{x}, t) = \int ds\, z(s) \tilde{P}^*(\psi, s; \mathbf{x}, t) \tag{9}$$

is the pdf of a particle being sampled for mixing. With this age-biased sampling, the C/D mixing model is further modified as follows

$$- \sum_{i=1}^{N} \sum_{j=1}^{N} \partial^2_{\psi_i \psi_j} (\langle \varepsilon_{ij} | \phi_k = \psi_k \rangle \tilde{P})$$
$$= \frac{C_D}{\tau} \left\{ \int d\psi' \int d\psi'' \tilde{P}^{**}(\psi') \tilde{P}^{**}(\psi'') T(\psi', \psi'' | \psi) - \tilde{P}^{**}(\psi) \right\} \tag{10}$$

The biasing function $z(s)$ in this study is chosen as a power law function of age

$$z(s) = \frac{\dfrac{1}{a} \left(\dfrac{s}{s_{\max}} \right)^{1/a - 1}}{\int ds \dfrac{1}{a} \left(\dfrac{s}{s_{\max}} \right)^{1/a - 1} P_s(s)} \qquad 1 \geqq a > 0, \tag{11}$$

where $s_{\max}$ is the maximum age among the particles in a grid cell. Figure 1a shows values of the numerator at various a values. When $a = 1$, this corresponds to the case of no sample bias, i.e., z is unity. For a values smaller than one, particles with old ages are sampled for mixing more frequently then those with young ages. We have chosen $a = 0.1$ to study the potential effects of strong sampling bias in turbulent shear flows. Implementation of biased sampling in the Monte-Carlo procedure is carried out as follows. First, particles in each grid cell are sorted out in sequential order according to their ages. To select a particle to be mixed from a grid cell containing n particles, a

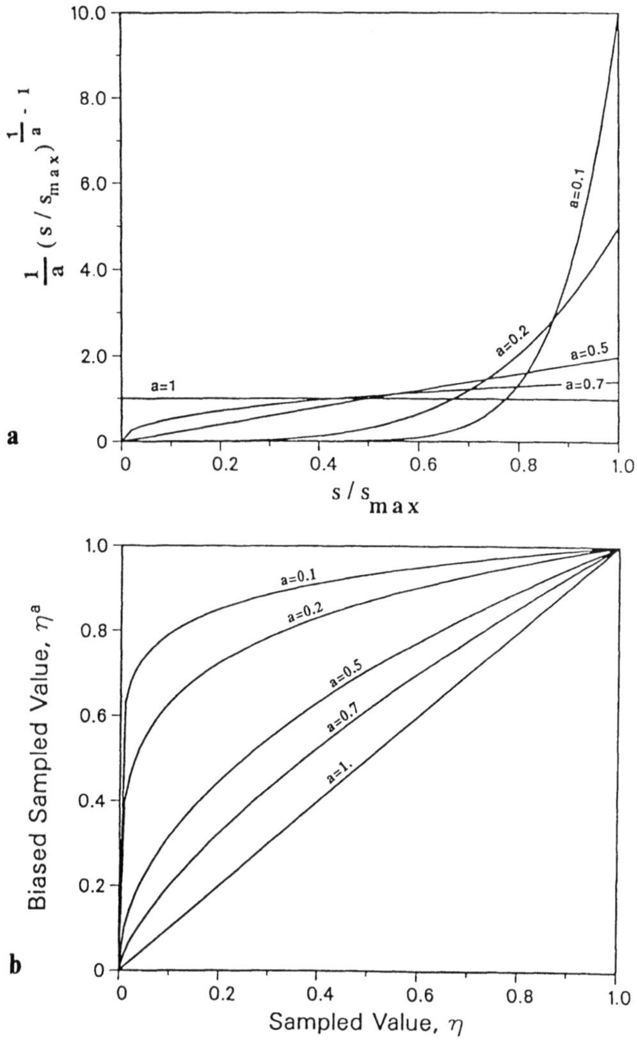

Fig. 1 a. Values of biasing function vs. normalized age s for various a values. Low a values correspond to high age-biased sampling. **b.** Biased sampled values as a function of η^a for various a values. η is a random variable uniformly distributed in the interval between 0 and 1

random number generator is called to yield a random number, η, between 0 and 1. The selected particle is chosen by the following formula

$$\textit{selected particle number} = n * \eta^a.$$

Values of the weighting factor η^a at various a values are plotted in Fig. 1b. It can be shown that the pdf of the random variable, $y \equiv \eta^a$, with η being uniformly distributed in the interval [0, 1] is

$$P(y) = \frac{1}{a} y^{1/a-1} \tag{12}$$

which is exactly the numerator in Eq. (11). When $a = 1$, the case of un-biased sampling is recovered.

2.3 Coalescence/Dispersion Model with Reaction Zone Conditioning

The Monte Carlo procedures for mixing and chemical reactions are usually carried out sequentially in separated steps during a time step. When combustion occurs, chemical reaction and mixing are intimately related to each other. To provide some physical insights, it is instructive to consider two extreme situations with fast and slow chemistry as compared to the mixing process (large and small Damköhler numbers). If chemical reactions occur at a much slower rate than that of mixing, the flow approaches the well-stirred reactor regime and no direct interaction between reaction and mixing is anticipated. On the other hand, if the chemical process becomes much faster than the mixing process in a small zone of the scalar space, the flamelet regime is approached and a thin reaction zone may exist in physical space between fluid parcels. This implies that mixed fluid parcels with scalar properties on either side (in scalar space with respect to mixture fraction) of the reaction zone will experience rapid chemical reactions, whereas mixing of parcels that are both on the same side of the reaction zone will not involve significant reaction. It follows that

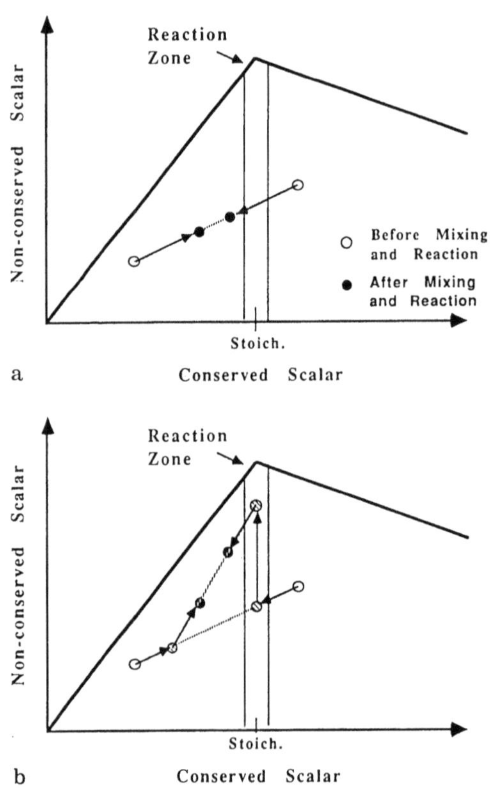

Fig. 2 a. The original Monte Carlo procedures used in simulating mixing and reaction processes by separated steps showing the influence of reaction zone is not taken into account. **b.** The proposed Monte Carlo procedures in simulating the combined effects of mixing and reaction by conditioning the mixing process on the existence of reaction zone. The vertical strip around the line $f = f_{\text{stoich}}$ denotes the reactive zone R in the composition space molecular mixing and reaction are then strongly coupled due to the strong variation of reaction rates in the scalar space.

Reaction zone conditioning for the mixing process refers to a Monte Carlo procedure which is developed to account for the combined effects of chemical reaction and mixing. This procedure does not introduce new variants of the mixing model. The idea is straightforward and can be applied to composition spaces of any dimension. Consider a convex scalar space and let R be the subset of the scalar space, where chemical reactions are considerably faster than those in the complement of R. Mixing of two particles, whose connecting line intersects R, must take into account the presence of R. This is done by performing mixing and reaction sequentially in a number of sub-steps within the mixing procedure, such that in at least one substep the mixed product passes through the fast reaction zone R. Figure 2a, b depicts the original and the reaction-zone conditioning Monte Carlo procedures in simulating both chemical and mixing processes. In the two-scalar composition domain, non-zero chemical reaction is confined only within a narrow region near the stoichiometric value of the mixture fraction. As demonstrated in Fig. 2a, the original process allows two particles be mixed through the fast reaction zone without any reaction, i.e., the process of combustion is in fact frozen. With conditioning on reaction zone, we show in Fig. 2b, the final mixed particles have indeed undergone chemical reaction reflecting the influence of fast reaction zone.

3. Monte Carlo Simulation

Monte Carlo simulation of joint pdfs with various mixing models has been performed for reacting and non-reacting turbulent jets. The flow configuration consists of a central jet of methane, a coaxial pilot flame and a coflowing air stream with velocities of 48 m/s, 24 m/s and 15 m/s respectively. Results for the following four mixing models will be presented: (A) Coalescence/Dispersion model and the original Monte Carlo procedure; (B) Coalescence/Dispersion model with age-biased sampling in the original Monte Carlo procedure; (C) Coalescence/Dispersion model with reaction zone conditioning; (D) Coalescence/Dispersion model with age-biased sampling and reaction zone conditioning.

To explore the important interactions between chemical reaction and the mixing process, a simple chemical reaction scheme consisting of two scalars is considered. One scalar, the mixture fraction f, is used to describe the mixing nature of the flow and the other scalar, temperature T, is used to indicate the progress of chemical reaction. A triangular composition domain is considered with the peak T situated at the stoichiometric point $f = 0.055$. This particular choice can be viewed as a simple representation of the composition domain for hydrocarbon fuels for which fast reactions occur only in a narrow zone around the stoichiometric point (Pope et al. 1989). Two extreme reaction cases are assumed: one with a very narrow strip of reaction zone and infinitely fast reaction and another one with a broad reaction zone and slow reaction. In the fast chemistry case, the increased rate of change of T was set to be 10^{10} K/ms uniformly inside the reaction zone until T reaches its maximum value at a fixed f. Outside the reaction zone, the reaction rate was set to zero. For the slow chemistry case, the increased rate of change of T was reduced to 1000 K/ms.

4. Results and Discussions

4.1 Nonreacting Flows

For nonreacting variable density flows, the scalar field can be described by a single conserved scalar, the mixture fraction. Density is directly determined from the mixture fraction as

$$\frac{1}{\rho} = \frac{f}{\rho_{\text{jet}}} + \frac{1-f}{\rho_{\text{air}}}.$$

Since no chemical reaction is involved, only mixing models with and without sampling bias will be investigated (i.e., Models A and B). Figure 3 presents the predicted mean and variance of the mixture fraction along the jet centerline indicating that age bias has little influence on the mean statistics. Typical radial profiles of the mean and variance at $x/D = 30$ are shown in Figs. 4 and 5. The differences between the different mixing models are noticeable but probably within the statistical error. Comparisons of higher moments allow us to check more subtle differences between the two mixing models. Skewness and flatness of the mixture fraction, shown in Figs. 6 and 7 respectively, exhibit insignificant differences. It is concluded that age bias has little influence on the statistics of the mixture fraction up to fourth moments for nonreacting turbulent jets.

Properties of the normalized age, s, on the other hand, show a strong dependence on the sampling procedure. As shown in Fig. 8, the mean value of s is reduced by bias sampling and is fairly uniform over the cross-section of the jet. The values of age variance with bias sampling, presented in Fig. 9, are drastically reduced to about ten

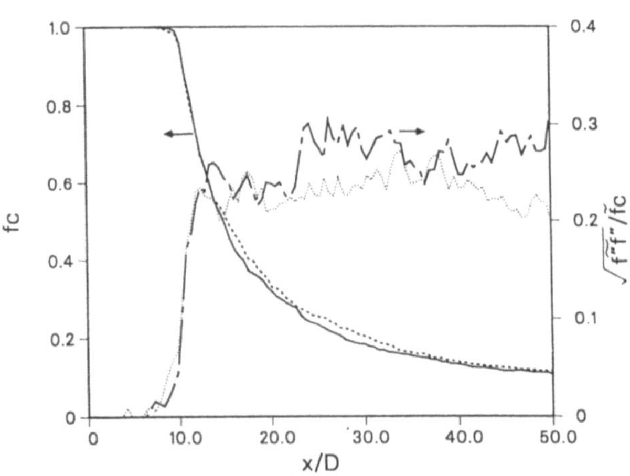

Fig. 3. Axial centerline profiles of the mean and variance of mixture fraction for turbulent nonreacting methane jets. Dashed and solid lines denote results with/without age-biased sampling

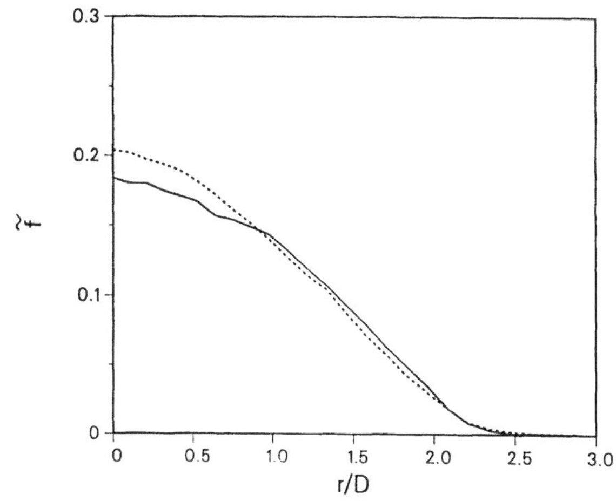

Fig. 4. Radial profiles of the mean mixture fraction at $x/D = 30$ for turbulent nonreacting methane jets. Dashed and solid lines denotes results with/without age-biased sampling

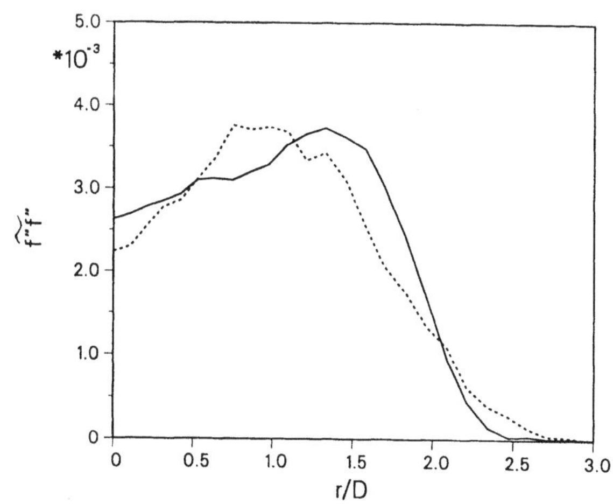

Fig. 5. Radial profiles of the variance of mixture fraction at $x/D = 30$ for turbulent non-reacting methane jets. Dashed and solid lines denote results with/without age-biased sampling

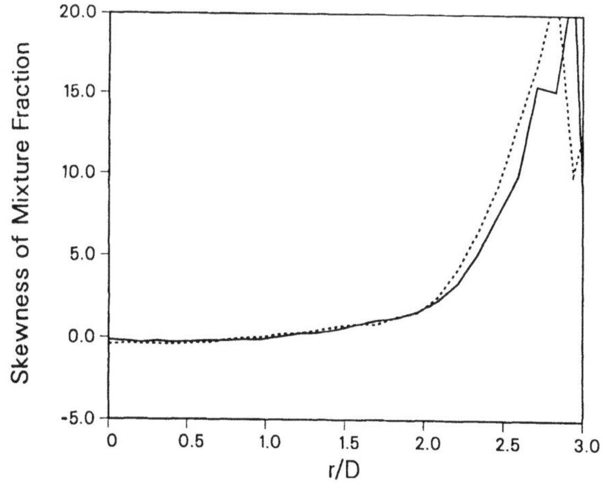

Fig. 6. Radial profiles of the skewness of mixture fraction (normalized third moment) at $x/D = 30$ for turbulent nonreacting methane jets. Dashed and solid lines denote results with/without age-biased sampling

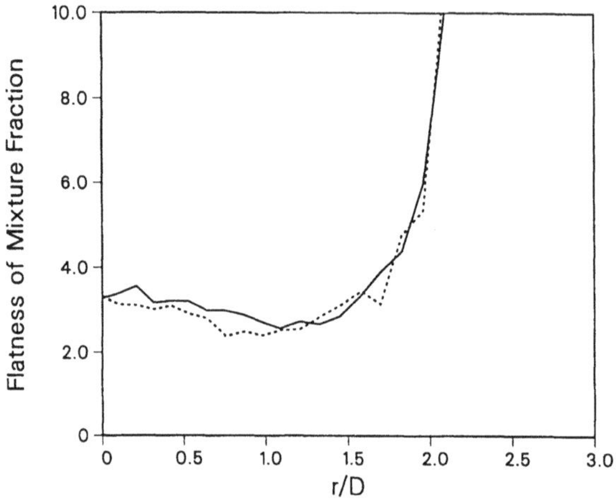

Fig. 7. Radial profiles of the flatness of mixture fraction (normalized fourth moment) at $x/D = 30$ for turbulent nonreacting methane jets. Dashed and solid lines denote results with/without age-biased sampling

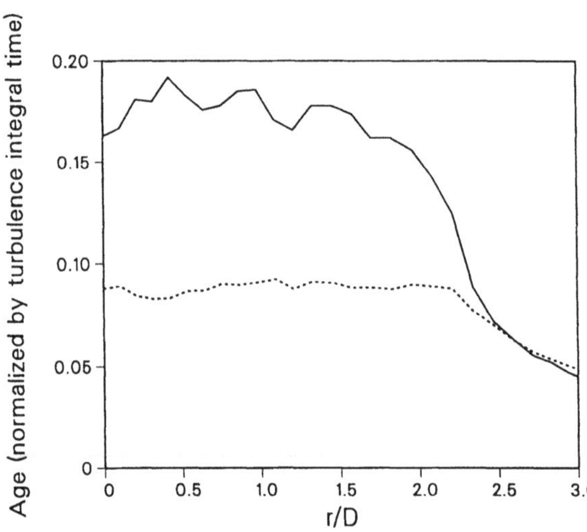

Fig. 8. Radial profiles of the mean normalized age s at $x/D = 30$ for turbulent nonreacting methane jets. Dashed and solid lines denote results with/without age-biased sampling

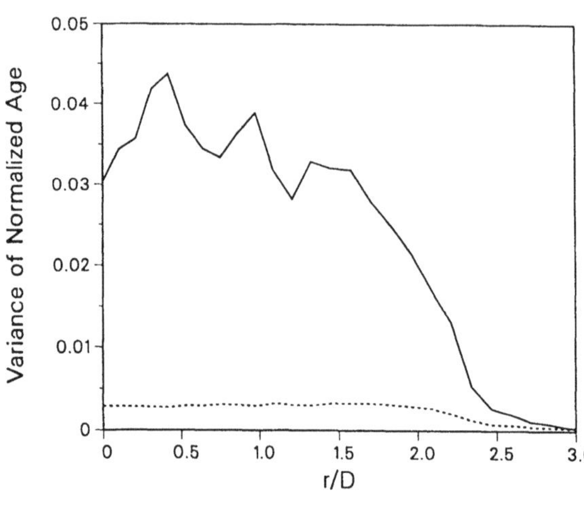

Fig. 9. Radial profiles of the variance of the normalized age s at $x/D = 30$ for turbulent nonreacting methane jets. Dashed and solid lines denote results with/without age-biased sampling

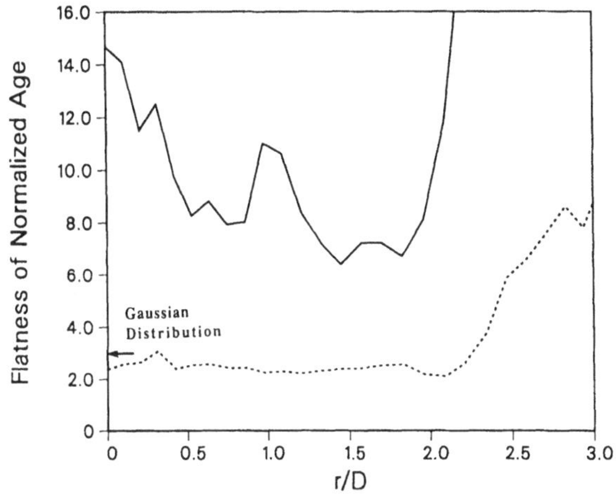

Fig. 10. Radial profiles of the skewness of normalized age s (normalized third moment) at $x/D = 30$ for turbulent nonreacting methane jets. Dashed and solid lines denote results with/without age-biased sampling

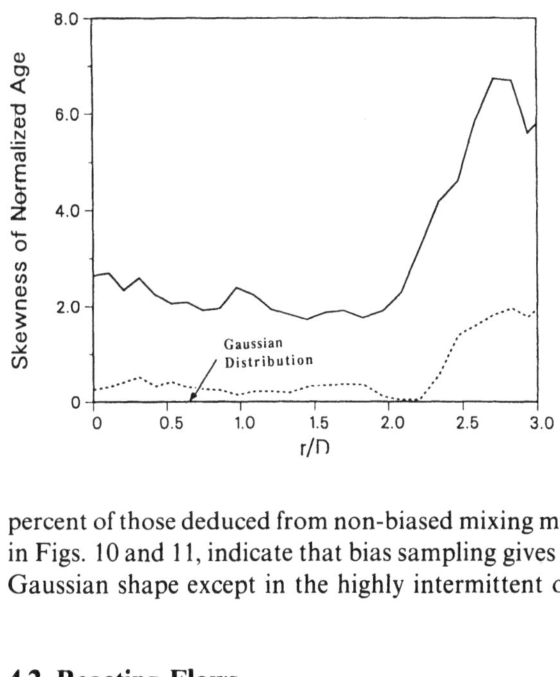

Fig. 11. Radial profiles of the flatness of normalized age s (normalized fourth moment) at $x/D = 30$ for turbulent nonreacting methane jets. Dashed and solid lines denote results with/without age-biased sampling

percent of those deduced from non-biased mixing model. Skewness and flatness of s, in Figs. 10 and 11, indicate that bias sampling gives the distribution of s close to the Gaussian shape except in the highly intermittent outer regions.

4.2 Reacting Flows

4.2.1 Slow Chemical Reaction

The reactions were confined to the quadrilateral subdomain in scalar space containing the vertex at $f_{\text{stoich.}}$ as shown in Fig. 12. The rate of temperature change was set to $\partial T/\partial t = 1000$ K/ms inside and zero outside this subdomain. The predicted joint pdf's for all four mixing/reaction models are presented in Figs. 12 and 13 in terms of scatter plots. In Fig. 12, comparison of the original fractional step method with and without age-biased sampling shows a slightly larger spread of the pdf for the biased case, but the difference is not significant. Conditioning the mixing process

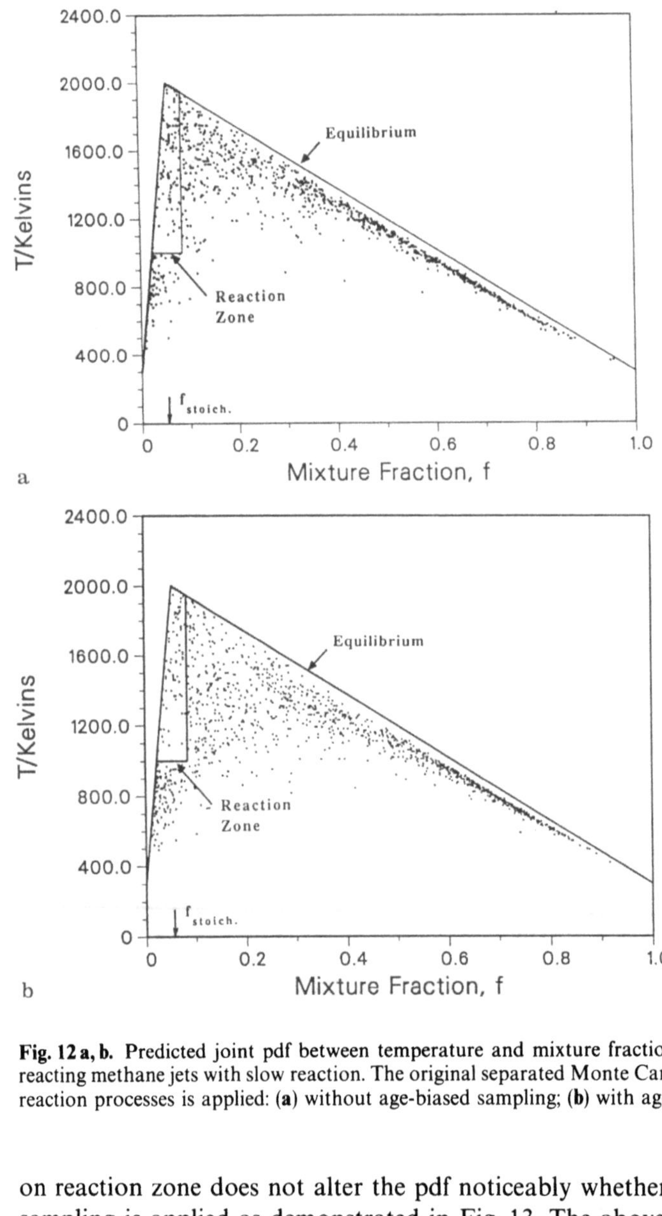

Fig. 12 a, b. Predicted joint pdf between temperature and mixture fraction at $x/D = 30$ for turbulent reacting methane jets with slow reaction. The original separated Monte Carlo procedure for mixing and reaction processes is applied: (**a**) without age-biased sampling; (**b**) with age-biased sampling

on reaction zone does not alter the pdf noticeably whether biased or non-biased sampling is applied as demonstrated in Fig. 13. The above comparisons indicate that flows with slow reactions behave essentially like non-reacting flows; thus, little influence of the sampling procedure on the statistics is observed.

4.2.2 Fast Chemical Reaction

Unlike the slow reaction model, fast reaction was confined in a very narrow region in the scalar space near the line $f = f_{\text{stoich.}}$ as presented in Fig. 14. The increased rate of

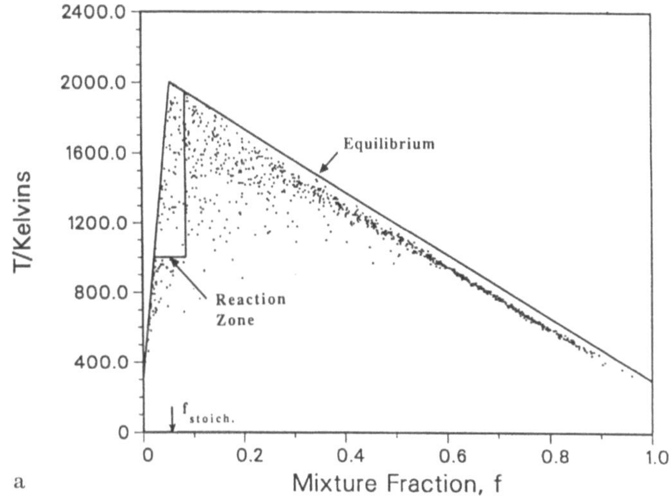

a

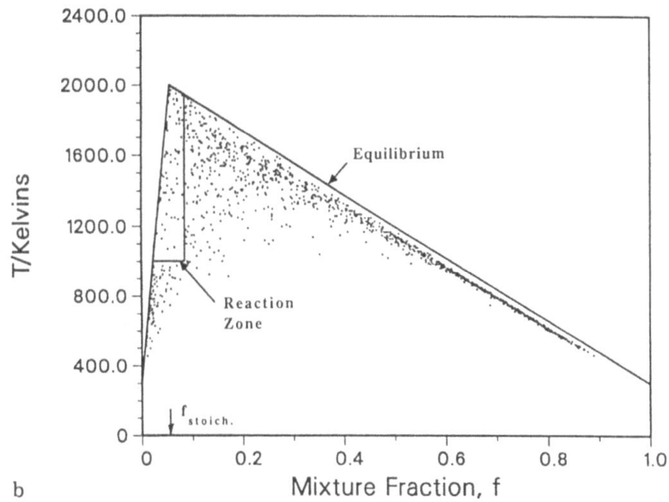

b

Fig. 13 a, b. Predicted joint pdf between temperature and mixture fraction at $x/D = 30$ for turbulent reacting methane jets with slow reaction. The proposed Monte Carlo procedure for the combined effects of mixing and reaction processes is applied: (**a**) without age-biased sampling; (**b**) with age-biased sampling

temperature change was set to $\partial T/\partial t = 10^{10}$ K/ms so that the reaction rate is nearly infinitely fast compared to the mixing process of the flow field. With such an artificial reaction model, one expects that once the jet is ignited by a pilot flame with a completely burned mixture, the temperatures should be maintained at the equilibrium values creating a perfect thin-flame-sheet combustion mode. If the model predicts temperatures below the equilibrium values, this would indicate that at least in this limiting case, the model is deficient.

The joint pdfs between T and f predicted by all four mixing/reaction models are shown in Figs. 14 and 15. As the reaction zone is very thin, it is expected that the

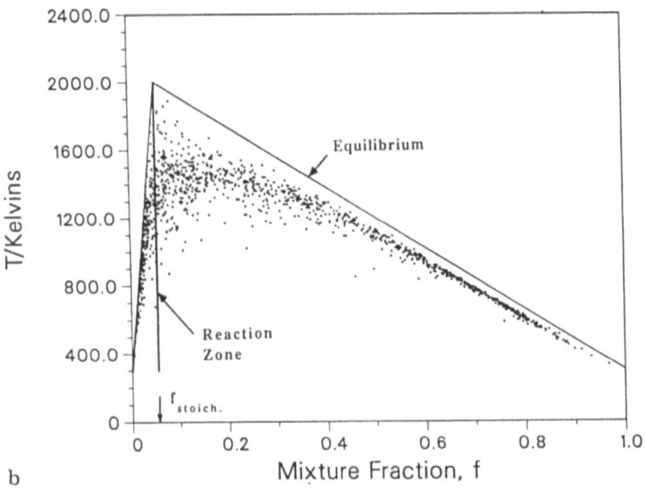

b

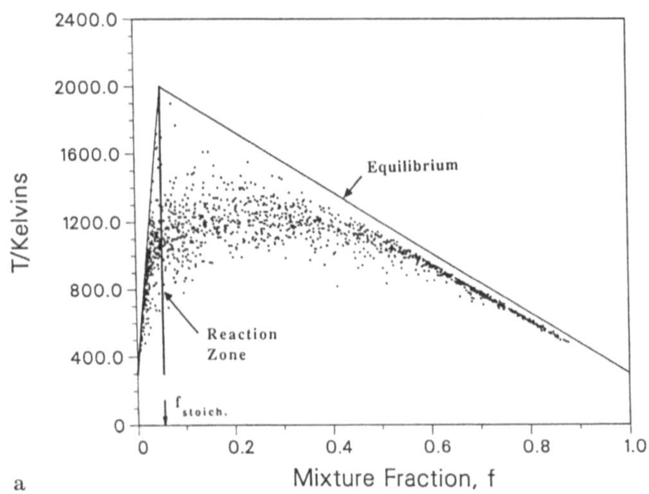

a

Fig. 14 a, b. Predicted joint pdf between temperature and mixture fraction at $x/D = 30$ for turbulent reacting methane jets with fast reaction. The original separated Monte Carlo procedure for mixing and reaction processes is applied: (**a**) without age-biased sampling; (**b**) with age-biased sampling

original modeling of reaction and mixing by separated steps will miss the fast reaction zone as the probability of mixed fluids being in the reaction zone is nearly zero. Consequently, as shown in Fig. 14, temperatures lower than equilibrium values were predicted suggesting that the model is deficient in the flame sheet regime. Age-biased sampling has little influence on the predicted temperatures and does not correct this deficiency. Reaction zone conditioning on the other hand has a dramatic effect on the predicted pdf as shown in Fig. 15. As both mixing and reaction processes are treated in a combined step, any reaction during mixing of two particles across the reaction zone is properly taken into account. The predicted temperatures indeed follow exactly the equilibrium line regardless of age-biased sampling. Only

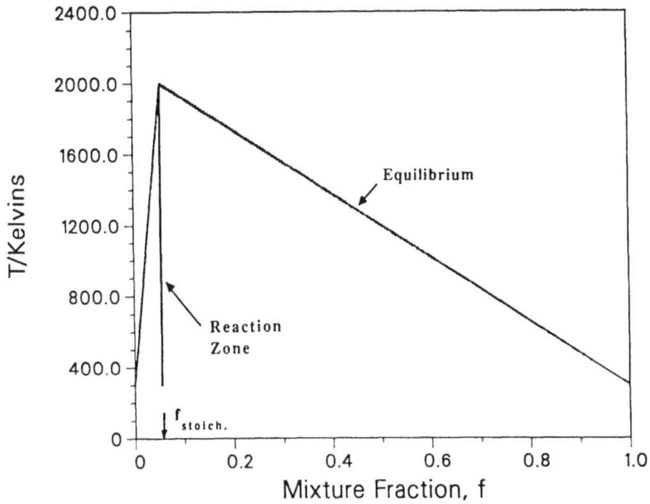

Fig. 15. Predicted joint pdf between temperature and mixture fraction at $X/D = 30$ for turbulent reacting methane jets with fast reaction. The proposed Monte Carlo procedure for the combined effects of mixing and reaction processes is applied without age-biased sampling. Age-biased sampling also produces equilibrium temperatures

the pdf for Model C (un-biased) is shown since Model D (biased) also produces equilibrium temperatures.

5. Conclusions

Modeling of the combined effects of chemical reaction on the joint scalar pdf has been shown to be essential if the reactive zone occupies a very small domain in the composition scalar space. Modifications to the coalescence/dispersion (C/D) model have been investigated, including age-biased sampling and reaction zone conditioning. Results of Monte Carlo simulation of turbulent reacting and nonreacting jet flows indicate that age-biased sampling has little influence on the scalar statistics except the age itself, and that modification of the mixing model by conditioning on reaction zone enables us to predict physically correct results in the flame-sheet combustion regime. Moreover, this reaction zone conditioning approach can incorporate any mixing models without fundamental changes.

References

Chen, J.-Y., Kollmann, W. (1988): Pdf modeling of chemical nonequilibrium effects in turbulent nonpremixed hydrocarbon flames. Twenty-second Symposium (International) on Combustion/The Combustion Institute, p. 645

Chen, J.-Y., Kollmann, W. (1989): Chemical models for pdf modeling of hydrogen-air nonpremixed turbulent flames. Combust. Flame, accepted for publication

Chen, J.-Y., Kollmann, W., Dibble, R. W. (1989): Pdf modeling of turbulent non-premixed methane jet flames. Combust. Sci. Tech. **64**, 315

Correa, S. M., Gulati, A., Pope, S. B. (1988): Assessment of a partial equilibrium/Monte Carlo Model for turbulent syngas flames. Combust. Flame, **72**, 159

Curl, R. L. (1963): Dispersed phase mixing: I. Theory and effects of simple reactors. A.I.Ch.E.J. **9**, 175

Haworth, D. C., Drake, M. C., Pope, S. B., Blint, R. J. (1988): The importance of time-dependent flame structures in stretched laminar flamelet models for turbulent jet diffusion flames. Twienty-second Symposium (International) on Combustion/The Combustion Institute, p. 589

Janicka, J., Kolbe, W., Kollmann, W. (1979): Closure of the transport equation for the probability density function scalar field. J. Non-Equilib. Thermodyn. **4**, 47

Kosaly, G. (1986): Theoretical remarks on a phenomenological model of turbulent mixing. Combust. Sci. Tech. **49**, 227

Kosaly, G., Givi, P. (1987): Modeling of turbulent molecular mixing. Combust. Flame **70**, 101

Pope, S. B. (1982): An improved turbulent mixing model. Combust. Sci. Tech. **28**, 131

Pope, S. B., Correa, S. M. (1986): Joint pdf calculations of a nonequilibrium turbulent diffusion flame. Twenty-first Symposium (International) on Combustion/The Combustion Institute, p. 1341

Pope, S. B., Norris, A., Masri, A. R. (1989): Mixing models for turbulent diffusion flames with finite rate kinetics, presented at the Twelfth Meeting of the Sandia Cooperative Group on the Aero-thermochemistry of Turbulent Combustion, Sandia Report SAND89-8220

A Lagrangian Intermittent Model for Turbulent Combustion; Theoretical Basis and Comparisons with Experiments

M. Gonzalez and R. Borghi

Université de Rouen—Laboratoire de Thermodynamique URA CNRS 230, Mont Saint Aignan, France

Abstract

A lagrangian turbulent combustion model in which the assumption of infinitely fast chemistry is relaxed has been developed. The allowance for a full spectrum of turbulent timescales instead of a single mean timescale related to the turbulent field is the main feature of the model. We present first the theoretical basis of the model and discuss on its essential entries which respectively depend on turbulence: a distribution of turbulent timescales, and chemical reactions: a chemical ignition delay. Second, numerical predictions on methane-air turbulent jet diffusion flames are compared with experimental data.

Introduction

The main information on turbulence required for understanding turbulence-chemistry interaction is provided by characteristic timescales. The specification of the relevant scale to be injected into any turbulent combustion model is complicated by the fact that fully turbulent flows display a wide spectrum of scales ranging from the Kolmogorov ones up to those which are representative of the largest structures.

The well-known hypothesis of infinitely fast chemical reactions greatly simplifies the problem since in that case the chemical and turbulent timescales are quite separated. Then, combustion is not altered by the exact shape of the spectrum of turbulent timescales and a single mean turbulent timescale, usually assumed to be proportional to k/ε is a sufficient information for the combustion model.

On the contrary, when some chemical characteristic timescale is of the same order of magnitude than the turbulent timescales (for example if the turbulent spectrum is wide enough or the chemical reactions are not too fast), strong interaction occurs between turbulence and combustion. The turbulent entry that is needed for the combustion model is not likely to be a single time and has to include the whole spectrum of turbulent timescales. The turbulent combustion models so far devised (the pdf models of Dopazo and O'Brien 1974, or Pope 1985, or the simpler "presumed pdf model" of Borghi and Moreau 1977) have neglected that point. We have studied a turbulent combustion model in which such a spectrum is accounted for, which is designed to handle partial extinction regimes, and which could be valid for thickened-wrinkled flames, while the Flamelet Model (Liew et al. 1981; Cant and Bray 1988) is devoted to the domain of wrinkled flames only.

Turbulent Shear Flows 7

Basic Principles of the Model

The first ideas on the model have been presented by Borghi and Gonzalez (1986); we give here the recent improvements and developments that have led to the design of a model now called M.I.L. (Modèle Intermittent Lagrangien).

The Origins of the M.I.L. Model

The starting point is the lagrangian I.E.M. model (see for instance Villermaux 1986) in which a set of lagrangian equations, with the diffusion term modelled as an exchange with the mean value, is used in order to study the composition evolution of a fluid particle (detailed discussion on I.E.M. and its extensions can be found in Borghi 1988):

$$\frac{dY_i}{dt} = \frac{\tilde{Y}_i - Y_i}{\tau} + \mathring{w}_i(Y_{J,J=1,\ldots,N}, T) \quad (1)$$

where: Y_i is the instantaneous mass fraction of species i and $\tilde{Y}_i$ its statistical mean value; $\mathring{w}_i$ is the instantaneous reaction rate for species i depending on the mass fractions of all other reacting species and on temperature; $1/\tau$ is the frequency of exchanges between fluid particles; in the I.E.M. model for homogeneous media, τ is unique and related to turbulence stating $\tau \propto k/\varepsilon$.

It is to be noticed that the modelled Eq. (1) allows to derive the equations classically used for $\tilde{Y}_i$, $\widetilde{Y''^2_i}$ and $\widetilde{Y''_i Y''_j}$ (see Borghi 1988).

The Hypothesis and Core of the Model

The M.I.L. model consists in an extension of I.E.M. in the sense that the exchange time τ, instead of being unique, is now assumed to obey a turbulent timescales distribution on which we shall discuss further.

But, in addition, it states also that combustion is sudden, i.e. with high global activation energy. The term "sudden combustion" relates to a chemical reaction that begins very slowly, that is with quasi-negligible consumption of reactants and low heat release, and, after a delay, terminates very rapidly, the most of the consumption of reactants, creation of products and heat release occurring in a quite short time interval compared to the "ignition delay". As allowance is made for a chemical time that is related to ignition phenomena and that may be of the same order of magnitude than the turbulent timescales, "sudden combustion" differs radically from "infinitely fast chemistry" which states any chemical time to be negligible in comparison to all other timescales.

Although M.I.L. can be written, in principle, for diffusion as well as premixed flames (with some modifications which do not alter its fundamental basis) we present the model in the case of diffusion flames. At the moment chemical reactions are assumed to be reduced to a one-step global reaction although a version with a moderately complex chemical scheme is in progress.

Let us now consider a burner feeded with oxidizer through an inlet A and with fuel through an inlet B (Fig. 1).

The behavior of the fluid particles inside the burner, from the entrance to the vicinity of a point P may be described on a phase plane (ϕ, Y_0). ϕ is an arbitrary inert species (defined, for example, as $\phi = 0$ in the fuel inlet and $\phi = 1$ in the oxidizer inlet) and Y_0 the instantaneous mass fraction of oxidizer.

With the hypothesis of sudden combustion and considering an homogeneous turbulent medium, the evolution of fluid particles is depicted by trajectories on the (ϕ, Y_0) phase plane; those trajectories, in agreement with Eq. (1) and sudden combustion are made with paths along the mixing lines AM and BM (M being the point where the mean value $(\tilde{\phi}, \tilde{Y}_0)$ is located) when weak reaction occurs, or paths along the equilibrium lines AS and BS (the point S corresponding to stoichiometric conditions) when reaction is very fast. The fluid particles jump from the former to the later when ignition occurs (Fig. 2). For example, a particle entering the burner through A travels along AM but may jump at a given time on AS and then follows AF via S (F represents the final state of the particle when time goes to infinity). A similar evolution is found for a particle entering through B and travelling on BM and BF.

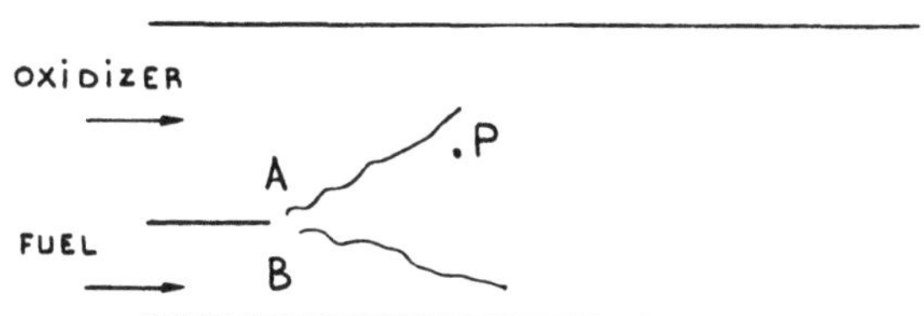

Fig. 1. Schematic picture of a diffusion flame burner

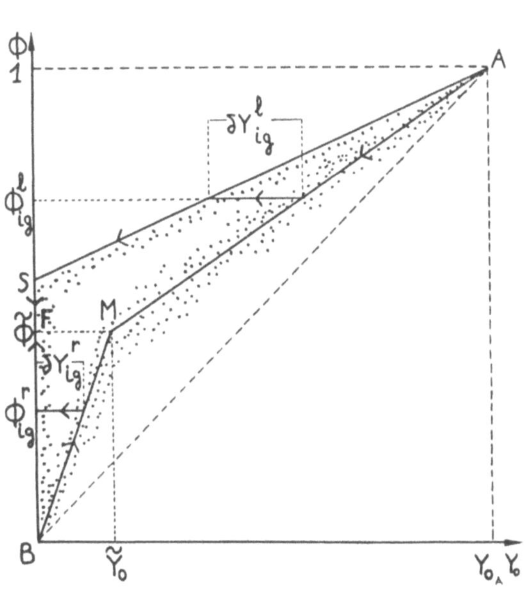

Fig. 2. Trajectories of fluid particles on the phase plane (ϕ, Y_0) in the case of a diffusion flame

Since combustion is sudden, the jumps from the mixing to the equilibrium lines are very fast and, consequently, the probability to find a fluid particle elsewhere than along *AM*, *BM*, *AS* or *BS* is quite negligible. That means that the projection of the joint p.d.f. $\tilde{P}_{\phi,Y_0}(\phi, Y_0)$ on the (ϕ, Y_0) plane is sustained only by *AM*, *BM*, *AS* and *BS*. As we are, in actual fact, interested in nonhomogeneous flows, the joint p.d.f. $\tilde{P}_{\phi,Y_0}(\phi, Y_0)$ is not so simple; it rather looks like two clouds of points respectively located around the mixing lines *AM*, *BM* and in vicinity of the equilibrium lines *AS*, *BS* (Fig. 2). A p.d.f. of that type has been found experimentally by A. R. Masri et al. 1988a on the plane mixture fraction/temperature.

The Mean Reaction Rate

The trajectories given by the M.I.L. model (Fig. 2) together with the equations of I.E.M. allow us to derive an expression for the mean reaction rate, for example the mean consumption rate of oxidizer. In that derivation, one has to consider the contribution of the particles burning in the vicinity of the equilibrium lines *AS* and *BS* $\tilde{\dot{w}}_0^{\text{eq}}$ as well as the contribution of each jump occurring between the mixing lines *AM* and *BM* and the equilibrium lines, $\tilde{\dot{w}}_0^{\text{jump}}$, and we write:

$$\tilde{\dot{w}}_0 = \tilde{\dot{w}}_0^{\text{eq}} + \tilde{\dot{w}}_0^{\text{jump}}. \tag{2}$$

Details of the derivation of $\tilde{\dot{w}}_0^{\text{eq}}$ and $\tilde{\dot{w}}_0^{\text{jump}}$ are given in the Appendix. We find:

$$\tilde{\dot{w}}_0^{\text{eq}} = \int_0^1 \tilde{P}_\phi(\phi) \int_{\tau_{\text{ig}}(\phi)}^{+\infty} \tilde{\dot{w}}_0^{\text{eq}} \cdot P_\tau(\tau) \cdot d\tau \cdot d\phi \tag{3}$$

and

$$\tilde{\dot{w}}_0^{\text{jump}} = -\int_{\tau_k}^{+\infty} \frac{\tilde{\phi} - \phi_{\text{ig}}^{\text{r}}(\tau)}{\tau} \cdot \delta Y_{0_{\text{ig}}}^{\text{r}} \cdot \tilde{P}_\phi(\phi_{\text{ig}}^{\text{r}}(\tau)) \cdot P_\tau(\tau) \cdot d\tau + \int_{\tau_k}^{+\infty} \frac{\tilde{\phi} - \phi_{\text{ig}}^{\text{l}}(\tau)}{\tau} \cdot \delta Y_{0_{\text{ig}}}^{\text{l}} \cdot \tilde{P}_\phi(\phi_{\text{ig}}^{\text{l}}(\tau)) \cdot P_\tau(\tau) \cdot d\tau \tag{4}$$

where: $\tau_{\text{ig}}(\phi)$ is the ignition delay,
$\tilde{P}_\phi(\phi)$ the p.d.f. of ϕ,
$P_\tau(\tau)$ the distribution of turbulent timescales,
$\dot{w}_0^{\text{eq}}$ is given by Eqs. (A.4) and (A.5),
τ_k is the Kolmogorov timescale,
$\phi_{\text{ig}}^{\text{r}}$ and $\phi_{\text{ig}}^{\text{l}}$ are the values of ϕ at the location of a jump respectively on the rich and lean side,
$\delta Y_{0_{\text{ig}}}^{\text{r}}$ and $\delta Y_{0_{\text{ig}}}^{\text{l}}$ represent the amounts of oxidizer consumed by an ignition respectively on the rich and lean side.

With $\tilde{\dot{w}}_0^{\text{eq}}$ and $\tilde{\dot{w}}_0^{\text{jump}}$ given respectively by Eqs. (3) and (4), $\tilde{\dot{w}}_0$ is found as a function of $\tilde{Y}_0$, $\tilde{\phi}$ and turbulence through $P_\tau(\tau)$, $\tilde{P}_\phi(\phi)$, and can now be used in the classical eulerian balance equation for $\tilde{Y}_0$.

The Essential Entries of the Model

The knowledge of three characteristics is needed:

—The p.d.f. of the inert scalar $\tilde{P}_\phi(\phi)$: it is a quite classical problem; we presume a β-function for $\tilde{P}_\phi(\phi)$ and compute its shape by means of $\tilde{\phi}$ and $\widetilde{\phi''^2}$.

—The distribution of timescales $P_\tau(\tau)$: at present, neither theoretical nor experimental studies have been carried out in order to determine $P_\tau(\tau)$.

Nevertheless, as a consequence of the hypothesis of sudden combustion, we may affirm that $P_\tau(\tau)$ is not likely to be modified by the chemical reaction and depends only on pure turbulence quantities; then, we express it as:

$$P_\tau(\tau) = f(\tau/\tau_t, \tau_k/\tau_t) \tag{5}$$

where

$$\tau_t \propto k/\varepsilon \quad \text{and} \quad \tau_k/\tau_t = Re_T^{-1/2}.$$

Following the experimental study of Bray et al. (1984) we assume an exponential form for f and write:

$$\begin{cases} P_\tau(\tau) = \dfrac{1}{\tau_t - \tau_k} e^{-(\tau - \tau_k/\tau_t - \tau_k)}; & \tau \geqq \tau_k \\ P_\tau(\tau) = 0; & \tau < \tau_k. \end{cases} \tag{6}$$

The function $P_\tau(\tau)$ is such that it verifies:

$$\int_{\tau_k}^{+\infty} P_\tau(\tau)\, d\tau = 1 \tag{7}$$

and

$$\int_{\tau_k}^{+\infty} \tau .\ P_\tau(\tau)\, d\tau = \tau_t. \tag{8}$$

—The ignition time $\tau_{ig}(\phi)$: it is a pure chemical quantity that may include complex chemistry effects related to conditions in the vicinity of the mixing lines AM and BM (Fig. 2.). It could be deduced from experiments or even computed knowing the full chemical mechanism. At present we estimate $\tau_{ig}(\phi)$ stating:

$$\tau_{ig}(\phi) = 1/\mathring{w}_0(\phi, \tilde{\phi}, \tilde{Y}_0) \tag{9}$$

where $\mathring{w}_0$ is the instantaneous consumption rate of oxidizer on the AM and BM lines assumed to be given by an Arrhenius law. Another method to compute $\tau_{ig}(\phi)$ that is now under investigation consists in integrating the I.E.M. equations for ϕ and Y_0 for all the sets $(\tilde{\phi}, \tilde{Y}_0, \tau)$ and to detect the values $\phi_{ig}(\tau)$ at which ignitions occur; it is then easy to build a library of $\tau_{ig}(\phi)$ with $\tilde{\phi}$ and $\tilde{Y}_0$ as parameters.

Comparisons with an Experiment

In order to test the M.I.L. model we have tried to compare numerical results with the experimental data of Masri et al. (1988b) on turbulent jet diffusion flames of methane near extinction. The experimental facility consists in a burner with a central jet of methane in a coflowing air stream (Fig. 3). The burner stability is ensured owen

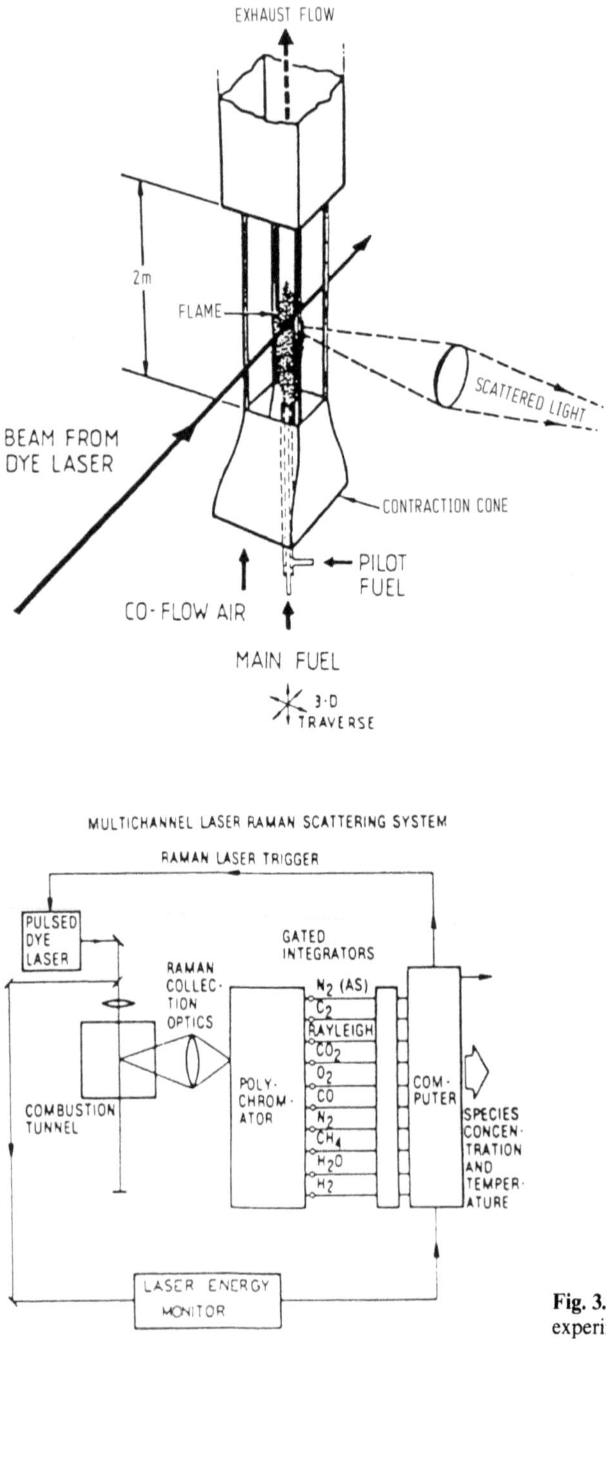

Fig. 3. Experimental setup of the experiment of A. R. Masri et al.

to a fully burnt gas flow provided by a pilot flame surrounding the methane jet. The experimental conditions are the following: coflow air: $U_e = 15\,\mathrm{m \cdot s^{-1}}$; pilot burnt gas: $U_{pb} = 24\,\mathrm{m \cdot s^{-1}}$, $T = 2600\,\mathrm{K}$; the flame characteristics are varied by varying the methane bulk velocity: $U_j = 36\,\mathrm{m \cdot s^{-1}}$ (flame *A*), $U_j = 41\,\mathrm{m \cdot s^{-1}}$ (flame *L*), $U_j = 48\,\mathrm{m \cdot s^{-1}}$ (flame *B*), $U_j = 55\,\mathrm{m \cdot s^{-1}}$ (flame *M*). The diameter of the central jet is $D_j = 7.2\,\mathrm{mm}$.

The flow field has been computed with a parabolic procedure including a standard $k - \varepsilon$ model (with the classical constants) for turbulence. The initial conditions for the mean and fluctuating velocities are those provided by A. R. Masri (personal communication; Fig. 4.).

Two types of calculation have been carried out: first, with the M.I.L. model working at the limit of infinitely fast chemical reaction by simply stating $\tau_{ig} < \tau_k$ (the value of τ_{ig} being, in that case, immaterial) and second, with finite chemistry, τ_{ig} being computed using Eq. (9) where $\mathring{w}_0$ is obtained from an Arrhenius law:

$$\mathring{w}_0 = -\rho K Y_0 Y_{CH_4} e^{-T_A/T}. \tag{10}$$

With, following Duterque et al. (1981):

$$K = 1.5 \times 10^{13}\,\mathrm{SI} \quad \text{and} \quad T_A = 20000\,\mathrm{K}$$

Figures 5 to 24 display comparisons between numerical and experimental data in the cases of flames *A* and *B* for the means and r.m.s. of mixture fraction and temperature. The corresponding radial profiles have been plotted at axial locations $x/D_j = 10$ and 20 for flame *A* and $x/D_j = 10$, 20 and 30 for flame *B*.

First of all, looking at the mean and r.m.s. of temperature, the outstanding feature is that large discrepancies are obtained with the model working at the limit of infinitely fast chemistry for flames *A* and *B* both. That is not surprising at all since the

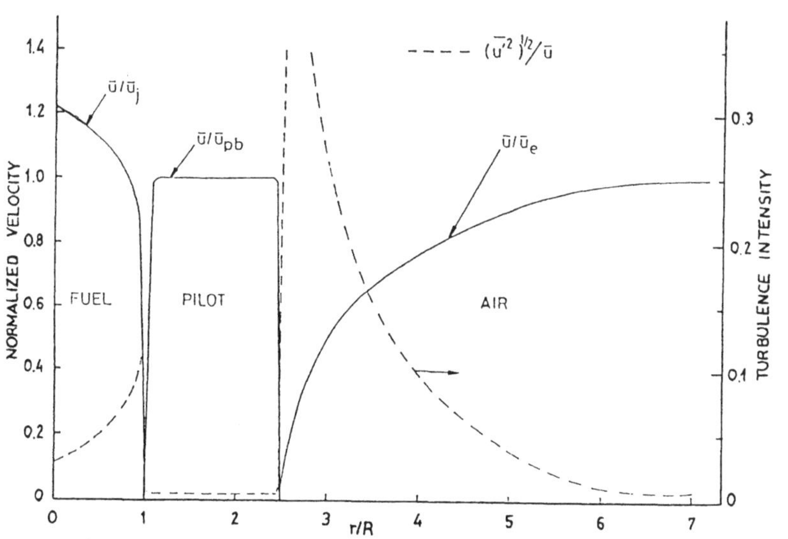

Fig. 4. Initial conditions for the mean and fluctuating velocities (from A. R. Masri)

flames under study are definitely not fast burning. Concerning mixture fraction, the differences are less distinct for that quantity is only indirectly influenced by chemical reaction.

For the flame A, the radial profiles of means and r.m.s. of mixture fraction and temperature computed in the case of finite chemistry (Figs. 5 to 12) agree rather well with the experimental data. Note that concerning mixture fraction, no difference is found at $x/D_j = 10$ between the numerical results obtained respectively with fast and finite chemistry (Figs. 5 and 6); at $x/D_j = 20$ the corresponding profiles display only a slight difference.

Concerning the flame B, the model with finite chemistry, gives satisfactory agreement for the mean temperature profiles at axial locations $x/D_j = 10, 20$ and 30 (Figs. 19 to 21). But looking at the r.m.s. value, noticeable discrepancies appear at

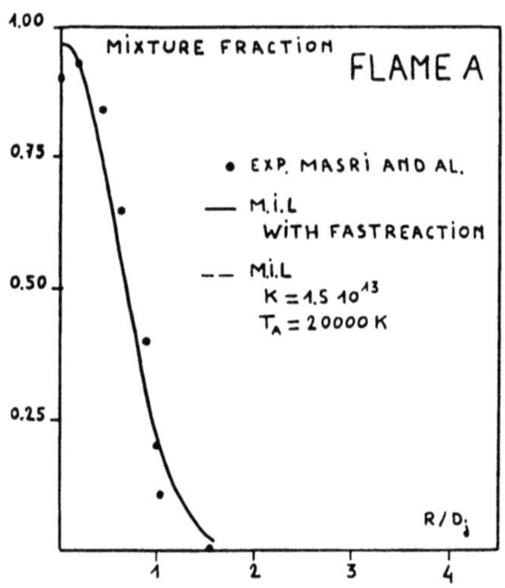

Fig. 5. Radial profile of mean mixture fraction at axial location $x/D_j = 10$; flame A

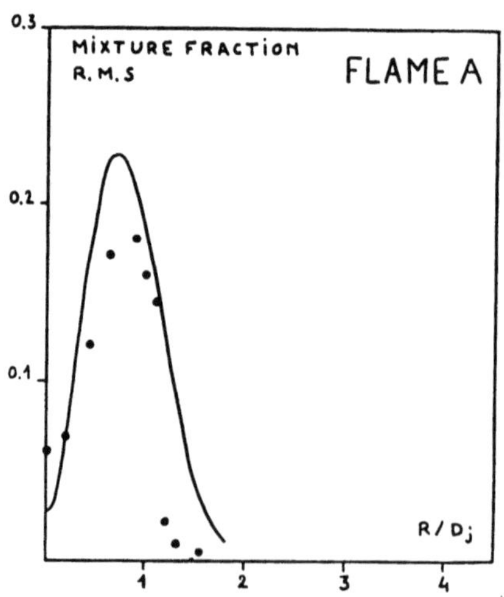

Fig. 6. Radial profile of mixture fraction r.m.s. at axial location $x/D_j = 10$; flame A

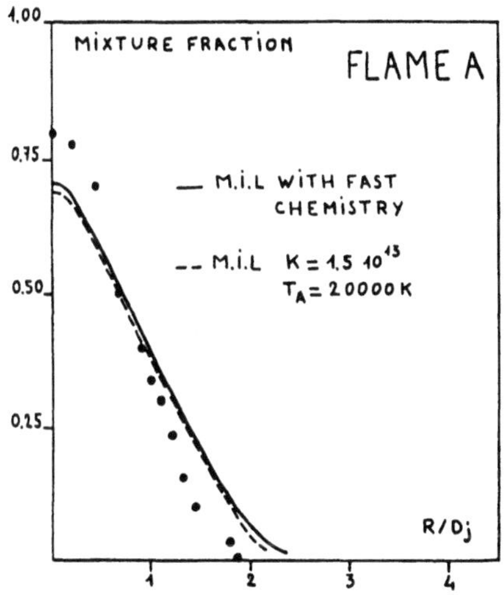

Fig. 7. Radial profile of mean mixture fraction at axial location $x/D_j = 20$; flame A

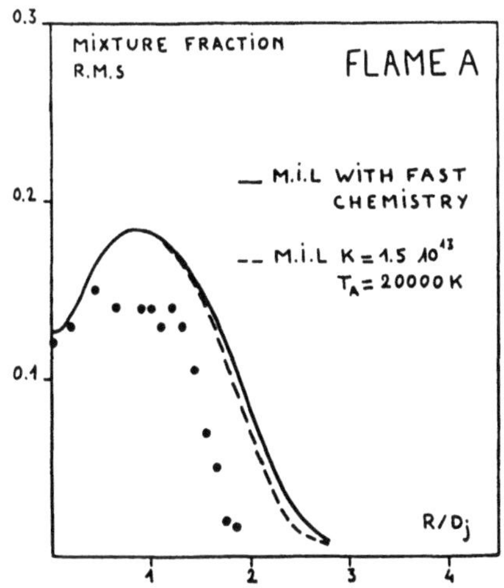

Fig. 8. Radial profile of mixture fraction r.m.s. at axial location $x/D_j = 20$; flame A

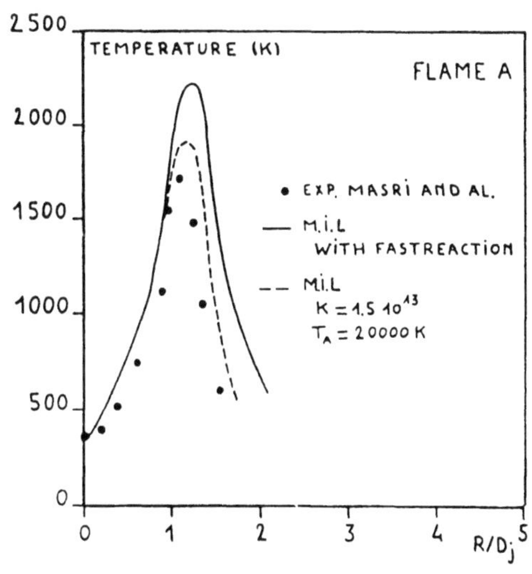

Fig. 9. Radial profile of mean temperature at axial location $x/D_j = 10$; flame A

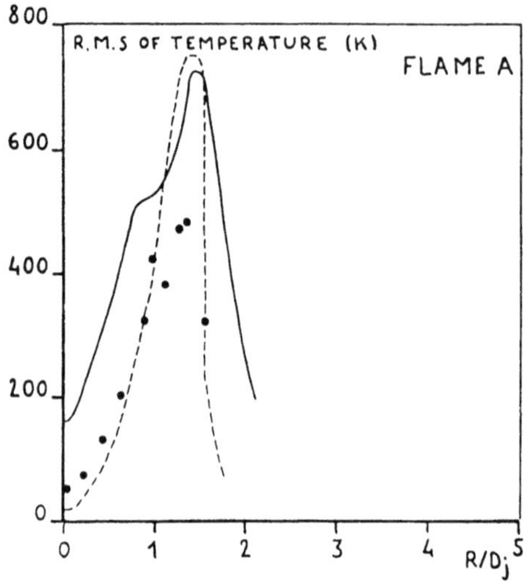

Fig. 10. Radial profile of temperature r.m.s. at axial location $x/D_j = 20$; flame *A*

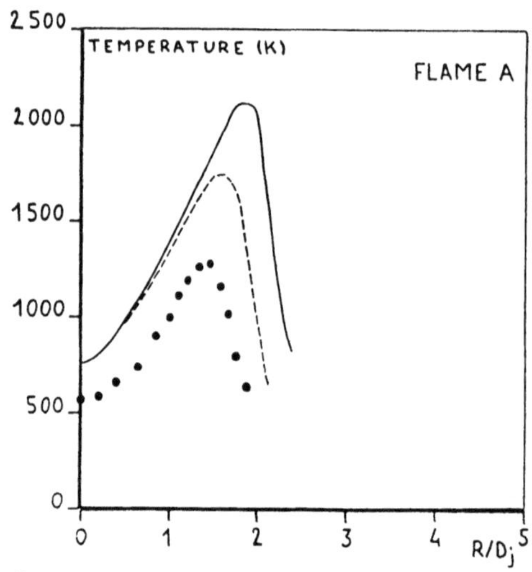

Fig. 11. Radial profile of mean temperature at axial location $x/D_j = 20$; flame *A*

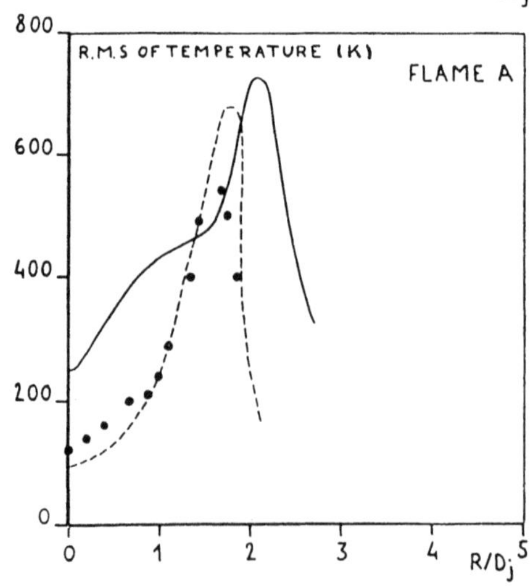

Fig. 12. Radial profile of temperature r.m.s. at axial location $x/D_j = 20$; flame A

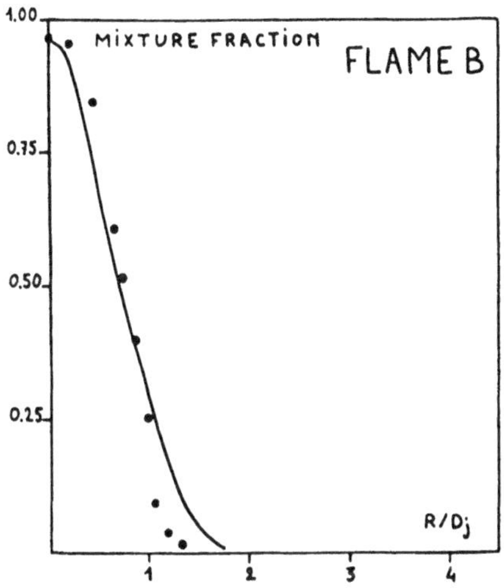

Fig. 13. Radial profile of mean mixture fraction at axial location $x/D_j = 10$; flame B

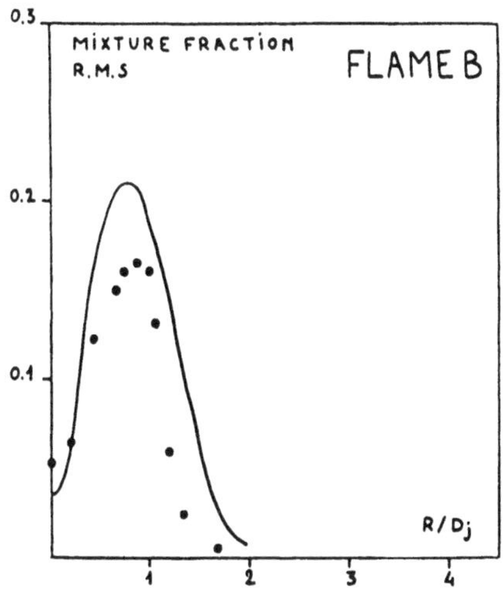

Fig. 14. Radial profile of mixture fraction r.m.s. at axial location $x/D_j = 10$; flame B

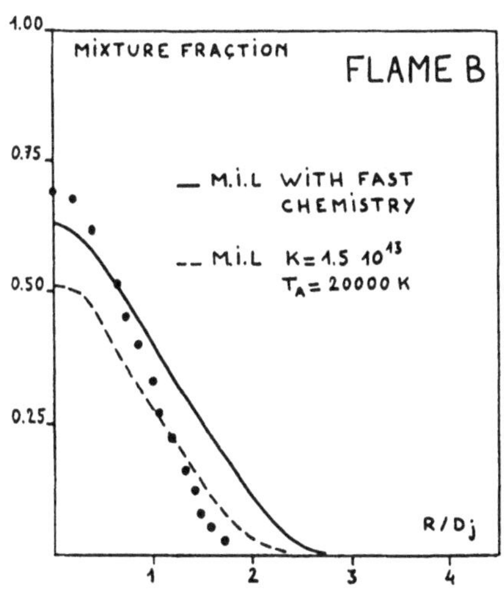

Fig. 15. Radial profile of mean mixture fraction at axial location $x/D_j = 20$; flame B

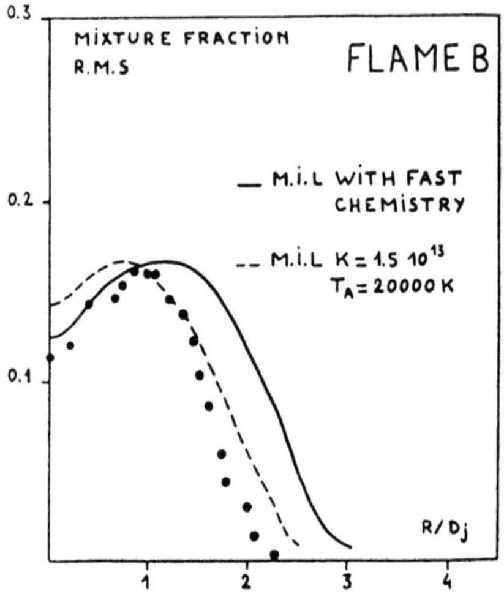

Fig. 16. Radial profile of mixture fraction r.m.s. at axial location $x/D_j = 20$; flame B

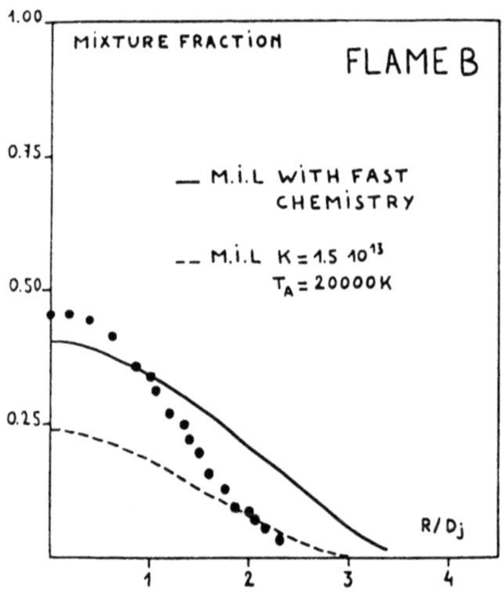

Fig. 17. Radial profile of mean mixture fraction at axial location $x/D_j = 30$; flame B

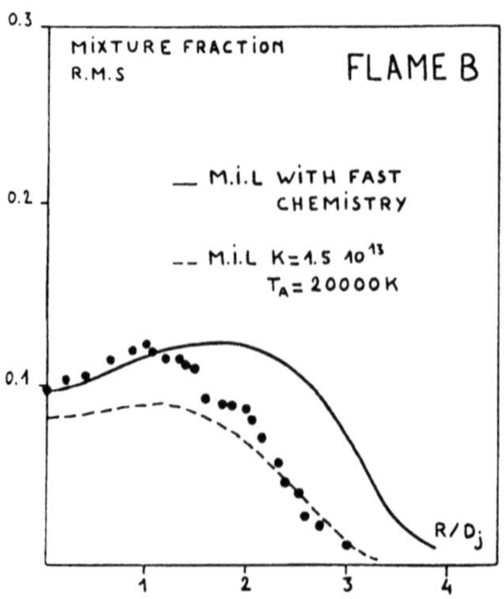

Fig. 18. Radial profile of mixture fraction r.m.s. at axial location $x/D_j = 30$; flame B

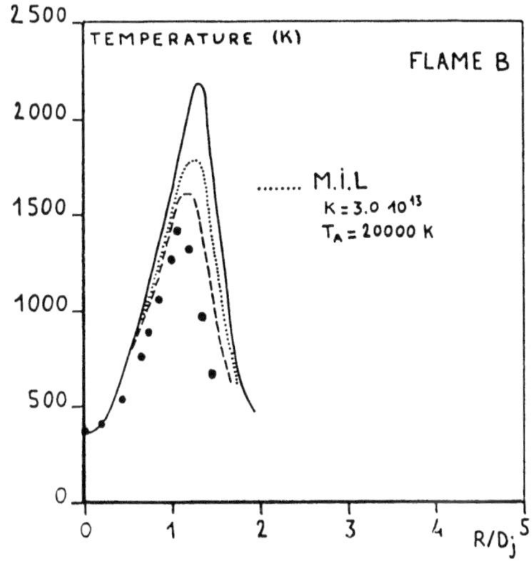

Fig. 19. Radial profile of mean temperature at axial location $x/D_j = 10$; flame B

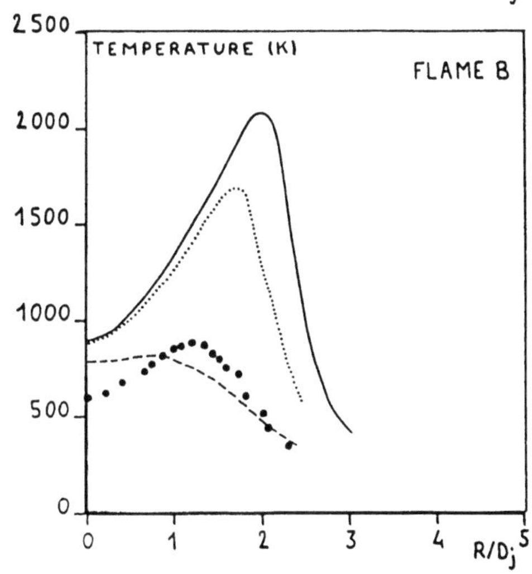

Fig. 20. Radial profile of mean temperature at axial location $x/D_j = 20$; flame B

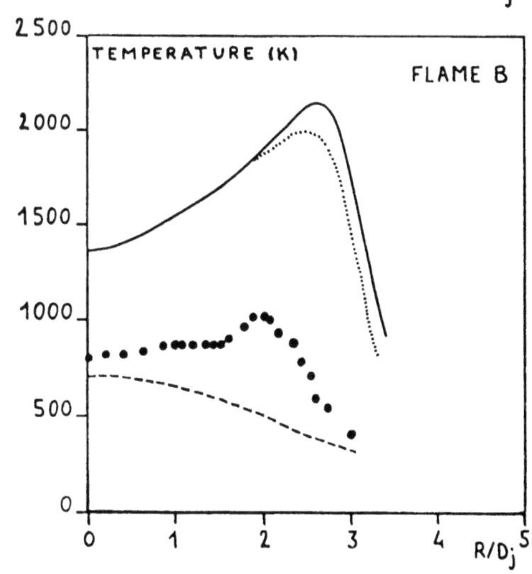

Fig. 21. Radial profile of mean temperature at axial location $x/D_j = 30$; flame B

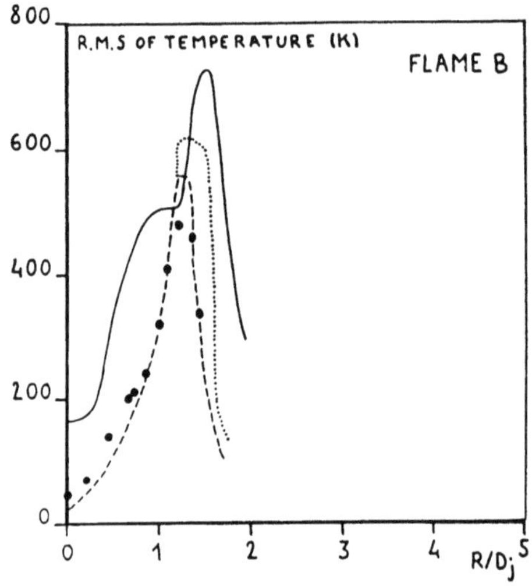

Fig. 22. Radial profile of temperature r.m.s. at axial location $x/D_j = 10$; flame B

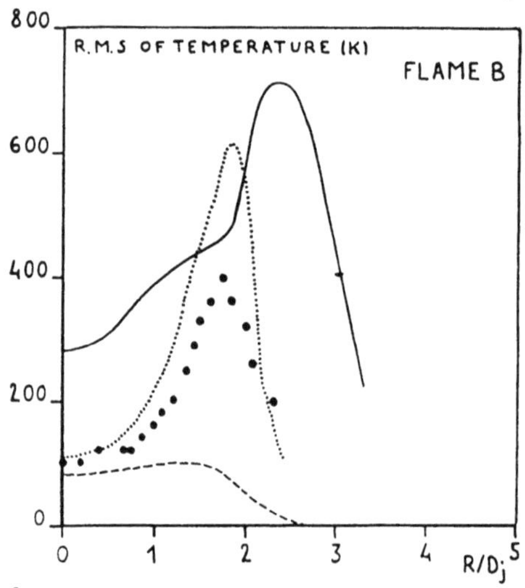

Fig. 23. Radial profile of temperature r.m.s. at axial location $x/D_j = 20$; flame B

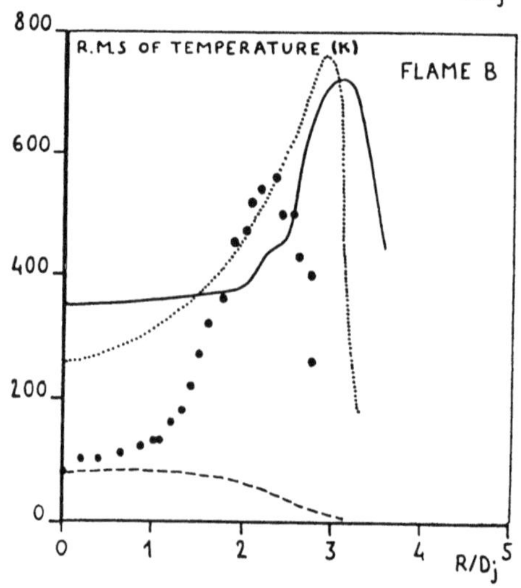

Fig. 24. Radial profile of temperature r.m.s. at axial location $x/D_j = 30$; flame B

$x/D_j = 20$ and 30 while at $x/D_j = 10$ the agreement is good (Figs. 22 to 24). The mean mixture fraction at $x/D_j = 30$ is also greatly underestimated (Fig. 17). Those results are due to the fact that the predicted flame is extinguished between $x/D_j = 10$ and $x/D_j = 20$ but no reignition is found further downstream as should be expected according to the experimental data.

We have tried a value $K = 3 \times 10^{13}$ SI (instead of 1.5×10^{13} SI) for the pre-exponential factor of the Arrhenius law (10) from which the ignition delay is deduced. In that case, the computed flame *B* is nowhere extinguished and at $x/D_j = 30$ the mean temperature is found to be close to the mean temperature at the limit of fast chemistry (Figs. 19 to 21). With values of *K* between 1.5×10^{13} SI and 3×10^{13} SI no intermediate solution has been found: either the flame is never extinguished or it is extinguished between $x/D_j = 10$ and $x/D_j = 20$, but never reignited.

Discussion

Our first impression is that, except the failure in predicting the reignition of flame *B*, the numerical results are not too bad generally speaking, considering the hypothesis that are made, particularly the one-step global reaction which, in that case, is far from reality. In addition, the turbulence model used is quite standard. Note also that concerning temperature we compute the Favre mean and r.m.s. while the corresponding measured quantities are conventionally averaged.

To our sense, the failure in the prediction of reignition may be understood looking at the axial evolution of the ignition time and of the mixing time. For simplicity we consider only one mixing time and one ignition time but, in actual fact, at each location, one has to account for a distribution of mixing times depending on turbulence and for a distribution of ignition times in the space of mixture fraction, related to chemistry and also to local conditions through $\tilde{\phi}$ and $\tilde{Y}_0$ (Eq. (9)). The mixing time as well as the ignition one increase with axial distance. Close to the

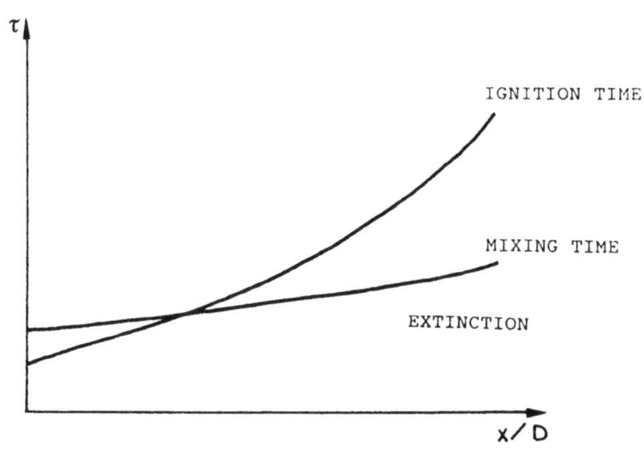

Fig. 25. Axial evolutions of the mixing time and ignition time in the case of an extinction with no reignition

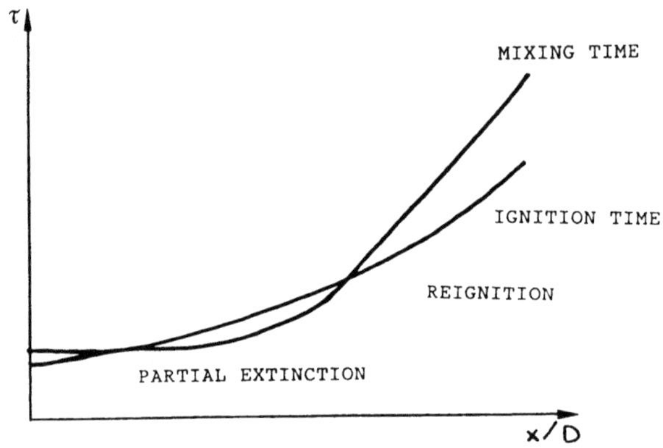

Fig. 26. Axial evolutions of the mixing time and ignition time in the case of a partial extinction and reignition

burner, owing to the pilot flame, the ignition time is sufficiently low in comparison with the mixing one and the flame is attached. Extinction may occur if the ignition time becomes greater than the mixing time. If they keep on behaving in the same way with axial distance no reignition can take place downstream (Fig. 25). A reignition can occur if: a) the flame is only partially extinguished; b) the ignition time and the mixing time evolve with axial distance in such a way that the ignition time, after being higher than the mixing one becomes smaller (Fig. 26). In our case, the curves Min $\{\tau_{ig}(\phi; x/D)\}$ and $\tau_t(x/D)$ (for example at the radial distance where the mean reaction rate is maximum) display only one cross-over point and consequently an extinction is predicted but no reignition is found.

Conclusion

We have designed a turbulent combustion model which includes a turbulent timescales distribution and is devoted to the study of non-infinitely fast combustion. Owing to an hypothesis of "sudden combustion", it allows us to derive the mean reaction rate. The model is likely to display the right parameters (τ_{ig}/τ_t, τ_{ig}/τ_k, τ_k/τ_t) to account for the action of turbulence on combustion and to deal with extinction regimes.

The first comparisons with the experiment of Masri et al. are encouraging since the model display the good trend for mean as well as r.m.s. values. Nevertheless, the difficulty in predicting reignition emphasizes the fact that in addition to taking account for a turbulent timescales distribution, the respective evolutions of the ignition delay and the turbulent timescale have to be well predicted.

Consequently, future work has to take two directions. First, studies on the turbulent timescales distribution: at the moment, as we lack on theoretical and experimental information, we can only guess and test different shapes depending on τ_k and τ_t. Second, the ignition delay: a library of ignition times including more chemistry should lead to a better behavior of the M.I.L. model.

Appendix

Derivation of the Mean Reaction Rate

The mean reaction rate near the equilibrium lines, denoted $\tilde{\mathring{w}}_0^{eq}$ is obtained from the I.E.M. equations for the oxidizer mass fraction Y_0 and the inert species ϕ:

$$\frac{dY_0}{dt} = \frac{\tilde{Y}_0 - Y_0}{\tau} + \mathring{w}_0 \tag{A.1}$$

$$\frac{d\phi}{dt} = \frac{\tilde{\phi} - \phi}{\tau} \tag{A.2}$$

From (A.1) and (A.2) we get:

$$\mathring{w}_0 = \frac{\tilde{\phi} - \phi}{\tau}\left[\frac{dY_0}{d\phi} - \frac{\tilde{Y}_0 - Y_0}{\tilde{\phi} - \phi}\right] \tag{A.3}$$

On the BS line (Fig. 2):

$$Y_0 = 0 \quad \text{and} \quad \frac{dY_0}{d\phi} = 0$$

give, with (A.3):

$$\mathring{w}_0^{eq} = -\frac{\tilde{Y}_0}{\tau} \tag{A.4}$$

While on AS:

$$Y_0 = \frac{Y_{0A}}{1 - \phi_s}(\phi - \phi_s)$$

gives:

$$\mathring{w}_0^{eq} = -\frac{\tilde{Y}_0(1 - \phi_s) - Y_{0A}(\tilde{\phi} - \phi_s)}{\tau(1 - \phi_s)} \tag{A.5}$$

Computing the mean reaction rate now requires the knowledge of the probability on AS and BS which, as a matter of fact, is related to the probability of jump at a given place on AM and BM. As any jump has the meaning of an ignition, it can be assumed that the probability of jump at each value of ϕ in the vicinity of the AM and BM lines is:

$$\alpha = \int_{\tau_{ig}(\phi)}^{+\infty} P_\tau(\tau)\, d\tau \tag{A.6}$$

where $P_\tau(\tau)$ is the distribution of exchange times and $\tau_{ig}(\phi)$ the ignition time.

The relation (A.6) just means that fluid particles experiencing exchanges at frequencies τ^{-1} higher than the reciprocal time τ_{ig}^{-1} cannot be ignited.

Now, integrating over ϕ and τ, we derive the expression for the mean reaction rate $\tilde{\mathring{w}}_0^{eq}$ in the vicinity of the equilibrium lines:

$$\tilde{\mathring{w}}_0^{eq} = \int_0^1 \tilde{P}_\phi(\phi) \int_{\tau_{ig}(\phi)}^{+\infty} \mathring{w}_0^{eq} \cdot P_\tau(\tau) \cdot d\tau \cdot d\phi \qquad (A.7)$$

Where: $\tilde{P}_\phi(\phi)$ is the p.d.f. of ϕ and $\tilde{\mathring{w}}_0^{eq}$ is given by (A.4) and (A.5),

It remains to estimate the contribution to the mean reaction rate due to the jumps (or ignitions) from the mixing to the equilibrium lines; the probability to find fluid particles between equilibrium lines and mixing lines is vanishingly zero, but the reaction rate tends to infinity, in such a way that their product, that gives the mean reaction rate, may remain finite.

For a given turbulent timescale τ, only one trajectory on the plane (ϕ, Y_0) is allowed for all fluid particles entering the burner in A and B. That trajectory consists in a "travel" on the AM (resp. BM) line and a sudden "jump" at a given value of ϕ, that is ϕ_{ig}^r on the rich side or ϕ_{ig}^l on the lean side. Therefore, in the case of a single timescale τ, the mean reaction rate is:

$$\tilde{\mathring{w}}_0 = \int_{\phi_{ig}^r}^{\phi_{ig}^l} \mathring{w}_0 \cdot \tilde{P}_\phi(\phi) \cdot d\phi \qquad (A.8)$$

or:

$$\tilde{\mathring{w}}_0 = \int_{\phi_{ig}^{r+}}^{\phi_{ig}^{l-}} \mathring{w}_0 \cdot \tilde{P}_\phi(\phi) \cdot d\phi - \int_{\delta\phi_{ig}^r} \mathring{w}_0 \cdot \tilde{P}_\phi(\phi) \cdot d\phi + \int_{\delta\phi_{ig}^l} \mathring{w}_0 \cdot \tilde{P}_\phi(\phi) \cdot d\phi \qquad (A.9)$$

The contribution to the mean reaction rate due to the "jumps" from the mixing to the equilibrium lines is represented by the two last terms of formula (A.9).

Owing to the fact that the variations on $\delta\phi_{ig}^r$ and $\delta\phi_{ig}^l$ of $\mathring{w}_0$ and $\tilde{P}_\phi(\phi)$ with respect to ϕ can be neglected, one writes:

$$\int_{\delta\phi_{ig}^r} \mathring{w}_0 \cdot \tilde{P}_\phi(\phi) \cdot d\phi = \mathring{w}_{0_{ig}}^r \cdot \tilde{P}_\phi(\phi_{ig}^r) \cdot \delta\phi_{ig}^r \qquad (A.10)$$

and

$$\int_{\delta\phi_{ig}^l} \mathring{w}_0 \cdot \tilde{P}_\phi(\phi) \cdot d\phi = \mathring{w}_{0_{ig}}^l \cdot \tilde{P}_\phi(\phi_{ig}^l) \cdot \delta\phi_{ig}^l \qquad (A.11)$$

Where $\mathring{w}_{0_{ig}}^r$ and $\mathring{w}_{0_{ig}}^l$ are the instantaneous reaction rates respectively at $\phi = \phi_{ig}^r$ and $\phi = \phi_{ig}^l$; they are easily derived from formula (A.3):

$$\mathring{w}_{0_{ig}}^r = \frac{\tilde{\phi} - \phi_{ig}^r}{\tau} \left[\frac{\delta Y_{0_{ig}}^r}{\delta\phi_{ig}^r} - \frac{\tilde{Y}_0 - Y_{0_{ig}}^r}{\tilde{\phi} - \phi_{ig}^r} \right] \qquad (A.12)$$

and

$$\mathring{w}_{0_{ig}}^l = \frac{\tilde{\phi} - \phi_{ig}^l}{\tau} \left[\frac{\delta Y_{0_{ig}}^l}{\delta\phi_{ig}^l} - \frac{\tilde{Y}_0 - Y_{0_{ig}}^l}{\tilde{\phi} - \phi_{ig}^l} \right] \qquad (A.13)$$

Where $Y_{0_{ig}}^r$ and $Y_{0_{ig}}^l$ are the values of Y_0 respectively at $\phi = \phi_{ig}^r$ and $\phi = \phi_{ig}^l$, $\delta Y_{0_{ig}}^r$ and $\delta Y_{0_{ig}}^l$ being the amounts of oxidizer consumed by a "jump" on the rich or on the lean side.

Now, stating $\delta\phi^{r}_{ig} \to 0$ and $\delta\phi^{l}_{ig} \to 0$, one derives from (A.12) and (A.13) combined respectively with (A.10) and (A.11):

$$\int_{\delta\phi^{r}_{ig}} \mathring{w}_0 \cdot \tilde{P}_{\phi}(\phi) \cdot d\phi = \tilde{P}_{\phi}(\phi^{r}_{ig}) \cdot \frac{\tilde{\phi} - \phi^{r}_{ig}}{\tau} \cdot \delta Y^{r}_{0_{ig}} \tag{A.14}$$

and

$$\int_{\delta\phi^{l}_{ig}} \mathring{w}_0 \cdot \tilde{P}_{\phi}(\phi) \cdot d\phi = \tilde{P}_{\phi}(\phi^{l}_{ig}) \cdot \frac{\tilde{\phi} - \phi^{l}_{ig}}{\tau} \cdot \delta Y^{l}_{0_{ig}} \tag{A.15}$$

$\delta Y^{r}_{0_{ig}}$ and $\delta Y^{l}_{0_{ig}}$ are simple algebraïc functions of ϕ^{r}_{ig}, ϕ^{l}_{ig}, $\tilde{\phi}$, $\tilde{Y}_0$, ϕ_s and Y_{0_A}:

$$\delta Y^{r}_{0_{ig}} = \frac{\tilde{Y}_0}{\tilde{\phi}} \cdot \phi^{r}_{ig} \tag{A.16}$$

$$\delta Y^{l}_{0_{ig}} = \left[\frac{Y_0 - \tilde{Y}_0}{1 - \tilde{\phi}} - \frac{Y_{0_A}}{1 - \phi_s}\right] \phi^{l}_{ig} + \left[1 + \frac{\phi_s}{1 - \phi_s}\right] Y_{0_A} - \frac{1}{1 - \tilde{\phi}} [Y_{0_A} - \tilde{Y}_0] \tag{A.17}$$

Now, since in actual fact the turbulent timescale τ is not unique but obeys a distribution $P_{\tau}(\tau)$, the final expression of $\tilde{\mathring{w}}_0^{\text{jump}}$ is found by integrating over τ in (A.14) and (A.15) and adding the two terms:

$$\begin{aligned} \tilde{\mathring{w}}_0^{\text{jump}} = & - \int_{\tau_k}^{+\infty} \frac{\tilde{\phi} - \phi^{r}_{ig}(\tau)}{\tau} \cdot \delta Y^{r}_{0_{ig}} \cdot \tilde{P}_{\phi}(\phi^{r}_{ig}(\tau)) \cdot P_{\tau}(\tau) \cdot d\tau \\ & + \int_{\tau_k}^{+\infty} \frac{\tilde{\phi} - \phi^{l}_{ig}(\tau)}{\tau} \cdot \delta Y^{l}_{0_{ig}} \cdot \tilde{P}_{\phi}(\phi^{l}_{ig}(\tau)) \cdot P_{\tau}(\tau) \cdot d\tau \end{aligned} \tag{A.18}$$

References

Borghi, R. (1988): Turbulent combustion modelling. Prog. Energy Combust. Sci. **14**, 245–292

Borghi, R., Gonzalez, M. (1986): Applications of lagrangian models to turbulent combustion. Combust. Flame **63**, 239–250

Borghi, R., Moreau, P. (1977): Turbulent combustion in a premixed flow. Acta Astronaut. **4**, 321–341

Bray, K. N. C., Libby, P. A., Moss, J. B. (1984): Scalar length scale variations in premixed turbulent flames. 20th Symposium (International) on Combustion, Ann Arbor

Cant, R. S., Bray, K. N. C. (1988): Strained laminar flamelet calculations of premixed turbulent combustion in a closed vessel. 22nd Symposium (International) on Combustion, Seattle

Dopazo, C., O'Brien, E. E. (1974): A probabilistic approach to the autoignition of reactive turbulent mixtures. Acta Astronaut. **1**, 1239–1255

Duterque, J., Borghi, R., Tichtinski, H. (1981) Study of quasi-global schemes for hydrocarbon combustion. Combust. Sci. Technol. **26**, 1–15

Liew, S. K., Bray, K. N. C., Moss, J. B. (1981): A flamelet model of turbulent non premixed combustion. Combust. Sci. Technol. **27**, 69–73

Masri, A. R., Bilger, R. W., Dibble, R. W. (1988a): Turbulent non premixed flames of methane near extinction: probability density functions. Combust. Flame **73**, 261–285

Masri, A. R., Bilger, R. W., Dibble, R. W. (1988b): Turbulent non premixed flame of methane near extinction: mean strucťure from Raman measurements. Combust. Flame **71**, 245–266

Pope, S. B. (1985): Pdf methods for turbulent reactive flows. Prog. Energy Combust. Sci. **11**, 119–192

Villermaux, J. (1986): Micromixing phenomena in stirred reactors. *Encyclopedia of Fluid Mechanics*, 707–768. Gulf Pub. Co., West Orange, MJ

Direct Simulation of a Passive Diffusion Flame in Sheared and Unsheared Homogeneous Turbulence

S. Elghobashi and K. K. Nomura

Mechanical Engineering Department, University of California, Irvine, CA 92717, USA

Abstract

The paper presents the results of direct numerical simulation of two homogeneous shear flows of different initial strain number (*St*) and a decaying isotropic turbulence with an infinite-rate chemical reaction between two species but without heat release. The simulations were performed using a $(96)^3$ computational grid and an initial Taylor's Reynolds number, Re_λ, of 25. The results show that shear flattens the reaction surface due to the augmented strain rates resulting from vortex stretching. The time evolution of the scalar pdf shows that the higher the strain number is the higher the scalar mixing rate. The directions of the instantaneous scalar flux and the instantaneous scalar gradient are uncorrelated, thus indicating that turbulent scalar transport is a nongradient mechanism.

Introduction

This paper is concerned with the direct numerical simulation of a chemically reacting, homogeneous turbulent shear flow. The rate of chemical reaction is infinitely fast so that the reaction is mixing-controlled. This type of reaction is predominant in most practical combustion problems where the fuel and oxidant are introduced separately into the combustion chamber to produce a turbulent "diffusion" flame. The assumption of instantaneous reaction is not far from reality in this case, because the time of recombination of the non-equilibrium radicals is usually orders of magnitude smaller than the smallest time-scale of turbulence. The Damköhler number, defined as the ratio of the characteristic flow time-scale to the reaction time-scale, is thus much greater than 1.

We further idealize the flow such that heat release due to chemical reaction is negligible and the fluid density remains constant, hence *passive diffusion flame*. Both fuel and oxidant have equal Schmidt numbers, and the binary chemical reaction proceeds to completion in a single step with a fixed stoichiometric ratio to form products. These assumptions allow a linear combination of the equations describing the conservation of the two nonpremixed reactants (fuel and oxidant) to yield an equation for a conserved scalar, denoted here by the mixture fraction F (Burke and Schumann (1928); Zeldovich (1949); Elghobashi (1977)).

The objective of this work is to examine the effects of shear on the turbulent mixing of scalars and the structure of the reaction surface. Homogeneous shear flow

Turbulent Shear Flows 7

is selected for this study because it is the simplest flow in which to study the interactions between the mean flow, turbulence, and chemical reaction.

Mathematical Description

Flow Description and Computational Domain

Figure 1 shows the mean velocity profile $U(z)$ with a uniform mean-gradient in the z-direction and the initial step profile of the mixture fraction, $F(z,t_0)$. Our computational domain is a finite cube with sides of length L, containing $(96)^3$ grid points. Initially, fuel is admitted into the upper half of the domain and oxidant into the lower half. The chemical reaction considered in this study is an isothermal reaction between two species, nitric oxide and ozone according to $NO + O_3 \rightarrow NO_2 + O_2$, diluted in an inert gas, N_2. The initial mass fraction of each of the reactants is taken to be 0.01 such that the density of the fluid remains essentially invariant throughout the flow. The Schmidt number for both reactants is assumed to be 0.7. The stoichiometric value of F, denoted by F_{st}, equals 0.385.

The assumption of infinite reaction rate implies that if the chemical reaction results in a flame it will be confined to a surface at which $F = F_{st}$ at any instant of time. On one side of the flame surface the fuel concentration vanishes, and on the other the oxidant concentration vanishes, whereas along the surface itself both vanish.

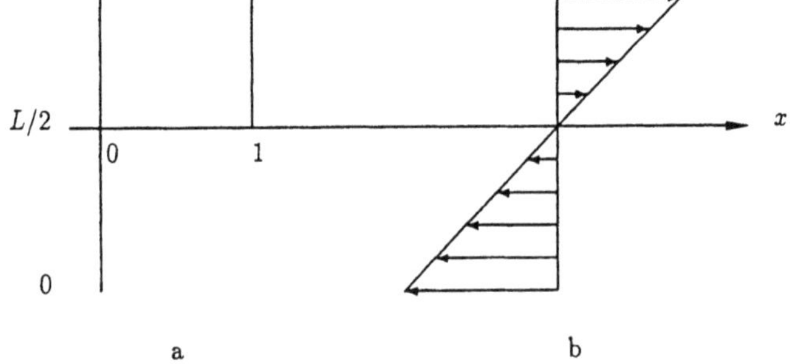

Fig. 1 a, b. Flow Representation: **a** Initial profile of the mixture fraction $F(z, t_0)$, **b** mean Velocity $U(z)$

Governing Equations

The governing equations are the instantaneous, three-dimensional Navier–Stokes, mixture fraction and continuity equations which can be written in dimensionless form as:

$$\frac{\partial u_i}{\partial t} + \frac{\partial}{\partial x_j}(u_j u_i) + Sx_3 \frac{\partial u_i}{\partial x_1} + Su_3 \delta_{i1} = \frac{1}{Re} \frac{\partial^2 u_i}{\partial x_j^2} - \frac{\partial p}{\partial x_i} \tag{1}$$

$$\frac{\partial f}{\partial t} + \frac{\partial}{\partial x_j}(u_j f) + Sx_3 \frac{\partial f}{\partial x_1} + su_3 = \frac{1}{ReSc} \frac{\partial^2 f}{\partial x_j^2} \tag{2}$$

$$\frac{\partial u_j}{\partial x_j} = 0 \tag{3}$$

In the above equations, u_i, f, and p are the deviations of the instantaneous velocity, mixture fraction, and pressure from their respective reference profiles, for example:

$$f(x, y, z, t) = F(x, y, z, t) - F_{ref}(z) \tag{4}$$

where $F_{ref}(z) = z \cdot (dF/dz)_{ref}$ and $(dF/dz)_{ref}$ is a fixed reference gradient for the mixture fraction. The nondimensional shear S and the nondimensional mixture fraction gradients s are defined as $S = (L/\Delta U)(dU/dz)_{ref}$, and $s = (L/\Delta F)(dF/dz)_{ref}$, where ΔU and ΔF are the differences in the mean velocity and mixture fraction at the top and bottom of the domain. The characteristic quantities used in nondimensionalizing Eqs. (1)–(3) are L, ΔU, ΔF and a reference density ρ_0. The dimensionless time, t, used in the presentation of the results is therefore defined as $t = (\tau \Delta U)/L$, where τ is the real time.

Initial and Boundary Conditions

The initial three-dimensional energy spectrum is prescribed by

$$E(k, 0) = (3v_0^2/4\pi)(k/k_p^2)e^{-(k/k_p)}, \tag{5}$$

where k_p is the wave number of the spectrum peak, and v_0 is the initial rms velocity ($v_0^2 = \langle u_i u_i \rangle_0 / 3$). The initial value of k_p is set equal to 6 and the initial Re_λ equals 25. The shape of the spectrum $E(k, 0)$ and the value of k_p determine the initial integral length scale l_0 according to:

$$l(t) = \frac{1}{2v^2(t)} \int \frac{E(k, t)}{k} dk \tag{6}$$

The value of v_0 is chosen to prescribe the initial strain number, St, defined as:

$$St = \frac{v_0}{S l_0} \tag{7}$$

which is the ratio of the strain rate of the energy-containing eddies to that of the mean flow. The spectrum of the dissipation rate of turbulence energy is calculated from

$$\varepsilon(k,t) = 8\pi^2 \nu k^2 E(k,t). \tag{8}$$

The computed initial dissipation spectrum $\varepsilon(k,0)$ has a peak at a wave number $k_d = 19$. The initial integral and Kolmogorov length-scales, normalized by L, equal respectively 0.0495 and 0.0024. The latter is about one quarter of the grid spacing indicating that the motions at the Kolmogorov scale are not being resolved in the present computations.

It should be mentioned that the step $F(z,t_0)$ is imposed only after the initial turbulence field has fully developed. This is measured by the value of the skewness of the velocity derivative. The skewness is initially zero and takes approximately 1 unit of dimensionless time to reach a value of 0.2 for the shear case and 0.5 for the isotropic case, at which we introduce the step profile $F(z,t_0)$ (Elghobashi et al. (1987)).

The boundary conditions for the dependent variables u_i, f, and p, are periodic in the x and y directions and shear-periodic in the z direction. Complete details of the numerical solution procedure are given by Gerz et al. (1989). It suffices here to state that the governing equations are discretized in an Eulerian framework using a second-order finite difference method on a staggered grid for all the terms except those corresponding to the mean advection (the third term in Eqs. (1) and (2)) where pseudospectral approximation is used. Integration in time is performed via the Adams–Bashforth scheme. Pressure is treated implicitly and obtained using a Poisson solver which includes the shear-periodic boundary conditions.

Results

Flow Structure

Results are presented for three flow conditions. Two homogeneous shear flows are considered, differing only in the initial value of the strain number, St. In one case, $St = 0.33$, referred to hereinafter as case L (low strain), and in the other, $St = 1$, referred to as case H (high strain). The third flow condition is that of decaying grid-turbulence (case I (isotropic) with the same initial conditions as case H.

Figure 2 shows the time development of the turbulence energy and its dissipation rate for cases L and H. The former case, with $St = 0.33$, has a larger eddy turnover time (ETT $= l/v$), i.e. the eddies take longer time before they break down into smaller ones, and thus its turbulence energy increases at a higher rate than the latter ($St = 1.$).

Figure 3 shows the time variation of the three length-scales normalized by their initial values l(integral), λ(Taylor), where $\lambda = [15\, \nu v^2/\varepsilon]^{1/2}$, and η (Kolmogorov), where $\eta = [\nu^3/\varepsilon]^{1/4}$, for the two cases. As shown in Fig. 2, case L experiences a continuous increase in the energy and dissipation, with the result of a slight decrease of η, and a slight increase of l and λ. In case H, all three scales initially increase up to time $t = 4$ when they begin to level off. In grid-turbulence (case I) all length-scales

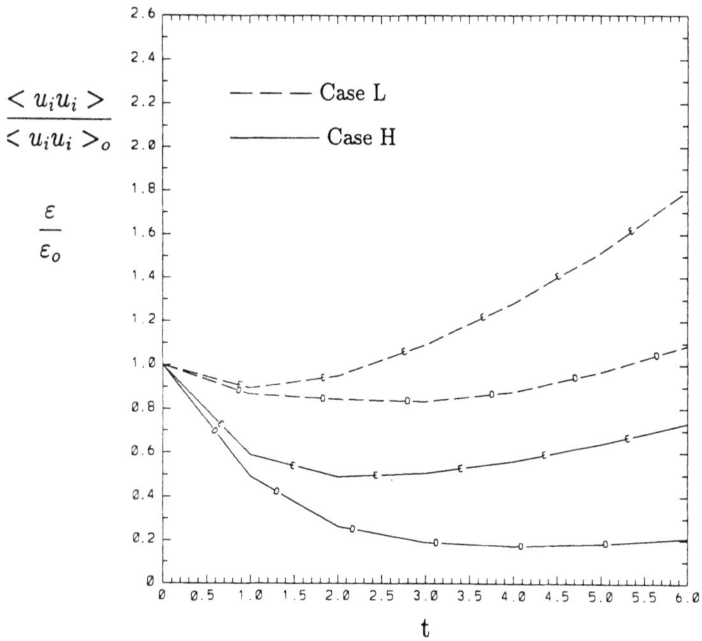

Fig. 2. Time development of normalized turbulent kinetic energy (-E-) and dissipation rate (-D-) for Case L (St = 0.33), and Case H (St = 1.0)

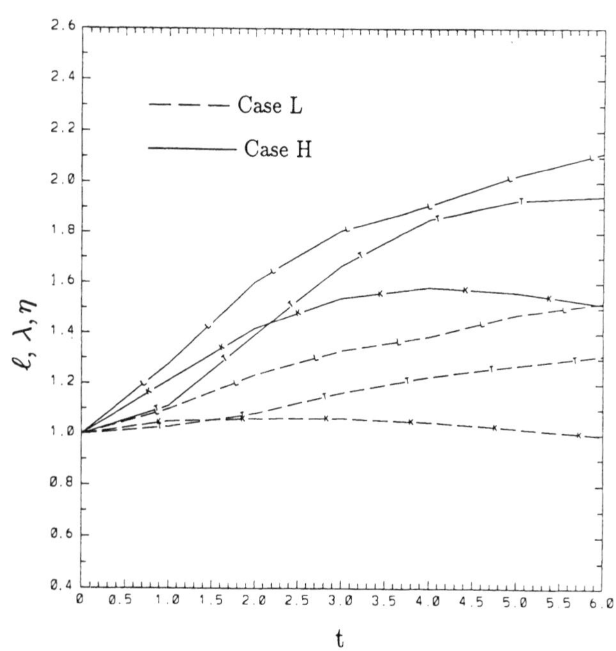

Fig. 3. Time development of normalized integral length-scale (-*L*-), Taylor micro-scale (-*T*-), and Kolmogorov length scale (-*K*-) for Case L (St = 0.33), and Case H (St = 1.0)

increase monotonically. The increase of l in both the shear and grid-turbulence cases is due to different reasons (Rohr et al. (1988)) which can be explained via Eq. (6). In the shear flow the rate of increase of energy $E(k, t)$ at low wave numbers, where the energy is transferred from the mean motion to the large-scale turbulence, is higher than the rate of increase of $v^2(t)$, since $v^2(t)$ is a measure of the energy content at all

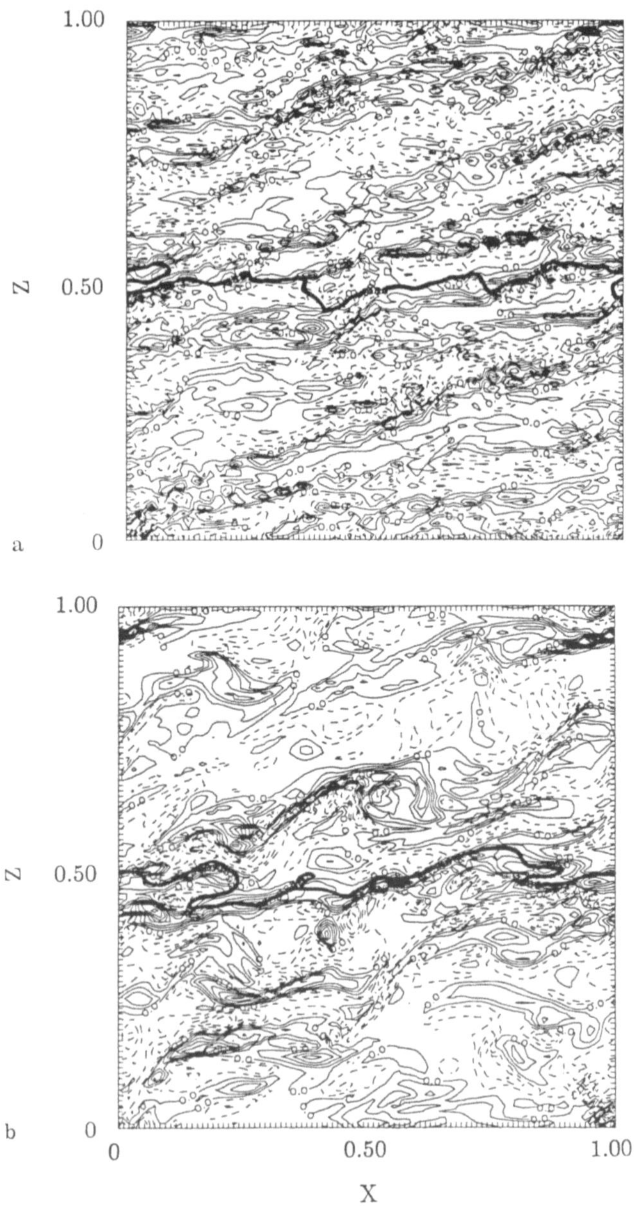

Fig. 4 a–c. Instantaneous contours of fluctuating vorticity (w_y) in the x–z plane with isoscalar contour line $F = F_{st}$ at time $t = 6$ for **a** Case L (St = 0.33), **b** Case H (St = 1.0) **c** Case I (isotropic)

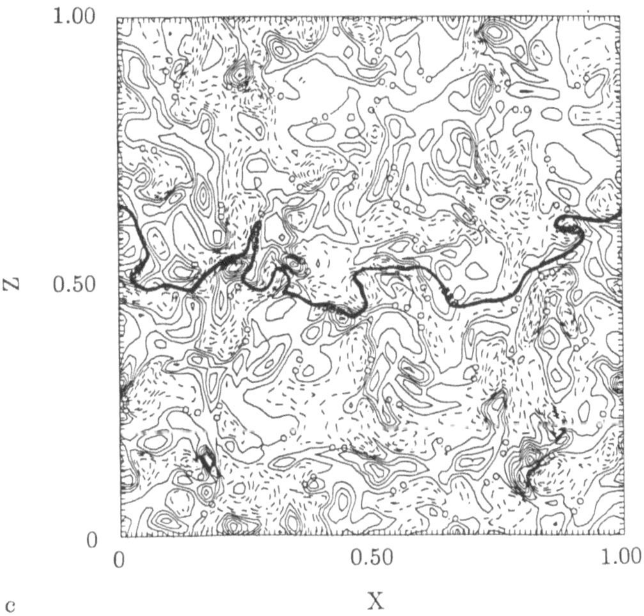

Fig. 4. (*Continued*)

scales. In grid turbulence the energy decays, but the decay of $E(k, t)$ at low wave numbers is less than that of $v^2(t)$ as indicated in Fig. 8 by the monotonic growth of the eddy turnover time ($= l/v$) in case I only. This indicates that the absolute size of the largest eddies grows only in shear flows, whereas the *apparent* growth of l in the decaying turbulence is due to the faster disappearance of the smaller scales.

The turbulence structure in the three cases can be compared by examining the instantaneous vorticity fields. Figure 4a–c shows contours of the instantaneous vorticity in the y-direction, $w_y = w_2 = u_{1,3} - u_{3,1}$, in a vertical ($x$–$z$) plane in the middle of the cube for cases L, H and I at time $t = 6$. This time is about six initial eddy turnover times for cases I and H and two turnover times for case L. The solid lines denote positive vorticity (counterclockwise rotation), and the dashed lines represent negative vorticity (clockwise rotation). The contours labeled by zero represent surfaces between counterrotating vortices. Superimposed on these contours is the isoscalar surface $F = F_{st}$, or the flame surface. The integral length-scale, which can be taken roughly as the size of a bundle of vorticity contours of the same sign, is shown to be smallest for case L (Fig. 4a) as explained earlier, largest for case H (Fig. 4b), and slightly less than the latter for the isotropic flow (Fig. 4c). It is also seen that the vortex structures have no particular orientation for the latter case (I), whereas they are inclined between 15° and 20° in the two shear cases. Following the behavior of these structures in time (not shown here) indicates that this inclination angle decreases monotonically from 45° at $t = 1$, due to the rotation of the principal axes of the strain tensor. Nomura and Elghobashi (1990) studied, for the present flows, the statistical distribution of the angle between the instantaneous local vorticity vector and each of the principal strain rate axes. They indicated that the vorticity tends to be aligned with the axis of the intermediate strain which is mostly extensional, and that in shear flow the magnitude of vorticity tends to be larger when there is strong alignment in agreement with Ashurst et al. (1987). This implies that vorticity is augmented due to stretching along the axis of the

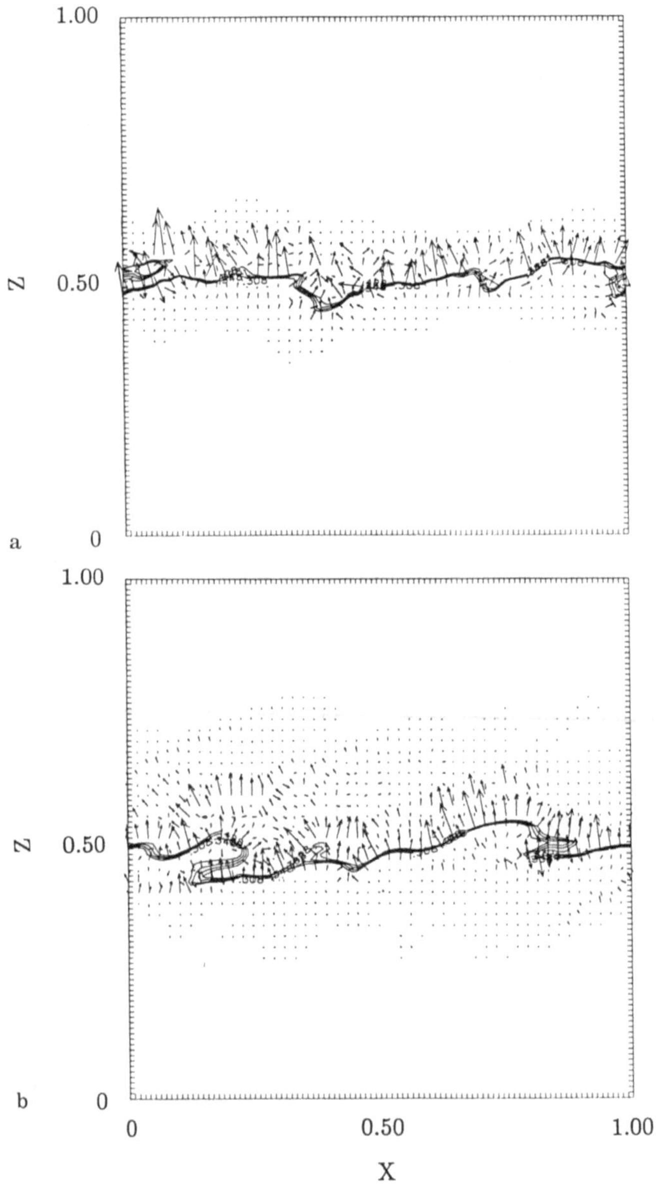

Fig. 5 a–c. Instantaneous scalar gradient vector (∇F) in x–z plane with isoscalar contour lines $F_{st} \pm 0.2F_{st}$ at time $t = 6$ for **a** Case L (St = 0.33), **b** Case H (St = 1.0), **c** Case I (isotropic)

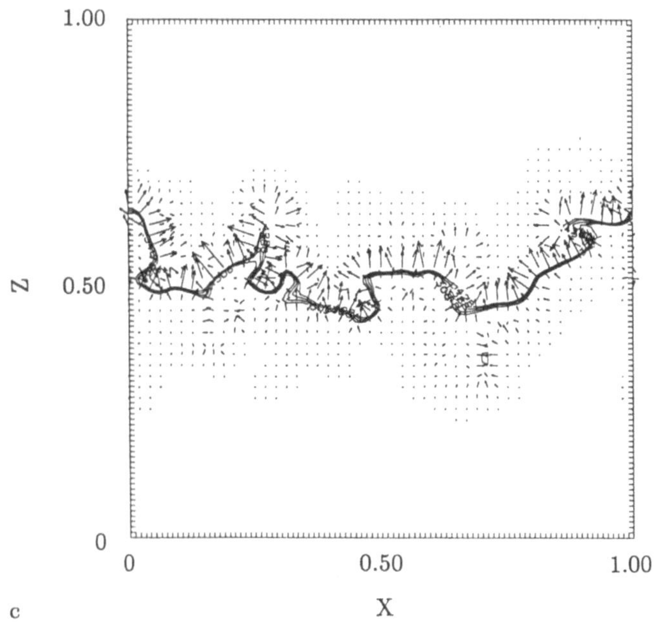

c

Fig. 5. (*Continued*)

intermediate strain. This in turn increases the strain rates (compressive and extensional) in the plane normal to the axis of the vortex tube, and hence increases the rate of scalar mixing due to the reduction of the diffusive scalar length-scale, $L_B = (\mathscr{D}/\gamma)^{1/2}$, where $\mathscr{D}$ is the molecular diffusivity of the scalar, and γ is a measure of the characteristic turbulence strain rates.

Flame Surface Structure and Scalar Transport

It should be emphasized here that we are considering an isothermal reaction for which a flame identified by a temperature peak does not occur. A more accurate name would be "stoichiometric isoscalar" but we will use the term "flame" for convenience. As discussed earlier, the instantaneous flame surface for infinite rate reactions coincide with the isoscalar surface F_{st}. Figure (5a–c) shows four surfaces covering the range $F_{st} \pm 0.2\, F_{st}$, projected onto a vertical (x–z) plane in the middle of the computational domain, for cases L, H, and I, respectively, at $t = 6$. Superimposed on each figure is the distribution of the local instantaneous scalar gradient vector (∇F) projected onto the x–z plane. As expected, the magnitudes of ∇F are greatest near the interface ($F = 0.5$) and therefore we see that the flame surface ($F_{st} = 0.385$) exists near peak values of ∇F. The magnitude of the maximum ∇F is nearly the same for cases H and I, whereas in case L, with smaller integral length-scales, the peak magnitude is slightly larger and occurs more frequently than in the other two cases. It is interesting to note that in zones with strong curvature the contours spread apart in a fan-like shape indicating smaller values of ∇F (shorter vectors).

Ashurst et al. (1987), in their numerical simulation of isotropic and homogeneous shear flows, showed that there is an increased probability for the scalar gradient to align itself in the direction of the most compressive strain, and the

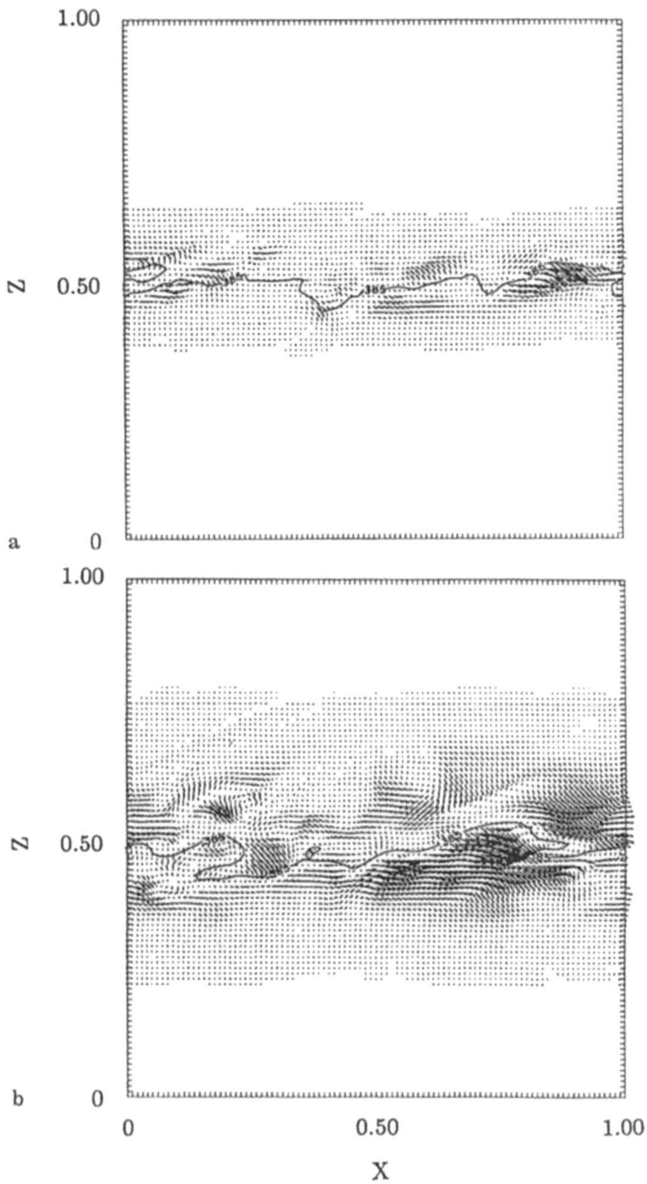

Fig. 6 a–c. Instantaneous turbulent scalar flux projected onto the center x–z plane with isoscalar contour line $F = F_{st}$ at time $t = 6$ for **a** Case L (St = 0.33), **b** Case H (St = 1.0), **c** Case I (isotropic)

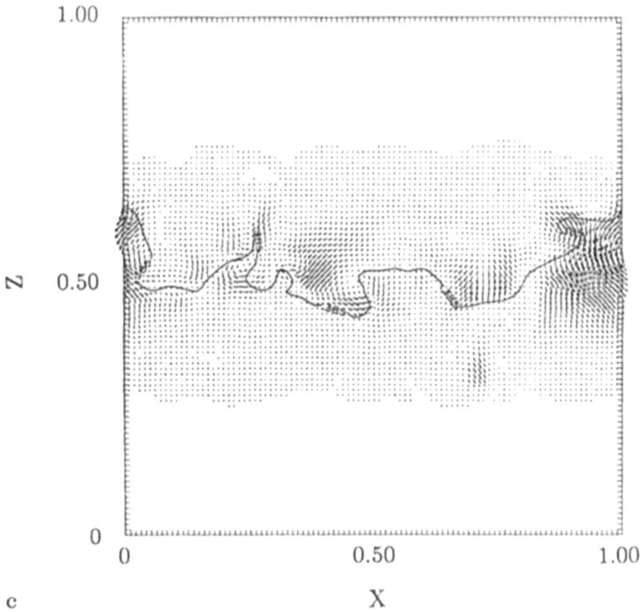

Fig. 6. (*Continued*)

average gradient is larger when it is pointing in that direction. Nomura and Elghobashi (1990) find the same alignment characteristics in the present homogeneous shear and decaying isotropic flows. Since the normal to the flame sheet coincides with the local scalar gradient vector, ∇F, the flame sheet will then tend to lie in the plane of the extensional and intermediate strain. As stated earlier, the intermediate strain is found to be predominantly extensional, and thus the flame sheet is primarily being stretched. The increasing strain rates in the shear flows cause greater rates of increase in surface area and in addition, greater scalar flux towards the flame sheet, thus enhancing the overall mixing in these cases.

Since ∇F tends to align normal to the intermediate strain which acts nearly parallel to the vorticity vector, it is expected that the flame surface should align itself with the vortex structures whose stretching is the primary mechanism of turbulent mixing. This is depicted in Fig. 4 which shows the superimposed flame surfaces (thickened for clarity) following the inclination of the sheared structures in cases L and H, and random structures in case I. It is evident that the characteristics of the flame surface in the shear flows are different from those in the isotropic case. In cases L and H, the large-scale distortions in the surface are flattened, whereas in the isotropic case, I, they are deeper and randomly oriented. Due to the presence of the mean shear, the most probable direction of the most extensive strain is initially at an angle of 45 degrees from the positive x-axis which then reduces with time to about 20 degrees (Nomura and Elghobashi 1990). The resulting effects on the flame are significant due to the amplification of strains by vortex stretching. The isoscalar surfaces tend to be stretched out by the extensional strains. The most compressive strains, acting normal to the extensional, flattens the flame surface folds which would otherwise be quite prominent as seen in the isotropic case.

In order to examine the interdependence of the instantaneous scalar flux on the scalar gradient we present in Fig. 6a–c the projection of the scalar flux on a vertical (x–z) plane, for cases L, H and I respectively, at $t = 6$. Thus each vector has a

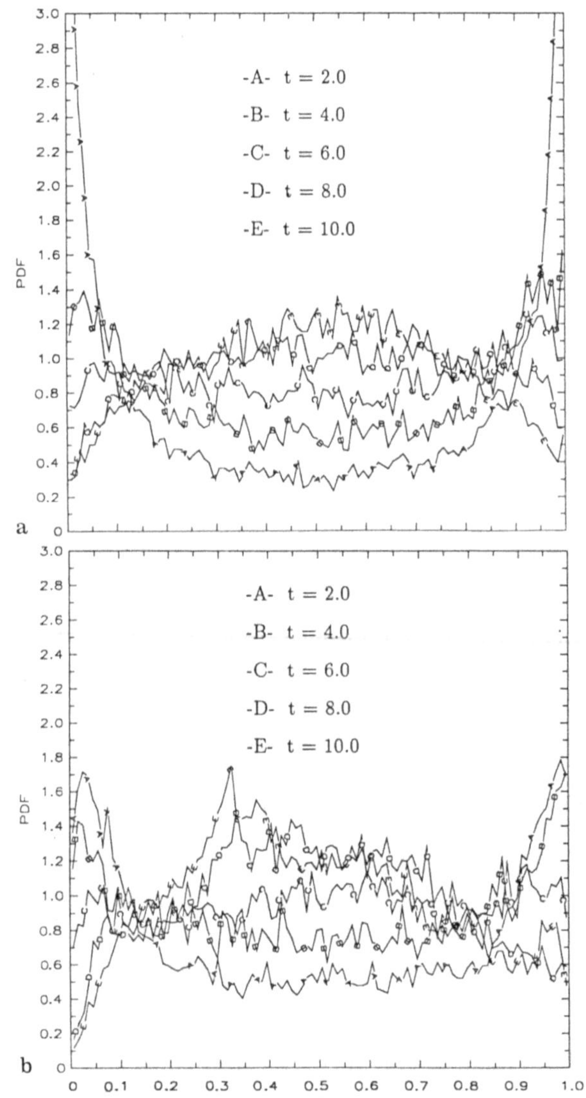

Fig. 7 a–c. Time development of mixture fraction (F) pdf for **a** Case L (St = 0.33), **b** Case H (St = 1.0), **c** Case I (isotropic)

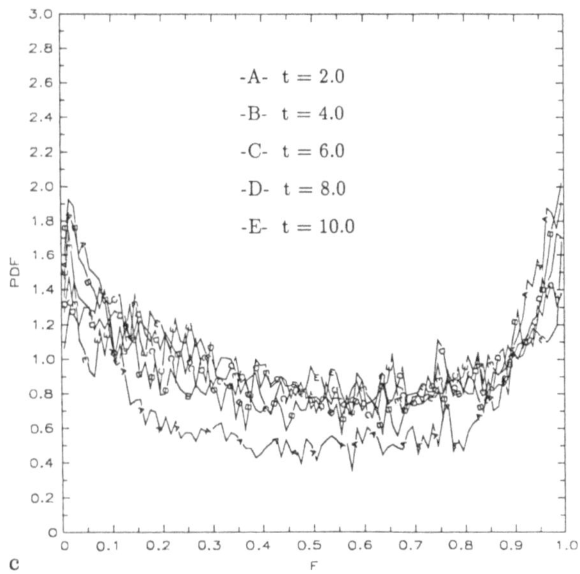

c

Fig. 7. (*Continued*)

magnitude equal to $[(uf')^2 + (wf')^2]^{1/2}$ and a slope equal to $[(wf')/(uf')]$. The stoichiometric isoscalar surfaces in Fig. 6a–c are those shown in Fig. 5a–c respectively. It is quite evident from comparing the two sets of Figs. (5a with 6a, etc.) that the direction of the instantaneous scalar flux is independent of the direction of the instantaneous scalar gradient (∇F). More specifically, there is no indication of either gradient or counter-gradient transport. The instantaneous scalar transport is a non-gradient process dominated mainly by the vorticity dynamics and the associated strain field. It is seen that case H, Fig. 6b, with the highest strain number ($St = 1.$) has the largest flux magnitudes, followed by case L and then case I.

Pdf of Mixture Fraction

The effect of the different strain numbers (St) on the mixing and reaction extent can also be seen by examining the probability density functions (pdf) of the mixture fraction. Figure 7 shows the time development of the pdf for the three cases within a small horizontal volume centered about the position of the initial interface between the fuel and oxidant. The initial two delta function distributions (at $t = 1$) are not shown. It is seen that the pdf for case H is the furthest developed at $t = 10$, whereas case I (isotropic) still retains some of its initial peaks at $F = 0$ and $F = 1$ indicating incomplete mixing. Complete mixing would result in a Gaussian pdf. Case L shows much better mixing than case I, but less extensive than case H.

The correspondence between the evolution of the pdf and that of the eddy turnover times can be seen in Fig. 8 which displays the time variation of the ETT for

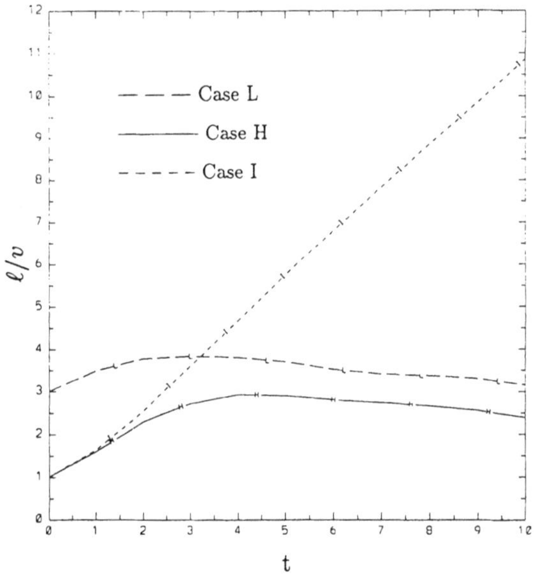

Fig. 8. Time development of eddy turnover time (ETT) for Case L (St = 0.33), Case H (St = 1.0), and Case *I* (isotropic)

the three cases. The ETT for the isotropic case increases monotonically as a result of growing length-scale and decaying energy. For the two shear cases there is an initial increase up to $t = 2.5$, followed by a leveling off. The nearly steady value of ETT for case H is less than that of case L. At $t \simeq 3$, the pdf's of the three cases are very similar at that time, a nearly flat distribution between $F = 0.1$ and 0.9. After that time the pdf of the isotropic case (Fig. 7c) stays nearly invariant, whereas the pdf for the other cases continue to develop indicating further mixing. At $t = 10$, case H with lower

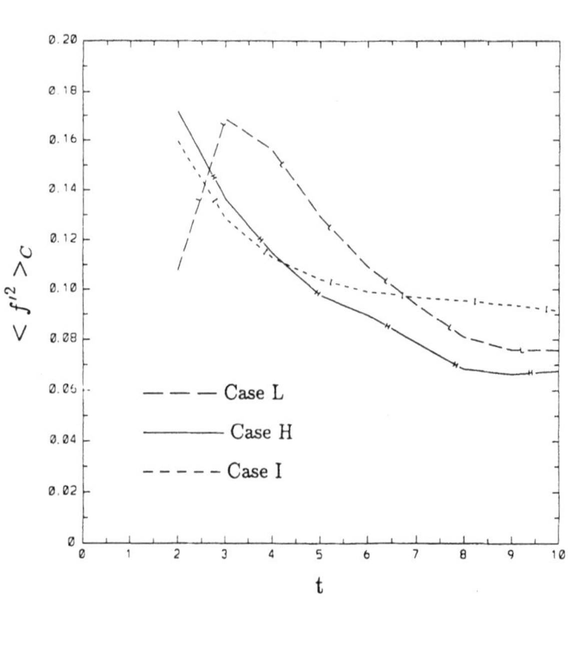

Fig. 9. Time development of scalar variance, $\langle f'^2 \rangle_C$ for Case *L* (St = 0.33), Case H (St = 1.0), and Case I (isotropic)

ETT shows a more developed pdf than that of case L. It is worth noting that for $t \leqq 3$, case L has a larger ETT than case I, thus resulting in higher mixing rate for the latter during the initial period.

The difference between the mixing rates of the three cases discussed above can also be seen in Fig. 9 which displays the time development of the scalar variance along the initial interface plane, $\langle f'^2 \rangle_C$. The instantaneous value f' is defined as:

$$f'(x, y, z, t) = F(x, y, z, t) - \langle F(x, y, z, t) \rangle_{x,y} \tag{9}$$

At $t = 10$, case H has the lowest value of $\langle f'^2 \rangle_C$, whereas case I has the highest, indicating the higher mixing rate in the shear case. Case L, with larger ETT, shows higher $\langle f'^2 \rangle_C$ than in H, indicating that the higher the strain number is the higher the mixing rate.

Velocity and Scalar Time-Scales

The time development of the velocity time-scale, $t_u = \langle u_i u_i \rangle / 2\varepsilon$, and the scalar time-scale, $t_f = [\langle f'^2 \rangle / \varepsilon_{f'}]_C$, is shown in Fig. 10 for cases I and H. The scalar dissipation rate, $\varepsilon_{f'}$, is defined as:

$$\varepsilon_{f'} = 2\mathscr{D} \langle f'_{,j} f'_{,j} \rangle. \tag{10}$$

The variation of t_u for the two cases is, as expected, similar to that of ETT in Fig. 8, i.e. t_u increases monotonically for case I and levels off for the shear case H at $t > 4$. However, t_f displays nearly the opposite behavior for cases I and H, i.e. it levels off, after an initial increase, for case I, whereas it shows an increase for the shear case at $t \geqq 8$. This increase in t_f can be explained by the fact that the higher strain rates in case

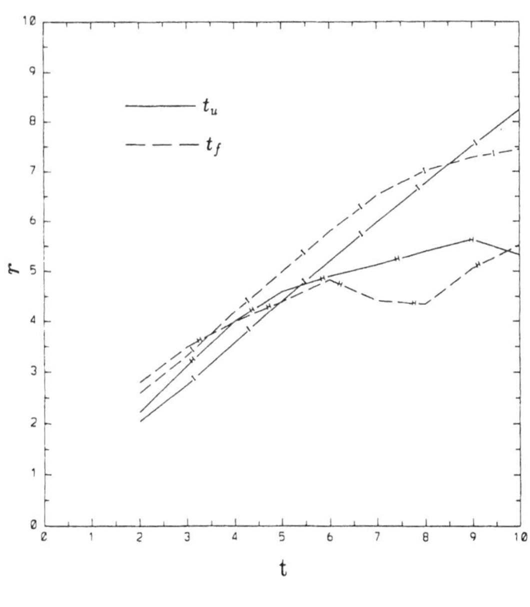

Fig. 10. Time development of the velocity time-scale, t_u, and scalar time-scale, t_f, for Case H (St = 1.0), and Case I (isotropic)

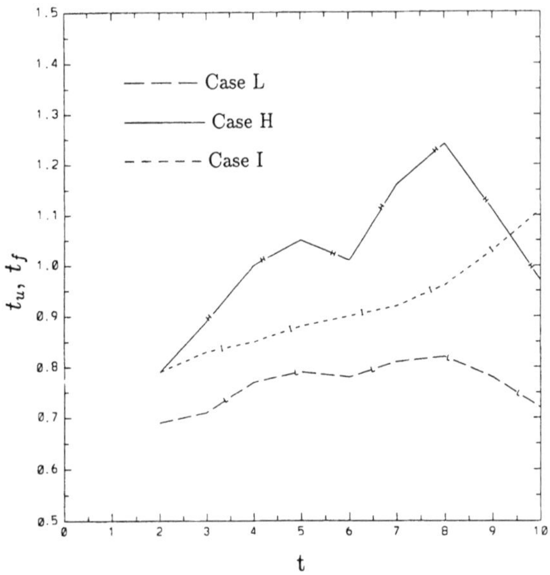

Fig. 11. Time development of the timescale ratio $r = t_u/t_f$ for Case L (St = 0.33), Case H (St = 1.0), and Case I (isotropic)

H eventually cause $\langle f'_{,j} f'_{,j} \rangle$ to diminish at a greater rate than that of the decaying amplitude f'. The levelling off of t_f in the decaying turbulence, case I, is due to the monotonic growth of the scalar length-scale ($L_B = \eta/Sc^{1/2}$). As a result, the decay rates of both the scalar fluctuation amplitude f' and the dissipation rate $\varepsilon_{f'}$ become limited *only* by the ability of molecular diffusion to reduce the scalar gradients.

The relative development of the scalar and velocity time-scales can be studied by examining the time variation of the time-scale ratio, $r = t_u/t_f$, shown in Fig. 11. Cases I and H have the same initial t_u and t_f. Case L has initially higher t_u and t_f values than case H. At $t \geqq 8$, both shear cases show a decaying r, mainly due to the increase of t_f as discussed earlier. The isotropic case exhibits an increasing r due to the monotonic increase of t_u and the levelling off of t_f as discussed above.

Concluding Remarks

The paper presents the results of direct numerical simulations on a $(96)^3$ grid and initial $Re_\lambda = 25$ of three homogeneous turbulent flows with an infinite rate chemical reaction but without heat release. One flow is that of decaying grid-generated turbulence. The other two are of homogeneous shear with two different initial strain numbers ($St = 0.33$ and 1.0).

The effects of shear on scalar mixing and structural features of the infinitely thin reaction-surface are studied by examining the instantaneous distributions of the conserved scalar, its gradient, variance, pdf and flux. The results show that:

- Shear flattens the reaction surface due to the augmented extensional and compressive strain rates.

- The higher the strain number is the higher the scalar mixing rate as evident from the time evolution of the scalar pdf (Fig. 7). It is also found that the time variation of the eddy turnover time is consistent with the development of the pdf of the mixture fraction.
- The directions of the instantaneous scalar flux and the instantaneous scalar gradient are uncorrelated, thus indicating that turbulent scalar transport is a nongradient mechanism.

Acknowledgements: The computations were performed on the Cray-2 of NAS at NASA Ames Research Center and on the Cray X-MP at the San Diego Supercomputer Center.

References

Ashurst, W. T., Kerstein, A. R., Kerr, R. M., Gibson, C. H. (1987): Alignment of vorticity and scalar gradient with strain rate in simulated Navier–Stokes turbulence. Phys. Fluids **30**, 2343–2353

Burke, S. P., Schumann, T. E. W. (1928): Diffusion flames. Ind. Eng. Chem. **20**, 998–1004

Elghobashi, S. E. (1977): Studies in the prediction of turbulent diffusion flames. Studies in Convection, (B. E. Launder ed.), **2**, 141–187

Elghobashi, S. E., Gerz, T., Schumann, U. (1987): Direct simulation of the initial development and the homogeneous limit of the thermal mixing layer. Sixth Symposium on Turbulent Shear Flows 511–516, Toulouse, France

Gerz, T., Schumann, U., Elghobashi, S. E. (1989): Direct numercial simulation of stratified homogeneous turbulent shear flows. J. Fluid Mech. **200**, 563–594

Nomura, K. K., Elghobashi, S. E. (1990): Direct simulation of an isothermal non-premixed flame in homogeneous turbulent shear flow. Paper AIAA 90-148, 28th Aerospace Sciences Meeting

Rohr, J. J., Itsweire, E. C., Helland, K. N., Van Atta, C. W. (1988): An investigation of the growth of turbulence in a uniform-mean-shear flow. J. Fluid Mech. **187**, 1–33

Zeldovich, Y. B. (1949): On the theory of combustion in initially unmixed gases. Zhur. Tekhn. Fiz. **19**, 1950. English Translation, N.A.C.A. Tech. Mem. No. 1296

Large Scale Structures in Reacting Mixing Layers

C. Lee, R. W. Metcalfe, and F. Hussain

Department of Mechanical Engineering, University of Houston, Houston, TX 77204-4792, USA

Abstract

The role of large-scale coherent structures in the enhancement of mixing and chemical reaction in turbulent free shear flows has been investigated by analyzing direct numerical simulations of temporally growing reacting mixing layers. The streamwise vortical modes are found to enhance the reaction rate by convoluting the reaction surface and enhancing the mixing of the two species, although the relative significance of this effect is very sensitive to the species diffusivity. For certain Schmidt numbers, the effects of flame shortening are balanced by flame sheet stretching by the vortex pairing process. The behavior of the helicity density, dissipation, and enstrophy production in the braid region has been examined. There does not appear to be a simple correlation between the amplitude of the helicity density and the position of some very strong coherent structures in this flow.

Introduction

Our present understanding of the role of turbulence in chemical reactions is severely limited. It has been argued frequently that large scale structures could be expected to play a significant role in turbulence production and transport processes (Hussain 1984), and this has been born out by recent numerical simulations (Riley et al. 1986). In order to develop better insight into these complex processes, we have analyzed the results of direct numerical simulations using education algorithms to characterize the large scale structures in the flow. This analysis suggests some of the mechanisms by which turbulent large-scale coherent structures enhance mixing and promote chemical reaction.

Methodology

A detailed direct numerical simulation of highly exothermic turbulent combustion would be prohibitively expensive even with present supercomputers. Therefore, the simulations considered here involve a number of simplifying approximations which permit the use of direct simulation techniques, but which capture some essential aspects of the physics. We are studying a binary, single-step, irreversible isothermal chemical reaction in an incompressible flow. This implies (a) no temperature effects

Turbulent Shear Flows 7

on the reaction rates and (b) no effects of the reactions on the flow field (cf. Riley and Metcalfe 1985, for a detailed discussion of these approximations). Using periodic boundary conditions in the streamwise direction, the simulation is of a temporally developing rather than spatially developing mixing layer. This is also an important simplification, since in order to treat the spatially growing layer numerically, inflow and outflow boundary conditions are required, which are very difficult to specify in a rigorous sense (Givi 1988; Buell and Huerre 1988). Additionally, in order to include a sufficiently long length of the flow so that the evolutionary dynamics can be captured and examined, an equivalent spatial simulation will require significantly more computational resources.

The governing equations of the flow are the incompressible Navier–Stokes equations, while those for reactant concentrations C_1 and C_2 satisfy the diffusion-reaction equations.

$$\frac{\partial}{\partial t}C_1 + \vec{\mathbf{u}}\cdot\nabla C_1 = -\alpha C_1 C_2 + \kappa_1 \nabla^2 C_1 \tag{3}$$

$$\frac{\partial}{\partial t}C_2 + \vec{\mathbf{u}}\cdot\nabla C_2 = -n\alpha C_1 C_2 + \kappa_2 \nabla^2 C_2. \tag{4}$$

Here the stoichiometric coefficient n, the reaction rate coefficient α, and the diffusivities κ_1 and κ_2 are all constants. Fourier series expansions are used for the dependent variables.

The initial reactant concentrations are defined by smooth functions to represent initially segregated species. The initially laminar mean velocity profile is defined as

$$U(z) = \frac{\Delta U}{2}\tanh\left(\frac{z}{\delta_i}\right)$$

where ΔU is the velocity difference across the layer and δ_i is the initial mean vorticity thickness (Fig. 1). The perturbation velocity is initialized with a component consisting of the linearly most unstable two-dimensional mode and its subharmonic (Michalke 1964) plus very low level three-dimensional random noise. By employing a passive Shvab-Zeldovich conserved scalar variable (Williams 1986), infinite reaction rate simulations can also be performed (see Riley and Metcalfe 1985 or Riley et al. 1986 for more details on this approach).

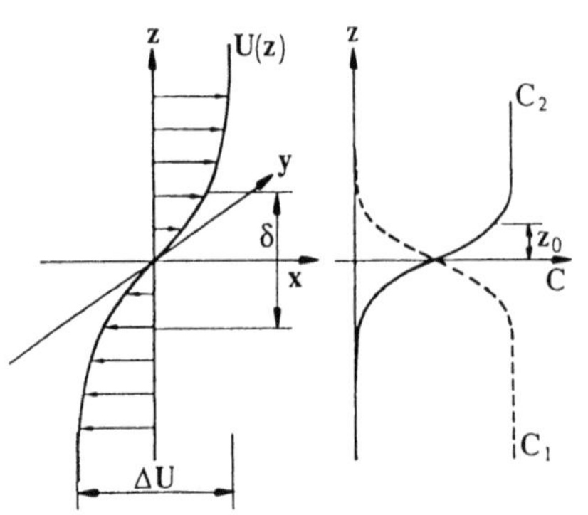

Fig. 1. Problem geometry and initial conditions

The velocity is nondimensionalized by ΔU, the mean velocity difference across the layer; time is nondimensionalized by $\lambda_f/\Delta U$, where λ_f is wavelength of the most unstable mode of the initial mean velocity profile; species concentrations are normalized by $C_\infty = C_{1\infty}$, the free-stream value of C_1. Important nondimensional numbers for our problems are the Reynolds number, $Re = \Delta U \delta/\nu$, where δ is the maximum slope thickness of initial base flow profile (Corcos and Sherman 1984), the Schmidt number, $Sc = \nu/\kappa$; and the first and second Damkohler numbers, $D_I = z_0 \alpha C_\infty/\Delta U$ and $D_{II} = z_0^2 \alpha C_\infty/\kappa$. D_I describes the ratio of convection time scale to the reaction time scale and D_{II} represents the ratio of the diffusion time scale to the reaction time scale where z_0 is initial concentration field thickness. Pseudospectral numerical methods are used to solve the equations (Canuto et al. 1988).

Numerical Results

The simulations reported here assume the stoichiometric coefficient to be 1 and the diffusivities of both the reactants and product to be the same. Table 1 contains a summary of the relevant simulation parameters. All cases were simulated with the same initial conditions for both the flow field and the concentration field. The computations use 64 × 64 × 64 computational grid points with execution times of approximately 6 seconds per time step on the NASA Ames Cray XMP-48 and NASA Lewis Cray XMP-24.

The time evolution of fundamental and subharmonic mode energy, and the energy of three-dimensional turbulence are shown in Fig. 2. The fundamental mode saturates at about $t = 1.6$ which corresponds to vortex roll-up and the subharmonic mode saturates at about $t = 4.4$. A three-dimensional perspective plot of the vorticity magnitude is shown in Fig. 3.

Our analysis of this simulation has focused on the role of the large scale structures in the evolution of the flow and their effect on the chemical reaction. We have found it useful in the past to apply an eduction scheme to characterize the large scale structures (Hussain 1986; Metcalfe et al. 1987b). This scheme, which is described in detail in the preceding reference, is based on detection of large-scale

Table 1. Simulation parameters

Run #.	α	κ	ν	Re	Sc	D_I	D_{II}
11	0.5	0.05	0.003	170	0.06	0.125	0.625
13	1.0	0.005	0.003	170	0.60	0.250	12.500
14	∞	0.005	0.003	170	0.60	∞	∞
15	∞	0.05	0.003	170	0.06	∞	∞
16	0.5	0.005	0.003	170	0.60	0.125	6.250
17	1.0	0.05	0.003	170	0.06	0.250	1.250
18	4.0	0.05	0.003	170	0.06	1.000	5.000
19	0.5	0.015	0.003	170	0.20	0.125	2.083
20	0.5	0.0015	0.003	170	2.00	0.125	20.833
21	∞	0.015	0.003	170	0.20	∞	∞

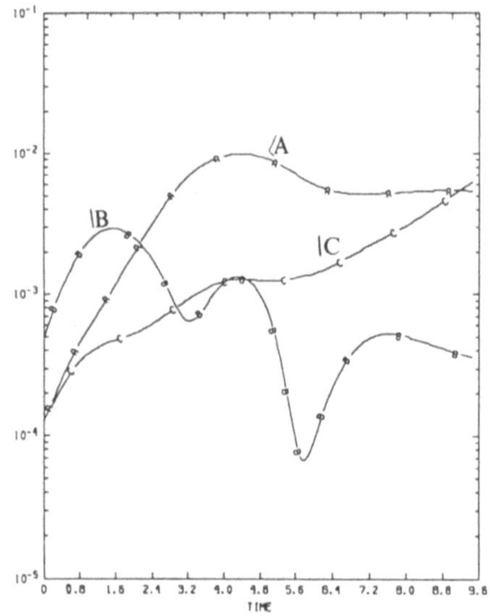

Fig. 2. A plot of the evolution of the modal energies as functions of time. Curve A is the subharmonic mode; Curve B is the fundamental mode; Curve C is the energy in the three dimensional modes

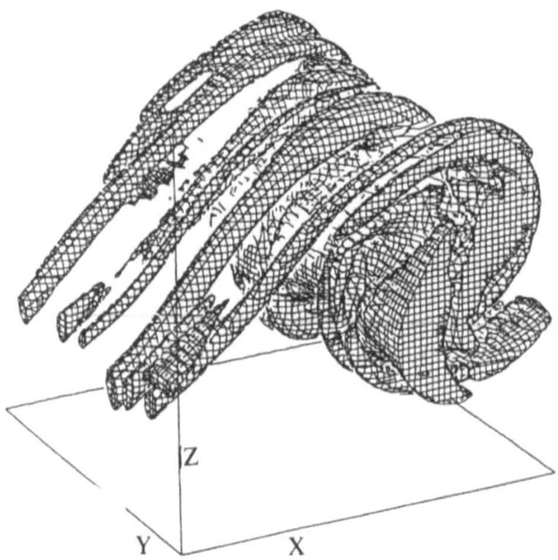

Fig. 3. Three dimensional perspective plots of the absolute value of the vorticity vector at $t = 4.8$

vorticity concentrations from smoothed vorticity maps, accepting structures of certain size and strength, and aligning the realizations through correlation of the vorticity distributions. The application of the eduction algorithm to the spanwise vorticity contours in our present simulation produce the coherent vorticity field shown in Fig. 4, which is at $t = 6.4$, shortly after pairing and subsequent saturation of the subharmonic. This is similar in both amplitude and shape to experimentally observed structures and results of previous numerical simulations (Metcalfe et al. 1987b). In some subsequent figures, the vorticity contour level corresponding to 25% of the peak value of this structure will be superimposed to provide a phase reference.

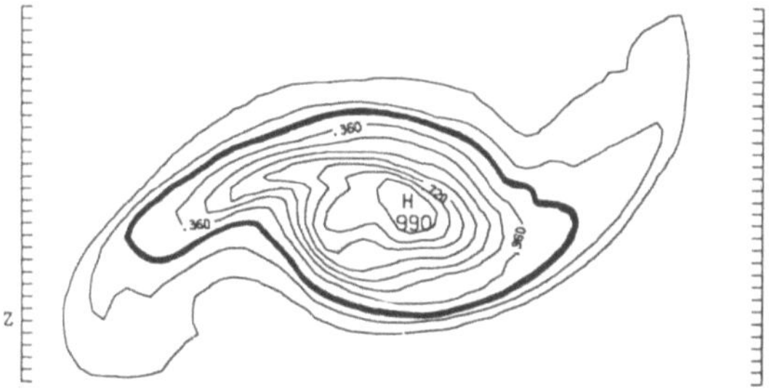

Fig. 4. Coherent spanwise vorticity contours at $t = 6.4$

During the roll up and pairing process, very stable counterrotating vortices ("ribs") develop in the streamwise direction. These structures have been extensively studied experimentally (Hussain 1984; Bernal and Roshko 1986), theoretically (Pierrehumbert and Widnall 1982; Lin and Corcos 1984; Riley et al. 1988) and numerically (Metcalfe et al. 1987a). A detailed analysis of the current simulations has shown that several of these structures form very early in the roll up process and maintain their strong, coherent, vortical structure throughout the entire course of the simulation, which was continued well past the saturation of the subharmonic mode. In this sense, these are the most coherent structures in this complex flow.

An important issue which has been the subject of much recent discussion in the literature is the role of helicity density ($u_i\omega_i$) in turbulent flows and whether it is useful in helping to develop an analytical definition of coherent structures (cf., e.g. Hussain et al. 1988). Moffatt (1985) has suggested that regions of helicity density and

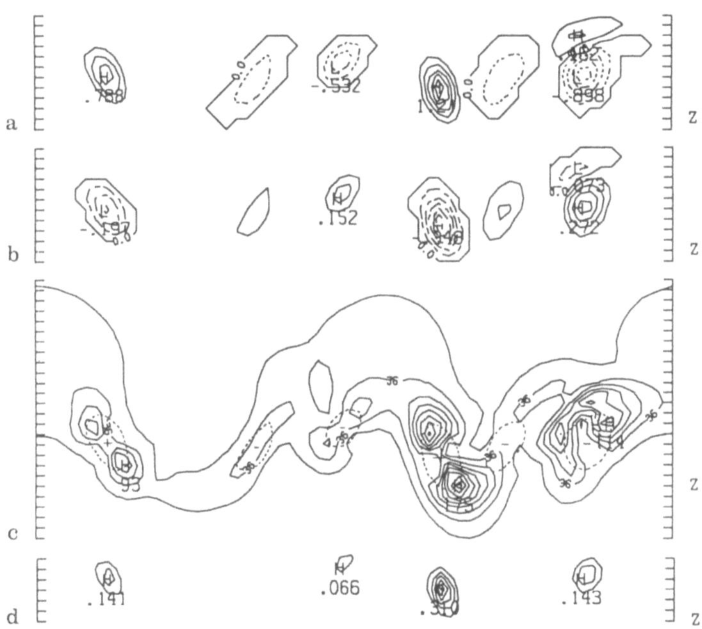

Fig. 5 a. Streamwise vorticity, **b** helicity density, **c** instantaneous energy dissipation, and **d** instantaneous enstrophy production contours in the y–z plane $x = 5\pi/8$ in the braid region at $t = 4.8$. The dashed line in **c** is a streamwise vorticity contour

dissipation should be spatially exclusive, so that regions of high helicity density will characterize coherent structures. For the ribs in the mixing layer, there is some indication of such behavior. First, a comparison of Figs. 5a and 5b shows that the helicity density correlates very strongly with the streamwise vorticity (ω_1) near the base of the rolls, and thus correctly corresponds to the location of the coherent structures. In addition, the regions of peak dissipation (Fig. 5c) tend to straddle the ribs (and hence the helicity density peaks) above and below the braids as the vortical motion of the ribs pumps fluid in a direction normal to that of the basic flow induced by the strain field of the spanwise vortices. Indeed, the effect of the strain field in sustaining the ribs is clearly shown in a plot of the enstrophy production ($\omega_i S_{ij} \omega_j$) (Fig. 5d) which has peaks coinciding with those of the ribs. However, as the stagnation point at the center of the braid is crossed (Fig. 6a), the helicity density goes to zero and changes sign. In fact, this is also the point of peak entrophy

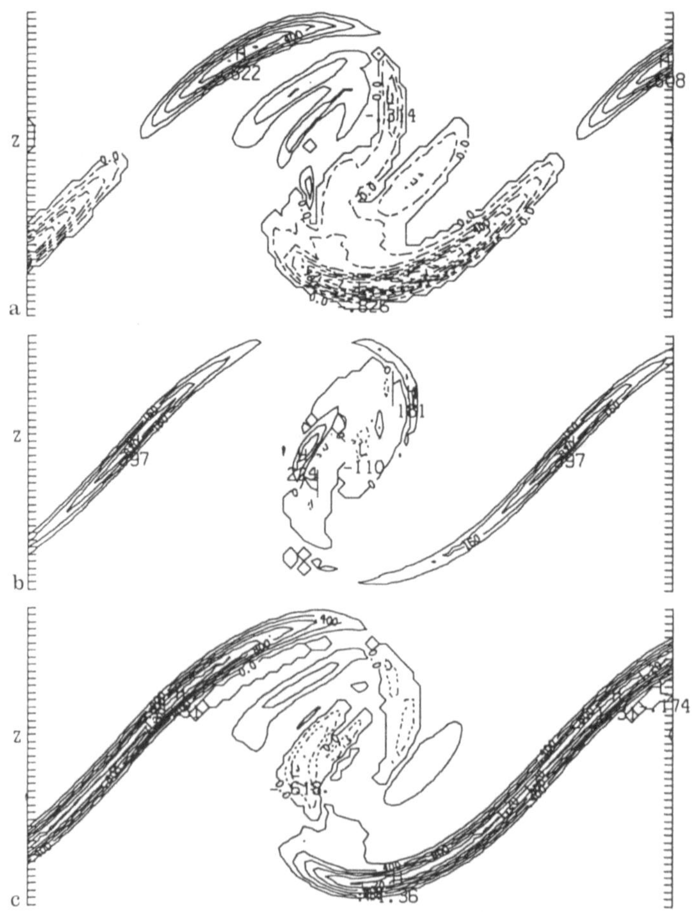

Fig. 6 a. Helicity density, **b** enstrophy production, and **c** streamwise vorticity contours in the x–z plane $y = 3\pi/2$ at $t = 4.8$

production (Fig. 6b). Thus, the strong correlation between the helicity density and the coherent rib structure breaks down near the stagnation point, where the rib itself remains strongly coherent and well defined (Fig. 6c) and where entrainment is presumed to be the maximum (Hunt and Hussain 1989).

The role of the vortex roll-up and pairing on the chemical reaction is two-fold:

(1) From the coherent structure interaction point of view, the vortex roll-up and pairing entrain the non-vortical fluid, i.e., the unreacted species on both sides of the free stream, into the chemical reaction zone. The entrainment is mostly a result of large-scale engulfment of the non-vortical fluid due to the Biot–Savart induction of the large-scale coherent structure near the interface (Hussain 1986).
(2) The chemical reaction zone is marked by very steep gradients of vorticity and reactant concentration across the braids of the vortices, whereas in the core of the vortices, the magnitude of the local gradient is small. The stretching of streamwise vortices (ribs) by the vortex roll-up and pairing enables them to counter the decay due to viscous diffusion and to retain their strength and elongate in the streamwise direction. These effects result in the stretching of the reaction zone and the convolution of the reaction surface, increasing product formation within the reaction zone.

Figure 7 shows the three dimensional perspective plot of the chemical reaction zone at $t = 4.8$ which shows the most intense reaction in the braid region. Note that flame shortening has reduced the reaction in the spanwise vortex cores (rolls). Figure 8 shows two contour plots of the reaction zone with different diffusivities on the same y–z plane. Note that the reaction enhancement due to the convoluting and lengthening of the flame sheet by the ribs is significantly higher with lower species diffusivity.

Figure 9 shows plots of the instantaneous product generation rate with respect to time for cases $D_1 = 0.125, 0.25, 1.0$ and ∞ with the same Schmidt number (i.e. same

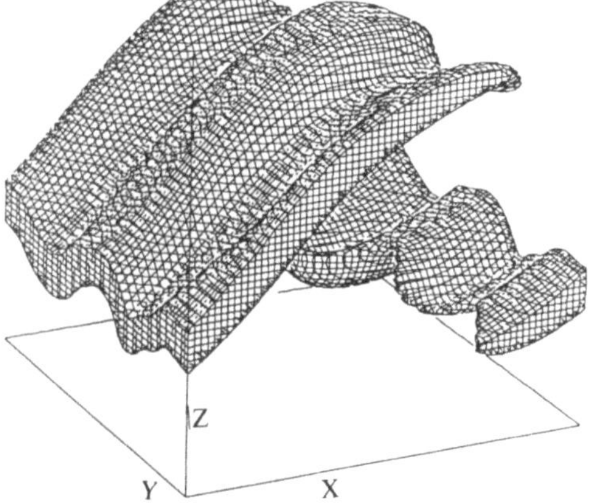

Fig. 7. A three dimensional perspective plot of the instantaneous reaction rate $\alpha C_1 C_2$ (reaction zone) at $t = 4.8$

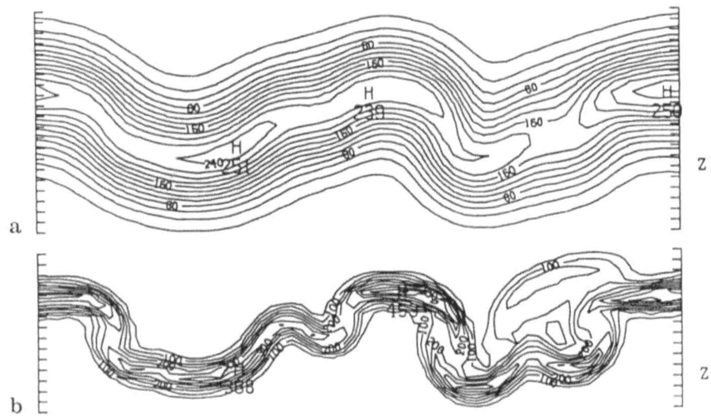

Fig. 8 a. Instantaneous reaction rate $\alpha C_1 C_2$ of Run 11 (Sc = 0.06) and **b** Run 16 (Sc = 0.6) in the y–z plane at $x = 5\pi/8$ at $t = 4.8$

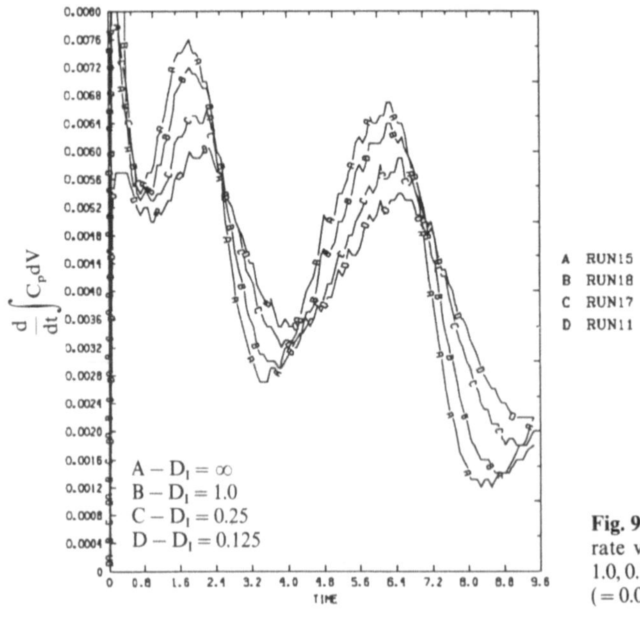

Fig. 9. Total product generation rate versus t for the cases $D_I = \infty$, 1.0, 0.25, and 0.125 with the same Sc (= 0.06)

diffusivity). Here there are two significant peaks in the instantaneous product generation rate. For the case of $D_I = 1.0$, the product rate curve reaches its first peak at about $t = 1.6$, corresponding to fundamental mode saturation (i.e. roll up location). The interdiffusion and reaction of the species within the cores results in their depletion there (flame shortening), so the product generation rate then declines. Then, as vortex pairing and the roll up of the subharmonic occur, new reactants are entrained (i.e. engulfed), so that the reaction rate again increases to a peak about $t = 6.4$, followed again by flame shortening. The effect of the species diffusivity on this

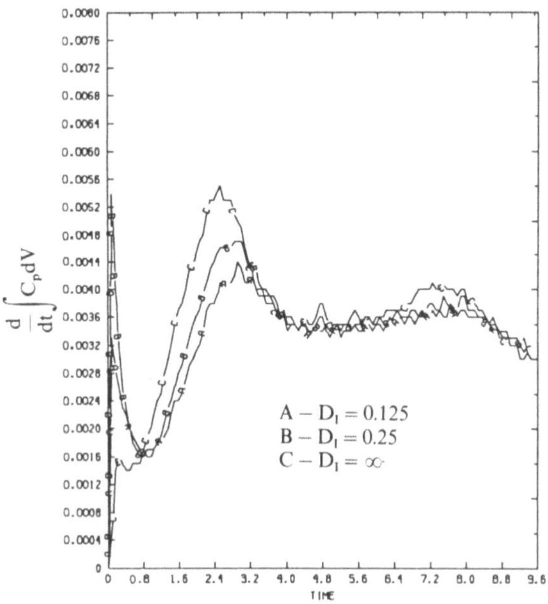

Fig. 10. Total product generation rate versus t for the cases $D_l = \infty$, 0.25, and 0.125 with the same Sc (= 0.6)

process is quite dramatic, as a comparison with Fig. 10 shows. In this case, the lower species diffusivity causes a delay in the onset of flame shortening, as can be seen by the shift in the first reaction rate peak from $t = 1.8$ to $t = 2.4$. During vortex pairing, flame shortening begins to take effect just as the enhanced stretching of the flame sheet caused by the pairing acts to increase the product generation. Consequently, the two effects approximately cancel out, and the second reaction rate peak is essentially annulled.

This phenomenon can be seen more clearly in Fig. 11, which shows the product generation rate for the cases $Sc = 0.06, 0.20, 0.60$ and 2.00 with the same $D_l (= 0.125)$. As Sc increases, the diffusion time scale becomes longer. The first flame shortening delays and then suppresses the growth of the product generation rate caused by pairing. Figure 12 shows contour plots of the reaction zone for the cases $Sc = 0.06$ and 0.60. When the diffusivity is low (Fig. 12b), significant reaction is still taking place in the vortex cores, while for high diffusivity, the core reaction rates are much lower (Fig. 12a).

Applying the eduction algorithm to the species and product concentration fields (Fig. 13a, b), using the vorticity field as the basis for spanwise alignment of the structures shows that even in this strongly three-dimensional flow, the product tends to correlate very highly with the vorticity. (In a purely two-dimensional flow with $Sc = 1$, the correlation would be exact). The instantaneous chemical reaction rate (Fig. 13c) shows a peak at the stagnation point in the braid where there is a peak in the strain field as the two reactants are being pumped into the braid. It is interesting

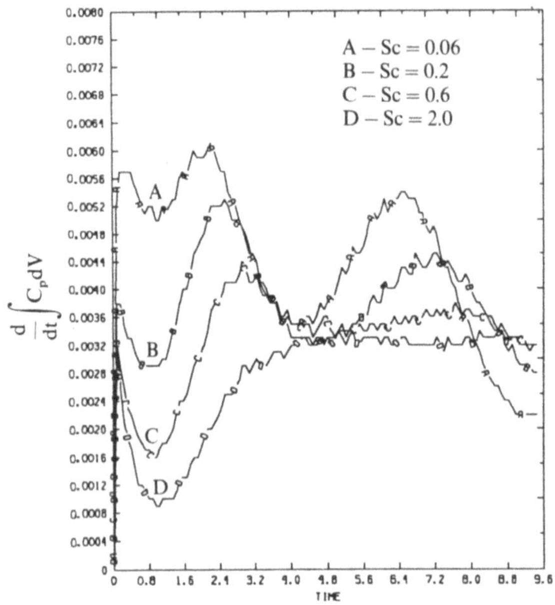

Fig. 11. Total product generation rate versus t for the cases Sc = 0.06, 0.2, 0.6, and 2.0 with the same $D_1 = (0.125)$

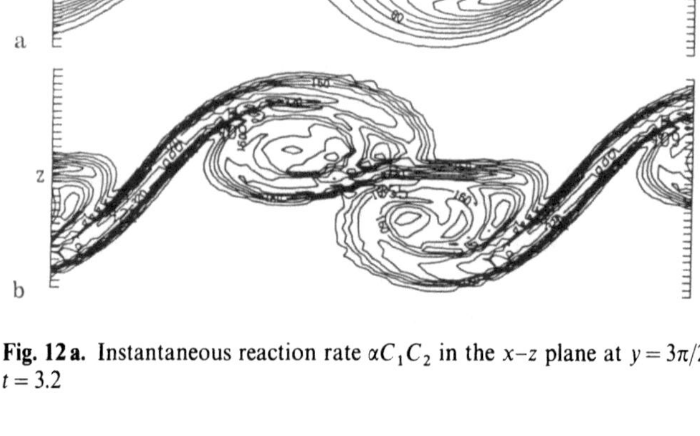

Fig. 12 a. Instantaneous reaction rate $\alpha C_1 C_2$ in the x–z plane at $y = 3\pi/2$ for Run 11 and **b** Run 16 at $t = 3.2$

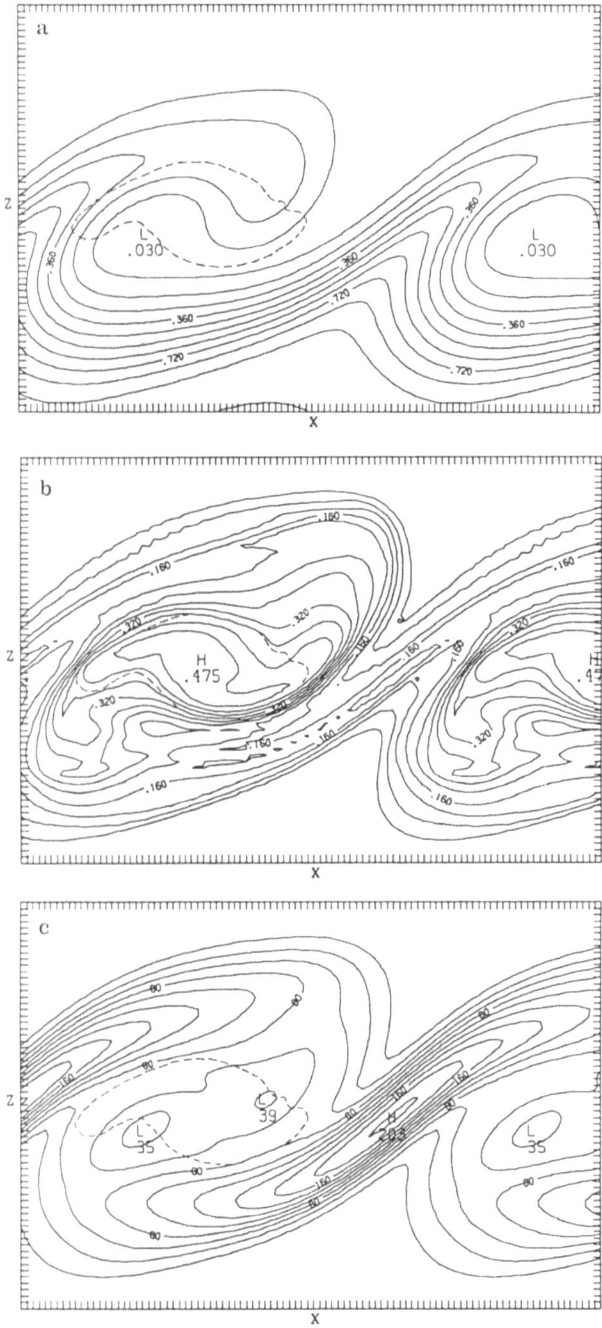

Fig. 13 a. Coherent concentration of reactant $\langle C_1 \rangle$. **b** Coherent concentration of product $\langle C_p \rangle$. **c** Coherent reaction rate $\langle \alpha C_1 C_2 \rangle$ at $t = 6.4$. The dashed line indicates the coherent spanwise vorticity to note that the eduction scheme identifies the locations of the two old vortex cores (which have subsequently merged, and do not show up in the vorticity eduction plot (Fig. 4)) as distinct and separate "cold" spots in the educed vortex core. These represent regions of species depletion due to flame shortening during the initial vortex roll up process before pairing takes place.

Conclusions

In this work, the effect of coherent structures in stretching the reaction zone and the entrainment of the unreacted species have been studied. A series of cases with varying Schmidt numbers, and first and second Damkohler numbers have been carefully examined. The vortex roll up and pairing continuously stretch the streamwise vortices which results in intense reaction in the braids. The streamwise vortices were shown to play an important role in product formation.

In order to resolve the details of the small scale streamwise vortices and to include a second pairing to simulate a more realistic turbulent chemically reacting mixing layer, this study will be extended to higher resolution. Also, the statistical aspects of turbulent flow, e.g., production, dissipation and helicity, mixedness, etc., need to be further studied in order to gain a more complete picture of the effect of turbulence on the chemical reaction. This work is currently in progress and should also lead to better insight into the formation and the evolution of the streamwise vortices.

Acknowledgements: This work is supported by ONR Contract Number N00014-87-K-0670. Computer time is supplied by the National Aerodynamic Simulation Program at NASA Ames Research Center.

References

Bernal, L. P., Roshko, A. (1986): Streamwise vortex structure in plane mixing layers. J. Fluid Mech. **170**, 499–525

Buell, J. C., Huerre, P. (1988): Inflow/outflow boundary conditions and global dynamics of spatial mixing layers. Proceedings of the 1988 Summer Program, Center for Turbulence Research, pp. 19–28

Cantwell, B., Coles, D. (1983): An experimental study of entrainment and transport in the turbulent near wake of circular cylinder. J. Fluid Mech. **136**, 321–374

Canuto, C., Hussaini, M. Y., Quarteroni, A., Zang, T. A. (1988): Spectral Methods in Fluid Mechanics, Springer-Verlag, New York

Corcos, G. M., Sherman, F. S. (1984): The mixing layer: deterministic models of a turbulent flow. part I. Introduction and the two-dimensional flow. J. Fluid Mech. **139**, 29–65

Givi, P. (1988): Model free simulations of turbulent reactive flows: a review. Progress in Energy and Combustion Science, in press

Hunt, J. C. R., Hussain, F. (1989): Private communication

Hussain, F. (1984): Coherent structures and incoherent turbulence. In Turbulence and Chaotic Phenomena in Fluids, IUTAM Symposium. (T. Tatsumi, ed.), pp. 453–460

Hussain, A. K. M. F. (1983): Coherent structures and incoherent turbulence. IUTAM Symposium on Turbulence and Chaotic Phenomena in Fluids, IUTAM, Kyoto, Japan, pp. 1–7

Hussain, A. K. M. F. (1986): Coherent structures and turbulence. J. Fluid Mech. 173, 303–356
Hussain, F., Moser, R., Colonius, T., Moin, P., Rogers, M. M. (1988): Dynamics of coherent structures in a plane mixing layer. Proceedings of the 1988 Summer Program, Center for Turbulence Research, pp. 49–56
Lin, S. J., Corcos, G. M. (1984): The mixing layer: deterministic models of a turbulent flow. Part 3. The effect of plane strain on the dynamics of streamwise vortices. J. Fluid Mech. **141**, 130–178
Metcalfe, R. W., Orszag, S. A., Brachet, M. E., Menon, S., Riley, J. J. (1987a): Secondary instability of a temporally growing mixing layer. J. Fluid Mech. **184**, 207–243
Metcalfe, R. W., Hussain, F., Menon, S., Hayakawa, M. (1987b): Coherent Structures in a Turbulent Mixing Layer: A Comparison Between Direct Numerical Simulations and Experiments. Turbulent Shear Flows, Vol 5, pp. 110–123
Michalke, A. (1964): On the inviscid instability of the hyperbolic tangent velocity profile. J. Fluid Mech. **19**, 543–556
Moffatt, H. K. (1985): Magnetostatic equilibria and analogous Euler flows of arbitrarily complex topology. Part 1. Fundamentals. J. Fluid Mech. **173**, 303–356
Orszag, S. A. (1971): Numerical simulation of incompressible flows within simple boundaries: accuracy. J. Fluid Mech. **49**, 75–112
Orszag, S. A., Pao, Y.-H. (1974): Numerical computation of turbulent shear flows. Adv. Geophys. **18A**, 225
Peyret, P., Taylor, T. D. (1983): Comp. Methods for Fluid Flows. Springer-Verlag, New York
Pierrehumbert, R. T., Widnall, S. E. (1982): The two- and three-dimensional instabilities of a spatially periodic shear layer. J. Fluid Mech. **114**, 50–82
Riley, J. J. Metcalfe, R. W. (1985): Direct numerical simulations of chemically reacting turbulent mixing layers. Paper in AIAA 23rd Aerospace Sciences Meeting
Riley, J. J., Metcalfe R. W., Orszag, S. A. (1986): Direct Numerical Simulations of Chemically Reacting Turbulent Mixing Layers. Phys. Fluids **29**, 406–422
Riley, J. J., Mourad, P. D., Moser, R. D., Rogers, M. M. (1988): Sensitivity of mixing layers to three-dimensional forcing, Proceedings of the 1988 Summer Program, Center for Turbulence Research, pp. 91–116
Williams, F. A. (1985): Combustion Theory. The Benjamin/Cummings Publishing Co., Menlo Park
Shtilman, L., Levich, E. Orszag, S. A., Pelz, R. B. Tsinober, A. (1985): On the role of helicity in complex fluid flows. Phys. Lett. **113A**, 32–37

Modelling and Numerical Simulation of Premixed Turbulent Combustion in a Boundary Layer

B. Rogg

University of Cambridge, Department of Engineering, Trumpington Street, Cambridge CB2 1PZ, UK

Abstract

In the present paper we model and numerically simulate premixed turbulent combustion in a boundary layer with injection. A laminar-flamelet approach is adopted. Attention is focused on the chemical-source-term closure for which a transport equation for the flame surface-to-area ratio, i.e., for the reciprocal of the length scale of the flamelet crossing process, is solved. Sensitivity analyses are performed to investigate the sensitivity of selected quantities, such as the maximum flame temperature, with respect to various model constants. The numerical results are compared to experimental data available in the literature.

1. Introduction

A particularly important regime of premixed turbulent combustion is the laminar-flamelet regime [1], the most prominent representative of which is combustion in Otto-engines. For flames in this regime the model of premixed turbulent combustion set forth by Bray, Champion, Moss and Libby,[1] see e.g. [2–5], and Pope's "Stochastic Flamelet Model" [6,7] have proved successful in describing several of the physical processes such as the phenomena of flame-generated turbulence and counter-gradient transport. Assuming that a turbulent burning velocity exists in the laminar-flamelet regime, from Damköhler onwards [8] this quantity has been modelled in terms of flamelet area. Since the flamelet area per unit volume, alternatively termed flamelet-surface density, is directly related to a characteristic scale of the wrinkled laminar flame, the prediction of this quantity is one of the most important topics of current research. For the Stochastic Flamelet Model first evolution equations have been given from which a mean value of the flamelet surface-to-volume ratio can be calculated [7]. Equivalently, to supplement their model, BCML have recently derived a transport equation for the probability of flamelet-burning [9]. As an alternative to the latter equation, modelled transport equations for the mean of the flame-surface density have been proposed by *Peters* [10, 11] and *Candel* [12].[2] Even though different authors may employ different

[1] Subsequently the shorthand "BCML" will be used for this collective of authors.

[2] The present paper is concerned with premixed turbulent combustion only; however, it appears appropriate to mention here the Coherent Flame Model developed by *Marble* and *Broadwell* for turbulent diffusion [13] flames. For a recent extension of this model see [14].

Turbulent Shear Flows 7

definitions for the flamelet surface-to-volume ratio or work with an equivalent of it, for convenience of notation subsequently this quantity will be termed σ and its governing partial differential equation will be referred to as the "σ-transport equation".

The objective of the present paper is to model and to numerically simulate premixed turbulent combustion in a boundary layer which arises from the injection of cold reactants into a hot stream of corresponding combustion products directed over a flat porous plate. This boundary-layer configuration, a schematic of which is shown in Fig. 1, was investigated experimentally and numerically by *Meunier* [15, 16], and his experimental data are used herein for comparison with numerical predictions. Unlike in his model where a presumed-pdf approach was used to determine the mean turbulent reaction rate, in the present paper the mean rate is calculated according to BCML, see e.g. [17], and the chemical-source-term closure is achieved through a modified version of *Peters*' transport equation for the flamelet surface-to-volume ratio. Sensitivity analyses are performed to investigate the sensitivity of selected quantities, such as the maximum mean temperature, with respect to various model constants, particularly to those appearing in the σ-transport equation. The numerical results are compared to experimental data available in the literature [15].

2 Modelling

2.1 The Flow-Field Model

Shown in Fig. 1 is a schematic of the geometry considered in the present paper, viz., the turbulent flow of hot combustion products over a flat plate. Through the plate, normal to it, cold reactants are injected into the boundary layer. As a consequence, in the boundary layer a turbulent flame is stabilised in which the reactants burn.

Conservation equations in boundary-layer form are adopted to describe the turbulent reacting flow under consideration. In particular, because the focus of

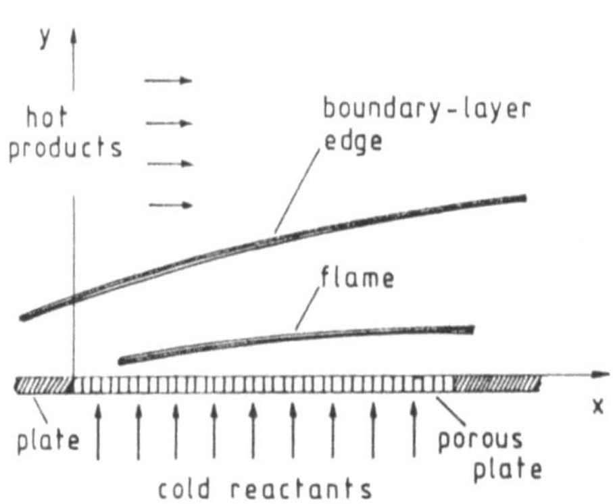

Fig. 1. Schematic of the turbulent boundary layer and the premixed flame

interest is the chemical-source-term closure, a conventional eddy viscosity gradient model of turbulent transport and a $k-\varepsilon$ turbulence model are adopted. Hence, by virtue of a parabolic turbulent-transport operator L,

$$L(\phi) \equiv \frac{\partial(\bar{\rho}\tilde{u}\phi)}{\partial x} + \frac{\partial(\bar{\rho}\tilde{v}\phi)}{\partial y} - \frac{\partial}{\partial y}\left(\frac{\mu_t}{\sigma_\phi}\frac{\partial \phi}{\partial y}\right),$$

the conservation equations for mass and momentum and the equations of the k-ϵ turbulence model can be written as

$$L(\phi) = S_\phi, \tag{1}$$

where ϕ is unity or any of the variables $\tilde{u}, \tilde{v}, k$ and ε, and where S_ϕ is a source term to be specified. In Eq. (1) the Favre-averaging procedure has been used; tildes identify Favre-averages, overbars time averages; x and y denote the longitudinal and transverse coordinates, respectively, and $\tilde{u}$ and $\tilde{v}$ the respective flow-velocity components; the σ_ϕ are model constants for which generally accepted values are adopted ($\sigma_u = \sigma_v = \sigma_k = 1.0$, $\sigma_\varepsilon = 1.3$); $\bar{\rho}$ is the density, μ_t the turbulent diffusivity, k the Favre turbulence kinetic energy, ε the Favre turbulence dissipation. In the source term of the k-equation changes of the kinetic turbulence energy due to the heat release have been taken into account by introducing a dilatation term [15, 18]. For the model constants appearing in the S_ε-term of Eq. (1) the values $C_{\varepsilon 1} = 1.44$, $C_{\varepsilon 2} = 1.92$ have been employed. Apart from appropriate boundary conditions the solution of these equations requires information about the variations of the density $\bar{\rho}$ and the turbulent viscosity μ_t. For μ_t the relationship $\mu_t = C_\mu \bar{\rho} k^2/\varepsilon$ is adopted with the generally accepted value $C_\mu = 0.09$. The mode of providing information about $\bar{\rho}$ depends on the approach to closure of the chemical source term and is briefly discussed in the following subsection.

2.2 The Thermochemistry Model

Difficulties arise in formulating a thermochemistry model that allows adequate simulation of the experiments reported in [15, 16] because in the experiments, firstly, the flow was nonadiabatic due to the heating of the reactants during their passage through the porous plate and, secondly, the temperature of the hot gas flowing over the plate was lower than the adiabatic flame temperature of the cold reactants. However, following *Meunier* et al. [15], herein these difficulties are overcome by including into the set of governing equations both the balance equation for the mean reaction-progress variable $\tilde{c}$ and the modelled energy equation in terms of mean temperature $\tilde{T}$, say. At the porous plate the boundary conditions for these equations are that $\tilde{T}$ be equal to the plate temperature and that $\tilde{c}$ attains a corresponding nonzero value; in the hot gas $\tilde{c}$ has a value of unity but $\tilde{T}$ there is subadiabatic, viz., equal to the temperature of the hot gas directed over the plate. The $\tilde{c}$-equation will be used to calculate the mean turbulent reaction rate, the energy equation together with the ideal-gas equation of state to calculate the mean density. Neglecting variations of the mixture molecular weight and treating the pressure as thermochemically constant, the latter equation reduces to

$$\bar{\rho} = \rho_f T_f/\tilde{T} = \rho_{ad} T_{ad}/\tilde{T}, \tag{2}$$

where here and below the subscripts f and ad are used to identify quantities in the cold reactants and their products at the *adiabatic* flame temperature, respectively.

2.2.1 Temperature and Reaction-Progress Variable

Upon introducing standard assumptions, such as equal and constant specific heats c_p for all species etc., it is straightforward to derive transport equations for the mean progress variable and the mean temperature, viz,

$$L(\tilde{c}) = \bar{\omega}, \tag{3}$$

$$L(\tilde{T}) = \bar{\omega}(T_{ad} - T_f). \tag{4}$$

Numerical values for the turbulent Prandtl numbers are $\sigma_c = 1.0$ and $\sigma_T = 0.7$. The mean turbulent reaction rate $\bar{w}$ can be expressed as, see e.g. [17],

$$\bar{w} = v_y w_y = \frac{g\bar{c}(1-\bar{c})}{\hat{L}_y |\bar{\sigma}_y|} \rho_f S_L I_0, \tag{5}$$

where v_y is the flamelet-crossing spatial frequency, w_y the mean reaction rate per crossing (that is, the average number of times per unit length that a flamelet crosses the mean contour), g a constant of order unity, and $\bar{c}$ the Reynolds mean of the reaction progress variable, various definitions of which have been examined recently [19, 20]. $\hat{L}_y$ is an integral length scale of the wrinkled flame, $|\bar{\sigma}_y|$ a geometrical factor, S_L the burning velocity of the unstrained laminar flame, and I_0 denotes an integral over the strain rate [19, 20].

The integral scale of the wrinkled flame, $\hat{L}_y$, may be viewed as the reciprocal of a suitable defined flame-surface density, σ. In the past various algebraic approximations to $\hat{L}_y$ have been explored none of which, however, describes properly its spatial variation through the flame; it is for this reason that (modelled) transport equations for $\hat{L}_y$ or, alternatively, σ are sought. In the present paper we define the Favre mean flame surface-to-volume ratio σ as

$$\sigma \equiv \frac{g}{\hat{L}_y |\bar{\sigma}_y|}, \tag{6}$$

so that Eq. (5) becomes

$$\bar{w} = \rho_f S_L \sigma \bar{c}(1-\bar{c}). \tag{7}$$

Since in the present paper the focus of interest is σ, in Eq. (7) and below *explicit* effects of strain on the turbulent mean reaction rate are neglected, i.e., I_0 is taken as unity. Effects of strain will enter the formulation implicitly through σ, the transport equation of which is discussed in the following subsection.

2.2.2 The Flamelet-Surface Density

The balance equation for the mean flamelet surface-to-volume ratio derived and modelled by *Peters* [10, 11] accounts for accumulation, convection, diffusion, production and dissipation of this quantity. Originally the equation was derived for constant density, turbulence intensity, integral length scale of the flow and turbulent kinematic viscosity ($= C_\mu k^2/\varepsilon$). To ascertain the extent to which this equation can be used to predict combustion in flows with variable properties is one of the topics of the present paper. For variable turbulence quantities the equation can be written as

$$L(\sigma) = C_{\sigma 1}\sigma\bar{\rho}\tilde{c}(1-\tilde{c})(\varepsilon/k) - C_{\sigma 2}\sigma^2\bar{\rho}k^{1/4}S_L^{1/2}, \qquad (8)$$

where $C_{\sigma 1}$ and $C_{\sigma 2}$ are empirical constants. For a statistically steady, one-dimensional flow with constant turbulence quantities, *Peters* [10, 11] determined numerical values for $C_{\sigma 1}$ and $C_{\sigma 2}$. Thus guidance is available to determine suitable values for these constants for the variable-properties case under consideration.

Equation (8) calls for some comments. In consistency with the balance equations for $\tilde{u}$, k, ε, $\tilde{T}$ and $\tilde{c}$, on the l.h.s. of (8) the L-operator represents accumulation, convection and diffusion of flame-surface density. The first term on the r.h.s. of Eq. (8) represents the increase of flamelet surface (production) due to the effects of strain as a consequence of the turbulent velocity fluctuations. The factor $\tilde{c}(1-\tilde{c})$ reflects the simple fact that flame surface cannot be increased by strain where no flame exists, i.e., in the reactants ($\tilde{c}=0$) and in the products ($\tilde{c}=1$), respectively. The second term on the r.h.s. of Eq. (8) describes the annihilation of flame surface (destruction, dissipation), e.g. mutual annihilation of adjacent surface elements as a consequence of reactant consumption or flamelet extinction due to excessive rates of strain.

It is interesting to note that the σ-dependence of the r.h.s. of Eq. (8) is similar as found by *Marble* and *Broadwell* for turbulent diffusion flames [13] and by *Candel* and coworkers for premixed flames [12], viz., production of flamelet surface proportional to σ and destruction proportional to σ^2.

3. Boundary Conditions and Numerical Method

At the wall the wall temperature and the thermochemical state of the reactants and their rate of injection F ($= \bar{\rho}\tilde{v}/\rho_f S_L$), and hence $\tilde{v}$, are specified; $\tilde{u}$ there is zero. The values of $\tilde{u}$, ε and μ_t near the wall are calculated using the wall functions described in Ref. [21]. At the boundary-layer edge the subadiabatic temperature of the hot gas and values for $\tilde{u}$ and k there are specified according to the experimental data reported in [15, 16]; $\tilde{c}$ there is taken as unity, $\partial\varepsilon/\partial y$ as zero, and $\tilde{v}$ is calculated from the continuity equation.

The origin of the x–y coordinate system is chosen such that injection starts at $x = 0$ cm. Since in [15, 16] measurements are reported for axial stations at $x = 5$ cm and $x = 19$ cm, the computational domain with respect to x is selected to start at $x = 5$ cm. Physically and mathematically consistent profiles are specified there for all dependent variables. In particular, transverse profiles according to the experimental data are specified for $\tilde{u}$, k, $\tilde{T}$ and $\tilde{c}$. The transverse initial profile for ε is calculated on the basis of the mixing-length hypothesis [22], the profile for $\tilde{v}$ is obtained by solving the continuity equation.

Boundary conditions for the σ-transport equation are that at the wall $\partial\sigma/\partial y$ be zero and $\sigma = 0$ at the boundary-layer edge. An initial profile for σ at $x = 5$ cm has been obtained by equating in Eq. (8) the production and the dissipation term.

The set of parabolic governing equations is solved by a space-marching method with respect to x starting at $x = 5$ cm. Successively, in each x-plane x^n, $n = 0, 1, \ldots$, the governing equations are discretized. The discretization with respect to x is of the backward-Euler type, that with respect to y is performed on a nonuniform mesh $\mathbf{M}$ of grid points,

$$\mathbf{M}^n = \{0 = y_1^n < y_2^n < \cdots < y_m^n = y_{max}\}. \tag{9}$$

In the y-direction a static adaptive-gridding procedure [23] combined with a piecewise monotonic cubic Hermite interpolation method [24] is employed that concentrates grid points in regions where the profiles of the dependent variables exhibit steep gradients and/or strong curvature. Therefore, the number of gridpoints, i.e. the subscript m in Eq. (9), may vary with x, but y_{max} is kept constant and is selected sufficiently large as to ensure that the conditions at the boundary-layer edge are asymptotically satisfied.

The marching scheme advances from one x-plane, x^n, to the next, x^{n+1}, in each plane solving the nonlinear system of difference equations

$$F(U^{n+1}, U^n; C) = f(U^{n+1}; C) - (U^{n+1} - U^n)/h^{n+1} \tag{10}$$

by a modified Newton method [25]. In Eq. (10), and below, h^{n+1} stands for $x^{n+1} - x^n, n = 0, 1, \ldots$; U^i denotes the solution of the nonlinear system in plane $x^i, i = n$ and $n + 1$, viz.,

$$U^i = (\tilde{u}_1^i, \tilde{v}_1^i, k_1^i, \varepsilon_1^i, \tilde{c}_1^i, \tilde{T}_1^i, \sigma_1^i, \tilde{u}_2^i, \tilde{v}_2^i, k_2^i, \varepsilon_2^i, \tilde{c}_2^i, \tilde{T}_2^i, \sigma_2^i, \\ \ldots, \tilde{u}_m^i, \tilde{v}_m^i, k_m^i, \varepsilon_m^i, \tilde{c}_m^i, \tilde{T}_m^i, \sigma_m^i), \tag{11}$$

and C is the parameter M-vector of the M model constants that appear in the governing equations,

$$C = (C_{\sigma 1}, C_{\sigma 2}, C_{\varepsilon 1}, C_{\varepsilon 2}, \cdots). \tag{12}$$

Thus, the linear system

$$(J(U_k^{n+1}) - I/h^{n+1})(U_{k+1}^{n+1} - U_k^{n+1}) \\ = -\omega_k^{n+1}(f(U_k^{n+1}; C) - (U_k^{n+1} - U^n)/h^{n+1}), \tag{13}$$

$k = 0, 1, \cdots$, is solved. Here U_k^{n+1} denotes the solution after k iterations, ω_k^{n+1} is the damping parameter, and $J(U_k^{n+1})$ and I are the Jacobian based on U_k^{n+1} and the identity matrix, respectively. The damping strategy [25] allows the Jacobians to be re-evaluated only periodically.

4. Sensitivity Analysis

For the reactive boundary-layer problem under consideration, sensitivity analysis is a useful tool to investigate quantitatively the dependence of the numerical predictions on the particular choice of the numerical values that have been assigned to the individual model constants appearing in the governing equations. Since the model constants of the k–ε model are relatively well established, in the present paper attention is focused on the σ-transport equation. In particular, we are interested in the sensitivity on the σ-equation model constants of the predicted values of the mean flame-surface density and the temperature, quantities that are a direct measure for the turbulent-burning rate.

In general, to predict the effect of the variation of the model constants contained in C on the dependent variables contained in U, where U and C are defined in Eqs. (11) and (12), respectively, the vector of first-order sensitivity coefficients $\partial U/\partial C_\alpha$, $\alpha = 1, \ldots, M$, must be calculated. The appropriate equation for the latter quantities is obtained by differentiating Eq. (10) with respect to C, viz.,

$$\frac{d}{dC_\alpha}(F(U^{n+1}, U^n; C) = \frac{\partial F}{\partial U^{n+1}} \frac{\partial U^{n+1}}{\partial C_\alpha} + \frac{\partial F}{\partial U^n} \frac{\partial U^n}{\partial C_\alpha} + \frac{\partial F}{\partial C_\alpha} = 0, \tag{14}$$

$\alpha = 1 \cdots, M$. Equation (14) can be rewritten as

$$\left(J(U^{n+1}) - \frac{I}{h^{n+1}} \right) \frac{\partial U^{n+1}}{\partial C_\alpha} = -\left(\frac{\partial F}{\partial C_\alpha} + \frac{I}{h^{n+1}} \frac{\partial U^n}{\partial C_\alpha} \right). \tag{15}$$

Since the first-order sensitivity coefficients at plane x^n are known, those at plane x^{n+1}, $\partial U^{n+1}/\partial C_\alpha$, can readily be calculated from Eq. (15) by solving a linear system of equations

5. Results and Discussion

Due to space limitations in the present volume, only a few results can be presented and discussed. These are results for one of the flames investigated by *Meunier* [15, 16], viz., a nitrogen-diluted, slightly lean propane/air flame, the reactants being injected with a rate of $F = 1.05$ percent. The data are: fuel-equivalence ratio 0.97, nitrogen mass fraction in the diluted reactive mixture 0.28, initial wall temperature (at $x = 5$ cm) 1000 K, initial free-stream temperature 1500 K and initial free-stream velocity 75.3 m/s. The conditions are such that wall temperature decreases and free-stream velocity increases slightly (approximately linearly between $x = 5$ cm and $x = 19$ cm by 10 percent); the free-stream temperature remains constant. Under these conditions it was found that a flame stabilizes, and experimental data are reported in [15, 16] for the two axial locations. Herein the data at $x = 19$ cm are used for comparison with the numerical predictions based on the σ-transport equation.

Shown in Fig. 2 are the measured and computed velocity profiles; the latter have been determined using the so-called slender-channel formulation[3]. It is seen that the agreement is good. Shown in Fig. 3 are the measured and calculated temperature profiles. The agreement, which is seen to be excellent, has been obtained for the following set of parameters: $C_{\sigma 1} = 7.5$, $C_{\sigma 2} = 0.075$ and $\sigma_\sigma = 1$. It is interesting to note that the ratio

$$r = C_{\sigma 2}/C_{\sigma 1} = 0.01 \tag{16}$$

is small, as was observed by *Peters* [11] for the non-dimensional, constant-property case.

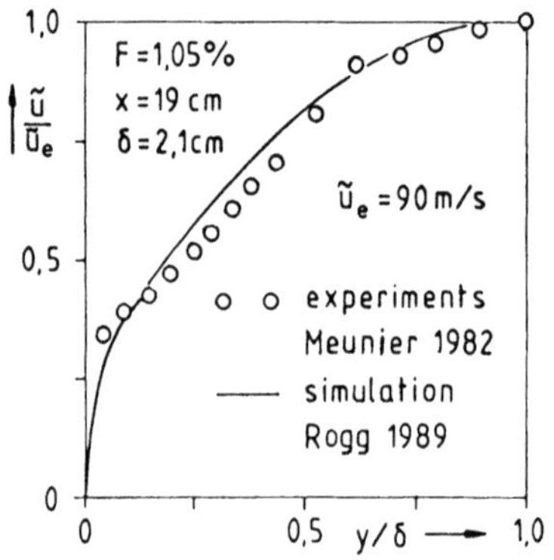

Fig. 2. Comparison between measured and predicted mean-velocity profiles

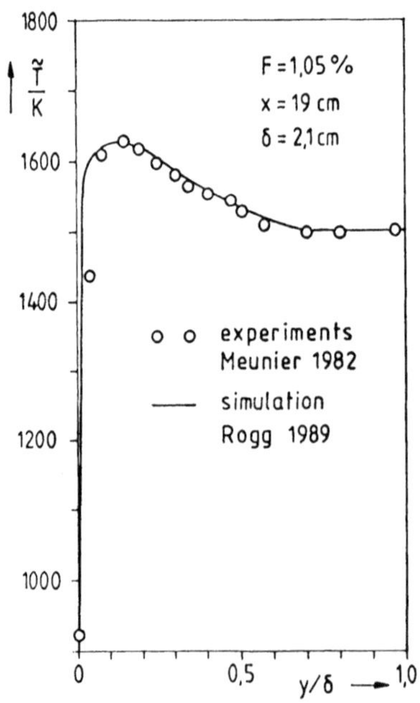

Fig. 3. Comparison between measured and predicted mean-temperature profiles

[3] See Acknowledgements

Shown in Figs. 4–6 are some results obtained from the sensitivity analysis. The abscissa in these figures is normalized such that $x^* \equiv (x - 5\,\text{cm})/(19\,\text{cm} - 5\,\text{cm})$. It is important to note that the relative sensitivity coefficients, which in the figures are denoted by

$$\beta_{i,j} = \frac{C_j}{U_i}\frac{\partial U}{\partial C_j} \approx \frac{\Delta U_j/U_i}{\Delta C_j//C_j}, \tag{17}$$

are local quantities, i.e., quantities that vary through the flame. Shown in Figs. 4 and 5 are the sensitivities of the mean flame-surface density with respect to the model parameter $C_{\sigma 1}$ and the ratio $r = C_{\sigma 2}/C_{\sigma 1}$ for the three contour lines where $\tilde{c}$ is 0.7, 0.8

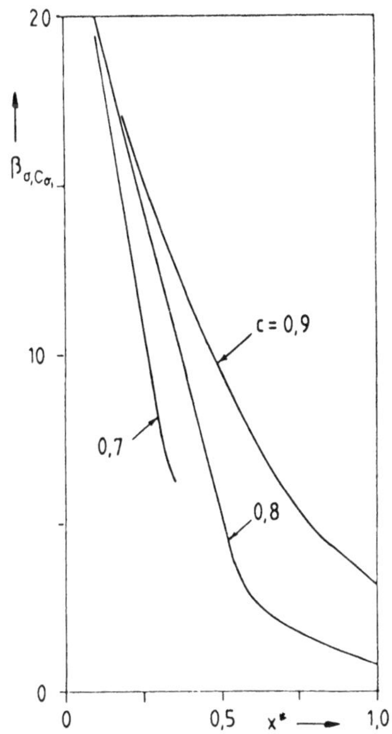

Fig. 4. Relative sensitivity coefficient of the mean flame-surface density with respect to $C_{\sigma 1}$ along the contours where $\tilde{c} = 0.7$, 0.8 and 0.9, respectively

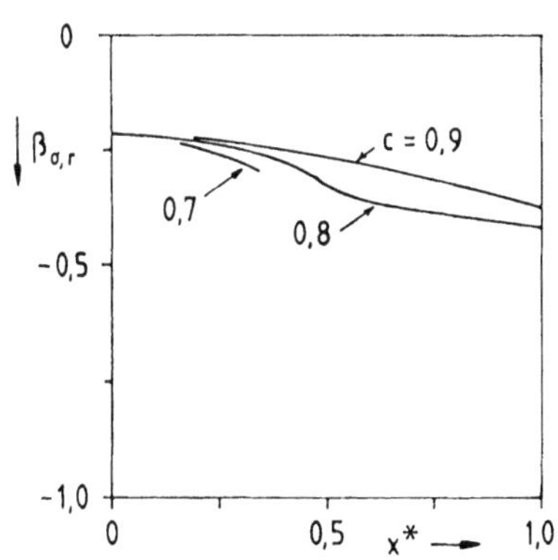

Fig. 5. As Fig. 4, but sensitivity with respect to the ratio $C_{\sigma 2}/C_{\sigma 1}$

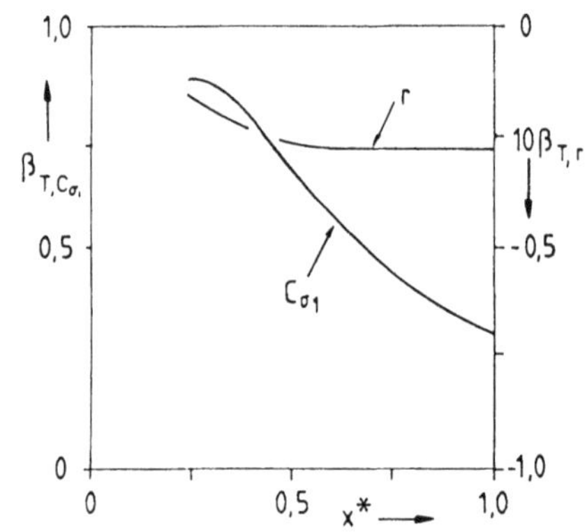

Fig. 6. Relative sensitivity coefficients of the mean temperature with respect to $C_{\sigma 1}$ and the ratio $C_{\sigma 2}/C_{\sigma 1}$ along the contours where $\tilde{T}$ attains its maximum

and 0.9, respectively. The sensitivity of σ with respect to $C_{\sigma 1}$ is remarkably high for small to intermediate values of x^*. Inspection of the temperature field (not shown here) reveals that these high sensitivities are related to the ignition process which takes place in the early parts of the boundary layer. The sensitivity of σ with respect to r (or $C_{\sigma 2}$) is seen to be fairly constant throughout the length of the boundary layer.

Shown in Fig. 6 is the sensitivity of the mean temperature *along the contour of maximum temperature* with respect to $C_{\sigma 1}$ and r. It is seen that throughout the flame the maximum temperature can be expected to be affected only little by changes in r, whereas the impact of changes in $C_{\sigma 1}$ on the maximum temperature increases with increasing x.

Conclusions

A transport equation for the mean flame surface-to-area ratio has been tested for premixed turbulent combustion in a boundary layer with injection. For this particular case model constants have been determined that lead to excellent agreement between numerical model predictions and experimental data available in the literature. Sensitivity analyses have been performed to indicate particular sensitivities of the predictions with respect to the empirical constants. Further computations for different geometries and different conditions and comparisons to corresponding experimental data are needed to further verify and establish the new transport equation.

Acknowledgements: The author appreciates stimulating discussions of this research with various colleagues. He would like to thank especially N. Peters who made available Eq. (8) and M. Champion who provided a copy of ref. [16]. P.A. Libby provided the idea of the slender-channel approximation, which, compared to the version of the paper presented at the Conference, has improved greatly the calculated velocity profiles. This work was supported by NATO through Grant No. 0102/89.

References

1. Peters, N.: Laminar Flamelet Concepts in Turbulent Combustion, 21st Symp. (Int.) on Comb., 1231–1250, The Combustion Institute, Pittsburgh (1986)
2. Bray, K. N. C., Libby, P. A., Moss, J. B.: Flamelet Crossing Frequencies and Mean Reaction Rates in Premixed Turbulent Combustion, Combust. Sci. and Tech. **41**, 143–172 (1984)
3. Bray, K. N. C., Libby, P. A., Moss, J. B.: Scalar Length Scale Variations in Premixed Turbulent Flames, 20th Symp. (Int.) on Comb., 421–427, The Combustion Institute, Pittsburgh (1984)
4. Champion, M., Libby, P.A.: Turbulent Premixed Combustion in a Boundary Layer, Combust. Flame **38**, 267–291 (1984)
5. Bray, K. N. C., Champion, M., Dave, N., Libby, P. A.: On the Thermochemistry of Premixed Turbulent Combustion in Variable Enthalpy Systems, Combust. Sci and Tech. **46**, 31–43 (1986)
6. Pope, S. B.: Ann. Rev. Fluid Mech **19**, 237–270, (1987)
7. Pope, S. B., Cheng, W. K.: The Stochastic Flamelet Model of Turbulent Premixed Combustion, to appear in the Proceedings of the 22nd Symp. (Int.) on Comb., The Combustion Institute, Pittsburgh (1988)
8. Damköhler, G.: Z. Elektrochem. **46**, 601 (1940)
9. Bray, K. N. C., Champion, M., Libby, P. A.: Private Communication (1988)
10. Franke, Ch., Peters N.: Flammenoberfläche und Brenngeschwindigkeit von turbulenten Vormischflammen, Proc. SFB 224, RWTH Aachen, 6–7 Oct. 1988, 126–147 (1988)
11. Peters, N.: Private Communication (1989)
12. Maistret, E., Veynante, D., Poinsot, T., Darabiha, N., Lacas, F., Candel, S.: Structure and Modeling of Turbulent Ducted Flames, Proc. Joint Meeting of the British and French Sections of the Combustion Institute, Rouen (France), 17–21 April 1989, 209–212 (1989)
13. Marble, F. E., Broadwell, J. E.: The Coherent Flame Model for Turbulent Chemical Reactions, Project SQUID Technical Report TRW-9-PU (1977)
14. Veynante, D., Candel, S. M., Martin, J. P. Coherent Flame Modeling of Chemical Reactions in a Turbulent Mixing Layer, in: Complex Chemical Reaction Systems, J. Warnatz and W. Jäger (Eds.), 386–398, Springer, Berlin/Heidelberg (1987)
15. Meunier, S., Champion, M., Bellet, J. C.: Premixed Combustion in a Turbulent Boundary Layer with Injection, Combust. Flame **50**, 231–238 (1983)
16. Meunier, S.: Combustion turbulente en couche limite avec injection, pariétale d'un mélange frais, Ph.D. thesis, University of Poitiers, France, (1982)
17. Cant, R. S., Bray, K. N. C.: Strained Laminar Flamelet Calculations of Premixed Turbulent Combustion in a Closed Vessel to appear in the Proceedings of the22nd Symp. (Int.) on Comb., The Combustion Institute, Pittsburgh (1988)
18. Bray, K. N. C.: ADARD CP 164, pp. II.2.1. (1975)
19. Cant, R. S., Rogg, B., Bray, K. N. C.: On Laminar Flamelet Modelling of the Mean Reaction Rate in a Premixed Turbulent Flame, Cambridge University Report CUED/A-THERMO/TR25, January 1989, to appear in Combust. Sci. Tech. (1989)
20. Cant, R. S., Rogg, B., Bray, K. N. C.: Flamelet Modelling in Premixed Turbulent Combustion, Proc. Joint Meeting of the British and French Sections of the Combustion Institute, Rouen (France), 17–21 April 1989, 191–195 (1989)
21. Patankar, S. V., Spalding, D. B.: Heat and Mass Transfer in Boundary Layers, Intertext Books, 2nd ed., London (1970)
22. Baker, R. J., Launder, B. E.: The Turbulent Boundary Layer with Foreign Gas Injection—II. Predictions and Measurements in Severe Streamwise Pressure Gradients, Int. J. Heat Transfer **17**, 293–306 (1974)
23. Giovangigli, V., Smooke, M. D.: Extinction of Strained Premixed Laminar Flames with Complex Chemistry, Combust. Sci. and Tech. **53**, 23 (1987)
24. Fritsch, F. N., Butland, J.: A Method for Constructing Local Monotone Piecewise Cubic Interpolants, SIAM J. Sci. Stat. Comput. **5**, 300–304 (1984)
25. Deuflhard, P.: A Modified Newton Method for the Solution of Ill-Conditioned Systems of non-linear Equations with Application to Multiple Shooting, Numer. Math. **22**, 289 (1974)

Investigation on the Combustion-Turbulence Interaction in Premixed Stagnation Flames of H_2-CH_4 Mixtures

Y. Liu, B. Lenze and W. Leuckel

Engler-Bunte-Institut, Lehrstuhl und Bereich Feuerungstechnik Universität Karlsruhe, FRG

Abstract

In the present paper measurements on both the influence of turbulence on premixed combustion and the effect of the flame front on the upstream turbulence conditions in the approach flow, using premixed stagnation flames as the experimental system, are presented.

The application of different H_2-CH_4 combustible mixtures (from pure CH_4- to nearly pure H_2-flames) enabled the laminar burning velocity S_L to be varied between 0.19 and 1.61 m/s. By use of a stagnation flame system and turbulence generating grids, turbulence intensities u'_0 from 0.12 to 2.90 m/s in the approach flow could be achieved. The resulting turbulent burning velocities S_T, determined by using a two component LDV technique, ranged from 0.6 to 18.4 m/s.

The results of precise measurements in more than 50 lean stagnation flames conform well to the correlation

$$S_T/S_L = 1 + 5.3 * S_L^{1/2} * u'_0/S_L$$

which we evaluated in our previous work for smaller S_L and u'_0 values.

The experimental results also show a strong affection of the upstream turbulence conditions by the presence of the flame front, i.e. a feedback effect of the combustion characteristics towards the approach flow. Due to this feedback effect, the turbulence intensity increases from the nearly constant value u'_0 in the unaffected approach flow up to the value u'_1 at the upstream edge of the reaction zone. This increase turns out to depend on the combustion intensity (S_T) and can be approximately described by the following relationship:

$$u'_1 = 2.2 * S_L^{1/2} * u'_0, (S_L > 0.25\,\mathrm{m/s})$$

Hence, the correlation of the turbulent burning velocity S_T and the turbulence intensity u'_1 at the beginning of the flame front turned out to be a linear function:

$$S_T/S_L = 1 + 2.4 * u'_1/S_L$$

This equation includes the observation that, in the case of lean flames, the combustion-turbulence interaction does depend only on the laminar burning velocity of the mixture, and not on the fuel type and mixture compositions individually.

Nomenclature

d	[mm]	diameter of perforation holes in the turbulence generating plate
D	[mm]	diameter of nozzle
PDF		probability density function
r	[mm]	radial coordinate
Re		Reynolds number
S	[m/s]	burning velocity

Turbulent Shear Flows 7

S_L	[m/s]	laminar burning velocity
S_T	[m/s]	turbulent burning velocity
u	[m/s]	axial velocity
$\bar{u}$	[m/s]	mean axial velocity
u'	[m/s]	axial turbulent fluctuation velocity
u'_0	[m/s]	constant u' in the unaffected approach flow
u'_1	[m/s]	u' at the upstream edge of the flame front
$u'_{max,1}$	[m/s]	u' maximum at flame front center
v	[m/s]	radial velocity
$\bar{v}$	[m/s]	mean radial velocity
v'	[m/s]	radial turbulent fluctuation velocity
x	[mm]	axial coordinate, from the stagnation plate surface upstream

1 Introduction

Since the introduction of the stagnation flame configuration as an experimental system, several investigators have used this burner system for fundamental studies of combustion-turbulence interactions in pre-mixed flames ([1, 2, 3, 5]). In comparison with conical Bunsen- and V-shaped flames the stagnation flame configuration shows significant advantages and is favored because of its flat flame front and the absence of any flow recirculation in the stabilization zone. Since the flame front is orientated normal to the approach flow, the turbulent burning velocity S_T can be clearly defined as the mean flow velocity entering the flame front, as was done in an analogous way for the definition of the laminar burning velocity.

In a previous experimental work [2] Liu and Lenze investigated the influence of turbulence on premixed combustion at different laminar burning velocities (S_L) by using a stagnation burner system. The application of CH_4–H_2 mixtures as fuel gases enabled to vary S_L from 0.18 m/s to 0, 72 m/s systematically. For this S_L-range, and for turbulence intensities u'_0 up to 1.2 m/s, the following correlation for lean flames has been established:

$$S_T/S_L = 1 + 5.3 * S_L^{1/2} * u'_0/S_L$$

The present experiments are aimed at investigating the combustion-turbulence interaction, i.e., both the influence of turbulence on the premixed combustion and the affection of the approach flow by the flame front, in a more extensive parameter (S_L and u'_0) range. By utilizing different H_2–CH_4–air combustion mixtures the laminar burning velocity S_L was varied between 0.19 and 1.61 m/s. By use of two stagnation burners of 40 mm and 80 mm diameter with turbulence generating grids placed at the burner exit, turbulence intensities in the approach flow u'_0 up to 2.9 m/s could be achieved. The turbulence and flame characteristics were measured by using a two-component Laser-Droppler-Velocimetry technique.

2 Experimental Set-up and Measuring Methods

The experimental set-up and the applied coordinate system are shown in Fig. 1. The concentric core the annular streams issue from a burner which consists of an internal nozzle and an annular orifice. Those two streams are directed towards a stagnation plate placed perpendicularly to the flow axis (see Fig. 1). The axially symmetrical core jet of the H_2–CH_4–air mixture is surrounded by a pure air annular co-flow which prevents the internal jet from forming a shear layer with the ambient air. Thus, a uniform velocity profile is preserved in the mixture jet right up to the stagnation plate. Two burners of $D = 40$ mm and 80 mm diameter of the internal nozzles, and 124 and 160 mm diameter of the annular orifices, were used in this work.

In order to generate the required turbulence intensities, perforated plates were placed at the exit of the internal nozzle. With the combination of different blockage ratios (geometric blockage $k = 0.3$, 0.5, 0.65) and various perforation hole diameters ($d = 3$, 6, 9, 12 and 15 mm) fluctuation velocities u' up to 2.9 m/s could be set in the approach flow. The mean exit velocity and the global stretch rate ($d\bar{u}/dx$) ranged from 2.5 to 12 m/s ($13300 < Re_D < 36000$) and from 70 to 440 $\sec^{-1}$, respectively.

Turbulent flame fronts were stabilized by a circular water-cooled stagnation plate of 240 mm diameter for both burners. The reaction zones of most of the flames were located in the axial distance range of $x = 10$ up to 40 mm, which means that the flame front has a distance from the plate surface of at least 10 mm which is much larger than the quenching distance. Therefore, the cooling of the stagnation plate can be considered to have no direct effect on the flame stabilization zone. The lower outer section of this stagnation plate was chamfered with an angle of $\alpha = 7°$, which allowed LDV measurements up to a distance of 5 mm from the stagnation point.

To measure mean flow velocities and turbulent fluctuations by LDV, magnesium oxide particles were introduced into the core flow. These particles were provided by a rotating brush-type particle generator and had diameters of between

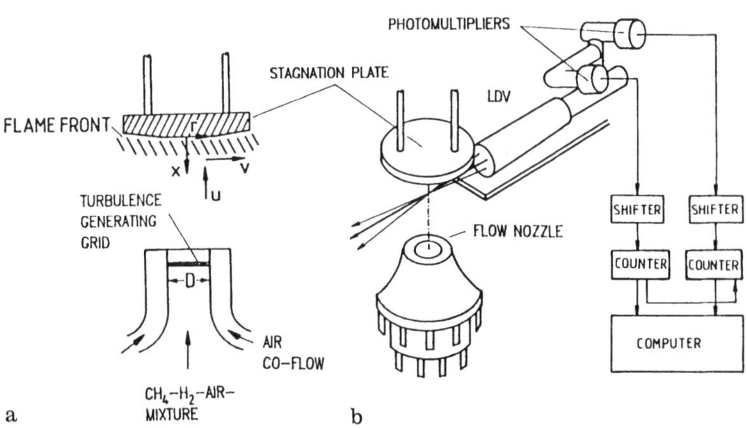

Fig. 1a, b. Experimental apparatus with stagnation plate oriented coordinate system

0.7 and 1.2 μm. In order to obtain both the axial and radial velocity components ($\bar{u}$, $\bar{v}$, u', v') as well as their probability density functions (PDF), a two-component LDV system, incorporating frequency shift and counter-type signal processors, was used. The measuring volume of the LDV had a length of about 1 mm and a diameter of about 80 μm. The typical data rate was between 2 and 5 kHz in the approach flow and decreases below 1 kHz in the reaction zone.

3 Results and Discussion

3.1 Turbulence Characteristics of Isothermal Stagnation Flows

The axial distributions of the mean velocities $\bar{u}$ and $\bar{v}$ and of the turbulence fluctuation velocities u' and v' in isothermal stagnation flows with different perforated plates but identical gas flow rate are presented in Fig. 2.

As to be expected, $\bar{u}$ decreases streadily towards the stagnation point ($x = 0$), while $\bar{v}$ remains around zero along the whole length of the flow centerline.

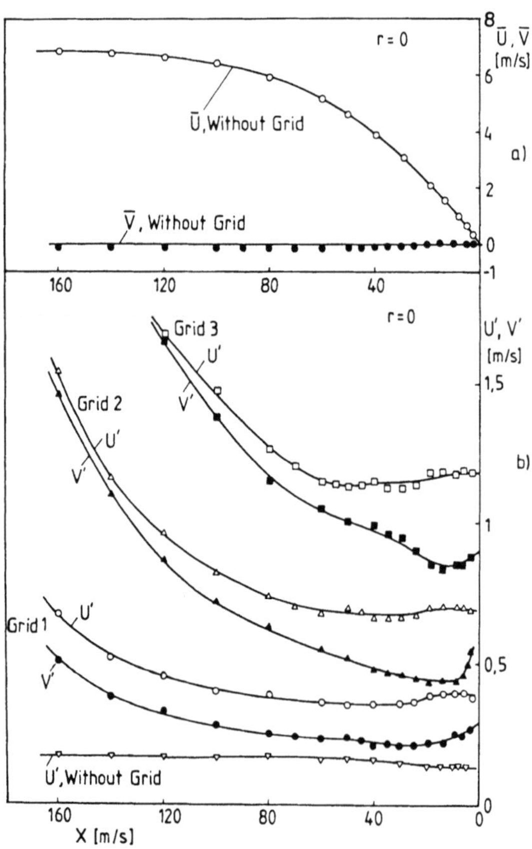

Fig. 2 a, b. Axial distributions of mean velocities $\bar{u}$ and $\bar{v}$ **a** and turbulence fluctuation velocities u' and v' **b** in isothermal stagnation flow for different turbulence generating grids

Different turbulence intensities can be generated by using perforated plates with the same gas mixture flow rate, as is shown in [2]. Those distributions of turbulence intensities have a distinct, common feature, namely that u' remains nearly constant over an axial distance range of about 60 mm upstream of the stagnation point (see also [2]). Also the measured radial profiles (not shown, see [1]) of $\bar{u}$ and u' in this region show that these remain approximately constant within a radius of 50 mm and 30 mm for nozzles of $D = 80$ mm and 40 mm, respectively (see also [1]). Such favourable flow characteristics, therefore, allow a clearly defined correlation of the flame front boundary and the approach flow, hence allowing an unambiguous determination of the turbulent burning velocity and its correlation with the turbulence characteristics.

3.2 Interaction Between the Approach Flow and the Flame Front

In order to illustrate the combustion-turbulence interaction in the stagnation flame system an example is taken of a particular flame (see Fig. 3). This flame A has been performed from a mixture of 6.3 vol% CH_4 and 8.4 vol% H_2, with a corresponding laminar burning velocity S_L of 0.7 m/s according to the experimental results of [3].

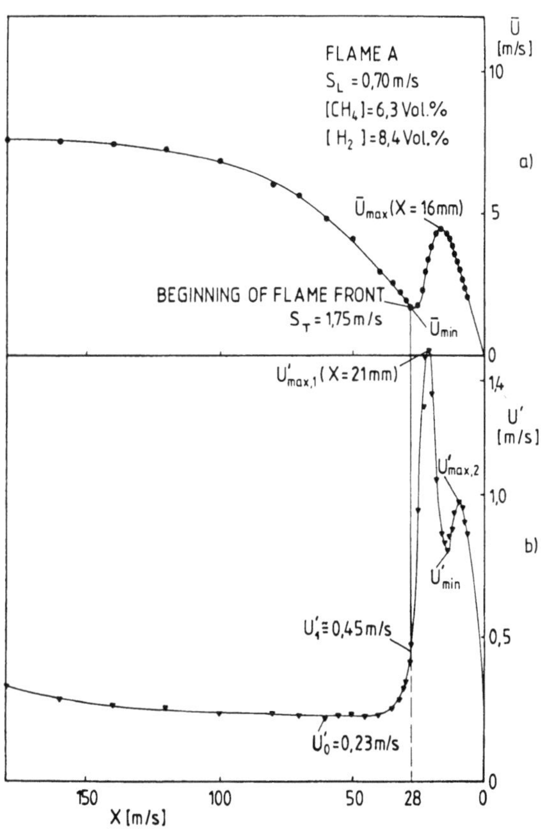

Fig. 3 a, b. Axial distribution of the mean velocity $\bar{u}$ **a** and the turbulence fluctuation velocity u', **b** for a nearly stoichiometric H_2–CH_4 flame

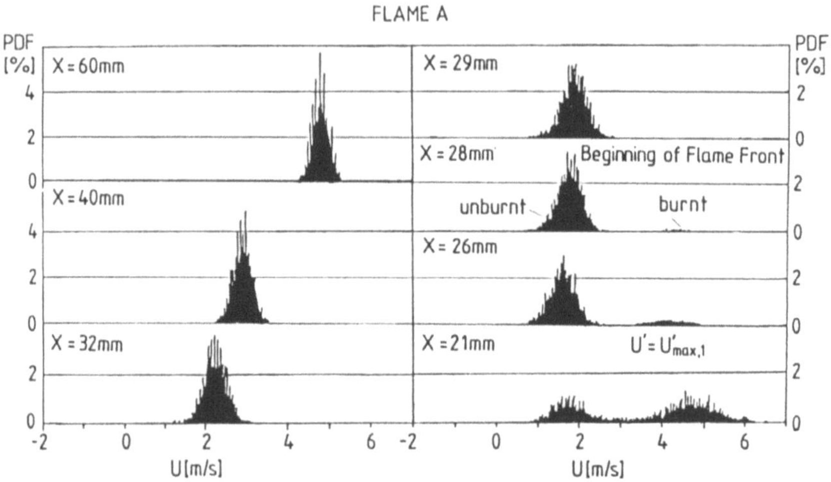

Fig. 4. Probability density functions of axial velocity u at various locations on the flow axis upstream and within the flame front

The axial distributions of $\bar{u}$ and u' and the probability density functions (PDF) of u at different locations along the flow axis of flame A are shown in Fig. 3 and 4, respectively. The mean axial velocity $\bar{u}$ progressively decreases towards the stagnation plate up to a particular location ($x = 28$ mm) where the trend of $\bar{u}$ deviates from its nearly linear decay. Looking at the corresponding PDF in Fig. 4 one observes that precisely at $x = 28$ mm a second peak of higher velocities (between 4 and 5 m/s) appears. This second peak increases steadily towards the flame front center, where at $x = 21$ mm both peaks are of about the same magnitude. Such bimodal u-distributions indicate the alternating presence of cold reactants and hot reacted gas which, due to the expansion caused by the combustion heat release, has higher velocities. The distinct point $x = 28$ mm, where the second peak in the PDF first appears and $\bar{u}$ deviates from its locally linear decay for the first time, can be considered as the beginning of the flame point. Due to the fact that the flame front is normal to the approach flow, the turbulent burning velocity S_T is, analogous to the definition of the laminar burning velocity, equal to $\bar{u}$ at this location ($S_T = \bar{u} = 1.75$ m/s).

Downstream of the beginning of the flame front $\bar{u}$ increases due to the above described intermittency effect of cold and hot gas, forms a local maximum and then drops back to zero at the stagnation point ($x = 0$).

Also in the presence of a flame u' remains nearly constant within a certain region in front of the flame (for instance between $x = 40$ and 70 mm in Fig. 3 for flame A). A comparison with the constant u'-distribution in the corresponding isothermal flow (see Fig. 2) suggests that the approach flow in this region is still unaffected by the flame front. Therefore, the constant u'_0, which is the turbulence intensity of the unaffected approach flow, is considered to be a flame-independent input parameter.

However, as the flow approaches the flame front, u' begins to increase and attains its first maximum $u'_{max,1}$ at the flame front center ($x = 21$ mm). Then u' decreases again and goes through a second local maximum before dropping back to zero at the stagnation point. By comparing the PDF's along the flow axis one can distinguish two steps of the u'-increase from the constant value u'_0 up to the value $u'_{max,1}$ within the flame front, namely a first step upstream and a second step downstream of the beginning of the flame front ($x = 28$ mm). In the second step, since the intermittency effect appears and increases steadily towards the flame front center, the steep u'-increase is mainly attributed to the bimodality of the u-distribution (see also [2]). Contrary to the second step no bimodality is visible in the first step. As the flow approaches the flame front, the u-distribution becomes broader but remains monomodal and of Gaussian shape up to the edge of the reaction zone ($x = 28$ mm). Since in a isothermal flow u' would remain constant up to a few millimeters distance from the stagnation point, also the increase in the first step is obviously due to the presence of the combustion front. It seems that a turbulent flame front with its fluctuations and instabilities (due to effects like gas expansion etc.) can represent a source of perturbation which also propagates up to some distance in the upstream direction, affecting the upstream turbulence by stimulating u' to increase. Also, the additional stagnation effect of the flame front may be the reason for turbulence generation upstream of the flame front, due to shear stresses of streamline divergence.

The turbulent burning velocity S_T of more than 50 stagnation flames as well as its corresponding turbulence characteristics have been determined and correlated. The needed laminar burning velocities S_L of CH_4–H_2–air mixtures are taken from the experimental results of [3].

The correlation of S_T and u'_0 (see Fig. 5) shows that, also in the more extended parameter range of the present results, the data conform well to the relationship

$$S_T/S_L = 1 + 5.3 * S_L^{1/2} * u'_0/S_L \tag{1}$$

which was found in the previous work [2].

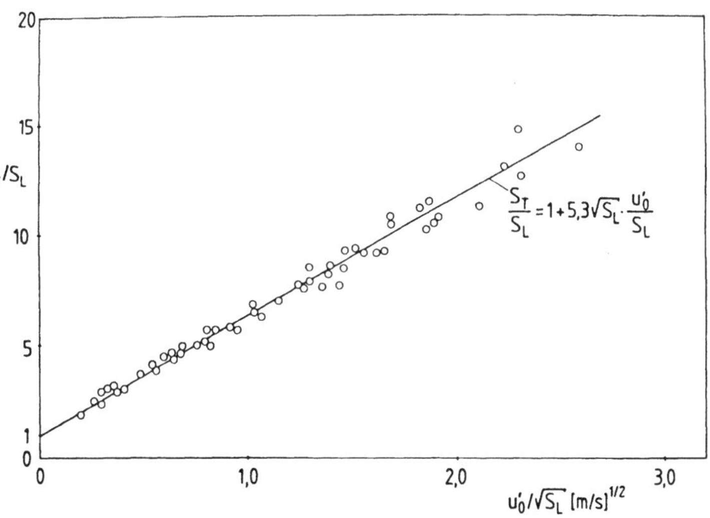

Fig. 5. Correlation of the turbulent burning velocity S_T and the flame-independent input turbulence intensity u'_0

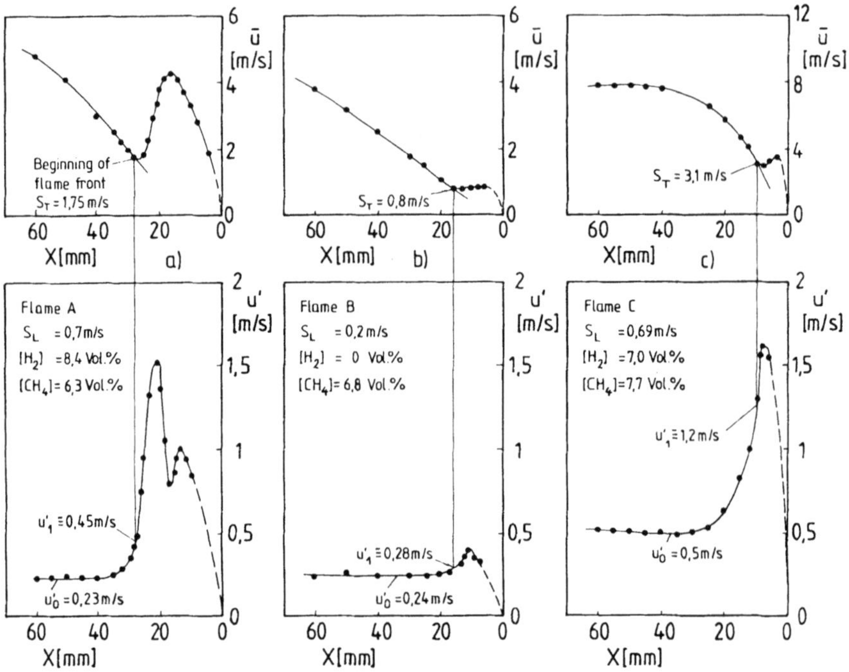

Fig. 6. Axial distributions of $\bar{u}$ and u' of three stagnation flames

Flame characteristics of two more flame examples, together with flame A, are presented in Fig. 6. As expected, the magnitude of the rise from u'_0 up to u'_1 increases with increasing S_L for given u'_0 (compare flame B and A) and with increasing u'_0 for a given S_L (compare flame A and C). Those results indicate that in both cases the u'_0/u'_1 rise increases with increasing S_T (compare Eq. (1)). By evaluating all the measured flames we found that this increase can be described in good approximation by the following relationship (see Fig. 7)

$$u'_1 = 2.2 * S_L^{1/2} * u'_0 \quad (S_L > 0.25\,\text{m/s}) \tag{2}$$

Finally, if we correlate the turbulent burning velocities S_T with the flame-dependent turbulence intensities u'_1 at the beginning of the flame front, this relationship turns out to be a linear function (Fig. 8):

$$S_T/S_L = 1 + 2.4 * u'_1/S_L \tag{3}$$

which is in qualitative agreement with the recent results by Said and Borghi who simulated the flame propagation with a "cellular automaton" ([4]) and found a linear correlation of the same form.

Although turbulent flames of different H_2:CH_4 ratios and various equivalence ratios (ranging from 0.65 to 1.0) have been measured in this work, the results can be described by the same relationships (Eqs. (1)–(3)). This fact again indicates that, in the case of lean flames, the mixture composition does not affect the combustion-turbulence interactions, as long as S_L remains constant (see also [2]).

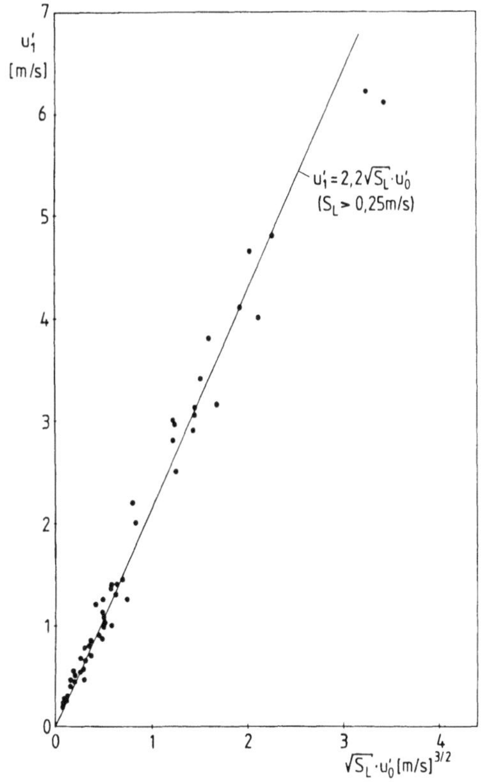

Fig. 7. Correlation of the *flame-dependent* u'_1 and the flame-independent u'_0

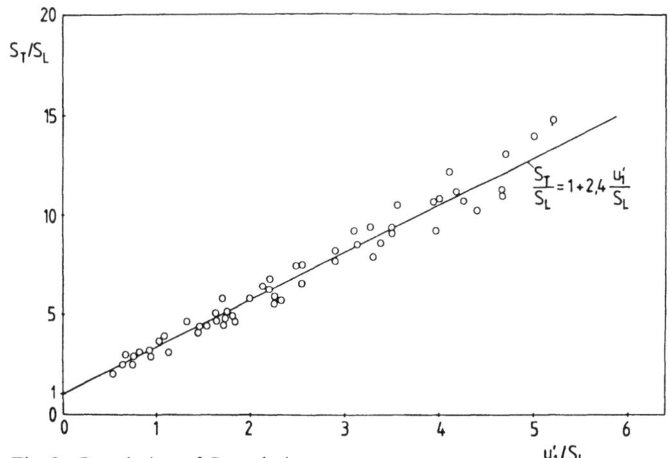

Fig. 8. Correlation of S_T and u'_1

Acknowledgement: These investigations were conducted within the German cooperative flame research program TECFLAM. The authors wish to express their gratitude to the Ministry of Research and Technology of the Federal Government of Germany/FRG for their financial support received. They also wish to thank cand.

chem. ing. S. Berbner and K. Hohlweck for their contribution to the flame experiments.

References

1. Cho, P., Law, C. K., Hertzberg, J. R. and Cheng, R. K.: Structure and Propagation of Turbulent Premixed Flames Stabilized in a Stagnation flow. 21st International Symposium on Combustion, Munich, 1986
2. Liu, Y. and Lenze, B.: The Influence of Turbulence on the Burning Velocity of Premixed Flames at Different Laminar Burning Velocities of CH_4–H_2 Mixtures. 22nd International Symposium on Combustion, Seattle, 1988
3. Liu, Y., Lenze, B., Leuckel, W.: Investigation on the Laminar and Turbulent Burning Velocities of Premixed Lean and Rich Flames of CH_4–H_2 Mixtures. 12th International Colloquium on the Dynamics of Explosions and Reactive Systems, Ann Arbor 1989
4. Said, R. and Borghi, R.: A Stimulation with a "Cellular Automaton" for Turbulent Combustion Modelling. 22nd International Symposium on Combustion, Seattle, 1988
5. Cho, P., Law, C. K., Cheng, R. K. and Shephered, I. G.: Velocity and Scalar Fields of Turbulent Premixed Flame in Stagnation flow. 22nd International Symposium on Combustion, Seattle, 1988

Coherent Flame Model for Non-Uniformly Premixed Turbulent Flames

D. Veynante, F. Lacas, E. Maistret* and S. M. Candel

Laboratoire E. M2. C., C. N. R. S., Ecole Centrale des Arts et Manufactures 92295 Chatenay–Malabry, France

Abstract

Modelling premixed turbulent flames is of importance in a variety of technological applications. In many industrial combustors, reactants may not be uniformly mixed and the local equivalence ratio depends on the flame element considered and changes with spatial location, time, degree of reaction, ... A way to model premixed turbulent flames in non uniformly mixed reactants is proposed. Our approach is based on the Coherent Flame Model due to Marble and Broadwell (1977). In this model as in various other flamelet descriptions, the reaction zone is viewed as a collection of laminar flame elements (flamelets) embedded in the turbulent flow. A first extension of the Coherent Flame Model is derived and tested but does not account for the reactions that may take place in the burned gases. An improved model is then proposed to deal with more complex situations in which unburned reactants remaining in the burned gases form a diffusion flame inside the hot stream. This configuration arises when rich and lean premixed flames coexist in the same flow. The temperature of combustion products is then sufficiently high to initiate and sustain a diffusion flame. Numerical calculations are provided to illustrate the different models.

Nomenclature

s	mass stoichiometric ratio
u_k	velocity component
V_{Di}	consumption rate per unit flame area
V_F	actual mass consumption of fuel
V_L	laminar flame speed
W_i	consumption rate of reactant i (per unit volume)
x_k	spatial coordinates
Y_0^f	mass fraction of oxidizer in fresh gases
Y_f^f	mass fraction of fuel in fresh gases
Y_o^b	mass fraction of oxidizer in burned gases
Y_f^b	mass fraction of fuel in burned gases
α	constant
β	constant
ε_s	strain rate
μ_t	turbulent viscosity
ν_t	kinematic viscosity

* Also with Groupe SNECMA, France

Turbulent Shear Flows 7

Σ flame surface density
Σ^{d} diffusion flame surface density
Σ^{p} premixed flame surface density
σ_i Schmidt number of specie i
σ_Σ Schmidt number of flame surface density
ρ density
ρ_f density in the fresh gases
Φ equivalence ratio

Introduction

Modelling premixed turbulent flames is of importance in a variety of technological applications such as dump combustors, afterburners and internal combustion engines. Many recent studies of turbulent combustion are based on flamelet concepts in which the reaction zone is viewed as a collection of laminar flame elements embedded in the turbulent stream (Williams, Carrier et al., Marble and Broadwell, Spalding, Peters, Clavin and Williams, Bray et al., Bray, Rogg). It is believed that this mode of combustion prevails if the flame thickness is small compared with the typical scales of the energy containing turbulent eddies. The main advantage of the flamelet concept is that it essentially decouples complex chemistry calculations from the turbulent flow field description. Chemical kinetics and transport properties are treated separately in a local analysis of a flame element and the resulting reaction rate is included in the calculation of the turbulent flow field. In practical situations, a data base (i.e. a "flamelet library") is first constructed from a local model and provides specific properties such as the consumption rates per unit of flame area or ignition and extinction conditions that are required in the computation of the entire flow field.

Flamelet models of premixed flames generally assume that the equivalence ratio is constant inside the flow field. But, in industrial combustors (afterburners, stratified charge internal combustion engines,...), reactants may not be uniformly mixed and the local equivalence ratio depends on the flame element considered and is a function of spatial location, time, degree of reaction,... Our aim is to propose a way to model premixed turbulent flame in non-uniformly mixed reactants. Our approach is based on the Coherent Flame Model due to Marble and Broadwell (1977). A feature which makes this flame model especially attractive is the use of the concept of flame surface density. The laminar flamelet analysis provides the local reaction rates per unit of flame area. The global reaction rate per unit volume in the flow is then written as the product of this local consumption rate by the flame surface density. An additional balance equation is used to determine the available surface density. These ideas were first proposed by Marble and Broadwell for non-premixed reactive flows. Applications have been carried out by Veynante et al., Lacas et al. and extensions to premixed configurations like turbulent ducted flames have been developed in parallel by Darabiha et al., Maistret et al. A report summarizing these studies is also available (Candel et al.).

We will first review essential aspects of the Coherent Flame Model and some results obtained with uniformly premixed reactants. The extension to non-uniformly premixed reactants is then presented. The last section deals with a more complex case which features lean and rich flames. A diffusion flame is formed in this case between the excess of fuel and oxidizer inside the burned gases. This configuration is observed in experiments performed by Phillips. A regime of laminar combustion involving triple flames is also analyzed by Buckmaster and Matalon. Turbulent configurations of this kind may be handled with an adaptation of a new model which simultaneously combines premixed and non premixed flame elements (see Veynante et al.) to the case of non uniformly premixed reactants.

Elements of the Basic Coherent Flame Model

The basic assumption of the model is that the chemical reaction takes place in thin sheets embedded in the turbulent flow (Fig. 1). Each flame element is mainly influenced by the local rate of strain acting in the plane of the flame. This strain modifies its structure and changes the local reaction rate but the reaction sheet conserves an identifiable structure and remains "coherent" inside the flow. As a consequence, the local consumption rate per unit of flame area may be obtained from an analysis of strained laminar diffusion flames as shown in Fig. 1c. The strain rate imposed by the turbulent motion has also the effect of increasing the available flame area. This production process is balanced by various destruction mechanisms such as flame shortening (mutual annihilation of adjacent flamelets by consumption of one of the reactants) or flame quenching (due, for example, to an excessively large strain rate).

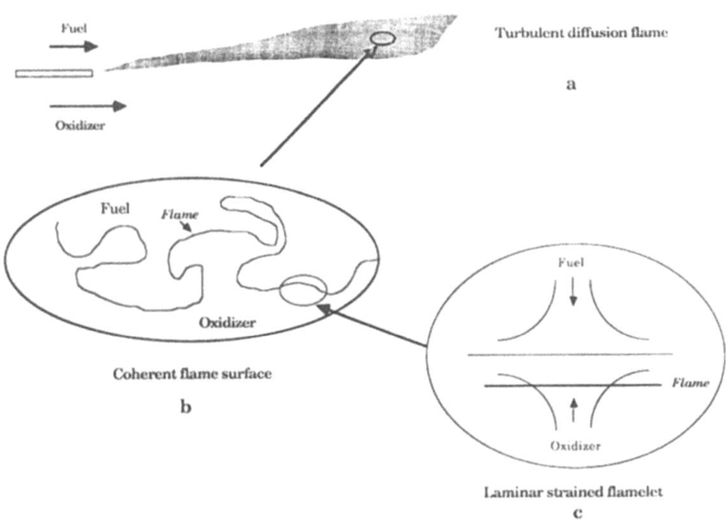

Fig. 1. Schematic representation of flamelet concept

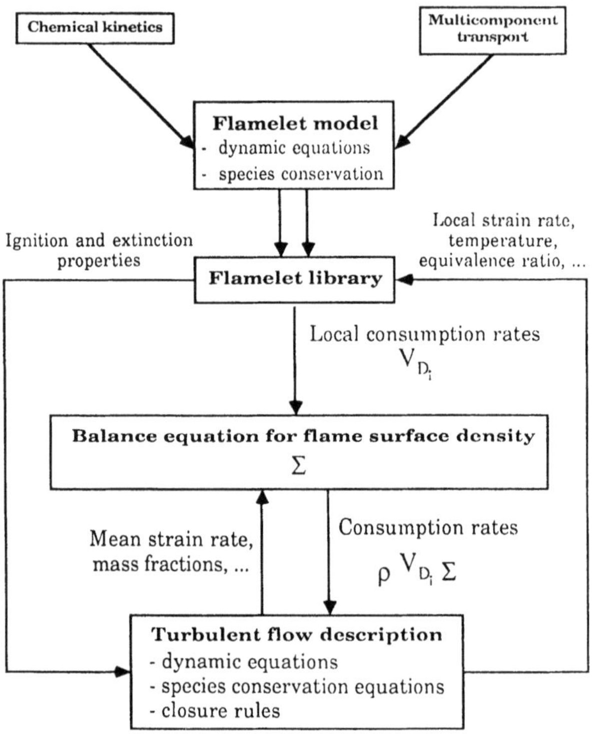

Fig. 2. Elements of the Coherent Flame Model

It follows from these general considerations that the mean consumption rate per unit volume at a point of the flow is determined as the product of the flame surface density at that point by the mass consumption rate per unit of flame area, V_{Di}, obtained from the analysis of local strained laminar flamelets. In its simplest form, the coherent flame description of turbulent reactive flows combines the three following elements summarized in Fig. 2 (details may be found in Marble and Broadwell, Veynante et al., Lacas et al. or Darabiha et al., Maistret et al.):

1) *a turbulent flow model* comprising a standard set of Reynolds or mass average dynamic equations, conservation equations for the main species and a set of closure rules for the turbulent fluxes. The global consumption rate of each reactant i, W_i, is determined as the product of the mean flame surface density, Σ, by the consumption rate per unit of flame area V_{Di} of the corresponding laminar strained flame:

$$\dot{W}_i = V_{D_i}\Sigma$$

2) *a local model of the laminar strained flame elements* providing the consumption rates per unit flame area, V_{Di}. Multiple species molecular transport and detailed chemistry may be incorporated into this description. This local model is used to build a flamelet library. An example of such an analysis carried on a premixed flame and taken from Darabiha et al., is displayed in Fig. 3. The local heat release per unit of flame area is plotted as a function of the strain rate for different values of the

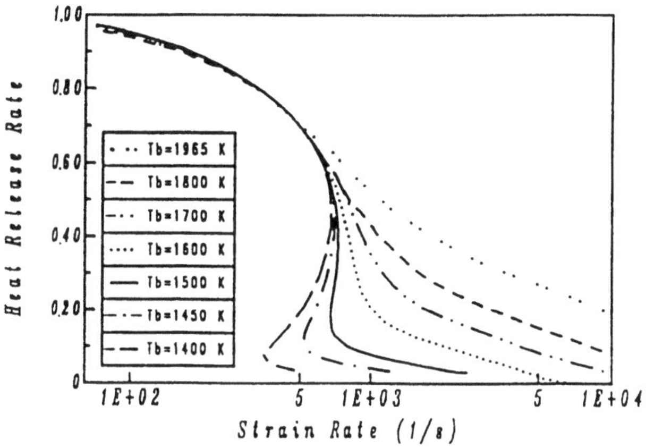

Fig. 3. Example of results from the local model for laminar strained Propane/Air flame elements ($\Phi = 0.75$); The local heat release per unit of flame area is plotted as a function of the strain rate for different values of the temperature of the burned gases (From Darabiha et al.). The vertical scale is divided by 10^8 J/m^2/s

temperature of the burned gases. Under certain conditions, the response curve exhibits turning points corresponding to extinction and ignition values of the strain rate.

3) *a balance equation for the flame area per unit volume (surface density)* accounting for transport, diffusion, production (due to strain rate) and destruction (due to mutual annihilation of adjacent flamelets, quenching,...) of flame area. This equation plays an essential role in the present flamelet description and employs local and global flow characteristics. The balance of flame surface looks like a classical transport equation. In its simplest form, the source term expresses the production of flame surface due to the strain rate ε_S and the destruction of flame area by consumption of the reactants. Considering premixed combustion and adopting the simplest assumptions, this equation may be written as:

$$\frac{\partial \Sigma}{\partial t} + \frac{\partial u_k \Sigma}{\partial x_k} = \frac{\partial}{\partial x_k}\left(\frac{\nu_t}{\sigma_\Sigma}\frac{\partial \Sigma}{\partial x_k}\right) + \alpha \varepsilon_s \Sigma - \beta \frac{V_L}{Y_f} \Sigma^2$$

Where Σ is the mean flame surface density, α and β are two constants and Y_j is the fuel mass fraction. The strain rate ε_S may be determined from velocity gradients of the mean flow (as proposed by Marble and Broadwell) or from the turbulent characteristics like the turbulent kinetic energy k and the dissipation $\varepsilon(\varepsilon_S = C_S\varepsilon/k)$.

Example of Results

An example of results obtained with the Coherent Flame Model is proposed in the case of a turbulent ducted flame. The experimental set up is shown in Fig. 4. A mixture of air and propane is injected through a long duct into a rectangular combustor of 300 mm length, 80 mm depth and 50 mm height. The upper and lower walls

Fig. 4. Experimental configuration used for tests

are made of ceramic material while the lateral walls are transparent artificial quartz windows which allow visualisation and optical measurements inside the chamber. In the inlet plane, a V-gutter flame holder is placed at the duct center producing a 50% blocage. Combustion is first ignited with a spark plug and then stabilised by hot gases recirculating behind the V-gutter.

While exact measurements of the local heat release (i.e. the source term of the species and enthalpy equations) are not available, many observations indicate that the light emitted by the combustion zone is related to the reaction intensity and hence to the heat release process. Certain chemical radicals like C_2 and CH appear almost exclusively in reactive zones and their concentration is always small. Hence, the self absorption of the light emitted by these radicals is not important and the light radiation is related monotonically to the heat release rate. This emission (516 nm for C_2 and 430 nm for CH) is easily collected through a lens by a photomultiplier with a narrow band interferential filter. Fig. 5 provides a comparison of the

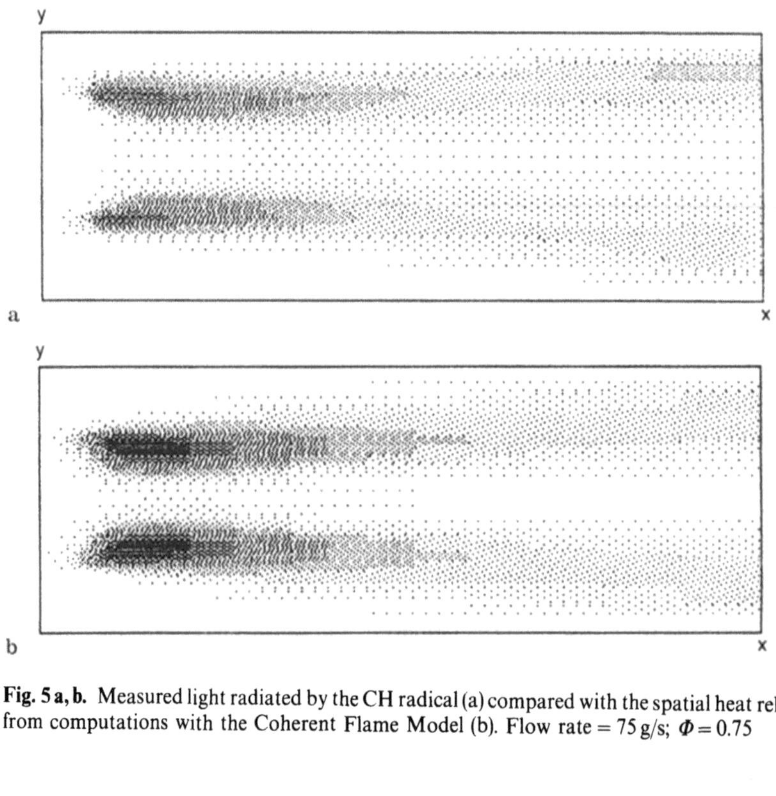

Fig. 5 a, b. Measured light radiated by the CH radical (a) compared with the spatial heat release obtained from computations with the Coherent Flame Model (b). Flow rate = 75 g/s; $\Phi = 0.75$

measured light from the CH radical with the spatial heat release rate obtained from a numerical simulation with a finite difference computational code using a $k-\varepsilon$ turbulent closure and the coherent flamelet model for the mean reaction term. The agreement is quite satisfactory.

A Simple Model for Nonuniformly Premixed Reactants

The present description of nonuniformly premixed turbulent flames is deduced from the basic form of the Coherent Flame Model in a simple situation. Consider a simple, one step, fast and irreversible reaction of the type:

$$\mathrm{F} + s\mathrm{O} \rightarrow \mathrm{P}$$

where F is the fuel, O the oxidizer and s the mass stoichiometric ratio.

The flame front separates burned and fresh gases. If one of the reactants, for example, the oxidizer O, is in excess, a part of the initial quantity of this reactant is mixed with burned gases. To introduce non-uniformly mixed reactants, one has to determine the local equivalence ratio of the reactants in fresh gases. This equivalence ratio may be defined by:

$$\Phi = \frac{s Y_{\mathrm{f}}^{\mathrm{f}}}{Y_{0}^{\mathrm{f}}}$$

where $Y_{\mathrm{f}}^{\mathrm{f}}$ and Y_{0}^{f} are the mass fractions of fuel and oxidizer in the fresh gases.

To determine $Y_{\mathrm{i}}^{\mathrm{f}}$, we write a balance equation for each reactant in the fresh gases. If V_{L} is the laminar flame speed (determined from the local flamelet analysis), this balance equation is:

$$\frac{\partial \rho_{\mathrm{f}} Y_{\mathrm{i}}^{\mathrm{f}}}{\partial t} + \frac{\partial \rho_{\mathrm{f}} u_{\mathrm{k}} Y_{\mathrm{i}}^{\mathrm{f}}}{\partial x_{\mathrm{k}}} = \frac{\partial}{\partial x_{\mathrm{k}}}\left(\frac{\mu_{\mathrm{t}}}{\sigma_{\mathrm{i}}}\frac{\partial Y_{\mathrm{i}}^{\mathrm{f}}}{\partial x_{\mathrm{k}}}\right) - \rho_{\mathrm{f}} V_{\mathrm{L}} \Sigma\, Y_{\mathrm{i}}^{\mathrm{f}}$$

The consumption term appearing in this last equation corresponds to the flame displacement ($V_{\mathrm{L}} dt$) during the time dt (Fig. 6). The volume fraction of fresh gases

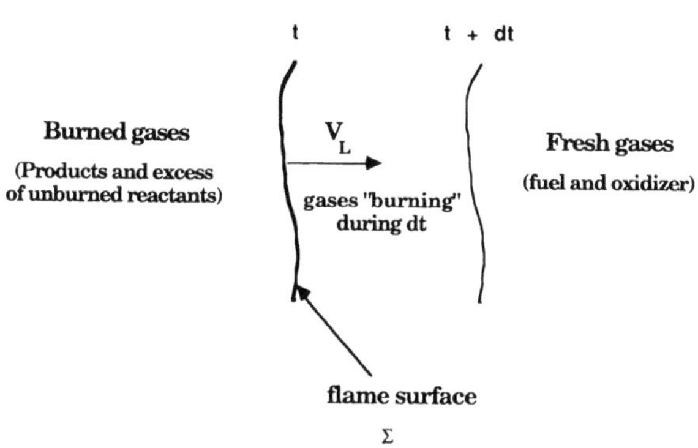

Fig. 6. Schematic representation of the consumption of fresh reactants. When the equivalence ratio is not equal to one, the excess of one of the reactants remains unburned in the burned gases

crossing the flame during this time is ($\Sigma\ V_L dt$). But, a part of the corresponding mass of reactants may be present in an unburned form in the burned gas stream. This is the case if for example one of the reactants is in excess. Accordingly, the balance equation for the reaction products is:

$$\frac{\partial \rho Y^{\mathrm{p}}}{\partial t} + \frac{\partial \rho u_{\mathrm{k}} Y^{\mathrm{p}}}{\partial x_{\mathrm{k}}} = \frac{\partial}{\partial x_{\mathrm{k}}}\left(\frac{\mu_{\mathrm{t}}}{\sigma_{\mathrm{p}}}\frac{\partial Y^{\mathrm{p}}}{\partial x_{\mathrm{k}}}\right) + V_{\mathrm{F}}(1+s)\,\Sigma$$

where V_F is the actual mass consumption of fuel F per unit of flame area.

The value of V_F depends on the laminar flame speed (and, accordingly, is related to the local laminar flamelet structure) and on the excess of one of the reactant (i.e. on the equivalence ratio Φ). If the fuel is in excess ($\Phi \geqq 1$), all the oxidizer reacts and the actual mass consumption is limited by the corresponding value of fuel:

$$V_{\mathrm{F}} = \frac{\rho_{\mathrm{f}} V_{\mathrm{L}} Y_0^{\mathrm{f}}}{s}$$

Of course, the excess of fuel remains unburned in the burned gases. This quantity may be expressed as the difference between the amount of available fuel that crosses the flame and the actual mass consumption of fuel:

$$\rho_{\mathrm{f}} V_{\mathrm{L}} Y_{\mathrm{f}}^{\mathrm{f}} \Sigma - V_{\mathrm{F}} \Sigma = \rho_{\mathrm{f}} V_{\mathrm{L}} Y_{\mathrm{f}}^{\mathrm{f}} \Sigma \left(\frac{\Phi - 1}{\Phi}\right)$$

On the other hand, when the oxidizer is in excess ($\Phi \leqq 1$), all the fuel reacts and a part of the oxidizer remains unburned. Accordingly, the mass consumption of fuel is:

$$V_{\mathrm{F}} = \rho_{\mathrm{f}} V_{\mathrm{L}} Y_{\mathrm{f}}^{\mathrm{f}}$$

The excess of oxidizer is given by:

$$\rho_{\mathrm{f}} V_{\mathrm{L}} Y_0^{\mathrm{f}} \Sigma - s V_{\mathrm{F}} \Sigma = \rho_{\mathrm{f}} V_{\mathrm{L}} Y_0^{\mathrm{f}} \Sigma (1 - \Phi)$$

By adding a balance equation for the flame density Σ, it is possible to compute non uniformly premixed turbulent flames. This balance equation may be written:

$$\frac{\partial \Sigma}{\partial t} + \frac{\partial u_{\mathrm{k}} \Sigma}{\partial x_{\mathrm{k}}} = \frac{\partial}{\partial x_{\mathrm{k}}}\left(\frac{\nu_{\mathrm{t}}}{\sigma_{\Sigma}}\frac{\partial \Sigma}{\partial x_{\mathrm{k}}}\right) + \alpha \varepsilon_{\mathrm{s}} \Sigma - \beta \frac{V_{\mathrm{L}}}{Y_{\mathrm{f}}^{\mathrm{f}}} \Sigma^2$$

where α and β are two constants and ε_S is the strain rate.

It is also possible to write balance equations for the unburned reactants, Y_i^b, in the burned gases (the reactants which are in excess). These equations are only useful if knowledge about burned gases is needed. One finds:

$$\frac{\partial \rho Y_{\mathrm{f}}^{\mathrm{b}}}{\partial t} + \frac{\partial \rho u_{\mathrm{k}} Y_{\mathrm{f}}^{\mathrm{b}}}{\partial x_{\mathrm{k}}} = \frac{\partial}{\partial x_{\mathrm{k}}}\left(\frac{\mu_{\mathrm{t}}}{\sigma_{\mathrm{f}}}\frac{\partial Y_{\mathrm{f}}^{\mathrm{b}}}{\partial x_{\mathrm{k}}}\right) + \max\left(0, \frac{\Phi - 1}{\Phi}\right) \rho_{\mathrm{f}} V_{\mathrm{L}} \Sigma Y_{\mathrm{f}}^{\mathrm{f}}$$

$$\frac{\partial \rho Y_0^{\mathrm{b}}}{\partial t} + \frac{\partial \rho u_{\mathrm{k}} Y_0^{\mathrm{b}}}{\partial x_{\mathrm{k}}} = \frac{\partial}{\partial x_{\mathrm{k}}}\left(\frac{\mu_{\mathrm{t}}}{\sigma_0}\frac{\partial Y_0^{\mathrm{b}}}{\partial x_{\mathrm{k}}}\right) + \max(0, 1 - \Phi) \rho_{\mathrm{f}} V_{\mathrm{L}} \Sigma Y_0^{\mathrm{f}}$$

A numerical test is provided in Fig. 7. The geometry is that used previously but we now assume that the flame holder separates two premixed propane/air streams

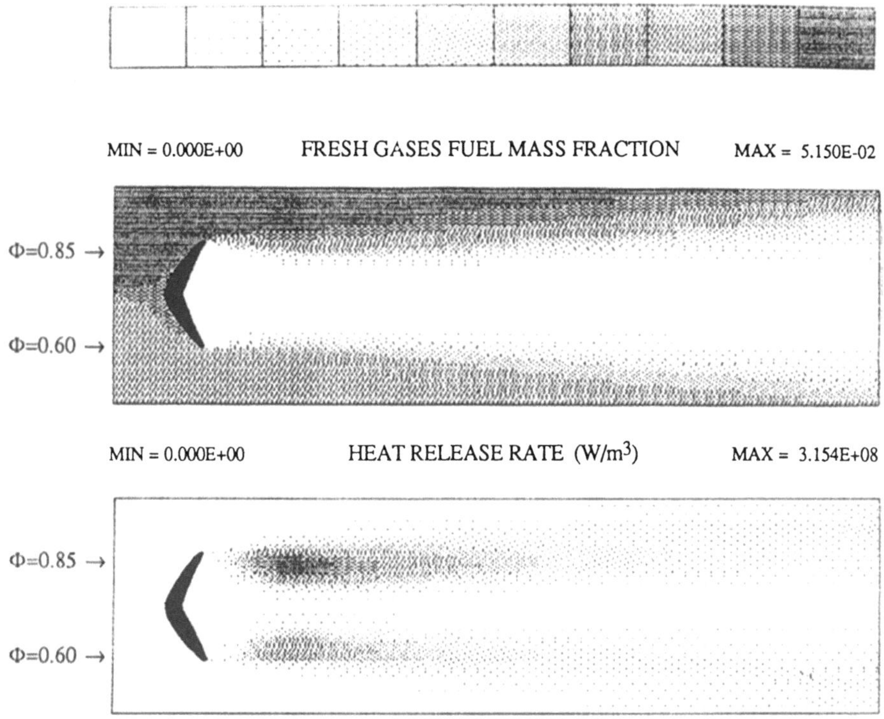

Fig. 7. Spatial distribution of fresh gases fuel mass fraction (a) and heat release rate (b) of a ducted flame. A flame holder separates two premixed propane/air streams with different values of the equivalence ratio ($\phi = 0.85$ and $\phi = 0.60$)

with different values of the equivalence ratio ($\Phi = 0.85$ and $\Phi = 0.60$). The calculated fuel mass fraction in the fresh gases is displayed in Fig. 7a. The corresponding heat release is shown in Fig. 7b. These distributions are clearly asymmetric. While experimental data for this configuration are not available, one may nevertheless conclude that the calculated flame structure is intuitively reasonable.

Model Combining Premixed and non Premixed Flame Elements

The unburned reactants in burned gases may also constitute a diffusion flame if both fuel and oxidizer are present, i.e. if simultaneous rich and lean premixed flames are present in the flow, because, in general, the temperature of the hot gases is sufficiently high to ignite and sustain the diffusion flame. Under these circumstances, premixed and non premixed flame elements are simultaneously present in the flow. Such a situation is already observed in experiments carried by Phillips and analysed mathematically by Buckmaster and Matalon. This phenomenon may be taken into account by introducing two flame surface densities: one for the area of diffusion flamelets (Σ^{d}), the other for the area of premixed flamelets (Σ^{p}). Of course, we assume in this analysis that the diffusion flame inside burned gases may be viewed as a collection of laminar flame elements. Let us note that we have previously derived a "combined" coherent flame model to describe the initial region of a turbulent diffusion flame (see Veynante et al. 1989). In this case, three surface densities are needed: one for premixed flames, one for diffusion flames and one for the non reactive contact surface between non premixed reactants.

To be consistent with this description, it is necessary to use two mass fractions (and two balance equations) for each reactant i: one represents the mass of the reactant i in a premixed form (in unburned gases) and the other the mass of the reactant i remaining in burned gases and burning in the diffusion flame. These equations are:

$$\frac{\partial \rho Y_{\mathrm{f}}^{\mathrm{f}}}{\partial t}+\frac{\partial \rho u_{\mathrm{k}} Y_{\mathrm{f}}^{\mathrm{f}}}{\partial x_{\mathrm{k}}}=\frac{\partial}{\partial x_{\mathrm{k}}}\left(\frac{\mu_{\mathrm{t}}}{\sigma_{\mathrm{f}}} \frac{\partial Y_{\mathrm{f}}^{\mathrm{f}}}{\partial x_{\mathrm{k}}}\right)-\rho_{\mathrm{f}} V_{\mathrm{L}} \Sigma^{\mathrm{p}} Y_{\mathrm{f}}^{\mathrm{f}}$$

$$\frac{\partial \rho Y_{0}^{\mathrm{f}}}{\partial t}+\frac{\partial \rho u_{\mathrm{k}} Y_{0}^{\mathrm{f}}}{\partial x_{\mathrm{k}}}=\frac{\partial}{\partial x_{\mathrm{k}}}\left(\frac{\mu_{\mathrm{t}}}{\sigma_{0}} \frac{\partial Y_{0}^{\mathrm{f}}}{\partial x_{\mathrm{k}}}\right)-\rho_{\mathrm{f}} V_{\mathrm{L}} \Sigma^{\mathrm{p}} Y_{0}^{\mathrm{f}}$$

$$\frac{\partial \rho Y_{\mathrm{f}}^{\mathrm{b}}}{\partial t}+\frac{\partial \rho u_{\mathrm{k}} Y_{\mathrm{f}}^{\mathrm{b}}}{\partial x_{\mathrm{k}}}=\frac{\partial}{\partial x_{\mathrm{k}}}\left(\frac{\mu_{\mathrm{t}}}{\sigma_{\mathrm{f}}} \frac{\partial Y_{\mathrm{f}}^{\mathrm{b}}}{\partial x_{\mathrm{k}}}\right)+\max\left(0, \frac{\Phi-1}{\Phi}\right) \rho_{\mathrm{f}} V_{\mathrm{L}} \Sigma^{\mathrm{p}} Y_{\mathrm{f}}^{\mathrm{f}}-V_{\mathrm{f}}^{\mathrm{d}} \Sigma^{\mathrm{d}}$$

$$\frac{\partial \rho Y_{0}^{\mathrm{b}}}{\partial t}+\frac{\partial \rho u_{\mathrm{k}} Y_{0}^{\mathrm{b}}}{\partial x_{\mathrm{k}}}=\frac{\partial}{\partial x_{\mathrm{k}}}\left(\frac{\mu_{\mathrm{t}}}{\sigma_{0}} \frac{\partial Y_{0}^{\mathrm{b}}}{\partial x_{\mathrm{k}}}\right)+\max(0, 1-\Phi) \rho_{\mathrm{f}} V_{\mathrm{L}} \Sigma^{\mathrm{p}} Y_{0}^{\mathrm{f}}-V_{0}^{\mathrm{d}} \Sigma^{\mathrm{d}}$$

This set of equations contains terms describing the consumption of reactants in the burned gases burning in the diffusion flame. $V_{\mathrm{i}}^{\mathrm{d}}$ represents the mass consumption rate of the reactant i in the diffusion flame elements. Of course, the equations for reactants in the fresh gases are the same as those previously proposed.

Two coupled equations are written, one for each flame surface density considered, Σ^{p} for the area of premixed flame and Σ^{d} for the diffusion flame:

$$\frac{\partial \Sigma^{\mathrm{p}}}{\partial t}+\frac{\partial u_{\mathrm{k}} \Sigma^{\mathrm{p}}}{\partial x_{\mathrm{k}}}=\frac{\partial}{\partial x_{\mathrm{k}}}\left(\frac{v_{\mathrm{t}}}{\sigma_{\Sigma}} \frac{\partial \Sigma^{\mathrm{p}}}{\partial x_{\mathrm{k}}}\right)+\alpha^{\mathrm{p}} \varepsilon_{\mathrm{s}} \Sigma^{\mathrm{p}}-\beta^{\mathrm{p}} \frac{V_{\mathrm{L}}}{Y_{\mathrm{f}}^{\mathrm{f}}} \Sigma^{\mathrm{p}^{2}}$$

$$\frac{\partial \Sigma^{\mathrm{d}}}{\partial t}+\frac{\partial u_{\mathrm{k}} \Sigma^{\mathrm{d}}}{\partial x_{\mathrm{k}}}=\frac{\partial}{\partial x_{\mathrm{k}}}\left(\frac{v_{\mathrm{t}}}{\sigma_{\Sigma}} \frac{\partial \Sigma^{\mathrm{d}}}{\partial x_{\mathrm{k}}}\right)+\alpha^{\mathrm{d}} \varepsilon_{\mathrm{s}} \Sigma^{\mathrm{d}}-\frac{\beta^{\mathrm{d}}}{\rho}\left(\frac{V_{\mathrm{f}}^{\mathrm{d}}}{Y_{\mathrm{f}}^{\mathrm{b}}}+\frac{V_{0}^{\mathrm{d}}}{Y_{0}^{\mathrm{b}}}\right) \Sigma^{\mathrm{d}^{2}}+\gamma^{\mathrm{d}} \frac{V_{\mathrm{L}}}{Y_{\mathrm{f}}^{\mathrm{f}}} \Sigma^{\mathrm{p}^{2}}$$

Except for the last term, this equation is the same as the one proposed by Marble and Broadwell for diffusion flames. The last term expresses the ignition of the diffusion flame in the burned gases. This term participates to the destruction of the premixed flame area. Conceptually, when the premixed flame area disappears because the premixed reactants are consumed, it is replaced by a diffusion flame sheet between the excess fuel and oxidizer in the burned stream.

Of course, this set of equations must be supplemented with expressions for the local consumption rate of each reactant in the diffusion flame sheet.

We now consider a test case of this combined model in the case of a two-dimensional mixing layer. Two parallel streams, each containing one of the two reactants, interact with different free stream velocities U_1 and U_2 (the velocity ratio

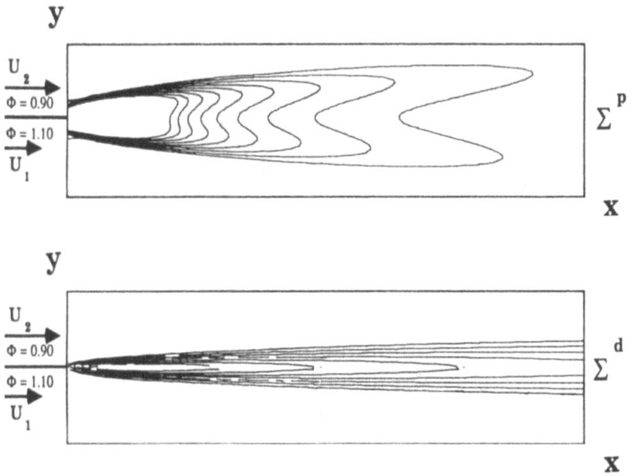

Fig. 8. Contours of constant densities of premixed flame (a) and diffusion flame area (b) in a mixing layer between two streams with different velocities (velocity ratio: 0.8) and different equivalence ratio ($\phi = 0.90$ and $\phi = 1.10$)

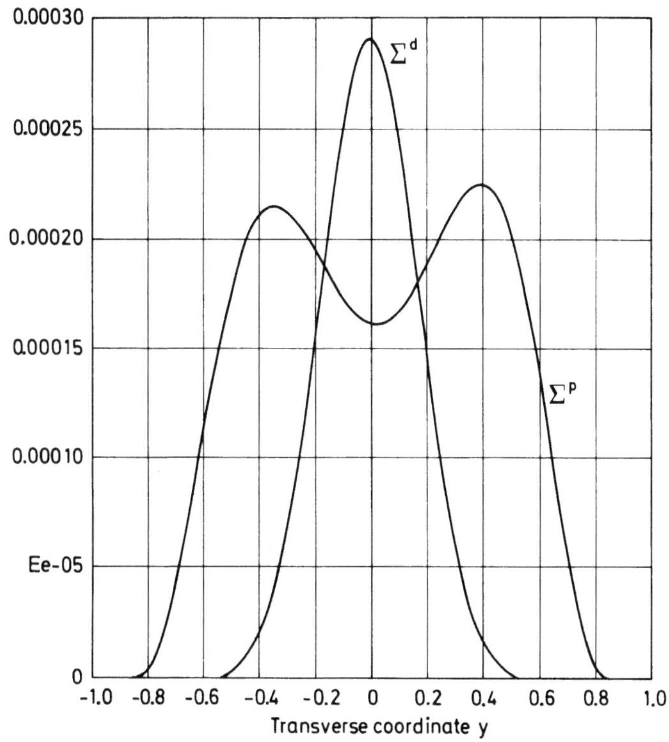

Fig. 9. Transverse profile of densities of premixed flame (Σ^p) and diffusion flame (Σ^d) in a well established mixing layer

U_1/U_2 is 0.8) and different equivalence ratios (0.90 and 1.10). Contours of constant densities of premixed flame and diffusion flame areas are displayed in Fig. 8. A diffusion flame is clearly apparent inside the mixing zone. A complementary view is given by the self similar transverse profiles of densities of premixed and diffusion flame areas in the well established mixing layer plotted in Fig. 9. The flame zone exhibits two premixed reaction fronts at the boundaries of the mixing layer separated by a diffusion front.

Conclusions

It is shown that the Coherent Flame Model of Marble and Broadwell may be extended to the case of non-uniformly premixed turbulent flames. A simple version is first provided but does not account for the possible ignition and stabilization of a diffusion flame inside the burned gases. A more complex model is then proposed to describe this important situation.

References

Bray, K. N. C., Libby, P. A., Moss, J. B. (1984): Comb. Sci. Tech., 143

Bray, K. N. C. (1986): In Complex Chemical Reaction Systems, J. Warnatz and W. Jaëger, eds., Springer Verlag, Heidelberg, pp. 356–375

Bray, K. N. C., Champion, M., Kostiuk, L. W., Libby, P. A. (1989): Seventh Symposium on Turbulent Shear Flows. Stanford

Buckmaster, J., Matalon, M. (1988): Anomalous Lewis number effects in tribrachial flames. 22nd Int. Symp. on Combustion. The Combustion Institute

Candel, S. M., Darabiha, N., Esposito, E., Giovangigli, V., Lacas, F., Laverdant, A., Maistret, E., Poinsot, T., Trouvé, A., Veynante, D., Zikikout, S. (1988): Internal report, Lab. E.M2.C., C.N.R.S. et Ecole Centrale des Arts et Manufactures. Châtenay Malabry, France

Carrier, G. F., Fendell, F. E., Marble, F. E. (1975): SIAM, J. Appl. Math. **28**, 00–00

Clavin, P., Williams, F. A. (1982): J. Fluid Mech. **116**, 215

Darabiha, N., Giovangigli, V., Trouvé, A., Candel, S. M., Esposito, E. (1987): Coherent flame description of turbulent premixed ducted flames. In Turbulent Reactive Flows, R. Borghi and S. N. B. Murthy, eds., Springer Verlag, New-York, pp. 591–637

Lacas, F., Zikikout, S., Candel, S. M. (1987): 23rd Joint Propulsion Conference. San Diego, AIAA Paper No 87–1782

Maistret, E., Darabiha, N., Poinsot, T., Veynante, D., Lacas, F., Candel, S., Esposito, E. (1989): Recent developments in the coherent flamelet description of turbulent combustion. Numerical Combustion, A. Dervieux and B. Larrouturou eds., Springer Verlag, Berlin pp. 98–117

Marble, F. E., Broadwell, J. E. (1977): The coherent flame model for turbulent chemical reactions. Project Squid Rep. TRW-9-PU

Marble, F. E., Broadwell, J. E. (1979): A theoretical analysis of nitric oxide production in a methane-air turbulent diffusion flame. EPA Tech. Rep.

Peters, N. (1984) Prog Energy Comb. Sci. **10**, 319

Peters, N. (1986): 21st Symp. (International) on Combustion. The Combustion Institute

Phillips, H. (1965): Flame in a buoyant methane layer. 10th Int. Symp. on Combustion. The Combustion Institute

Rogg, B. (1989): Seventh Symposium on Turbulent Shear Flows. Stanford

Spalding, D. B. (1978): Seventeenth Symposium (International) on Combustion. The Combustion Institute

Veynante, D., Candel, S. M., Martin, J. P. (1986): In Complex Chemical Reaction Systems, J. Warnatz and W. Jaëger, eds. Springer Verlag, Heidelberg. pp. 386–398

Veynante, D., Lacas, F., Candel, S. M. (1989): A new flamelet combustion model combining premixed and non premixed turbulent flames. A.I.A.A. 27th Aerospace Sciences Meeting. Reno (Nevada)

Williams, F. A. (1975): AGARD Conf. Proc. 164

Index of Contributors

Borghi R. 293
Bradbury L. J. S. 169
Buch K. A. 17

Candel S. M. 367
Chen J.-Y. 277

Dahm W. J. A. 17

Elghobashi S. 313

Fric T. F. 225

Gerz T. 27
Gibson C. H. 3, 63
Gonzalez M. 293

Hancock P. E. 253
Henningson D. S. 155
Hunt J. C. R. 101
Hussain F. 9, 331

Johnson A. E. 253
Johnston J. P. 241
Jones W. P. 271

Kawaguchi Y. 119
Kikkawa Y. 119
Kim J. 87, 155
Kollmann W. 277
Kurita N. 209

Lacas F. 367
Lai Y. G. 81
Lasheras J. C. 195
Lee C. 331
Lee M. J. 87, 101
Lenze B. 357
Leuckel W. 357
Liu Y. 357

Maistret E. 367
Martin J. E. 195
Meiburg E. 195
Melander M. V. 9
Metcalfe R. W. 331
Miyata M. 209

Nagano Y. 47
Nakamura I. 209
Nomura K. K. 313

O'Brien J. E. 137

Plesniak M. W. 241

Reshotko E. 137
Rogg B. 345
Roshko A. 225

Schumann U. 27
So R. M. C. 81
Sohn K. H. 137
Suzuki H. 119
Suzuki K. 119

Tagawa M. 367

Veynante D. 367

Williamson C. H. K. 173

MIX
Papier aus verantwortungsvollen Quellen
Paper from responsible sources
FSC® C105338

If you have any concerns about our products,
you can contact us on
ProductSafety@springernature.com

In case Publisher is established outside the EU,
the EU authorized representative is:
Springer Nature Customer Service Center GmbH
Europaplatz 3, 69115 Heidelberg, Germany

Printed by Libri Plureos GmbH
in Hamburg, Germany